Preface to the English Edition

This is a translation of my book "Gunron" (Group Theory), published by the Iwanami Shoten.

This volume contains the translation of the entire first volume of the two-volume set of the Japanese edition. The second part will be translated soon.

For the translation, I have corrected a few of the mistakes and cleared up some of the obscurities of the original version. But, the main body of the text has been unchanged. Since the publication of the Japanese edition in 1977–78, the theory of finite simple groups has continued its remarkable progress reaching its zenith in 1980 when the classification of finite simple groups was almost completed. In view of the importance of this tremendous accomplishment, it would be desirable to incorporate these most recent advances into this book. However, I feel that the time has not yet come to include this progress in a book such as this one. So, I have regretfully kept the main body of the text unchanged, except for the few paragraphs added at the end of Chapter 3, § 5, and for several places in Chapter 6. Also, a few papers are added to the bibliography at the end of the first volume.

It is a pleasure to acknowledge my indebtedness to several persons.

Professor Koichiro Harada, Dr. Yoshimi Egawa, and Dr. Hiroshi Suzuki have given me many valuable comments on the Japanese edition of the first volume. I thank them for their careful reading.

This translation could not have been done without the assistance of Kazuko Suzuki. I thank Kazuko for her fine effort and numerous suggestions. If at least the worst blunders have been avoided and the translation is a bit more easily readable, this accomplishment is due entirely to her devoted help.

Finally, I am grateful to Springer-Verlag for publishing this translation and to the editorial staff and the production department of Springer for friendly cooperation.

October 1980   Michio Suzuki

Preface

The concept of a group is one of the most fundamental in modern mathematics. Although we can find its primordial forms in such things as the congruence of geometric figures and the motions in space which were studied even in ancient times, it was not until the first half of the nineteenth century that this idea was clearly defined and recognized as a subject of serious mathematical thought. At this time, it was quite prominent in the research concerning general polynomial equations of degree greater than four, particularly in the works of Abel and Galois. In fact, the group consisting of the permutations of the roots of the given polynomial was the key to resolving the problem of its algebraic solvability. Later in the nineteenth century, the notion of the congruence of geometric figures was generalized, and this clarified an important relationship among the various geometries and the transformation groups of their geometric objects. Furthermore, the works of Lie on the theory of continuous groups gave added significance to the concept of a group, and by the end of that century, it had been firmly established as one of the most important ideas of modern mathematics. During these years, a generalization of the concepts of transformation groups and permutation groups led to the study of abstract groups, and Burnside's classical "Theory of Groups of Finite Order" (Second Edition), published in 1911, presents the theory of abstract groups as it was in the early part of the twentieth century. Thereafter, group theory developed slowly but surely until finally, in 1955, the publication of a few basic papers triggered a sudden explosion of advances which has been especially remarkable during the last fifteen years or so.

One of my main aims was to present an introduction to this recent progress in the theory of finite simple groups. I have tried to keep the preliminaries down to a bare minimum. The first chapter contains a fairly detailed discussion of the basic ideas underlying the theory of groups, while the second presents the fundamental theorems and methods. After laying this groundwork, I went on to discuss several specific branches of group theory. However, because of the tremendous diversification, it is extremely difficult—if not impossible—to present a comprehensive survey of all the various parts of group theory. Such topics as the theory of permutation groups and the representation theory of finite groups are so important that they must be included in any book dealing with groups, although I was unfortunately unable to discuss them in any great detail because of the space limitations. In spite of these and other omissions, the book became rather bulky, so it has been published in two parts. The first volume ends with Chapter 3 which discusses the properties of several particular classes of groups. The fourth chapter concerns commutators, a concept which I consider to be perhaps the most distinctly group-theoretical, and some of their applications. Up until the middle of Chapter 4, I have mainly dealt with general group theory, whereas the rest of the book treats almost exclusively the theory of finite groups. Chapter 5 discusses the methods and theorems pertaining to finite groups, and the last chapter serves as an introduction to the recent progress in the theory of finite simple groups.

To the reader who has some knowledge about the rudiments of matrices, determinants, and elementary number theory, this book should present no difficulty. Nevertheless, for the sake of convenience, I have collected some of the necessary materials after the Preface. There is a small number of places in the text where some outside references are needed, and there, I have included those references.

With few exceptions, each section ends in a collection of exercises which serve to present the interesting concepts and theorems which are not discussed in the text. As a rule, each exercise is supplemented by a "Hint" which either literally contains a set of hints or provides an outline of a proof. Often, it also presents some additional remarks. Of course, the hints are only suggestions, and it may well be possible to come up with better proofs, so I urge the readers to work on the exercises without feeling compelled to follow the given hints. I have placed each exercise so that all of the necessary preliminaries (for the intended solution) will hopefully have been discussed in earlier sections. Also, I have not referred to the results of any exercise later on in the text (although some exercises do depend on the results of earlier ones). The reader may thus proceed without solving a single exercise and still not experience any great disadvantage as far as the understanding of the proofs of the theorems is concerned.

It gives me great pleasure to express my indebtedness to the numerous people without whom this undertaking could not have been achieved. I learned mathematics under the guidance of Professor Shokichi Iyanaga and Professor Kenkichi Iwasawa, and I am very grateful for all their kind counsel and encouragement. The influence of my many friends who supplied me with new ideas in the proofs or preprints of their unpublished papers is evident throughout this book. I sincerely thank all those friends who helped me directly or indirectly. And finally, I wish to thank Mr. Hideo Arai of the Iwanami Shoten, Publishers for his invaluable help concerning everything related to the publication of this book.

January 1977   Michio Suzuki

Remarks

Here, we have collected several remarks for the readers' convenience.

(1) Let A be a set. If B is a subset of A, we will use the notation

$$B \subset A,$$

and this does *not* exclude the possibility that $A = B$. If B is a proper subset (which is different from A), and if we wish to emphasize this difference, we will use the notation

$$B \underset{\neq}{\subseteq} A.$$

(2) Let A and B be two sets. The notation

$$f: A \to B$$

will mean that f is a mapping from the set A to the set B. If C is another set, and if

$$g: B \to C,$$

then the *composite* of f and g is defined as a mapping from A into C such that an element a of A is mapped to

$$g(f(a)).$$

The composite of f and g is usually denoted by $g \circ f$, but in this book, we will also write it as

$$fg,$$

and it is often called the *product* of f and g. Thus, we have

$$(fg)(a) = g(f(a)).$$

For this reason, we often use the exponential notation so that the image of an element x by a mapping σ is written as x^σ instead of as $\sigma(x)$. Therefore, we have

$$x^{\sigma\tau}=(x^\sigma)^\tau$$

for any mappings σ and τ, provided the product is defined.

The following important theorem holds.

Theorem. *The composition of mappings satisfies the associative law. To put it more precisely, the following property is satisfied. Let A, B, C, and D be sets, and let*

$$f: A \to B, g: B \to C, \text{ and } h: C \to D$$

be three mappings. Then, the mappings fg and gh are defined, and we have the associative law:

$$(fg)h = f(gh).$$

Proof. If $a \in A$, then both sides of the above formula map the element a to

$$h(g(f(a))).$$

Thus, $(fg)h$ and $f(gh)$ coincide. $\square$

(3)   We will follow the convention that the set of natural numbers is the set of positive integers $\{1, 2, 3, \ldots\}$ which excludes zero.

We have collected here some of the basic properties of integers which will be used frequently throughout this book.

(1)   **The Division Theorem.** *Let m be an arbitrary integer, and let n be a natural number. Then, there is a unique pair of integers (q, r) which satisfies the properties*

$$m = qn + r \quad \text{and} \quad 0 \leq r < n.$$

(2)   Two natural numbers m and n are said to be *relatively prime* if they have no common prime factors. If m and n are relatively prime, then there is a pair of integers (u, v) which satisfies the equation

$$um + vn = 1.$$

(3)   Let m and n be two natural numbers such that m and n are relatively prime. If m divides the product nl of n and another integer l, then m divides l.

(4) An integer greater than 1 can be written as a product of prime numbers. This decomposition is unique if we disregard the order of the factors in the product.

(5) Let m be a fixed natural number. Two integers a and b are said to be *congruent* modulo m if their difference $a - b$ is a multiple of m. We write this relation as

$$a \equiv b \pmod{m}.$$

The congruence (relation) satisfies the following properties for any integer c:

$$a \equiv b \pmod{m} \Rightarrow a + c \equiv b + c \pmod{m}$$

$$ac \equiv bc \pmod{m}.$$

If an integer n is relatively prime to m, the congruence

$$xn \equiv b \pmod{m}$$

has a solution, and all of the solutions are congruent modulo m.

(The following sketch may convince the reader of the validity of the proposition (1). On the number line, mark off all of the multiples of the natural number n, and consider the number line to be divided into the small intervals of size n. Then, any integer m will fall into one of these small intervals. So, there is a pair (q, r) which satisfies the conditions of (1). If (q', r') is another pair which satisfies the same conditions, then n must divide $r - r'$, and $r - r'$ is an integer which lies between $-(n-1)$ and $+(n-1)$. So, we must have $r = r'$. Thus, the pair (q, r) is unique.

In (1), if m and n are relatively prime, then n and r are also relatively prime. So, by induction on n, (2) can be proved using (1). We have $(2) \Rightarrow (3) \Rightarrow (4)$. The first part of (5) follows easily from the definition of the congruence relation, while the remaining assertion of (5) may be proved by using (2) and (3).)

Some basic knowledge of the theory of general fields will facilitate the reading of this book. A *field* is a set of elements which is equipped with the operations of addition, subtraction, multiplication, and division by nonzero elements. The set of rational numbers and the set of real numbers are examples of fields. At first, the reader may think of a field as always being the field of rational numbers, but many of the theorems are valid for all fields in general, and the theory will be all the more enriched if the reader bears this in mind.

In a general field F, we denote the addition and the multiplication by $a + b$ and ab, respectively. The field F contains two special elements u and e such that for any $a \in F$,

$$a + u = a \quad \text{and} \quad ae = a.$$

The element u is called the zero (element) of F, and e is the identity of F. It can be proved that the field F contains a unique zero and a unique identity (see Theorem 1.2, Chapter 1). These elements correspond to 0 and 1 in the usual field of numbers. Thus, they will be written as $u=0$ and $e=1$. In any field, we will assume that $1 \neq 0$.

In the field F, we have the usual rules:

$$a+b=b+a, \quad ab=ba,$$
$$(a+b)+c=a+(b+c), \quad (ab)c=a(bc),$$
$$(a+b)c=ac+bc.$$

Furthermore, for any $a \in F$ and $b \neq 0$, we have elements $-a$ and b^{-1} such that

$$a+(-a)=0, \quad b(b^{-1})=1.$$

In this book, we will suppose that the axiom of choice is valid. As is well known, the axiom of choice is independent of the axioms of set theory. Thus, Mathematics without the axiom of choice is possible. However, we will apply the axiom of choice, or one of its equivalent forms (Zorn's lemma), whenever it is convenient.

Let P be a set in which a relation $\leqq$ is defined for some pair of elements (a, b). The relation $\leqq$ is an *order* if the following properties are satisfied:

$$a \leqq b, b \leqq c \Rightarrow a \leqq c,$$
$$a \leqq b, b \leqq a \Rightarrow a = b,$$
$$a \leqq a.$$

The set P in which an order relation is defined for some pairs is called a *partially ordered set*. A partially ordered set may contain a pair (a, b) such that a and b do not satisfy the order relation. If we have either $x \geqq y$ or $y \geqq x$ for any pair of elements x and y of P, then P is said to be *linearly ordered*. A partially ordered set M is called an *inductive set* if an upper bound of any linearly ordered nonempty subset is in M.

The lemma of Zorn may be stated as follows.

A nonempty inductive set contains at least one maximal element.

For the proof of the equivalence of Zorn's lemma and the axiom of choice, the readers are referred to [IK] pp. 61–63.

The symbol [IK] stands for the textbook by Iyanaga and Kodaira which is listed in the Bibliography at the end of this volume. Similarly, textbooks of Akizuki–Suzuki and Asano–Nagao are referred to as [AS] and [AN], respectively. The numeral in brackets, such as in Brauer [1], refers to the paper [1] of Brauer in the bibliography.

The preceding remarks should be sufficient to prepare the reader. But, for a finer understanding of this book, a small working knowledge of linear algebra and the theory of finite fields will be necessary.

From linear algebra, we need the definition of a vector space, the properties of the bases of a vector space, and the uniqueness of its dimension ([IK] pp. 279–283). Later, we will also require some knowledge of the eigenvalues of a matrix and the theory of linear equations ([IK] pp. 383–387, pp. 344–347).

There exist fields consisting of a finite number of elements. If F is a finite field, the number of its elements is a power of a prime number p. For any prime power p^n, there is a unique field consisting of exactly p^n elements. This field will be denoted by $GF(p^n)$. If x is an element of the finite field $GF(p^n)$, then we have

$$x + x + \cdots + x \,(p \text{ times}) = px = 0.$$

Also, for any elements x and y of $GF(p^n)$,

$$(x + y)^p = x^p + y^p \quad \text{and} \quad (xy)^p = x^p y^p.$$

This implies that the mapping

$$\sigma : x \to x^p$$

is an automorphism of $GF(p^n)$. The fixed field of σ (which consists of those elements left fixed by σ) is the subfield $GF(p)$. Thus, $GF(p^n)$ is a Galois extension of $GF(p)$, and its Galois group is generated by σ([AS] I pp. 199–201).

Contents

List of Notation

Chapter 1

Basic Concepts

§1. The Definition of a Group and Some Examples

Let us consider the set consisting of all the integers $\{0, \pm 1, \pm 2, \ldots \}$. The sum $m + n$ of any two integers m and n is also an integer, and the following two rules of addition hold for any arbitrary integers m, n, and p:

(1) $m + n = n + m$.
(2) $(m + n) + p = m + (n + p)$.

Furthermore, for any two given integers m and n, the equation

$$m + x = n$$

has a unique solution, $x = n - m$, which is also an integer.

Similar situations often occur in many different fields of mathematics, and they do not necessarily concern only the integers. Consider, for instance, the set of all nonsingular 2×2 matrices (that is, all 2×2 matrices A such that the determinant of A is not zero). Let A and B be any two 2×2 matrices,

$$A = \begin{pmatrix} a_1 & a_2 \\ a_3 & a_4 \end{pmatrix} \quad \text{and} \quad B = \begin{pmatrix} b_1 & b_2 \\ b_3 & b_4 \end{pmatrix}.$$

Then, the product of A and B is also a 2×2 matrix. This product, denoted by the juxtaposition of the letters A and B, is defined as

$$AB = \begin{pmatrix} a_1 b_1 + a_2 b_3 & a_1 b_2 + a_2 b_4 \\ a_3 b_1 + a_4 b_3 & a_3 b_2 + a_4 b_4 \end{pmatrix}.$$

With respect to the binary operation of the multiplication of matrices, we note that, in general, AB is not necessarily equal to BA, but the associative law $(AB)C = A(BC)$ is valid for any three 2×2 matrices A, B, and C. If the 2×2 matrix A is nonsingular, then the equations $AX = B$ and $YA = B$ have unique solutions, $X = A^{-1}B$ and $Y = BA^{-1}$, respectively, where X and Y are both 2×2 matrices.

A given set of elements together with an operation satisfying the associative law is said to be a group or to form a group if any linear equation has a unique solution which is in the set. Thus, the totality of nonsingular 2×2 matrices together with multiplication is said to be a group, as is the set of all the integers with addition.

In the set of all the integers, multiplication, as well as addition, is defined, and for any three integers m, n, and p, the associative law of multiplication is satisfied: $(mn)p = m(np)$. However, the equation $mx = n$ need not have an integral solution x. So, the set of all integers does not form a group with respect to multiplication. On the other hand, the set of all integers does form a group with respect to addition, as we saw earlier. Thus, we can say that a given set is a group only after we specify the operation defined on the set; to be precise, we should say that a given set is a group with respect to the operation specified on the set.

An operation defined on a set G is, like the addition or multiplication of integers or the multiplication of matrices, a rule to assign a third element c of G to any pair of elements (a, b) where $a, b \in G$. Thus, an operation is nothing but a mapping from the direct product set $G \times G$ consisting of all ordered pairs (a, b) of elements a and b of G into the set G. In the case of the addition of integers, the third element c is the sum of two integers; while in the case of multiplication, it is the product. As can be seen from these examples, the exact meaning of an operation varies from one example to another, but we often denote $c = ab$, for the sake of convenience, and call c the *product* of two elements, a and b.

We will now state the formal definition of a group in terms of the product, but the readers are cautioned again that the product xy merely signifies the element c determined by the operation specified on the set G; so, for example, if G is the set of all integers, it need not be the usual product defined on the set of integers. In this particular case, since the set forms a group with respect to addition, the product xy is really the sum $x + y$.

Definition 1.1. A nonempty set G on which a product operation is defined is said to be or to form a **group** (with respect to the given product operation) if the following two conditions are satisfied:

(i) For any three elements a, b, and c of G, the associative law holds:

$$(ab)c = a(bc).$$

(ii) For two arbitrary elements a and b, there exist elements x and y of G which satisfy the equations

$$ax = b \quad \text{and} \quad ya = b.$$

The following properties of a group are important.

Theorem 1.2. (ii)′ *There is a unique element e in G such that, for all $g \in G$,*

$$ge = eg = g.$$

(ii)″ *For any element $a \in G$, there is a unique element $a' \in G$ such that*

$$aa' = a'a = e,$$

where e is the element of G defined in (ii)′.

(iii) *The solutions x and y of the equations $ax = b$ and $ya = b$ are unique, and we have*

$$x = a'b \quad and \quad y = ba'$$

where a' is the element associated with the element a in (ii)″.

Proof. Since the set G is not empty, we take an element a of G. By the property (ii), there are solutions $x = e$ and $y = e'$ of the equations

$$ax = a \quad and \quad ya = a.$$

Also, if g is an arbitrary element of G, there are elements u and v of G such that

$$au = g = va;$$

so, we have

$$ge = (va)e = v(ae) = va = g.$$

The second equality, $(va)e = v(ae)$, follows from the associative law (i). Similarly, we obtain $e'g = g$. Since the element g is arbitrary, we may take $g = e$ to get $e'e = e$. On the other hand, the element e satisfies $ge = g$ for any $g \in G$, so $e'e = e'$. Therefore, we have $e' = e'e = e$. This proves that an arbitrary solution of the equation $ax = a$ is equal to a solution of $ya = a$. Thus, the uniqueness of the element e is proved, and (ii)′ holds.

The proof of (ii)″ is similar. By (ii), there are elements a' and a'' of G such that

$$aa' = e = a''a.$$

Using (i) and (ii)′, we have

$$a'' = a''e = a''(aa') = (a''a)a' = ea' = a'.$$

Hence the proof of the uniqueness of a' is similar to that of the uniqueness of e in (ii)'.

The proof of (iii). If $ax = b$, then the left multiplication of a' of (ii)" gives us

$$a'b = a'(ax) = (a'a)x = ex = x.$$

Thus, a solution of $ax = b$ is $x = a'b$, and it is unique; similarly, the solution of $ya = b$ is uniquely determined to be $y = ba'$. $\square$

Corollary. *A nonempty set G with an operation is a group if the conditions* (i), (ii)', *and* (ii)" *are satisfied.*

Definition 1.3. The element e defined in (ii)' is called the **identity** of G; the element a' defined in (ii)" is called the **inverse** of a.

The identity of a group is usually denoted by 1, while the inverse of an element a is customarily denoted by a^{-1}.

Theorem 1.4. *We have* $(a^{-1})^{-1} = a$ *and* $(ab)^{-1} = b^{-1}a^{-1}$; *that is, the inverse of the product is the product of the inverses of the factors, taking the product in the reverse order.*

Proof. The first equality follows from (ii)" (the uniqueness of the inverse). The second one is proved by the equality

$$(ab)(b^{-1}a^{-1}) = ((ab)b^{-1})a^{-1} = (a(bb^{-1}))a^{-1}$$
$$= (ae)a^{-1} = aa^{-1} = e$$

and the uniqueness of the inverse. Notice the change in the order of the factors from ab to $b^{-1}a^{-1}$. $\square$

The *product* of the elements $a_1, a_2, \ldots, a_n$ $(n \geq 3)$ of G is defined inductively by

$$a_1 \cdots a_n = (a_1 \cdots a_{n-1})a_n.$$

(1.5) *For* $n + 1$ *arbitrary elements* $a_0, a_1, \ldots, a_n$ *of a group G, we have* $a_0(a_1 \cdots a_n) = a_0a_1 \cdots a_n.$

Proof. Set $a = a_0, a_1 \cdots a_{n-1} = b$, and $a_n = c$. Then by the above definition

$$a_1 \cdots a_n = bc,$$

so the left side of the equation in (1.5) is equal to $a(bc)$. By the associative law, we have $a(bc) = (ab)c$. On the other hand,

$$ab = a_0(a_1 \cdots a_{n-1}),$$

so by the inductive hypothesis, $ab = a_0 a_1 \cdots a_{n-1}$. From the definition, we get $(ab)c = a_0 a_1 \cdots a_n$; thus, (1.5) holds. $\square$

Theorem 1.6. *The* **general associative law** *holds in any group. That is, if $x_1, x_2, \ldots, x_n$ are n arbitrary elements of a group, then the product of $x_1, \ldots, x_n$ is uniquely determined irrespective of the ways the product is taken, provided that the order of factors is unchanged.*
For example, $(xy)((z(uv))w) = x(((y(zu))v)w)$.

Proof. Suppose that, as the last step in computing a product of the elements $x_1, x_2, \ldots, x_n$ of G, we take the product of two elements a and b of G, where a is a product of $x_1, \ldots, x_p$ and b is a product of $x_{p+1}, x_{p+2}, \ldots, x_n$. Since $p \geq 1$, by the inductive hypothesis, we find that a is equal to the product $x_1 \cdots x_p$ and b is the product $x_{p+1} \cdots x_n$. If we set $a = a_0$, $x_{p+i} = a_i$ $(i = 1, 2, \ldots, n - p)$, then, by (1.5), we have

$$a_0(a_1 \cdots a_{n-p}) = a_0 a_1 \cdots a_{n-p};$$

the left side is $(x_1 \cdots x_p)(x_{p+1} \cdots x_n)$, while the right side is $x_1 \cdots x_n$. Thus, the general associative law holds. $\square$

The inverse of a product of n elements is given by

(1.7)
$$(a_1 \cdots a_n)^{-1} = a_n^{-1} \cdots a_1^{-1}.$$

On the right side, the factors are arranged in decreasing order of indices.
If $a_1 = a_2 = \cdots = a_n$, then we use the power notation (for $a = a_1$)

$$a_1 a_2 \cdots a_n = a^n.$$

If $n = -m$ is a negative integer, then we define

$$a^n = (a^{-1})^m;$$

also, we define $a^0 = 1$ (where 1 is the identity of G). Then, by (1.7), we have $(a^{-1})^m = (a^m)^{-1}$.

(1.8) *The formulas*

$$a^m a^n = a^{m+n} \quad and \quad (a^m)^n = a^{mn}$$

hold for any element a of a group G and any pair of integers m and n.

Proof. We consider several cases according to the signs of m and n; but in each case, the formula is a particular case of the general associative law (1.6).

$\square$

Definition 1.9. We say that two elements a and b of a group are **commutative**, or commute, if $ab = ba$. A group is said to be **abelian** (so named after N. H. Abel (1802–1829)) or commutative, if any two elements commute.

(1.10) *The* **general commutative law** *holds in an abelian group. In general, let* $a_1, a_2, \ldots, a_n$ *be n mutually commuting elements in a group. (The group need not be abelian.) Then the product of these elements is uniquely determined and does not depend on the way the product is taken or on the order of the factors. Thus, if $(\sigma(1), \sigma(2), \ldots, \sigma(n))$ is an arrangement of $(1, 2, \ldots, n)$ in some order, then we have*

$$a_{\sigma(1)} a_{\sigma(2)} \cdots a_{\sigma(n)} = a_1 a_2 \cdots a_n.$$

Proof. By (1.6) we have

$$a_{\sigma(1)} a_{\sigma(2)} \cdots a_{\sigma(n)} = a_{\sigma(1)} \cdots (a_{\sigma(i)} a_{\sigma(i+1)}) \cdots a_{\sigma(n)}.$$

By assumption, $a_{\sigma(i)} a_{\sigma(i+1)} = a_{\sigma(i+1)} a_{\sigma(i)}$ holds, so we can move any a_k, except the last, one place to the right. By repeating this process, we can move a_n to the position furthest to the right, and we have

$$a_{\sigma(1)} a_{\sigma(2)} \cdots a_{\sigma(n)} = a_{\tau(1)} \cdots a_{\tau(n-1)} a_n,$$

where $(\tau(1), \ldots, \tau(n-1))$ is an arrangement of $(1, 2, \ldots, n-1)$ in some order. By the inductive hypothesis, we get (1.10). $\square$

(1.11) *If a and b are two commuting elements of a group, then the formula*

$$(ab)^n = a^n b^n$$

holds for any integer n.

Proof. By the assumption, we have $ab = ba$. So, for the inverses, we get $a^{-1} b^{-1} = b^{-1} a^{-1}$ by (1.4). Then, (1.11) is a particular case of (1.10) for any (positive or negative) value of n. $\square$

Definition 1.12. The number of elements in a group G is called the **order** of G and is denoted by $|G|$. If $|G|$ is finite, then G is said to be a finite group; otherwise G is an infinite group.

So far, we have regarded the operation defined on G as multiplication and have used the multiplicative notation. But this is only for convenience, and we may use other notations. For example, in the group consisting of all the integers, it is natural to use the additive notation and talk about the sum $a + b$ of two elements. Even for an arbitrary abelian group G, we sometimes use the additive notation: we call the operation addition and let $a + b$ represent the sum of two elements. In this case, we say G is an *additive group*. An additive group, in this book, is always understood to be abelian. We denote the identity of an additive group by 0, and the inverse of an element a by $-a$.

EXAMPLE 1. The totality of integers forms an additive group with respect to the usual addition. Because of its importance, we follow the customary notation and use the special symbol $\mathbb{Z}$ to denote this additive group.

EXAMPLE 2. An arbitrary field F forms an additive group with respect to the addition defined on F. We use the notation F^+ to emphasize this additive structure. The set $F^* = F - \{0\}$ of nonzero elements of F forms a group with respect to multiplication. The identity of this multiplicative group (in the sense of (1.3)) is the identity of the field F. We adopt the definition of a field according to which the multiplicative group F^* of F is abelian. A set which satisfies all the axioms for fields except possibly the commutativity of multiplication is said to be a *skew-field*. If K is a skew-field, the multiplicative group $K^* = K - \{0\}$ is not necessarily a commutative group. By a famous theorem of Wedderburn (see [AS] p. 200), there exists no finite skew-field which is not commutative.

EXAMPLE 3. Let $\mathbb{R}$ denote the field consisting of all rational numbers. The following also holds when we consider $\mathbb{R}$ to be the field consisting of all real numbers. Then the set of all nonsingular 2×2 matrices with coefficients in $\mathbb{R}$ forms a group. Not only the set of all the nonsingular 2×2 matrices, but also the totality of 2×2 matrices having determinant 1 forms a group with respect to the multiplication of matrices. This follows immediately from the formula

$$\det(AB) = \det A \cdot \det B.$$

The identity of these groups of matrices is the *identity matrix*

$$I = \begin{pmatrix} 1 & 0 \\ 0 & 1 \end{pmatrix}.$$

EXAMPLE 4. We define two subsets G and H of the group of all nonsingular 2×2 matrices discussed above (Example 3) as follows: Let G be the totality of matrices of the form

$$\begin{pmatrix} r & s \\ 0 & 1 \end{pmatrix} \qquad (r, s \in \mathbb{R}, r \neq 0),$$

and let H be the set of all matrices of the following form:

$$\begin{pmatrix} 1 & n \\ 0 & 1 \end{pmatrix} \quad (n \in \mathbb{Z}).$$

We claim that both G and H form groups with respect to matrix multiplication. First, the proof that both sets G and H are closed under multiplication is trivial:

$$\begin{pmatrix} r & s \\ 0 & 1 \end{pmatrix}\begin{pmatrix} r' & s' \\ 0 & 1 \end{pmatrix} = \begin{pmatrix} rr' & rs' + s \\ 0 & 1 \end{pmatrix}.$$

The identity matrix I is contained in both G and H, and becomes the identity of G and of H. Furthermore, we have

$$\begin{pmatrix} r & s \\ 0 & 1 \end{pmatrix}^{-1} = \begin{pmatrix} r^{-1} & -r^{-1}s \\ 0 & 1 \end{pmatrix},$$

so the inverse of an element of G is also contained in G. Thus, the set G is a group by Cor. to Theorem 1.2. Similarly, the set H is a group. In this example, H is contained in the set G; thus the set H is one of many examples of a subgroup. We will define subgroups in the next section.

EXAMPLE 5. Let R_n $(n > 2)$ be a regular n-gon in the plane P. The motions of P (congruences of P) which leave R_n invariant are the rotations of degree $2\pi k/n$ $(k = 0, 1, \ldots, n-1)$ around the center O of the n-gon R_n, and the reflections with respect to either the perpendicular bisectors of the sides of R_n or the bisectors of the interior angles at the vertices. If n is odd, the reflection with respect to the bisector of an interior angle coincides with the reflection with respect to the perpendicular bisector of opposite side, so for any n there are exactly $2n$ motions which leave R_n invariant. Any motion σ on P moves a point p of P to a point q of P which is denoted by $q = \sigma(p)$. If ρ is another motion on P, then the point $\sigma(p)$ is moved to the point $\rho(\sigma(p))$. The transformation which sends each point p to $\rho(\sigma(p))$ is also a motion on P which we call the *product* of σ and ρ and denote by $\sigma\rho$. This defines an operation. The transformation on P which sends each point of P to itself is a motion, and it is the identity with respect to the operation defined above. If σ is a motion, the inverse transformation σ^{-1} is also a motion. If σ, ρ, and τ are three motions, then both $(\sigma\rho)\tau$ and $\sigma(\rho\tau)$ are the motion which sends a point p to $\tau(\rho(\sigma(p)))$; thus, the associative law (ii) is satisfied. Thus, the totality of motions of P forms a group with respect to the operation defined above; also the set of $2n$ motions of P which leave the n-gon R_n invariant forms a finite group of order $2n$ with respect to the same operation. The group consisting of these $2n$ motions is said to be the *dihedral group* of order $2n$ and is usually denoted by D_{2n}.

Exercises

1. Assume that a set G with an operation satisfying the associative law satisfies the following two conditions (a) and (b):

    (a)   There exists an element e of G such that $ge = g$ for all $g \in G$.

    (b)   For any element a of G, there exists an element a' such that $aa' = e$.

Then, show that G is a group with respect to the given operation.

    (*Hint.* The conditions (a) and (b) are about one half of the conditions (ii)′ and (ii)″; so prove the remaining half. Apply (b) to the element a'. There exists an element a'' such that $a'a'' = e$. The associative law applied to $aa'a''$ yields $a = ea''$; so for any $a \in G$ we obtain $ea = a = ea'' = a''$, proving (ii)′ and (ii)″.)

2. Assume the following (a)′ in place of (a) in Exercise 1:

    (a)′   There exists an element e of G such that $eg = g$ for all $g \in G$.

Is a set G satisfying (a)′ and (b) a group? If not, how is the structure of G related to that of a group?

    (*Hint.* The set G need not be a group. If we set $H = Ge$, H is a group. Let U be the set of elements x in G such that $xe = e$. Then any element of G can be written as hu ($h \in H$, $u \in U$), and this expression is unique provided that an element h of H is written as he. If $h_1, h_2 \in H$ and $u_1, u_2 \in U$, then the product in G is given by $(h_1 u_1)(h_2 u_2) = (h_1 h_2)u_2$.)

§2. Subgroups

The concept of subgroups is one of the most basic ideas in group theory.

Definition 2.1. A nonempty subset H of a group G is said to be a **subgroup** of G if the following conditions are satisfied:

    (i) $a, b \in H \Rightarrow ab \in H$;

    (ii) $a \in H \Rightarrow a^{-1} \in H$.

(2.2) *Let H be a subgroup of a group G. Then, the following statements hold:*

    (a) *The subgroup H contains the identity of G.*

    (b) *Let a and b be two elements of H. As elements of G, they determine the product ab, and this product is an element of the subgroup H. Thus, the product of two elements of H is defined in H, and so, with respect to this operation, H forms a group.*

    (c) *The identity of the group H coincides with the identity of the group G. Let a be an element of H. Then, the inverse of a in the group H coincides with the inverse of a in the group G.*

Proof. By definition, H is nonempty, so there is an element a contained in the subgroup H. By (ii), we have $a^{-1} \in H$. From (i), $aa^{-1} \in H$. But $aa^{-1} = 1$, so aa^{-1} is the identity of G. Thus, the assertion (a) holds.

The operation on H referred to in (b) is the one naturally induced from the operation defined on G and satisfies the associative law. The subgroup H contains the identity of G which is clearly the same as the identity of the operation on H. By the condition (ii), the inverse of any element of H with respect to the operation defined on G is also contained in H. The uniqueness of the inverse implies that this inverse is also the inverse with respect to the same operation on H. Thus, the assertions (b) and (c) follow. $\square$

In Example 3 of §1, the totality of matrices having determinant 1 forms a subgroup of the group of all nonsingular matrices. We also note that in Example 4, the group H forms a subgroup of G, while in Example 5, the dihedral group is a subgroup of the group consisting of all the motions on P.

The relation that H is a subgroup of a group G will be denoted by:

$$H \subset G \quad \text{or} \quad G \supset H.$$

This is the same notation used to denote a subset; in this book, we mainly consider subsets which are also subgroups, so we hope no confusion will arise. The readers are cautioned, however, that even if a group H is a subset of a group G, H may not be a subgroup of G (Example 2, §1). So, in the case that H is a subset of G, it is necessary to check whether or not H is indeed a subgroup of G.

(2.3) *For any group G, the set G itself is a subgroup of G; the subset $\{1\}$ consisting of only the identity 1 is a subgroup of G. If H is a subgroup of G and K is a subgroup of H, then K is a subgroup of G. If H and K are subgroups of G and if K is a subset of H, then K is a subgroup of H.*

Proof. The proofs are relatively easy, so we will only prove that if K is a subgroup of H, then K satisfies the condition (i) of the definition of a subgroup of G. Let x and y be two elements of K. Then x and y are also elements of H, so the product xy with respect to the operation defined on G is an element of H since H is a subgroup of G. This product xy coincides with the product of the elements x and y with respect to the operation defined on the group H (see (2.2)). Since K is a subgroup of H, we have $xy \in K$. Thus, the condition (i) holds. $\square$

Definition 2.4. Among the subgroups of G, the subgroups G and $\{1\}$ are said to be **trivial**. A subgroup H is said to be a **proper subgroup** of G if $H \neq G$. If M is a proper subgroup of G and if $M \subset H \subset G$ for a subgroup H of G implies $G = H$ or $H = M$, then M is said to be a **maximal subgroup** of G.

If we need to emphasize that H is a proper subgroup of G, we use the notation

$$H \subsetneqq G.$$

We stress again that a maximal subgroup is a subgroup which is maximal among proper subgroups. A group may contain no maximal subgroup (Exercise 3).

(2.5) *Let H and K be two subgroups of a group G. The intersection $H \cap K$ of H and K is a subgroup of G. In general if H_λ is a subgroup for each $\lambda \in \Lambda$, then their intersection $\bigcap H_\lambda$ is a subgroup of G.*

Proof. Set $L = \bigcap H_\lambda$. We will prove that L satisfies the conditions (i) and (ii) of Definition 2.1. Let $a, b \in L$. By the definition of L, we have $a, b \in H_\lambda$ for all $\lambda \in \Lambda$. Since H_λ are subgroups by assumption, we have $ab \in H_\lambda$ for all $\lambda \in \Lambda$; whence $ab \in \bigcap H_\lambda = L$. This proves (i). The condition (ii) can be proved similarly. □

From now on, we will denote the subgroup formed by the intersection of H and K by $H \cap K$:

$$H \cap K = \{x \mid x \in H \quad \text{and} \quad x \in K\}.$$

We use the notations $H \cap K \cap L$ and $\bigcap H_\lambda$ to denote not only the intersection subsets, but also the subgroups formed by these subsets.

The problem of surveying the subgroups of a given group G is one of the most important problems in group theory. If a subset S of G is given, three subgroups defined from S are of particular interest:

$\langle S \rangle$:   the subgroup of G generated by S;
$N_G(S)$: the normalizer of S in G;
$C_G(S)$: the centralizer of S in G.

Before defining these subgroups, we will prove the following lemma.

(2.6) *Let S be a subset of G, and let H be the set of all elements of G which can be written as the product of a finite number of elements of S or of the inverses of elements of S; that is :*

$$x \in H \Rightarrow x = u_1 u_2 \cdots u_n$$

where $u_i \in S$ or $u_i^{-1} \in S$ for every i. Then H forms a subgroup of G which contains S.

Proof. We will verify the conditions (i) and (ii) of (2.1) for H. If x and y are elements of H, then we can write

$$x = u_1 u_2 \cdots u_n, \quad \text{and} \quad y = v_1 v_2 \cdots v_m$$

where $u_i \in S$ or $u_i^{-1} \in S$ for all i, and similarly, $v_j \in S$ or $v_j^{-1} \in S$ for all j. Hence the product

$$xy = u_1 u_2 \cdots u_n v_1 v_2 \cdots v_m$$

is an element which belongs to H, so the condition (i) is satisfied. The condition (ii) follows immediately from the formula (1.7). It is also trivial to show that H contains S. Thus (2.6) has been proved. $\square$

Definition 2.7. The subgroup H of G defined in (2.6) is said to be the subgroup generated by the subset S, and is denoted by $\langle S \rangle$. The set S is said to be a **generating set** or a **set of generators** of H.

If subsets S_λ $(\lambda \in \Lambda)$ are given, the notation

$$\langle S_\lambda \, (\lambda \in \Lambda) \rangle \quad \text{or} \quad \langle S_1, S_2, \ldots \rangle$$

will denote the subgroup generated by the union of the subsets S_λ, while the notation $\{S_\lambda \, (\lambda \in \Lambda)\}$ will denote the collection of subsets S_λ $(\lambda \in \Lambda)$.

(2.8) *The subgroup $\langle S \rangle$ coincides with the intersection of all subgroups of G which contain the subset S.*

Proof. Let L be the intersection of all subgroups of G which contain S, and let $H = \langle S \rangle$. Since H is a subgroup of G which contains S, we obviously have $L \subset H$. Let K be any subgroup of G containing S. Then, K contains any element s of S and s^{-1} as well. It follows from the definition and repeated applications of (2.1) (i) that all the elements of H are contained in K. Hence $H \subset K$. Since this holds for any subgroup K of G which contains S, we conclude that H is contained in L: $H \subset L$. Since we saw earlier that $L \subset H$, we have $H = L$. $\square$

Corollary. *The subgroup $\langle S \rangle$ is the smallest subgroup of G which contains the subset S.*

Next, we will define the conjugate subset.

Definition 2.9. Let S be a subset of a group G, and let g be a fixed element of G. The subset consisting of all elements of the form $g^{-1}sg$ $(s \in S)$ is said to be the **conjugate** of S by the element g, and is denoted by $g^{-1}Sg$ or S^g:

$$S^g = g^{-1}Sg = \{g^{-1}sg \,|\, s \in S\}.$$

The subset S^g is said to be conjugate to S via g; S^g is conjugate to S in G if g is an element of G. The set of all conjugates of an element x is called the **conjugacy class** of x in G.

(2.10) *The following formula holds for any $g, h \in G$:*

$$S^{gh} = (S^g)^h.$$

Let H be the set of all elements g satisfying $S^g = S$. Then H is a subgroup of G.

Proof. The elements of S^{gh} can be written as $(gh)^{-1}s(gh)$ $(s \in S)$. Since $(gh)^{-1} = h^{-1}g^{-1}$ by (1.4), the general associative law gives us:

$$(gh)^{-1}s(gh) = h^{-1}(g^{-1}sg)h.$$

Hence we have $S^{gh} = (S^g)^h$.

By definition, $H = \{g \,|\, S^g = S\}$. So, if $a, b \in H$, $S^a = S = S^b$. From the first half of this proposition, we get

$$S^{ab} = (S^a)^b = S^b = S,$$

and so, the condition (2.1) (i) holds. If $a \in H$, then $c = a^{-1}$ satisfies the equality

$$S = S^1 = S^{ac} = (S^a)^c = S^c.$$

Thus, H satisfies (2.1) (ii), too. Therefore, H is a subgroup of G. □

Definition 2.11. The subgroup H defined in (2.10) is said to be the **normalizer** of S in G and is denoted by $N_G(S)$:

$$N_G(S) = \{g \,|\, g \in G, \, S^g = S\}.$$

Finally, we will define the centralizers.

(2.12) *The totality of the elements of G which commute with all the elements of a subset S forms a subgroup of the group G.*

Proof. Let $H = \{g \,|\, gs = sg \text{ for all elements } s \text{ of } S\}$. We have

$$gs = sg \Leftrightarrow s = g^{-1}sg.$$

Using the notation $s^g = g^{-1}sg$, we use (2.10) to get $s^{gh} = (s^g)^h$. Thus, (2.12) follows from (2.10).

Thus, (2.12) follows from (2.10). □

Definition 2.13. The subgroup defined in (2.12) is called the **centralizer** of S in G and will be denoted by $C_G(S)$.

The subgroups defined in Definitions 2.7, 2.11, and 2.13 are very important in the study of groups. Throughout this book, we will always use the notation introduced in this section to denote the corresponding subgroups or subsets. For example, S^g is always the conjugate of S by the element g, and $C_G(S)$ always denotes the centralizer of S in G.

Definition 2.14. The centralizer of the group G itself is said to be the **center** of G.

The center of a group will be the subject of many investigations, so we use a special symbol $Z(G)$ to denote the center of G:

$$Z(G) = \text{the center of } G = C_G(G).$$

The next theorem and its proof reveal the most significant aspect of the concept of conjugacy.

(2.15) *If H is a subgroup of G, then any conjugate of H is also a subgroup of G.*

Proof. Let H^g be a conjugate of H by an element g of G. An element of H^g can be written as $g^{-1}ug$ $(u \in H)$. If we choose another element $g^{-1}vg$ $(v \in H)$ of H^g, then the general associative law gives us:

(2.16)
$$(g^{-1}ug)(g^{-1}vg) = g^{-1}u(gg^{-1})vg$$
$$= g^{-1}uvg.$$

By assumption, H is a subgroup of G, so $u, v \in H$ implies that $uv \in H$. Hence the formula (2.16) shows that H^g satisfies (2.1)(i).

From the formula (1.7), we get:

(2.17)
$$(g^{-1}ug)^{-1} = g^{-1}u^{-1}g.$$

(We can also get (2.17) from (2.16) by substituting u^{-1} for v.) Hence, H^g satisfies (2.1)(ii) as well. So, H^g is a subgroup of G. □

Definition 2.18. For any subgroup H of G, a conjugate H^g is called a **conjugate subgroup** of H.

Theorem 2.19. *For any subset S of G and any element g of G, we have :*

$$\langle S^g \rangle = \langle S \rangle^g,$$
$$N_G(S^g) = N_G(S)^g,$$
$$C_G(S^g) = C_G(S)^g.$$

Proof. By induction beginning with (2.16), we have

$$(g^{-1}u_1 g)(g^{-1}u_2 g) \cdots (g^{-1}u_n g) = g^{-1}(u_1 u_2 \cdots u_n)g.$$

From the above formula, together with (2.17) and Definition 2.7, we have $\langle S^g \rangle = \langle S \rangle^g$.

An element x of G belongs to $N_G(S^g)$ if and only if $(S^g)^x = S^g$. By (2.10),

$$(S^g)^x = S^g \Leftrightarrow S^{gx} = S^g \Leftrightarrow S^{gxg^{-1}} = S.$$

The last equality is equivalent to saying that $gxg^{-1} \in N_G(S)$, which in turn is equivalent to saying that $x \in N_G(S)^g$. This proves that $N_G(S^g) = N_G(S)^g$. The last equality, $C_G(S^g) = C_G(S)^g$, may be proved similarly. $\square$

Corollary. $N_G(S) \subset N_G(C_G(S))$.

The proof of this containment follows immediately from the definitions of both sides and the formula (2.19).

We will now study the subgroups generated by a single element.

Theorem 2.20. *Let g be an element of a group G. Then, the subgroup of G generated by g coincides with the totality of integral powers of g: $\{g^n\}$ ($n = 0, \pm 1, \pm 2, \ldots$). And, we always have one of the following two cases:*

(1) If $n \neq m$, then $g^n \neq g^m$. In this case, $\langle g \rangle$ contains infinitely many elements.

(2) For some natural number k, we have $g^k = 1$, and the k elements $1, g, g^2, \ldots, g^{k-1}$ are distinct so that $\langle g \rangle$ consists of exactly these k elements. In this case,

$$g^n = g^m \Leftrightarrow m \equiv n \pmod{k}.$$

Thus, $g^n = 1$ if and only if $n \equiv 0 \pmod{k}$.

Proof. It follows from Definition 2.7 that each element of $\langle g \rangle$ can be written as the product $u_1 u_2 \cdots u_n$, where each u_i is either g or g^{-1}. Hence $\langle g \rangle$ coincides with the powers of g (cf. (1.8)).

In order to prove that we always have one of the two cases described above, we assume that (1) does not hold. Then, there are integers m and n such that $m \neq n$ but $g^m = g^n$. Without loss of generality, we may assume $m > n$. By (1.8), we get

$$g^{m-n} = g^m g^{-n} = g^m (g^n)^{-1} = 1.$$

Thus, there is a natural number l such that $g^l = 1$. Let k be the smallest natural number satisfying $g^k = 1$. By the division algorithm, any integer m can be written as

$$m = qk + r \qquad 0 \le r < k$$

for some integers q and r. Again by (1.8) we get

$$g^m = g^{qk}g^r = (g^k)^q g^r = g^r.$$

Thus, $\langle g \rangle = \{1, g, g^2, \ldots, g^{k-1}\}$ and, by the minimality of k, $g^m = 1$ if and only if $m \equiv 0 \,(\text{mod } k)$. Since $g^m = g^n \Leftrightarrow g^{m-n} = 1$, we have

$$g^m = g^n \quad \text{if and only if} \quad m \equiv n \,(\text{mod } k).$$

Therefore, $1, g, g^2, \ldots, g^{k-1}$ are distinct elements of $\langle g \rangle$; thus, (2) is proved. $\qquad \square$

Definition 2.21. When the case (2) holds for an element g of G, g is said to be an element of finite order and the natural number k satisfying the properties of (2) is called the **order** of g. If the case (1) holds, the order of g is infinite.

(2.22) *Let a and b be two commuting elements having orders m and n respectively. The order of the product ab is a divisor of mn. If m and n are relatively prime, then the order of ab is equal to mn.*

Proof. By (1.8) and (1.11) we get

$$(ab)^{mn} = a^{mn}b^{mn} = (a^m)^n(b^n)^m = 1.$$

Thus, the order l of ab divides mn by (2.20). So, $l \le mn$.

Next, suppose that m and n are relatively prime. If l is the order of ab, then

$$1 = (ab)^{lm} = a^{lm}b^{lm} = (a^m)^l b^{lm} = b^{lm},$$

so the order n of b divides lm. Since n and m are relatively prime by assumption, n divides l. Similarly, m divides l; so, mn divides l. This proves that $mn \le l \le mn$, or $mn = l$. $\qquad \square$

If two elements a and b do not commute, then the order of the product ab can be arbitrary large (cf. Exercise 6, §3).

Definition 2.23. A group generated by a single element is said to be a **cyclic group**.

Theorem 2.20 and (2.22) are the main theorems concerning the structure of cyclic groups.

Theorem 2.24. *A group G, $G \neq \{1\}$, with no nontrivial subgroup is a finite cyclic group of prime order.*

Proof. Let G be a group different from $\{1\}$ which does not contain any nontrivial subgroup. If g is a nonidentity element of G, then $\langle g \rangle$ is a subgroup of G different from $\{1\}$. So, by the assumption that G does not contain any nontrivial subgroup, $G = \langle g \rangle$. If $|G|$ is not finite, $\langle g^2 \rangle$ is a proper subgroup of G (because $\langle g^2 \rangle$ does not contain g^n for any odd integer n), and this contradicts the assumption that G has no nontrivial subgroup. Hence $|G|$ is finite. Let $|G| = k$. Suppose that k is not prime, and let $k = lm$ with $l > 1$ and $m > 1$. Then, for $g^l = h$, we have:

$$h \neq 1 \quad \text{and} \quad h^m = (g^l)^m = g^{lm} = g^k = 1.$$

Thus, $\langle h \rangle$ has order m. This implies that $\{1\} \neq \langle h \rangle \neq G$, which is contrary to the assumption. Hence $|G|$ is prime. $\square$

The converse of Theorem 2.24 holds, too (cf. Corollary to (3.3)).
In general, the problem of surveying all of the possible subgroups of a given group G is a very difficult problem, but for the particular group $\mathbb{Z}$, this problem has an easy solution.

(2.25) *For each natural number k, let H_k be the totality of multiples of k. Then H_k is a subgroup of $\mathbb{Z}$. Conversely, any nontrivial subgroup of $\mathbb{Z}$ is one of these subgroups H_k ($k > 1$).*

Proof. It is obvious that H_k is a subgroup of $\mathbb{Z}$. Let H be a nontrivial subgroup of $\mathbb{Z}$. If an integer m is contained in H, then so is $-m$. Hence H contains a natural number. Let k be the smallest natural number contained in H. Then any multiple of k is contained in H, so $H_k \subset H$. On the other hand, the division algorithm states that any $m \in H$ can be written as $m = qk + r$ ($0 \leq r < k$). As $m, k \in H$, we have $r \in H$. By the minimality of k, we have $r = 0$; so, $m = qk$ and $H_k = H$. Since H is nontrivial, $H \neq \mathbb{Z}$ and $k > 1$. $\square$

(2.26) *The center $Z(G)$ of a group G is a commutative subgroup of G. Let H be a subgroup of $Z(G)$. Then, the subgroup $\langle H, x \rangle$ generated by H and an element x of G is abelian.*

Proof. The first part will be obvious because it follows almost immediately from the second. We will prove the second assertion in two steps.
    (a) *Let $K = \langle H, x \rangle$, and let k be an element of K. Then, we will prove that k can be written as hx^m ($h \in H$ and $m \in \mathbb{Z}$).*

By Definition 2.7, k can be expressed as a product of the form $u_1\, u_2\, \cdots\, u_n$ where $u_i \in H$, $u_i = x$, or $u_i = x^{-1}$. Since $H \subset Z(G)$ by our assumption, $u_1, u_2, \ldots, u_n$ commute with each other. Thus, all the factors of k contained in H can be collected to the left, and the remaining factors are either x or x^{-1}. Let h be the product of all the factors of k which are elements of H. Then, $k = hx^m$ for some integer m.

(b) *If $h_1, h_2 \in H$, then $(h_1\, x^m)(h_2\, x^n) = (h_2\, x^n)(h_1\, x^m)$.*

Proof. Using the general associative law and the fact that $H \subset Z(G)$, we can prove (b) as follows:

$$(h_1\, x^m)(h_2\, x^n) = ((h_1\, x^m)h_2)x^n = (h_2(h_1\, x^m))x^n$$
$$= h_2(h_1(x^m x^n)) = h_2(h_1(x^n x^m))$$
$$= h_2(x^n(h_1\, x^m)) = (h_2\, x^n)(h_1\, x^m).$$

From (a) and (b), it follows that K is an abelian subgroup. $\square$

A family $\{H_\lambda\}$ ($\lambda \in \Lambda$) of subgroups H_λ is said to be *linearly ordered* if, for any indices $\lambda, \mu \in \Lambda$, either $H_\lambda \subset H_\mu$ or $H_\mu \subset H_\lambda$. The following lemma is indispensable in applying Zorn's lemma to a family of subgroups.

(2.27) *Let $\{H_\lambda\}$ ($\lambda \in \Lambda$) be a linearly ordered family of subgroups of G. Then, the union of the subgroups H_λ is a subgroup of G.*

Proof. Let H be the union of the subgroups H_λ. An element of H is, then, contained in some subgroup H_λ. Hence, if x and y are elements of H, then $x \in H_\lambda$ and $y \in H_\mu$ for some $\lambda, \mu \in \Lambda$. By assumption, the family $\{H_\lambda\}$ is linearly ordered, so we have $H_\lambda \subset H_\mu$ or $H_\mu \subset H_\lambda$. Both cases can be handled similarly, so let us assume $H_\lambda \subset H_\mu$. Then x, $y \in H_\mu$. Since H_μ is a subgroup, $xy \in H_\mu$ and $x^{-1} \in H_\mu$. Thus, both xy and x^{-1} are elements of H. This proves that H is a subgroup of G. $\square$

Exercises

1. Prove that the following is both a necessary and sufficient condition for a nonempty subset H of a group G to be a subgroup of G:

$$a,\, b \in H \Rightarrow ab^{-1} \in H.$$

2. (a) Show that the condition (i) of Definition 2.1 alone is sufficient for a nonempty subset H of a finite group G to be a subgroup of G.

(b) Let the order of every element of a group G be finite. Prove that the condition (i) of (2.1) is sufficient for a nonempty subset to be a subgroup.

(c) Show that, in general, the condition (i) alone is not sufficient for a nonempty subset to be a subgroup.

3. Let $\mathbb{R}$ be the additive group consisting of all rational numbers. Prove that $\mathbb{R}$ contains no maximal subgroup.

(*Hint.* (a) Let M be a maximal subgroup of $\mathbb{R}$. Show that $M \neq \{0\}$ and that a rational number r and a natural number k exist such that $r \notin M$ but $kr \in M$. For example, the following proposition may be used to prove (a):

(b) For any nonzero rational numbers r and s, there exist nonzero integers m and n such that $mr = ns$.

(c) Let $M_k = \{s \in \mathbb{R} \mid ks \in M\}$, where k is the natural number defined in (a). Show that M_k is a subgroup of $\mathbb{R}$.

(d) $M_k = \mathbb{R}$ (because $M_k \supsetneq M$ and M is maximal).

(e) We get a contradiction: $M = \mathbb{R}$.

If the concept of factor groups (to be defined in §4) were available, then we could get (d) immediately by applying Theorem 2.24 to the factor group $\mathbb{R}/M$.)

4. For a given subset X of a group G, let $\mathscr{H}$ be the set of subgroups H satisfying $H \cap X = \varnothing$ (the empty set). The set $\mathscr{H}$ becomes a partially ordered set by defining $H < K$ if and only if H and K are members of $\mathscr{H}$ and H is a subgroup of K. Show that, if $\mathscr{H}$ is not empty, $\mathscr{H}$ is inductively ordered, so $\mathscr{H}$ has at least one maximal element by Zorn's lemma.

Pick a subgroup H_0 satisfying $H_0 \cap X = \varnothing$, and let $\mathscr{H}_0$ denote the subset of $\mathscr{H}$ consisting of the members which contain H_0. Show that $\mathscr{H}_0$ is also inductively ordered, and has a maximal element. (For the inductively ordered sets and Zorn's lemma, see [IK] Chapter 1, §9.)

5. Suppose that a group G contains a proper subgroup H and a finite subset S such that $G = \langle H, S \rangle$. Show that there exists a maximal subgroup which contains H. In particular, if a group G is generated by a finite number of elements, then G has at least one maximal subgroup.

(*Hint.* Choose a finite subset T of G so that $G = \langle H, T \rangle$ and $|T|$ is minimal. For an element t of T, set $T_0 = T - \{t\}$. Then $\langle H, T_0 \rangle = K$ is a proper subgroup of G and satisfies $K \cap \{t\} = \varnothing$. Apply Exercise 4 to the family of those proper subgroups of G which contain K but do not contain t. A maximal element of this family is a maximal subgroup of G.)

§3. Cosets

Definition 3.1. Let H be a subgroup of a group G, and let x be an element of G. The subset of G consisting of the products hx ($h \in H$) is said to be a **right coset** of H in G and is denoted by Hx. A **left coset** xH is defined similarly. The number of distinct right cosets of H is called the **index** of H in G and denoted by $|G : H|$.

For any subset S of G, S^{-1} denotes the set of the inverses of the elements in S:

$$S^{-1} = \{s^{-1} \mid s \in S\}.$$

If S is a right coset of H, then $S = Hx$ for some $x \in G$. By Theorem 1.4, the inverse of an element hx of S is $x^{-1}h^{-1}$, so S^{-1} coincides with the left coset $x^{-1}H$. Similarly, $(yH)^{-1} = Hy^{-1}$. Thus, the number of distinct right cosets of H is equal to the number of distinct left cosets of H. We may define the index of H using left cosets. In the following, we will state most of the results for right cosets, but the left cosets have similar corresponding properties.

We call a right coset of H simply a coset of H. The following omnibus lemma contains the basic properties of cosets and is useful in many applications.

(3.2) *Let H be a subgroup of G.*

   (i)   *Every element g of G is contained in exactly one coset of H. This coset is Hg.*

   (ii) *Two distinct cosets of H have no common element.*

  (iii) *The group G is partitioned into a disjoint union of cosets of H.*

  (iv) *The function $h \to hx$ is a one-to-one correspondence between the elements of the set H and those of the coset Hx. In particular, if H is a finite subgroup, then any coset of H has the same number of elements as H.*

   (v) *Two elements x and y of G are contained in the same coset of H if and only if $xy^{-1} \in H$.*

Proof. (i) By (2.2), the subgroup H contains the identity 1 of G, so Hg contains $1 \cdot g = g$. Conversely, if a coset Hx contains g, there is an element h of H such that $hx = g$. Hence

$$Hg = H(hx) = (Hh)x = Hx.$$

The proposition (i) has been proved. The propositions (ii) and (iii) are corollaries of (i).

The equality $hx = h'x$ holds if and only if $h = h'$. This proves (iv). The coset of H containing the element y is Hy by (i). Since $x \in Hy$ is equivalent to $xy^{-1} \in H$, the assertion (v) holds. $\square$

Theorem 3.3. $|G| = |H| \cdot |G : H|$. *Thus, if G is a finite group, the order, as well as the index, of a subgroup is a divisor of the order of G.*

Proof. By (3.2) (iii, iv), the set G is partitioned into a disjoint union of $|G : H|$ subsets, each containing $|H|$ elements. Counting the number of elements in G, we get $|G| = |H| \cdot |G : H|$. $\square$

The above theorem of Lagrange is one of the basic results in finite group theory; it is very important in that this theorem of Lagrange puts a severe constraint on the subgroups of a given finite group. There are many useful corollaries of Theorem 3.3.

Corollary 1. *A finite cyclic group of prime order contains no nontrivial subgroup.*

This is the converse of Theorem 2.24, and follows immediately from Theorem 3.3.

Corollary 2. *The order of an element of a finite group G divides the order $|G|$.*

Proof. The order of an element g is equal to the order of the subgroup $\langle g \rangle$, whence the assertion follows from Theorem 3.3.   □

Corollary 3. *If the order of a finite group G is n, then every element x of G satisfies $x^n = 1$.*

Proof. The order of x divides n by Corollary 2, so the assertion follows from Theorem 2.20.   □

Let p be a prime number. The residue classes modulo p which contain integers prime to p forms a finite group of order $p - 1$ with respect to the multiplication of residue classes. This is a prototype of the factor group which will be discussed in the next section. Corollary 3 applied to this group gives us

$$n^{p-1} \equiv 1 \quad (\text{mod } p)$$

for any integer n which is prime to p. This is the theorem of Fermat in elementary number theory. For an arbitrary modulus m, we get the theorem of Euler by replacing $p - 1$ by the value $\varphi(m)$ of the Euler function.

Definition 3.4. Let H be a subgroup of a group G. A subset T of elements of G is said to be a **right transversal** of H if T contains exactly one element of each right coset of H. A **left transversal** is defined similarly.

If $U = \{u_\lambda\}$ $(\lambda \in \Lambda)$ is a right transversal of H, then Hu_λ $(\lambda \in \Lambda)$ are distinct cosets of H and G is the union of these cosets. This is equivalent to saying, for any $g \in G$, there exists a unique $\lambda \in \Lambda$ such that $g = hu_\lambda$ $(h \in H$ and $u_\lambda \in U)$. Clearly, we have $|U| = |G : H|$.

(3.5) *For two subgroups H and K of G such that $H \supset K$, we have*

$$|G : K| = |G : H| \cdot |H : K|.$$

If $U = \{u_\lambda\}$ is a right transversal of H in G and $V = \{v_\mu\}$ is a right transversal of K in H, then $VU = \{v_\mu u_\lambda\}$ is a right transversal of K in G.

Proof. Since $|U| = |G:H|$ and $|V| = |H:K|$, it suffices to prove the second part. Let g be an element of G. Then there is a pair (h, λ) such that $g = hu_\lambda$, $h \in H$, and $u_\lambda \in U$. The element h of H belongs to a coset Kv_μ, whence $g \in Kv_\mu u_\lambda$. Thus, the set VU contains a right transversal of K in G. It suffices to show that two distinct elements of VU are not contained in the same coset of K. Suppose that $(v_\mu u_\lambda)(v_\sigma u_\rho)^{-1} \in K$ (cf. (3.2) (v)). From

$$(v_\mu u_\lambda)(v_\sigma u_\rho)^{-1} = v_\mu u_\lambda u_\rho^{-1} v_\sigma^{-1} \in K \quad \text{and} \quad v_\mu, v_\sigma \in H,$$

we get $u_\lambda u_\rho^{-1} \in H$. Since U is a right transversal, we conclude $\lambda = \rho$, whence it follows that $u_\lambda u_\rho^{-1} = 1$ and $v_\mu v_\sigma^{-1} \in K$. Similarly, since V is a right transversal of K, we get $v_\mu = v_\sigma$. Thus, two different elements of VU never belong to the same coset of K. Therefore, VU is a right transversal.   □

If $K = \{1\}$, then $|G:K| = |G|$; so, (3.5) is a generalization of Theorem 3.3.

Various methods of enumeration are quite important in finite group theory. As an example of one of them, we will state a result on the number of conjugates of a subset.

Theorem 3.6. *Let S be a subset of group G, and let $\mathfrak{S}$ be the set of conjugates of S. Then we have $|\mathfrak{S}| = |G : N_G(S)|$. If $U = \{u, v, \cdots\}$ is a right transversal of $N_G(S)$, then*

$$\mathfrak{S} = \{S^u, S^v, \ldots\}$$

and $S^u, S^v, \ldots$ are distinct conjugates of S.

Proof. If $S^x = S^y$ for $x, y \in G$, we have $S^{xy^{-1}} = S$ by (2.10), whence $xy^{-1} \in N_G(S)$. So, by (3.2) (v), x and y are contained in the same coset of $N_G(S)$. Conversely, if x and y are contained in the same coset, we get $S^x = S^y$ by reversing the above argument. Thus, the number of distinct conjugates of S is equal to the index of $N_G(S)$. The remaining part of Theorem 3.6 follows immediately.   □

Corollary 1. *The number of conjugates of any subset is a divisor of the order $|G|$.*

Proof. This is clear from Theorems 3.6 and 3.3.   □

Corollary 2. *The conjugacy class containing x has exactly $|G : C_G(x)|$ elements.*

Proof. Corollary 2 is a special case of Theorem 3.6. □

Definition 3.7. Let H and K be two subgroups of a group G. The set of elements of the form hxk ($h \in H$, $k \in K$) is called a **double coset** with respect to H and K:

$$HxK = \{hxk \mid h \in H, k \in K\}.$$

(3.8) *Let H and K be two subgroups of G, and let g be an element of G.*

  (i) *There is a unique double coset containing g; this double coset is HgK.*
  (ii) *Two distinct double cosets never have a common element.*
  (iii) *The set G is the union of disjoint double cosets with respect of H and K.*
  (iv) *A double coset with respect to H and K is a union of right cosets of H as well as of the left cosets of K. The double coset HxK contains exactly $|K : K \cap H^x|$ right cosets of H.*

Proof. If 1 is the identity of G, then $g = 1 \cdot g \cdot 1$ belongs to HgK. If an element g belongs to a double coset HxK, then $g = hxk$ for some $h \in H$ and $k \in K$. Hence

$$HgK = H(hxk)K = HxK,$$

whence (i) follows. The propositions (ii) and (iii) are corollaries of (i).

The part (iv) can be proved as follows. Right cosets of H which are contained in HxK can be written as Hxk for some $k \in K$. For $k_1, k_2 \in K$, $Hxk_1 = Hxk_2$ if and only if $k_1 k_2^{-1} \in K \cap H^x$. (The notation is as in the preceding section.) By (3.2) (v), we have

$$(K \cap H^x)k_1 = (K \cap H^x)k_2.$$

Conversely, if $(K \cap H^x)k_1 = (K \cap H^x)k_2$, by tracing the preceding proof backwards, we get $Hxk_1 = Hxk_2$. Thus, the number of right cosets of H in HxK is the index of $K \cap H^x$ in K. □

Theorem 3.9. *Let H and K be subgroups of G. Suppose that $|G : H|$ is finite. Then we have*

$$|G : H| = \sum_{x \in X} |K : K \cap H^x|$$

where the set X is a complete system of representatives from distinct double cosets with respect to H and K.

Proof. By (3.8) (ii, iii) we have a partition of G into disjoint union of distinct double cosets with respect to H and K: $G = \bigcup HxK$. Count the number of right cosets of H in both sides of the above equality; (3.8) (iv) gives the result. □

Theorem 3.9 is an important result on which one of the proofs of the fundamental theorem of Sylow will be based.

Product of Subgroups

Definition 3.10. Let U and V be two subsets of a group G. The set $UV = \{uv \mid u \in U, \ v \in V\}$ consisting of the products of elements $u \in U$ and $v \in V$ is said to be the **product** of U and V.

The associative law of multiplication gives us $(UV)W = U(VW)$ for any three subsets $U, V,$ and W.

(3.11) *For any two subgroups H and K of finite orders, the following equality holds:*

$$|HK| = |H| \cdot |K|/|H \cap K|.$$

Proof. The product HK is a double coset with respect to H and K, so by (3.8) (iv), we have

$$|HK| = |H| \cdot |K : K \cap H|.$$

Applying Theorem 3.3 to the group K, we get

$$|K : K \cap H| = |K|/|H \cap K|,$$

whence (3.11) is proved. □

The product of two subgroups is not necessarily a subgroup (Exercise 6 (b)). We have the following theorem.

(3.12) *Let H and K be subgroups of a group G. The following two conditions are equivalent:*

(1) *The product HK is a subgroup of G;*
(2) *We have $HK = KH$.*

Proof. $(1) \Rightarrow (2)$. If HK is a subgroup of G, HK contains both H and K, whence $KH \subset HK$. On the other hand, the inverse of an element x of HK is also contained in HK. So, $x^{-1} \in HK$ and $x \in (HK)^{-1} = KH$. Thus, $HK \subset KH$ and (2) holds.

$(2) \Rightarrow (1)$. Let $L = HK$. The inverse of any element x of L belongs to KH. Since $KH = L$, L satisfies (2.1) (ii). Furthermore,

$$LL = HKHK = H(KH)K = H(HK)K \subset HK,$$

which implies the condition (2.1) (i). Hence L is a subgroup of G. $\square$

(3.13) *Let H and K be two subgroups of a group G. If the index $|G : H|$ is finite, the following propositions hold :*

(i) *The index $|K : H \cap K|$ is also finite and*

$$|K : H \cap K| \leq |G : H|.$$

(ii) *The equality in* (i) *holds if and only if $G = HK$.*
(iii) *If $|G : K|$ is also finite and relatively prime to $|G : H|$, then $G = HK$.*

Proof. By (3.8) (iv), HK contains $|K : H \cap K|$ right cosets of H. Clearly HK is a subset of G; we get (i) and (ii).

By (i), $|K : H \cap K|$ is finite. So, under the assumptions of (iii), the index $|G : H \cap K|$ is finite. Set $|G : H| = m$, $|G : K| = n$, and $|K : H \cap K| = k$. By (3.5), we have

$$|G : H \cap K| = m|H : H \cap K| = nk.$$

Hence m divides nk. But, by our assumption, m and n are relatively prime. Hence m divides k; in particular, $m \leq k$. On the other hand, we have $k \leq m$ by (i). Thus, $m = k$ and $G = HK$ by (ii). $\square$

Corollary 1. *If $H_1, \ldots, H_n$ are subgroups of finite indices in G, then the intersection $H_1 \cap H_2 \cap \cdots \cap H_n$ is also a subgroup of finite index in G.*

Proof. If $n = 2$, $|H_2 : H_1 \cap H_2|$ is finite by (3.13) (i). So, by (3.5), $|G : H_1 \cap H_2|$ is finite. If $n \geq 3$, we can use induction on n to prove the assertion. $\square$

Corollary 2. *Let H be a subgroup of finite index in G. Let D be the intersection of all conjugate subgroups of $H : D = \bigcap H^x$. Then D is also a subgroup of finite index.*

Proof. By Theorem 3.6 and (3.5), the number of conjugate subgroups of H is finite; the assertion follows from Corollary 1. $\square$

As we will see in the next section, D is what we call a normal subgroup of G. Corollary 2 says that a subgroup of finite index contains a normal subgroup of finite index.

If a group G is a product of two subgroups H and K, the structure of G is restricted by the structures of H and K. How the properties of H and K reflect on G is an interesting problem, on which many results have been obtained. The readers are referred to Chapter 13 of Scott [1] for many interesting theorems in this area.

The following is a useful theorem called **the Dedekind law**.

Theorem 3.14. *Let U and V be two subsets, and let L be a subgroup of a group G. If U is a subset of L, we have*

$$UV \cap L = U(V \cap L).$$

Proof. The right side is a subset of UV and, at the same time, it is a subset of L because L is a subgroup of G. Thus, we have

$$UV \cap L \supset U(V \cap L).$$

Suppose that an element l is contained in $UV \cap L$. Then $l \in L$ and $l = uv$ ($u \in U, v \in V$). By assumption, L is a subgroup which contains U, so $v = u^{-1}l \in L$. This implies that $v \in V \cap L$ and $l = uv \in U(V \cap L)$. Thus, $UV \cap L \subset U(V \cap L)$, and Theorem 3.14 is proved. $\square$

In Theorem 3.14, if U, V, and UV are subgroups of G, $UV \cap L$ is not only a subgroup, but also the product of the subgroups U and $V \cap L$; this extra information is useful in many occasions.

The following lemma is often useful in studying groups which are products of their subgroups.

(3.15) *Let G be a product of the subgroups H and $K : G = HK$. For any conjugate subgroup $H^x (x \in G)$, there is an element k of K such that $H^x = H^k$.*

Proof. Since $G = HK$, the element x of G can be written as a product $x = hk$ ($h \in H, k \in K$). By (2.10), we have

$$H^x = H^{hk} = (H^h)^k = H^k. \qquad \qquad \square$$

Corollary 1. *If a group G is a product of proper subgroups H and K, then H and K are not conjugate in G.*

Proof. If K were conjugate to H in G, then by (3.15), there would exist an element k of K such that $K = H^k$. Then,

$$G = G^k = (HK)^k = H^k K^k = KK = K,$$

contrary to the assumption that K is a proper subgroup of G. (The third equality in the above formula follows from (2.16).) $\square$

Corollary 2. *For any proper subgroup H of a group G, we have $G \neq HH^x$.*

EXAMPLE. As an application of the preceding discussion, we shall prove that, *for any prime p, a finite group of order p^2 is abelian* (cf. Chapter 2, Theorem 1.4).

Let G be a group of order p^2. By Cor. 2 of Theorem 3.3, the order of any element of G is a divisor of $|G|$, so it is 1, p, or p^2. If G contains an element g of order p^2, then $G = \langle g \rangle$ and G is a cyclic group. Obviously, G is abelian.

Next, assume that the order of every element except the identity is p. In this case, every element $g, g \neq 1$, generates a cyclic group of order p. We will prove the following proposition:

If H and K are distinct subgroups of order p, then $H \cap K = \{1\}$ and $G = HK$.

Proof. By (2.5), $H \cap K$ is a subgroup of H, so $H \cap K = H$ or $H \cap K = \{1\}$ by Cor. 1 of Theorem 3.3. If $H \cap K = H$, then $H \subset K$. Since H and K have the same order, they must be equal, and this is contrary to our assumption. Thus, $H \cap K = \{1\}$. By (3.11), we get $|HK| = p^2$, whence $G = HK$.

We will now arrive at a contradiction by assuming that G is nonabelian. If G is nonabelian, then, there is a pair of elements x and y such that $xy \neq yx$. Let $z = x^{-1}y^{-1}xy$, $H = \langle x \rangle$, and $K = \langle y \rangle$. The subgroups H and K are of order p and $y \notin H$, so $H \neq K$. Applying Cor. 2 of (3.15) and the above proposition to H, we get $H = H^y$. Thus, $z = x^{-1}x^y \in H$. Similarly,

$$z = (y^{-1})^x y \in K^x K = K.$$

So, $z \in H \cap K = \{1\}$. This gives us

$$z = x^{-1}y^{-1}xy = 1, \quad \text{or} \quad xy = yx,$$

and this contradicts the definitions of x and y. □

The above proposition also implies that every element x of G, $x \neq 1$, is contained in a uniquely determines subgroup of order p. So, G contains exactly $p + 1$ subgroups of order p. (Each subgroup of order p contains exactly $p - 1$ nonidentity elements, so the number of such subgroups is $(p^2 - 1)/(p - 1) = p + 1$.) The nontrivial subgroups of G are only these $p + 1$ subgroups.

As to the existence of the groups of order p^2, the totality of p^2-th roots of unity clearly forms a cyclic group of order p^2. The set of matrices with complex coefficients of the form

$$\begin{pmatrix} \omega & 0 \\ 0 & \zeta \end{pmatrix} \qquad \omega^p = \zeta^p = 1$$

forms an abelian group of order p^2, but this group is not cyclic. Thus, there are two classes of abelian groups of order p^2.

The element $z = x^{-1}y^{-1}xy$, used so effectively in the above proof, is the **commutator** of x and y, which often appears in group theory. We will study commutators in detail in Chapter 4. The method used in the last part of the preceding proof has many other applications. As an example, we state the following lemma, which can be proved by the same method.

(3.16) *If two subgroups H and K of a group G satisfy the conditions*

$$H \cap K = \{1\}, \quad H \subset N_G(K), \quad \text{and} \quad K \subset N_G(H),$$

then every element of H commutes with every element of K.

Exercises

1. Let H_k be the subgroup of $\mathbb{Z}$ defined in (2.25). Show that $|\mathbb{Z} : H_k| = k$ and that the integers $0, 1, 2, \ldots, k - 1$ form a transversal of H_k in $\mathbb{Z}$.

2. Find the number of left cosets of K which are contained in the double coset HxK. (This is analogous to (3.8) (iv).)

3. (a) Let H be a finite subgroup of a group G. Show that the number of right cosets of H contained in a double coset HxH is equal to the number of left cosets of H contained in HxH.

(b) Let G and H be the groups defined in §1, Example 4. Set

$$x = \begin{pmatrix} n & 0 \\ 0 & 1 \end{pmatrix}$$

where n is a natural number. Compute the number of right cosets of H contained in the double coset HxH; do the same for the left cosets of H in HxH.

(*Hint*. The number of right cosets is one, while the number of left cosets is n. To prove this, use (3.8) (iv) and Exercise 2. This example shows that it is not always possible to choose a transversal T of a subgroup H such that T satisfies the property that two distinct members lie in different right cosets as well as in different left cosets of H. However, if $|G : H|$ is finite, it is possible to choose such a transversal (Zassenhaus [1], p. 12).)

4. Let H be a proper subgroup of a finite group G. Show that there exists an element of G which is not conjugate to any element of H.

(*Hint*. An upper bound for the number of elements conjugate to some element of H is obtained from Theorem 3.6. The conclusion of this exercise does not necessarily hold for infinite groups (Chapter 3, §6).)

5. Let H be a nonempty subset of a group G. Show that H is a subgroup of G if and only if H satisfies $HH \subset H$ and $H^{-1} \subset H$. In this case, we always have the equalities:

$$HH = H \quad \text{and} \quad H^{-1} = H.$$

6. Let G denote the finite group of order $2n$ defined in §1, Example 5. Let σ be the reflection with respect to the bisector of the interior angle at a vertex P, and let τ be the reflection with respect to the perpendicular bisector of a side PQ through P. Set $S = \langle \sigma \rangle$ and $T = \langle \tau \rangle$. Prove the following propositions:

(a) $|S| = |T| = 2$;
(b) If n is odd, the subset ST is not a subgroup of G;
(c) The product $\sigma\tau$ is a rotation around the center of the n-gon and sends the vertex P into Q;
(d) The subset ST is not a subgroup even if n is even.

(*Hint.* Use (3.3) for (b), and (c) for (d).)

7. Suppose that a group $G = HK$ is a product of two subgroups H and K. Prove that, for any conjugate subgroup H^x of H and any conjugate K^y of K, we have $G = H^x K^y$ and there is an element z of G such that $H^x = H^z$ and $K^y = K^z$.

(*Hint.* Apply (3.15) to get $G = H^x K$; apply (3.15) again to get $G = H^x K^y$.)

§4. Normal Subgroups; Factor Groups

Definition 4.1. A subgroup H of a group G is said to be a **normal subgroup** of G if $Hx = xH$ for any element x of G; in this case, we write

$$H \lhd G.$$

The condition can also be expressed as $x^{-1}Hx = H$, so the normal subgroup can be defined as a subgroup which coincides with all conjugate subgroups, or a subgroup H such that $N_G(H) = G$.

(4.2) *Let H be a normal subgroup of a group G. For any subgroup K of G, the product HK is a subgroup of G. Thus, $HK = \langle H, K \rangle$ for any $K \subset G$.*

Proof. If H is normal in G, then $HK = KH$ for any subgroup K. Hence (3.12) proves that HK is a subgroup of G. $\square$

(4.3) *Let H and K be normal subgroups of a group G. Then HK and $H \cap K$ are also normal subgroups of G. In general, let H_λ ($\lambda \in \Lambda$) be a family of normal subgroups of G. Then the intersection $\bigcap H_\lambda$ is a normal subgroup; the union $\prod H_\lambda = \langle H_\lambda \rangle$ is normal, too.*

From the second definition of a normal subgroup ($H^x = H$ for all x of G), we get (4.3) by applying the following general proposition.

(4.4) *For any subsets H and K of a group G, we have*

$$(HK)^x = H^x K^x \quad \text{and} \quad (H \cap K)^x = H^x \cap K^x$$

for any $x \in G$. In general, for any family H_λ ($\lambda \in \Lambda$) of subsets of G, the formulas

$$(\prod H_\lambda)^x = \prod H_\lambda^x, \qquad (\cap H_\lambda)^x = \cap H_\lambda^x$$

hold for any $x \in G$.

Proof. The formula (2.16) gives us $(uv)^x = u^x v^x$ for any elements x, u, and v. So, by definition, we have $(HK)^x = H^x K^x$. Using induction on the number of factors, we get

$$(H_1 H_2 \cdots H_n)^x = H_1^x H_2^x \cdots H_n^x \qquad (x \in G).$$

Even if the index set is infinite, $\prod H_\lambda$ consists of the products of a finite number of elements. Then, we have $(\prod H_\lambda)^x = \prod H_\lambda^x$.

The assertions about the intersections follow from the definition. $\square$

The trivial subgroups of any group, that is, G and $\{1\}$, are normal; every subgroup of an abelian group is normal.

From a given subgroup H, we can construct two normal subgroups, $\langle H^x \rangle$, the subgroup generated by all conjugates of H, and $\cap H^x$, the intersection of all conjugate subgroups of H. That these subgroups are normal can be proved by applying Theorem 2.19 and (4.4).

Definition 4.5. A group G is said to be a **simple group** if $G \neq \{1\}$ and G contains no nontrivial normal subgroup.

Theorem 4.6. *An abelian simple group is a cyclic group of prime order; conversely, any finite cyclic group of prime order is an abelian simple group.*

Proof. An abelian simple group contains no nontrivial subgroup. So, by Theorem 2.24, it is a finite cyclic group of prime order. The converse statement follows from Cor. 1 of Theorem 3.3. $\square$

An arbitrary finite group may be considered to be a layer of various simple groups (cf. for example, Theorem 5.9). Thus, the study of the structure of finite simple groups and the ultimate classification of all finite simple groups is the fundamental problem in finite group theory. In the last 20 years, finite group theory has developed tremendously and the classification problem is expected to be solved in the near future.

EXAMPLE 1. The center $Z(G)$ of any group G is a normal subgroup. In fact, any subgroup of $Z(G)$ is normal in G.

EXAMPLE 2. For any subset S of G, we have

$$C_G(S) \lhd N_G(S)$$

by the corollary of Theorem 2.19. A subset S is said to be *normal*, if $S^x = S$ for all $x \in G$. If S is a normal subset of G. then $C_G(S) \lhd G$.

(4.7) *Let H be a normal subgroup of a group G, and let $\bar{G}$ denote the set of all cosets of H. For any two elements X and Y of $\bar{G}$, we define their product XY as the subset of G obtained by taking the product of the two subsets X and Y of G. Then XY is a coset of H. With respect to this multiplication on $\bar{G}$, the set $\bar{G}$ forms a group.*

Proof. Let X and Y be two elements of $\bar{G}$. Then, there are elements x and y of G such that $X = Hx$ and $Y = Hy$. By assumption, H is normal so that $Hx = xH$ for any $x \in G$. Whence

$$XY = (Hx)(Hy) = H(xH)y = H(Hx)y = Hxy.$$

This proves that XY is a coset of H. If $Z \in \bar{G}$, then by the associative law for the product of subsets, we have $(XY)Z = X(YZ)$. Thus, the multiplication defined on $\bar{G}$ satisfies the associative law. By definition, we get

(4.8) $(Hx)(Hy) = Hxy;$

so, the coset which contains the identity 1 of G, namely H, is the identity of $\bar{G}$, and the inverse of Hx is the coset Hx^{-1}. Then, by the corollary of Theorem 1.2, $\bar{G}$ forms a group. $\square$

Definition 4.9. The group $\bar{G}$ which was defined in (4.7) is called the **factor group** of G by H, or modulo H, and is written

$$\bar{G} = G/H.$$

The mapping $x \to Hx$ from G into $\bar{G}$ is called the **canonical mapping**.

The formula (4.8) is the most important property of the canonical mapping and from it we get

(4.10) $(Hx)^{-1} = Hx^{-1}.$

The following theorem is obvious from the definitions.

(4.11) *The order of the factor group G/H is equal to the index of the normal subgroup H :*

$$|G/H| = |G : H|.$$

The relationship between the factor group G/H and the group G will be stated in a precise form by the Homomorphism Theorem to be discussed in the next section. We will be content to prove the Correspondence Theorem, which is frequently referred to when we study the factor group G/H.

Let H be a normal subgroup of a group G, and consider a subgroup U which contains H. Since $H \subset U$, Hu is a subset of U for any $u \in U$. So, U is a union of the cosets of H which are contained in U. Let $\bar{U}$ be the set of cosets of H contained in U. Then, by (4.8) and (4.10), $\bar{U}$ satisfies conditions (i) and (ii) of (2.1), and $\bar{U}$ is a subgroup of $\bar{G}$. As $H \lhd G$, H is a normal subgroup of U. Therefore, the factor group U/H can be defined, and is in fact the subgroup $\bar{U}$ of $\bar{G}$.

The converse of the above discussion holds; that is, given a subgroup $\bar{U}$ of $\bar{G}$, as defined above, it is possible to find the subgroup U of G such that $U \supset H$ and $\bar{U} = U/H$. This constitutes the essential point of the Correspondence Theorem.

Theorem 4.12. *Let H be a normal subgroup of a group G, and let $\bar{G} = G/H$. For any subgroup $\bar{V}$ of $\bar{G}$, there corresponds a subgroup V of G such that*

$$H \subset V \quad \text{and} \quad \bar{V} = V/H.$$

The subgroup V consists of those elements of G which are contained in some element of $\bar{V}$ and is uniquely determined by $\bar{V}$. Thus, between the set $\bar{\mathscr{S}}$ of subgroups of $\bar{G}$ and the set $\mathscr{S}$ of subgroups of G which contain H, there exists a one-to-one correspondence, $\bar{V} \leftrightarrow V$.

Proof. For a given subgroup $\bar{V}$ of $\bar{G}$, define

$$V = \{v \in G \,|\, Hv \in \bar{V}\}.$$

Then, by (4.8) and (4.10), the set V satisfies conditions (i) and (ii) of (2.1). Therefore, V is a subgroup of G. The coset H is the identity of $\bar{G}$. Then, from (2.2), we get $H \in \bar{V}$ and $H \subset V$. As shown before, the set of cosets of H which are contained in V coincides with $\bar{V}$; thus, $\bar{V} = V/H$. The remaining assertions are obvious. $\square$

As in the above proof, the image of a subset S of G under the canonical mapping is often denoted by adding the bar on top, like $\bar{S}$. This is called the *bar-convention* (which we sometimes do not honor).

(4.13) *The one-to-one correspondence given by the Correspondence Theorem satisfies the following properties.*

Let $\bar{U}_1$ and $\bar{U}_2$ be subgroups of $\bar{G}$, and let U_i be the subgroup of G corresponding to $\bar{U}_i (i = 1, 2)$.

(i)
$$\bar{U}_1 \supset \bar{U}_2 \Leftrightarrow G \supset U_1 \supset U_2 \supset H$$

and
$$|\bar{U}_1 : \bar{U}_2| = |U_1 : U_2|.$$

(ii) $\bar{U}_1$ *is conjugate to* $\bar{U}_2$ *in* $\bar{G}$ *if and only if* U_1 *is conjugate to* U_2 *in* G.

(iii)
$$\bar{U}_1 \vartriangleright \bar{U}_2 \text{ if and only if } U_1 \vartriangleright U_2.$$

Proof. (i) The first statement is obvious. The image of a coset $U_2 x$ under the canonical mapping is $\bar{U}_2 \bar{x}$, whence we have

$$|\bar{U}_1 : \bar{U}_2| \leq |U_1 : U_2|.$$

On the other hand, if the images of two elements u and v of U are contained in the same coset of $\bar{U}_2$, we get $\bar{u}\bar{v}^{-1} \in \bar{U}_2$ from (3.2) (v). Thus, $uv^{-1} \in U_2$. Hence, again by (3.2) (v), we have $U_2 u = U_2 v$. Combined with the previously proved inequality, we obtain $|\bar{U}_1 : \bar{U}_2| = |U_1 : U_2|$.

(ii) For any subgroup U and for any element x of G, the image of $x^{-1}Ux$ under the canonical mapping is $\bar{x}^{-1}\bar{U}\bar{x}$. So, the assertion is obvious.

(iii) This follows immediately from (ii).   □

Definition 4.14. A normal subgroup H of a group G is said to be a **maximal normal subgroup** of G if the following two conditions are satisfied:

(i) H is different from G; and
(ii) $H \subset K \subset G$, where K is a normal subgroup of G, implies that either $K = G$ or $K = H$.

Similarly, a normal subgroup M of G is said to be a **minimal normal subgroup** of G if $M \neq \{1\}$ and the only normal subgroup N of G satisfying $M \supset N \supset \{1\}$ is either $\{1\}$ or M.

A group may contain no maximal or minimal normal subgroup, but a finite group contains at least one maximal and one minimal normal subgroup.

Theorem 4.15. *Let H be a normal subgroup of a group G. If H is maximal normal subgroup of G, the factor group G/H is a simple group. Conversely, if the factor group G/H is simple, then H is a maximal normal subgroup of G.*

Proof. If H is a proper normal subgroup of G such that the factor group $\bar{G} = G/H$ is not simple, then $\bar{G}$ contains a nontrivial normal subgroup $\bar{N}$. By the Correspondence Theorem, there is a subgroup N of G such that $H \subset N$

and $\bar{N} = N/H$. Then, by (4.13) (iii), $N \lhd G$, and we have $H \subsetneq N \subsetneq G$ because $\bar{N}$ is not trivial. Hence H is not a maximal normal subgroup of G.

Conversely, if H is not a maximal normal subgroup, then there is a normal subgroup N of G such that $H \subset N$ and $H \neq N \neq G$. By (4.13) (iii), $\bar{N} = N/H$ is a normal subgroup of $\bar{G}$ which is nontrivial because $H \neq N \neq G$. Hence $\bar{G}$ is not simple. $\square$

A similar theorem concerning minimal normal subgroups also holds, but it is a little more complicated to state (cf. Chapter 2, (4.14), Corollary 3).

Exercises

1. (a) Prove that a subgroup H of a group G is a normal subgroup of G if and only if $x^{-1}Hx \subset H$ for all $x \in G$.

(b) Show that an element y of G need not be an element of $N_G(H)$ even if it satisfies $y^{-1}Hy \subset H$. Also prove that, in general, the subset $\{y \mid y \in G, y^{-1}Hy \subset H\}$ is not necessarily a subgroup of G.

(*Hint.* Choose G, H, and x as in §3, Exercise 3(b), and set $y = x^{-1}$. Then the conjugate subgroup H^y of H is a proper subset of H. If H is a finite subgroup, then $|H^y| = |H|$, and the situation in (b) does not occur.)

2. Show that if S is a normal subset of G, the subgroup $\langle S \rangle$ is a normal subgroup of G.

3. (a) Prove that any subgroup of index 2 is normal.

(b) Let G be a finite group, and let p be the smallest prime divisor of the order $|G|$. Show that any subgroup of index p is normal (Theorem of Ore).

(*Hint.* (a) follows immediately from the definition of normal subgroup. As for (b), supposing that a subgroup H of index p is not normal, we can pick a conjugate K, $K \neq H$. Prove that $G = HK$, and use Cor. 2 of (3.15) to get a contradiction. Or, suppose that G and H provide a counterexample of minimal order. Then by induction, $H \cap K \lhd G$ and the factor group $G/H \cap K$ is of order p^2. Apply §3, Example, to get a contradiction.)

4. Show that the following eight matrices form a group of order 8 with respect to matrix multiplication:

$$\pm \begin{pmatrix} 1 & 0 \\ 0 & 1 \end{pmatrix}, \ \pm \begin{pmatrix} 0 & 1 \\ -1 & 0 \end{pmatrix}, \ \pm \begin{pmatrix} i & 0 \\ 0 & -i \end{pmatrix}, \ \pm \begin{pmatrix} 0 & i \\ i & 0 \end{pmatrix} \qquad (i = \sqrt{-1}).$$

Verify that this group Q contains exactly one element of order 2. Also show that all subgroups of Q are normal, although Q is not abelian.

(*Hint.* Use Exercise 3(a) and Theorem 3.3. There are exactly 3 subgroups of order 4, all of which are cyclic. This group Q is called the *quaternion group*, and is connected with the quaternion skew-field of Hamilton. The quaternion skew-field is isomorphic to the algebra consisting of the totality of 2×2 matrices of the form

$$\begin{pmatrix} \alpha & \beta \\ -\bar{\beta} & \bar{\alpha} \end{pmatrix} \quad (\alpha, \beta \in \mathbb{C})$$

where $\bar{\alpha}$ denotes the usual complex conjugate of α.)

§5. Homomorphisms; Isomorphism Theorems

Definition 5.1. A function f defined on a group G to a group G' (not necessarily distinct from G) is said to be a **homomorphism** from G into G' if it satisfies

$$f(xy) = f(x)f(y) \quad \text{for all } x, y \in G.$$

If a homomorphism f from G into G' induces a one-to-one correspondence and is surjective, f is said to be an **isomorphism**. If there is an isomorphism from G onto G', we say that G and G' are **isomorphic** and write $G \cong G'$.

If a homomorphism f from G into G' induces a one-to-one correspondence, f may be considered to be an isomorphism from G onto $f(G)$. In this case, f is said to be *an isomorphism into G'*, or a *monomorphism*. If $f(G) = G'$, f is called an *epimorphism*.

Let f be a homomorphism from G into G'. The subset

$$H = \{x \mid x \in G, f(x) \text{ is the identity of } G'\}$$

is called the **kernel** of f and is denoted by $\operatorname{Ker} f$.

For a given group G, the set of all groups which are isomorphic to G is called the *isomorphism class* of G. In group theory, we often study the properties of an isomorphism class; that is, the properties which are common to all the groups in an isomorphism class.

EXAMPLE 1. Let G be a cyclic group, and let g be a generator of G. Define a function f from $\mathbb{Z}$ into G by the formula

$$f(n) = g^n \quad (n \in \mathbb{Z}).$$

By (1.8), the function f is a homomorphism from $\mathbb{Z}$ to G, and it is clearly an epimorphism. If G is an infinite cyclic group, the function f is an isomorphism, while if G is a finite cyclic group of order k, then the kernel of f is the subgroup H_k defined in (2.25).

EXAMPLE 2. The function $A \to \det A$ which assigns the determinant of A to a nonsingular matrix A is a homomorphism from the group of nonsingular matrices onto the multiplicative group F^* of the ground field F, and its kernel is the totality of matrices having determinant 1.

EXAMPLE 3. Let G be the group defined in §1, Example 4. If f is the function

$$\begin{pmatrix} r & s \\ 0 & 1 \end{pmatrix} \to r$$

defined on G into the multiplicative group $\mathbb{R}^*$ of nonzero rational numbers, the multiplication rule for elements of G shows that f is a homomorphism of G onto $\mathbb{R}^*$. The kernel of f is the subgroup consisting of matrices

$$\begin{pmatrix} 1 & s \\ 0 & 1 \end{pmatrix}.$$

The function which sends the above matrix to the rational number s is a homomorphism of the subgroup $\mathrm{Ker}\, f$ onto the additive group $\mathbb{R}^+$, and we have $\mathrm{Ker}\, f \cong \mathbb{R}^+$.

EXAMPLE 4. Suppose that a group G and an element g of G are given. Let H be a subgroup of G. Define a function f from H into G by the formula

$$f(x) = g^{-1}xg \quad (x \in H).$$

By (2.16), the function f is a homomorphism; it is also an isomorphism from H onto the conjugate subgroup H^g of H. Thus, we have $H^g \cong H$. This is the essence of conjugation.

EXAMPLE 5. Let H be a normal subgroup of a group G, and let f be the canonical mapping from G onto the factor group G/H. By (4.8), the mapping f is a homomorphism from G onto G/H. This is the main significance of the canonical mapping, or *the canonical homomorphism* as it is sometimes called. The kernel of the canonical mapping coincides with the normal subgroup H.

The condition for homomorphism, $f(xy) = f(x)f(y)$, means that the image of the product of two elements x and y in G is the product of the images $f(x)$ and $f(y)$ in G'. Thus, when G' is an additive group, the condition for homomorphism becomes

$$f(xy) = f(x) + f(y).$$

For example, the multiplicative group of positive real numbers is isomorphic to the additive group of all real numbers by the log function. We will now prove some elementary properties of homomorphisms.

(5.2) *Let f be a homomorphism from a group G into a group G'. The following propositions hold :*

(i)   *The homomorphism f maps the identity of G onto the identity of G';
similarly, we have*

$$f(x^{-1}) = f(x)^{-1}.$$

(ii) *The kernel of f is a subgroup of G.*

(iii) *Let U be a subgroup of G. The image $f(U) = \{f(x) \,|\, x \in U\}$ is a
subgroup of G'. For a subgroup U' of G', the inverse image*

$$f^{-1}(U') = \{x \,|\, x \in G, f(x) \in U'\}$$

is a subgroup of G.

(iv) *For two elements x and y of G, $f(x) = f(y)$ if and only if x and y lie in
the same right coset of the kernel of f. In particular, if f is surjective,
then f is an isomorphism if and only if the kernel of f is $\{1\}$.*

(v)   *If two subsets R and S of G are conjugate in G, then the images $f(R)$
and $f(S)$ are conjugate in G'.*

(vi) *If $H \lhd G$, then $f(H) \lhd f(G)$.*

Proof. (i) Let $e = f(1)$, where 1 is the identity of G. When $x = y = 1$ in the
formula for homomorphism, we have

$$e = f(1) = f(1)f(1) = ee.$$

This proves that e is the identity of G'. When $y = x^{-1}$, we get

$$e = f(1) = f(x)f(x^{-1});$$

that is, $f(x^{-1}) = f(x)^{-1}$.

(ii) It follows from (i) and the definitions that the kernel satisfies the
conditions (i) and (ii) of (2.1).

(iii) This proof is similar to those of (2.15) and of the Correspondence
Theorem (cf. Examples 4 and 5).

(iv) If $f(x) = f(y)$ for two elements x and y of G, then the definition of
homomorphism and (i) give us

$$f(xy^{-1}) = f(x)f(y^{-1}) = f(x)f(y)^{-1}$$
$$= f(x)f(x)^{-1} = e.$$

By definition, xy^{-1} lies in the kernel of f. So, by (3.2) (v), x and y are contained
in the same right coset of the kernel. Conversely, if x and y lie in the same right
coset of the kernel, we get $f(x) = f(y)$; thus, (iv) holds.

(v) If $R = S^x$ ($x \in G$), then $f(R)$ is conjugate to $f(S)$ by the element $f(x)$ of
G'. Finally, (vi) is proved similarly.   □

When a group G is isomorphic to a group G', the isomorphism f between them gives a one-to-one mapping of elements which preserves the multiplicative structure. Thus, as far as the multiplication of their elements is concerned, the two groups G and G' are regarded as being the same. If a property of groups is defined using multiplication (for example, the commutativity or the simplicity of groups), then the property holds not only for an individual group G, but also for any group in the isomorphism class of G.

We sometimes denote a property of groups by a symbol, such as P, and call it property P. For example, P may be the property of being an abelian group, or of being an infinite group; or, P may be the property that a group G consists of matrices with coefficients in the field of rational numbers. In each of these cases, if a group G has the property P, we call G a P-group.

A property P of groups is said to be a *group theoretical property*, if the following condition is satisfied:

If a group G satisfies the property P, then any group which is isomorphic to G satisfies P.

Among the three properties stated above, the first two are group theoretical, but the last one is not. A group theoretical property is a property of isomorphism classes. The concepts of subgroups, conjugates and cosets are defined in Definitions 2.1, 2.9, and 3.1, respectively, based on the operation defined on the elements of a group, so propositions concerning these concepts are group theoretical.

The next theorem is called the Homomorphism Theorem, and it is one of the most basic theorems in group theory.

Theorem 5.3. *Let f be a homomorphism from a group G onto a group G'. Then,*

(1) the kernel K of f is a normal subgroup of G.

Furthermore,

(2) there is an isomorphism g from G/K onto G' such that $f = g \circ \theta$, where θ is the canonical homomorphism from G onto G/K. In this case, we say that the diagram

is commutative.

Proof. (1) For any element x of G, the image of any element of xK under f is equal to $f(x)$; whence $xK \subset Kx$ by (5.2) (iv). A similar containment holds for x^{-1}, too. Therefore, we have

$$x^{-1}K \subset Kx^{-1}.$$

Multiplying both sides by x, we get $Kx \subset xK$. So, $Kx = xK$ holds for any $x \in G$, and K is a normal subgroup of G by Definition 4.1.

(2) Define a function g from G/K to G' by

$$g(Kx) = f(x).$$

By (5.2) (iv), the function g is well defined and does not depend on the choice of a representative from Kx. From (4.8) and the fact that f is a homomorphism, we know that

$$g(KxKy) = g(Kxy) = f(xy)$$
$$= f(x)f(y) = g(Kx)g(Ky).$$

Hence, g is a homomorphism from G/K onto G'. If the coset Kx lies in the kernel of g, then

$$1 = g(Kx) = f(x),$$

so that $x \in \text{Ker } f$. Thus, g is an isomorphism by (5.2) (iv). It is clear from the definition that $f = g \circ \theta$ holds; so the diagram is commutative. $\square$

Corollary. *Let N be a normal subgroup of a group G, and let θ be the canonical homomorphism from G onto G/N. Let f be a homomorphism from G into a group G'. There is a homomorphism g from G/N into G' such that $f = g \circ \theta$, if and only if $N \subset \text{Ker } f$. In this case, we have $f(G) \cong (G/N)/(K/N)$ where $K = \text{Ker } f$.*

Proof. If there is a homomorphism g satisfying $f = g \circ \theta$, we have $f(N) = 1$; so $N \subset \text{Ker } f$. Conversely, suppose $N \subset \text{Ker } f$. Set $K = \text{Ker } f$ and $\bar{G} = G/N$. We define a function g from $\bar{G}$ into G' by

$$g(Nx) = f(x)$$

As in the proofs of (5.2) (iv) and Theorem 5.3 (2), the function g is uniquely determined independent of the choice of a representative x from Nx, and g is proved to be a homomorphism from $\bar{G}$ into G'. Clearly, $f = g \circ \theta$ by definition, and we have $\text{Ker } g = \bar{K}$. From the Homomorphism Theorem, we get

$$G/K = f(G) \cong g(\bar{G}) \cong \bar{G}/\bar{K} = (G/N)/(K/N). \quad \square$$

(5.4) *Let f be a homomorphism from G into G'.*

(i) *For any subgroup U of G, the restriction of f on U is a homomorphism from U into G'.*

(ii) *If g is a homomorphism from G' into a group G'', then the composite mapping h, where $h = g \circ f$, is a homomorphism from G into G''.*

Proof. In both cases, the proofs are easy. For example, let h be the composite mapping of f and g. Then, we have $h = g \circ f$, and

$$h(xy) = g(f(xy)) = g(f(x)f(y))$$
$$= g(f(x))g(f(y)) = h(x)h(y).$$

So, h is a homomorphism. □

The restriction of f on U is denoted by f_U, while the composite mapping $h = g \circ f$ is called the *product* of f and g, and is denoted by $h = fg$.

Next, we will generalize the Correspondence Theorem.

Theorem 5.5. *Let f be a homomorphism from a group G onto a group G', and let K be the kernel of f. The mapping f induces a one-to-one correspondence φ between the set of subgroups of G which contain K and the set of subgroups of G'. The function φ is*

$$U \to \varphi(U) = \{f(u) \mid u \in U\},$$

and the inverse function φ' is given by

$$U' \to \varphi'(U') = \{u \mid f(u) \in U'\}.$$

Furthermore, the correspondence φ satisfies the following additional properties for two subgroups U and V of G containing K :

(1)   *$U \supset V$ if and only if $\varphi(U) \supset \varphi(V)$; in this case, we have*

$$|U : V| = |\varphi(U) : \varphi(V)|.$$

(2)   *$U \rhd V$ if and only if $\varphi(U) \rhd \varphi(V)$; and in this case*

$$U/V \cong \varphi(U)/\varphi(V).$$

(3)   *U and V are conjugate in G if and only if $\varphi(U)$ and $\varphi(V)$ are conjugate in G'.*

Proof. The Homomorphism Theorem gives us an isomorphism g from G/K onto G' such that $f = g \circ \theta$, where θ is the canonical homomorphism from G onto G/K. By Theorem 4.12 and (4.13), θ induces a one-to-one correspondence $U \to \bar{U}$ between the subgroups of G containing K and the subgroups of G/K; and, this function satisfies the properties (1), (2), and (3). The isomorphism g induces a one-to-one correspondence ψ of subgroups. So, the function φ defined by $\varphi(U) = \psi(\bar{U})$ induces a one-to-one correspondence between the subgroups of G containing K and those of G'. The function φ satisfies the properties (1), (2), and (3) because the function ψ is induced by the isomorphism g and (1), (2), and (3) are group theoretical properties. $\square$

Corollary (The First Isomorphism Theorem). *Let f be a homomorphism from a group G onto a group G'. Let N' be a normal subgroup of G' and set $N = f^{-1}(N')$. Then,*

$$N \lhd G \quad \text{and} \quad G/N \cong G'/N'.$$

As an application of the preceding theorem, we will study cyclic groups.

(5.6) *The factor groups and subgroups of a cyclic group are cyclic.*

Proof. The assertion about factor groups is trivial to prove. In general, if a subset S of a group G generates G, then the factor group $\bar{G}$ is generated by

$$\bar{S} = \{\bar{s} \mid s \in S\}.$$

This is easily seen from the fact that the canonical mapping is a homomorphism (Example 5).

Let U be a subgroup of a cyclic group G. Assume $G \neq U \neq \{1\}$. By Example 1, there is a homomorphism f from $\mathbb{Z}$ onto G. The subgroup $f^{-1}(U)$ is a nontrivial subgroup of $\mathbb{Z}$. So, by (2.25) $f^{-1}(U) = H_k$. The subgroup H_k is, by definition, a cyclic group generated by the natural number k, and f induces a homomorphism from H_k onto U. Thus, U is isomorphic to a factor group of a cyclic group (Theorem 5.3). So, by the first part of the assertion, we know that U is cyclic. $\square$

Corollary. *Let G be a cyclic group of finite order d. For each positive divisor k of d, G contains a unique subgroup of index k (as well as one of order k). If g is one of the generators of G, then we have*

$$|G : \langle g^k \rangle| = k$$

for any divisor k of d, where $d = |G|$. Conversely, if a finite group G of order d contains at most one subgroup of order k for each positive divisor k of d, then G is cyclic.

Proof. Apply the Correspondence Theorem to the homomorphism $n \rightarrow g^n$ from $\mathbb{Z}$ onto G. Then the first half follows from (2.25) and the equivalence

$$H_k \supset H_d \Leftrightarrow k \mid d.$$

In order to prove the converse statement, let $\psi(k)$ be the number of elements of G whose order is exactly k. If an element x of G is of order k, then

$\langle x \rangle$ is a subgroup of order k. By assumption, all the elements of order k of G are contained in $\langle x \rangle$. Hence we have

$$\psi(k) = 0 \quad \text{or} \quad \psi(k) = \varphi(k),$$

where $\varphi(k)$ denotes the number of elements of order k in the cyclic group of order k. Thus, φ is the Euler function of elementary number theory. By a basic property of the Euler function, we have

$$\sum_{k \mid d} \varphi(k) = d.$$

(This is a property of the cyclic group of order d: the left side can be considered to be the sum of the number of elements of order k.) On the other hand, the definition of ψ gives us:

$$\sum_{k \mid d} \psi(k) = d.$$

Thus, we must have $\psi(k) = \varphi(k)$ for all divisors k of d. In particular, $\psi(d) = \varphi(d) > 0$. So, G has an element of order exactly d. This means that G is cyclic. $\square$

As an application on the preceding corollary, we will consider a finite group G which is a subgroup of the multiplicative group F^* of a field F. The solutions of the equation $x^d = 1$ are the d-th roots of unity, and there are at most d such roots of unity in F. The corollary proves that *any finite subgroup of F^* is cyclic.*

Theorem 5.7. *Let H be a normal subgroup of a group G, and let K be any subgroup of G. Then, the following hold :*

- (i)   *HK is a subgroup of G;*
- (ii) *$H \cap K$ is a normal subgroup of K;*
- (iii) *$HK/H \cong K/H \cap K$ (**The Second Isomorphism Theorem**);*
- (iv) *the formula $|HK : K| = |H : H \cap K|$ holds provided at least one side is finite.*

Proof. Consider the canonical mapping from G onto the factor group G/H. Let f be the restriction of the canonical mapping on K. Then f is a homomorphism from K into G/H. By definition, f is a homomorphism from K onto HK/H. Let K_0 be the kernel of f. The Homomorphism Theorem (5.3) proves that

$$HK/H \cong K/K_0.$$

But, the kernel of the canonical homomorphism is K (Example 5), whence we get $K_0 = H \cap K$. This proves (ii) and (iii).

The statement (i) had already been proved in (4.2), but since we showed in the above argument that $f(K) = HK/H$, HK is proved again to be a subgroup by Correspondence Theorem.

If $|HK : K|$ is finite, (3.13) applied to HK proves the proposition (iv). If $|H : H \cap K|$ is finite, then the number of left cosets of K in HK is finite as in (3.8) (iv). Thus, (iv) holds in either case. □

The Jordan-Hölder Theorem

Definition 5.8. Let G be a group. If a finite chain of subgroups of G

$$(\mathscr{G}): G_0 = G \supset G_1 \supset \cdots \supset G_r = \{1\}$$

satisfies the property that, for each $i = 1, 2, \ldots, r$, G_i is a maximal normal subgroup of G_{i-1}, we say that $(\mathscr{G})$ is a **composition series** of length r of G. The set of factor groups

$$\{G_0/G_1, G_1/G_2, \ldots, G_{r-1}/G_r\}$$

is said to be the set of **composition factors**.

By Theorem 4.15, each composition factor is a simple group.

A group need not have a composition series. In fact, the additive group $\mathbb{Z}$ does not have a composition series. Let $G_0 = \mathbb{Z}$ and, for each $i > 0$, let G_i be a maximal normal subgroup of G_{i-1}. Then for any natural number r, $G_r \neq \{0\}$ so that we cannot reach the trivial subgroup in finite number of steps. On the other hand, every finite group has a composition series.

The Jordan-Hölder Theorem holds for any group with a composition series.

Theorem 5.9. *Let G be a group with a composition series*

$$(\mathscr{G}): G_0 = G \supset G_1 \supset \cdots \supset G_r = \{1\}.$$

Let $(\mathscr{H}): H_0 = G \supset H_1 \supset \cdots \supset H_s = \{1\}$ be any composition series of G. Then the length of $(\mathscr{H})$ is equal to the length of $(\mathscr{G})$; that is, $r = s$. Furthermore, there is a one-to-one correspondence between the composition factors of $(\mathscr{G})$ and the composition factors of $(\mathscr{H})$ such that the corresponding factor groups are isomorphic.

Proof. Two composition series $(\mathscr{G})$ and $(\mathscr{H})$ are said to be *equivalent* if $r = s$ and there is a one-to-one correspondence between two sets of composition factors satisfying the property stated in the theorem. Theorem 5.9 asserts that any two composition series of G are equivalent.

We will prove the following proposition by induction on r.

Let G be a group with a composition series $(\mathcal{G})$ of length r, and let H_1 be any maximal normal subgroup of G. Then G has a composition series whose first term is H_1, and any composition series $(\mathcal{H})$ of G with the first term H_1 is equivalent to $(\mathcal{G})$.

If $r = 1$, then G is a simple group and the proposition is obviously true. Now as our inductive hypothesis, assume that the proposition holds for groups with a composition series of length i, where $i < r$. If H_1 coincides with the first term G_1 in $(\mathcal{G})$, then the series $G_1 \supset G_2 \supset \cdots \supset G_r = \{1\}$ forms a composition series of length $r - 1$ of G_1. The assertion follows immediately. Finally, assume that $H_1 \neq G_1$. Then $G_1 H_1$ is a normal subgroup of G and is strictly larger than H_1. Since H_1 is maximal normal, we have $G_1 H_1 = G$. Let $K_2 = H_1 \cap G_1$. By (4.3), K_2 is a normal subgroup of G. By the Second Isomorphism Theorem, we have

$$G_1/K_2 \cong G/H_1 \quad \text{and} \quad H_1/K_2 \cong G/G_1.$$

By Theorem 4.15, the right side of each formula is a simple group; hence, so is the other side. In particular, K_2 is a maximal normal subgroup of G_1. By the inductive hypothesis, there is a composition series of G_1 with K_2 as the first term: $G_1 = K_1 \supset K_2 \supset \cdots \supset K_t = \{1\}$. Furthermore, this series is equivalent to the given composition series $G_1 \supset G_2 \supset \cdots \supset G_r = \{1\}$ of G_1. Let $(\mathcal{K})$ be the chain of subgroups

$$(\mathcal{K})\colon G \supset G_1 \supset K_2 \supset \cdots \supset K_t = \{1\}.$$

Then, $(\mathcal{K})$ is a composition series of G which is equivalent to $(\mathcal{G})$. In particular, we have $r = t$. Define another chain

$$(\mathcal{K'})\colon G \supset H_1 \supset K_2 \supset \cdots \supset K_t = \{1\}.$$

As before, $(\mathcal{K'})$ is also a composition series of G, but its first term is H_1. From the isomorphisms

$$G_1/K_2 \cong G/H_1 \quad \text{and} \quad H_1/K_2 \cong G/G_1,$$

we know that $(\mathcal{K'})$ is equivalent to $(\mathcal{K})$. Since $r = t$, we know from the inductive hypothesis that the proposition we are trying to prove holds for composition series of H_1 (since their length is less than r). If $(\mathcal{H})$ is any composition series with H_1 as the first term, $(\mathcal{H})$ is equivalent to $(\mathcal{K'})$. This proves that $(\mathcal{H})$ is equivalent to $(\mathcal{G})$. $\square$

We will consider generalizations of the Jordan-Hölder Theorem in Chapter 2, §3.

Exercises

1. Prove (5.6) directly (without using Theorem 5.5).
   (*Hint.* Imitate the proof of (2.25).)

2. Let G be a group with a composition series, and let N be a normal subgroup of G. Show that there is a composition series of G having N as a term.
   (*Hint.* When N is a maximal normal subgroup of G, this has been proved as the proposition used in the proof of (5.9). We will prove a more general theorem later (Chapter 2, §3).)

3. A subgroup H of a group G is said to be a *composition subgroup* of G if there is a finite chain of subgroups

$$G_0 = G \supset G_1 \supset G_2 \supset \cdots \supset G_r = H$$

such that for all i ($i = 1, 2, \ldots, r$), G_i is a maximal normal subgroup of G_{i-1}. The above series is called a *composition series of length r between G and H*, and the set of factor groups

$$\{G_0/G_1, G_1/G_2, \ldots, G_{r-1}/G_r\}$$

is said to be the set of composition factors.
   (a) Prove that if H is a composition subgroup of G, its composition factors are all simple groups and the set of their isomorphism classes is uniquely determined.
   (b) Prove that if H is a composition subgroup of G and if K is a composition subgroup of H, then K is a composition subgroup of G.
   (c) Show that if H is a composition subgroup of G and if N is a maximal normal subgroup of G, then either $H \subset N$, or $H \cap N$ is a maximal normal subgroup of H. In the latter case, prove that $H/H \cap N \cong G/N$.
   (d) Prove that if H and K are composition subgroups of G, then so is $H \cap K$, and that the set of simple groups in the composition factors between G and $H \cap K$ is the union of the simple groups in the composition factors between G and H and between G and K.
   (e) Let H and K be composition subgroups of G such that $H \subset K$. If $H \neq K$, then the length of a composition series between G and K is smaller than the length of a composition series between G and H. If $K \neq G$, then the length of a composition series between K and H is smaller than the length of a composition series between G and H. Prove the above assertions.
   (*Hint.* Use induction on the length of a composition series between G and H. The subgroup generated by two composition subgroups is also a composition subgroup (Wielandt), but the proof of this requires a method to be discussed later (Chapter 2, §3, Exercise 3).)

§6. Automorphisms

Definition 6.1. Let G be a group. An isomorphism of G onto itself is called an **automorphism** of G.

The identity function $x \to x$ $(x \in G)$ is certainly an automorphism of G.

(6.2) *An automorphism preserves any group theoretical property. Namely, if σ is an automorphism of a group G, then for any two subgroups H and K of G such that $H \supset K$, we have*

$$\sigma(H) \cong H, \qquad |\sigma(H) : \sigma(K)| = |H : K|,$$

and $K \lhd H$ implies $\sigma(K) \lhd \sigma(H)$ and $H/K \cong \sigma(H)/\sigma(K)$. In general,

$$N_G(H^\sigma) = N_G(H)^\sigma, \qquad C_G(H^\sigma) = C_G(H)^\sigma, \qquad \langle S^\sigma \rangle = \langle S \rangle^\sigma.$$

We write the image of H by an automorphism σ either as H^σ or $\sigma(H)$.

Let σ and ρ be two automorphisms of a group G. The *product* $\sigma\rho$ is by definition the composite $\rho \circ \sigma$ of σ and ρ, and is an automorphism by (5.4). Due to this notation, the image is often denoted by x^σ, so we have

$$x^{\sigma\rho} = (x^\sigma)^\rho \quad \text{for any } x \in G.$$

(6.3) *Let A be the totality of automorphisms of a group G. Then A forms a group under the multiplication defined above.*

Proof. The composition of functions satisfies the associative law. The identity function is the identity of A. If σ is an automorphism, its inverse function is defined and is easily proved to be an automorphism of G as well (the inverse σ^{-1} is defined as usual: $\sigma^{-1}(x) = y$ when $\sigma(y) = x$). By Cor. of Theorem 1.2, A forms a group. $\square$

Definition 6.4. The totality A of automorphisms of a group G equipped with the multiplication defined by composition is said to be **the group of automorphisms** of G; we write $A = \text{Aut } G$. If g is an element of G, then the function

$$x \to g^{-1}xg = x^g$$

is an automorphism of G (§5, Example 4), which is called the **conjugation** by g, or the **inner automorphism** by g; we use the symbol i_g to denote the conjugation by the element g.

By (2.10), we have $x^{gh} = (x^g)^h$. So, the definition of the product of automorphisms proves that

(6.5)
$$i_{gh} = i_g i_h.$$

(The image by the conjugation i_g is often simply written x^g, instead of the more correct notation $i_g(x)$ or x^{i_g}.)

The formula (6.5) shows that the function $g \to i_g$ from G into the group Aut G is a homomorphism. Therefore, the set of inner automorphisms $\{i_g\}$ is a subgroup of Aut G, which is written

$$\text{Inn } G,$$

and is called **the group of inner automorphisms** of G. If G is abelian, then Inn $G = \{1\}$.

Theorem 6.6. *The group of inner automorphisms of G is isomorphic to the factor group $G/Z(G)$, where $Z(G)$ is the center of G :*

$$\text{Inn } G \cong G/Z(G).$$

Furthermore, Inn *G is a normal subgroup of* Aut *G.*

Proof. The function $g \to i_g$ is a homomorphism from G onto Inn G, so Inn $G \cong G/K$ where K is the kernel of the above homomorphism. An element g of G lies in the kernel K if and only if $i_g = 1$, that is $g^{-1}xg = x$ for all $x \in G$. Thus, we have $K = Z(G)$, by Definition 2.14.

The last assertion follows immediately from the formula

(6.7)
$$\sigma^{-1} i_g \sigma = i_{\sigma(g)}$$

which holds for any $g \in G$ and $\sigma \in$ Aut G. The formula (6.7) is proved as follows: for any $x \in G$,

$$(\sigma^{-1} i_g \sigma)(x) = i_g \sigma(\sigma^{-1}(x)) = \sigma(g^{-1} \sigma^{-1}(x)g)$$
$$= \sigma(g^{-1})x\sigma(g) = i_{\sigma(g)}(x). \quad \square$$

Corollary. *If $Z(G) = \{1\}$, then* Inn *$G \cong G$.*

If $Z(G) = \{1\}$, we may identify Inn G with G. Thus, we consider the function $g \to i_g$ as an injection of G into Aut G, and regard G as a subgroup of Aut G. The formula (6.7) is now written as

$$\sigma^{-1} g \sigma = \sigma(g) = g^\sigma.$$

In other words, the conjugation by σ induces an automorphism of G (which is identified with the normal subgroup Inn G), but the induced automorphism coincides with σ. Thus, the identification of g with i_g does not cause any confusion as to the meaning of g^σ.

The factor group Aut $G/$Inn G is called the *outer automorphism class group* of G, we denote it by Out G.

We often encounter the problem of determining the structure of Aut G for a given group G, but this problem is usually rather difficult. If G is a nonabelian finite simple group, the famous **Schreier conjecture** is the following: the composition factors of Out G consist of cyclic groups of (various) prime orders. The conjecture is still open; it has been verified to be true for all the known finite simple groups. The eventual classification of finite simple groups would certainly solve the Schreier conjecture. But at present, there seems to be no promising lead for a more direct approach.

We will give an upper bound of the order $|$Aut $G|$ of the group of automorphisms when the order n of a finite group G is fixed.

(6.8)   *We have $|$Aut $G| < n^a$ where $a = [\log_2 n]$.*

Proof. Choose a set of generators $X = \{x_1, \ldots, x_m\}$ in such a way that the number m of elements in X is smallest. We claim that an automorphism σ of G is uniquely determined by knowing the m elements $\{x_1^\sigma, x_2^\sigma, \ldots, x_m^\sigma\}$. First, by (2.6), any element u of G is written as a finite product $u_1 u_2 \cdots u_r$ where $u_i \in X$ or $u_i \in X^{-1}$. By the definition of an automorphism,

$$u^\sigma = u_1^\sigma u_2^\sigma \cdots u_r^\sigma,$$

so u^σ is uniquely determined by

$$\{x_1^\sigma, x_2^\sigma, \ldots, x_r^\sigma\} \qquad ((x_i^{-1})^\sigma = (x_i^\sigma)^{-1}).$$

Hence, σ is determined by the subset $\{x_1^\sigma, \ldots, x_m^\sigma\}$. Thus, $|$Aut $G| < n^m$.

Define a chain of subgroups H_i by $H_0 = \{1\}$ and

$$H_i = \langle x_1, \ldots, x_i \rangle \qquad (i = 1, 2, \ldots, m).$$

Clearly, $H_{i-1} \subset H_i$. If $H_{j-1} = H_j$ for some j, then the element x_j could be removed from the set X, so that the remaining set still generates the group G. This contradicts the minimality of m. Thus, $H_0, H_1, \ldots, H_m$ are all distinct and we have $|H_i : H_{i-1}| \geq 2$. By (3.5),

$$|G| = |H_m : H_{m-1}| \cdot |H_{m-1} : H_{m-2}| \cdots |H_1 : H_0| \geq 2^m;$$

so, $m \leq [\log_2 n]$ and (6.8) holds. $\square$

Interest in (6.8) stems from the fact that (6.8) gives a uniform upper bound by a simple function of n. There is a finite group of order 2^m such that its group of automorphisms is of order

$$(2^m - 1)(2^m - 2)(2^m - 2^2) \cdots (2^m - 2^{m-1}).$$

See (9.14). So, the upper bound in (6.8) is very close to the maximum of $|\operatorname{Aut} G|$ when $n \to \infty$.

The above proof of (6.8) shows that the integer a may be replaced by the number of prime factors of n. Furthermore, we can see that the minimum number of generators of any finite group of order n is at most a. Since there is a finite group of order 2^m which cannot be generated by any set of $m - 1$ elements, the above proposition is the best possible. But, it has been conjectured that any finite simple group can be generated by two elements. All the known finite simple groups are, in fact, generated by two elements.

We study next the group of automorphisms of cyclic groups.

(6.9)   *The group of automorphisms of a cyclic group is abelian.*

Proof. Let G be a cyclic group and let g be a generator of G. For any automorphism σ of G, $\sigma(g) = g^n$ for some integer n. If ρ is another automorphism, $\rho(g) = g^m$ for some integer m. Hence we have

$$\rho\sigma(g) = (g^n)^m = g^{nm} = (g^m)^n = \sigma\rho(g).$$

Thus, $\rho\sigma$ and $\sigma\rho$ map the generator g to the same element of G. Since any element of G is a power of g, $\rho\sigma$ and $\sigma\rho$ have the same effect on any element of G. So, we have $\rho\sigma = \sigma\rho$. $\square$

Let $G = \langle g \rangle$ be the cyclic group generated by g. For any integer n, the function σ defined by

$$\sigma \colon g^k \to g^{kn}$$

is a homomorphism of G into itself. If the function σ is an automorphism of G, g^n must be a generator of G. Hence there is an integer l such that $(g^n)^l = g$. If G is infinite, we have $nl = 1$. This implies that either $n = 1$ or $n = -1$. Clearly the function $g^k \to g^{-k}$ defines the nonidentity automorphism of G. On the other hand, if G is a finite cyclic group of order d, we have

$$nl \equiv 1 \pmod{d}.$$

Hence the residue class of $n \pmod{d}$ is prime to d. Conversely, if the integer n is prime to the order d of G, then $\sigma \colon g^k \to g^{nk}$ is an automorphism of G. We have proved the following theorem.

(6.10) *The group of automorphisms of an infinite cyclic group is finite and is of order 2. If G is a finite cyclic group of order d, then* Aut *G is isomorphic to the multiplicative group of integers* mod *d consisting of classes prime to d. In particular, its order is equal to the value of the Euler function* $\varphi(d):|$Aut $G| = \varphi(d)$.

The structure of the multiplicative group of integers mod d has been studied in elementary number theory. Suppose, for example, that d is a power of a prime number p. If $d = p^m$ and $p > 2$, it is a cyclic group of order $p^{m-1}(p-1)$. But, if $p = 2$ and $m > 2$, it is not cyclic.

The group of automorphisms appears often in the study of groups. One of the reasons is the following theorem.

Theorem 6.11. *Let H be a subgroup of a group G. We have*

$$C_G(H) \lhd N_G(H),$$

and the factor group $N_G(H)/C_G(H)$ *is isomorphic to a subgroup of* Aut *H.*

Proof. The first part has been proved in §4, Example 2. Let u be an arbitrary element of $N_G(H)$. By definition, we have $H = H^u$, so the conjugation by the element u induces an automorphism θ_u of H. In fact, θ_u is the restriction of i_u on H. The element u is an element of $N_G(H)$, but not necessarily an element of H. So, θ_u need not be an inner automorphism of H. Since we have

$$\theta_{uv} = \theta_u \theta_v \qquad (u, v \in N_G(H))$$

(cf. the proof of (6.5)), the function $\theta: u \to \theta_u$ defines a homomorphism from $N_G(H)$ into Aut H. Let K be the kernel of θ. Then, by the Homomorphism Theorem (5.3), the image of θ is isomorphic to $N_G(H)/K$. The kernel K of θ consists of elements u such that $\theta_u = 1$. Thus, by definition, K coincides with $C_G(H)$. (The first part is proved again at the same time.) $\square$

This theorem shows that Aut H controls the factor group $N_G(H)/C_G(H)$. For example, if H is cyclic, then $N_G(H)/C_G(H)$ is always abelian (by (6.9)).

Definition 6.12. Let H be a subgroup of a group G. We call H a characteristic subgroup of G if every automorphism of G maps H onto itself; we write

$$H \text{ char } G.$$

Thus, H is characteristic if and only if $H^\sigma = H$ for all $\sigma \in$ Aut G.

By definition, H is normal if and only if $H^g = H$ for all inner automorphisms of G. Thus, a characteristic subgroup is always normal; the converse is, however, not true in general (Example 5 at the end of this section).

(6.13) (i) *Let H be a subgroup of a group G. If $H^\sigma \subset H$ for all $\sigma \in$ Aut G, then H is a characteristic subgroup of G.*

(ii) *The trivial subgroups of G are characteristic; the center $Z(G)$ is a characteristic subgroup of G. In general,*

$$H \text{ char } G \Rightarrow C_G(H) \text{ char } G.$$

(iii) *Every subgroup of a cyclic group is characteristic.*

(iv) *If a subgroup H of a finite group G is the unique subgroup of its order, then H is characteristic.*

Proof. (i) By assumption, $H^\sigma \subset H$ for all $\sigma \in$ Aut G. Hence $\sigma^{-1}(H) \subset H$, and this proves $H \subset H^\sigma$. So, we have $H = H^\sigma$ for all $\sigma \in$ Aut G. Hence, H is a characteristic subgroup of G.

(ii) It is obvious that $G^\sigma \subset G$ and $\{1\}^\sigma = \{1\}$. So, the trivial subgroups are characteristic. Let H be a characteristic subgroup of G. Choose an element g of $C_G(H)$. For any $\sigma \in$ Aut G and $x \in H$, we have

$$g\sigma^{-1}(x) = \sigma^{-1}(x)g.$$

Applying the automorphism σ to both sides, we get

$$\sigma(g)x = x\sigma(g).$$

Since this holds for any $x \in H$, we conclude that $\sigma(g) \in C_G(H)$. Hence $C_G(H)^\sigma \subset C_G(H)$ for all $\sigma \in$ Aut G; by (i), $C_G(H)$ is characteristic.

(iii) An automorphism of a cyclic group maps any element g to its power g^n; so, it leaves every subgroup invariant. Thus, by (i), every subgroup is characteristic.

(iv) Since $|H^\sigma| = |H|$, (iv) follows easily from (i). (If $|G:H|$ is finite and if H is the only subgroup with index equal to $|G:H|$, then H is characteristic. This follows from

$$|G:H^\sigma| = |G:H|;$$

G may be infinite here.)   □

(6.14) *Let H and K be subgroups of a group G. Assume that H is a characteristic subgroup of K.*

(i) K *char* $G \Rightarrow H$ *char* G.

(ii) $K \lhd G \Rightarrow H \lhd G$.

Proof. (i) Let σ be an automorphism of G. Since K char G, we have $K^\sigma = K$. Thus, the restriction $\rho = \sigma_K$ of σ on K is an automorphism of K. By assumption H char K; so, we have $H^\rho = H$. The function ρ is the restriction of σ, so

$H^\rho = H^\sigma = H$. The last equality holds for all $\sigma \in$ Aut G; so, H is characteristic.

(ii) The preceding proof is valid for $\sigma = i_g$, the inner automorphism of G,
and shows that H is invariant by all inner automorphisms of G. Thus, H is
normal in G. □

EXAMPLE 1. In general, $H \lhd K$ and $K \lhd G$ do not imply $H \lhd G$. Let H be the
group defined in §1, Example 4: let K be the subgroup of G defined in §5,
Example 3. Then we have $H \lhd K$ and $K \lhd G$, but H is not a normal subgroup of G.

Endomorphisms

Definition 6.15. Let G be a group. A homomorphism from G into itself is
called an **endomorphism** of G. A subgroup H of G is said to be a **fully invariant
subgroup** of G if H is invariant by any endomorphism θ of G; that is, if
$\theta(H) \subset H$ for any endomorphism θ.

It follows easily from the definition that trivial subgroups are fully invariant and that any fully invariant subgroup is characteristic. But, a characteristic subgroup need not be fully invariant. There are examples of groups
for which the centers are not fully invariant (Chapter 2, §4, Exercise 2).

Let σ and ρ be endomorphisms of a group G. The composite $\rho \circ \sigma$ is also
an endomorphism of G; we define the *product* $\sigma\rho$ to be the composite $\rho \circ \sigma$.
The sets G^σ and G^ρ are subgroups of G. If the elements of G^σ commute with
any element of G^ρ (in this case, we say that the subgroup G^σ *commutes
elementwise* with G^ρ), we can define the *sum* of σ and ρ as follows: Let τ be the
function $x \rightarrow x^\sigma x^\rho$ defined on G. We have

$$(xy)^\tau = (xy)^\sigma (xy)^\rho = x^\sigma y^\sigma x^\rho y^\rho$$
$$= x^\sigma x^\rho y^\sigma y^\rho = x^\tau y^\tau.$$

(The third equality is the consequence of the elementwise commutativity.)
The above formula shows that τ is an endomorphism of G. We define τ to be
the sum and write $\tau = \sigma + \rho$. When G^σ and G^ρ commute elementwise, we say
that σ and ρ are *summable*. The sum of endomorphisms is defined only when
they are summable; in this case we have

$$\sigma + \rho = \rho + \sigma.$$

If any pair of the three endomorphisms σ, τ, and ρ are summable, then $\sigma + \tau$
and ρ are summable. Similarly, σ and $\tau + \rho$ are summable, and we have

$$(\sigma + \tau) + \rho = \sigma + (\tau + \rho).$$

Furthermore, the formulas

$$\sigma(\tau + \rho) = \sigma\tau + \sigma\rho, \qquad (\sigma + \tau)\rho = \sigma\rho + \tau\rho$$

can be proved easily. We will just prove the second. Since σ and τ are summable, G^σ and G^τ commute elementwise. Then for any endomorphism ρ, $G^{\sigma\rho}$ and $G^{\tau\rho}$ commute elementwise, and $\sigma\rho$ and $\tau\rho$ are summable. For any element x of G, we have

$$x^{\sigma\rho + \tau\rho} = x^{\sigma\rho} x^{\tau\rho} = (x^\sigma x^\tau)^\rho$$
$$= (x^{\sigma + \tau})^\rho = x^{(\sigma + \tau)\rho},$$

and the formula $\sigma\rho + \tau\rho = (\sigma + \tau)\rho$ is proved.

If G is an abelian group, any two endomorphisms are summable. Thus, the set of all endomorphisms forms a ring with respect to the multiplication and the addition defined above. The identity function is the identity 1 for multiplication. We express the identity element for addition as 0, which is the endomorphism sending every $x \in G$ into the identity of G.

Definition 6.16. The ring consisting of all endomorphisms of an abelian group G is called the **endomorphism ring** of G and we denote it by $E(G)$.

Definition 6.17. An endomorphism σ of a group G is said to be **normal** if σ commutes with all inner automorphisms of G.

(6.18) *Let σ and ρ be endomorphisms of a group G.*

(i) *If σ and ρ are normal, so is the product $\sigma\rho$. Furthermore, if they are summable, then $\sigma + \rho$ is also normal.*

(ii) *If σ is a normal automorphism, then there is a homomorphism z from G into the center $Z(G)$ such that*

$$g^\sigma = z(g)^{-1} g \qquad \text{for any} \quad g \in G.$$

Conversely, if z is a homomorphism from G into $Z(G)$, the above formula defines a normal endomorphism of G.

(iii) *If $Z(G) = \{1\}$, the only normal automorphism is the identity.*

Proof. (i) Let τ be any inner automorphism of G. Since σ and ρ are normal, we have $\sigma\tau = \tau\sigma$ and $\rho\tau = \tau\rho$ by definition of normality. Hence, $(\sigma\rho)\tau = \tau(\sigma\rho)$ and $\sigma\rho$ is normal. If σ and ρ are summable, then $\sigma\tau$ and $\rho\tau$ (and similarly $\tau\sigma$ and $\tau\rho$) are summable. So,

$$(\sigma + \rho)\tau = \sigma\tau + \rho\tau = \tau\sigma + \tau\rho = \tau(\sigma + \rho),$$

and this shows that $\sigma + \rho$ is also normal.

(ii) Since σ commutes with any inner automorphism i_g, for any x of G, we have

$$\sigma i_g = i_g \sigma, \quad \text{or} \quad g^{-1}\sigma(x)g = \sigma(g)^{-1}\sigma(x)\sigma(g).$$

Thus, $g^\sigma g^{-1}$ commutes with every element of G^σ. Since $G^\sigma = G$, $g^\sigma g^{-1}$ belongs to the center of G. Set

$$g^\sigma = z(g)^{-1}g, \qquad z(g) \in Z(G).$$

Then, z is a function from G into $Z(G)$. The formula $(gh)^\sigma = g^\sigma h^\sigma$ gives us

$$z(gh) = z(g)z(h),$$

so z is a homomorphism. Conversely, if z is a homomorphism from G into $Z(G)$, then the formula $g^\sigma = z(g)^{-1}g$ defines an endomorphism of G. By definition we have $\sigma i_g = i_g \sigma$ for any g; so σ is normal. (The function σ is one-to-one if and only if z has no fixed point $\neq 1$ on $Z(G)$. This is the condition for σ to be an automorphism when G is finite.)

(iii) This is a corollary of (ii). $\quad\square$

EXAMPLE 2. An endomorphism σ of a group G may be a monomorphism without being an automorphism. For example, let σ be the function on $G = \mathbb{Z}$ defined by $\sigma(n) = kn$. For any k, σ is a monomorphism; but if $k > 1$, σ is not an automorphism.

EXAMPLE 3. An endomorphism σ may be an epimorphism without being an automorphism. Let $G = \mathbb{C}^*$ be the multiplicative group of the nonzero complex numbers, and let σ be the function defined by $\sigma(x) = x^n$. This is an endomorphism of G. Since any complex number has its n-th root in $\mathbb{C}$, σ is an epimorphism. However, the kernel of σ is the set of n-th roots of unity. So, σ is not an automorphism if $n > 1$. (A group is said to be *anti-Hopfian* if it has an endomorphism onto itself which is not an automorphism. There are anti-Hopfian groups generated by finitely many elements, but their construction is difficult (Schenkman [2], p. 173.)

EXAMPLE 4. For the additive group $\mathbb{Z}$ of integers, the endomorphism ring $E(\mathbb{Z})$ is isomorphic to the ring of usual integers.

EXAMPLE 5. Let $\mathbb{R}$ be the field of rational numbers, and set $G = \mathbb{R}^+$, the additive group of $\mathbb{R}$. *In this case,*

$$E(G) \cong \mathbb{R} \quad \text{and} \quad \text{Aut } G \cong \mathbb{R}^*.$$

Furthermore, G has no nontrivial characteristic subgroup.

Proof. For each $t \in R$, the function $x \to tx$ is an endomorphism of G. Conversely, let an endomorphism ρ correspond to $r = \rho(1)$. We claim $\rho(x) = rx$ for all rational numbers x. If n is a natural number, n is an n fold addition of 1. So, we get $n = 1 + 1 + \cdots + 1$ and

$$\rho(n) = n\rho(1) = rn.$$

Similarly, n fold addition of $1/n$ gives 1, so $n\rho(1/n) = \rho(1) = r$. Hence we have $\rho(1/n) = r/n$. For any positive rational number p/q (where p and q are positive integers), we have

$$\rho(p/q) = \rho(p(1/q)) = p\rho(1/q) = p(r/q) = rp/q.$$

Thus, $\rho(x) = rx$ for any rational number x. This implies that ρ is uniquely determined by $r = \rho(1)$.

If $\sigma \in E(G)$ corresponds to $s = \sigma(1)$, then the definition of operations in $E(G)$ gives us

$$(\rho + \sigma)(1) = \rho(1) + \sigma(1) = r + s,$$

$$(\rho\sigma)(1) = \sigma(r) = sr.$$

Hence the function $\rho \to \rho(1)$ becomes an isomorphism from the endomorphism ring $E(G)$ onto $\mathbb{R}$.

An element of $E(G)$ lies in Aut G if and only if it has an inverse in $E(G)$. Thus, we have Aut $G = \mathbb{R}^*$. Finally, let H be a characteristic subgroup of G which contains $r \neq 0$. If s is a nonzero rational number, the function σ defined by $\sigma(x) = sx$ is an automorphism of G. Hence H contains $\sigma(r) = sr$; thus, H contains all rational numbers and $H = G$. This proves that G contains no characteristic subgroup except the trivial ones. (For any field, the additive group contains no nontrivial characteristic subgroup.) □

Exercises

1. Let G be a group with $Z(G) = \{1\}$. Show that the centralizer (in Aut G) of Inn G is $\{1\}$ and in particular, that $Z(\text{Aut } G) = \{1\}$.

2. Suppose that $Z(G) = \{1\}$ and Inn G char Aut G. Let $A = \text{Aut } G$. Show that any automorphism of A is inner.

(*Hint.* As $Z(G) = \{1\}$, we have $Z(A) = \{1\}$. We may suppose that $G \subset A \subset A_1 = \text{Aut } A$. By (6.14) (ii), $G \lhd A_1$. The conjugation by an element α of A_1 induces the automorphism α in A (6.7) and an automorphism β of G in G. Then $\beta \in A$ and $\alpha\beta^{-1}$ commutes with all elements of G; that is, $\alpha\beta^{-1} \in C_{A_1}(G) = C$. Prove $C \lhd A_1$ and $A \cap C = \{1\}$. Apply (3.16) and Exercise 1.)

3. Let G be a nonabelian simple group. Show that any automorphism of Aut G is inner.

(*Hint.* Show Inn G char Aut G, and use Exercise 2. This proposition is one of the few theorems which are valid for any simple group. A group X is said to be a *complete group* if $Z(X) = \{1\}$ and any automorphism of X is inner. Exercises 1 and 3 show that the group of automorphisms of a nonabelian simple group is complete. For any group X, define $X_0 = X$ and $X_1, X_2, \ldots$ by

$$X_i = \text{Aut } X_{i-1} \qquad (i = 1, 2, \ldots).$$

If X is a finite group with $Z(X) = \{1\}$, then the sequence of groups $\{X_i\}$ reaches a complete group in a finite number of steps (theorem of Wielandt [1]).)

4. Let σ be a normal endomorphism of a group G. Set $\sigma(G) = H$ and $\sigma(g) = z(g)^{-1}g$ for any $g \in G$.

(a) Show that z is a homomorphism from G into $C_G(H)$.

(b) Show that H is a normal subgroup of G such that $G = HC_G(H)$, and $H \cap C_G(H) = Z(H) \subset Z(G)$.

(c) Show that both H and $C_G(H)$ are invariant by σ. Prove that the restriction ρ of σ on $C_G(H)$ is a homomorphism from $C_G(H)$ into $Z(H)$, and that for any element x of $Z(H)$, we have $x = \zeta(x)\rho(x)$ where ζ is the restriction of z on H.

(d) Conversely, suppose that $G = HC_G(H)$, and that a homomorphism $\zeta: H \to Z(H)$ and a homomorphism $\rho: C_G(H) \to Z(H)$ are given so that, for any element x of $Z(H)$, the formula $x = \zeta(x)\rho(x)$ is satisfied. Prove that, for an element g of G, the formula

$$\sigma(g) = \rho(c)\zeta(h)^{-1}h,$$

where $g = hc$, $h \in H$ and $c \in C_G(H)$, defines a normal endomorphism from G into H.

§7. Permutation Groups; G-sets

Definition 7.1. Let X be a set. A one-to-one function from X onto X is called a **permutation** on the set X. We denote by $\Sigma(X)$ the set of all permutations on X.

A function p defined on the set X is a permutation if and only if the following three conditions are satisfied:

(i) $p(x) \in X$ for all $x \in X$;
(ii) Given $a \in X$, there is an element $x \in X$ such that $p(x) = a$;
(iii) $p(u) = p(v) \Rightarrow u = v$.

If p and q are permutations on a set X, the composite $q \circ p$ is also a permutation on X which is defined to be the *product* of p and q; we write the product of p and q as the juxtaposition pq. It follows from the definition that the associative law holds;

$$p(qr) = (pq)r.$$

The identity function 1 on X is the identity with respect to the operation just defined. If p is a permutation on X, the inverse function p^{-1} is defined by the formula

$$p^{-1}(x) = y \quad \text{when} \quad p(y) = x,$$

and clearly p^{-1} is also a permutation. Thus, the set $\Sigma(X)$ of permutations on X forms a group under the operation defined above. We call $\Sigma(X)$ the **symmetric group** on X.

(7.2) *Let θ be a one-to-one function from a set X onto a set Y. For a permutation p on X, we define a function $\theta(p)$ on Y by the formula*

$$\theta(p)(y) = \theta(p(\theta^{-1}(y))) \qquad (y \in Y).$$

Then $\theta(p)$ is a permutation on Y. Furthermore, the function $\theta: p \to \theta(p)$ is an isomorphism from the symmetric group $\Sigma(X)$ onto the symmetric group $\Sigma(Y)$ on Y.

Proof. It is easy to prove that the function $\theta(p)$ satisfies the conditions (i), (ii), and (iii) for $\theta(p)$ to be a permutation on Y. Since the proofs are similar, we will prove (ii) as a sample.

(ii) Let $y' \in Y$ be given. Since the function θ is surjective, there is an element $x' \in X$ such that $\theta(x') = y'$. By assumption, p is a permutation. So, there is an element $x \in X$ such that $p(x) = x'$. If $y = \theta(x)$, then $x = \theta^{-1}(y)$ and

$$y' = \theta(x') = \theta(p(x)) = \theta(p(\theta^{-1}(y))) = \theta(p)(y).$$

Thus, $\theta(p)$ satisfies the condition (ii).

If p and q are permutations on X, then for any $y \in Y$,

$$\theta(pq)(y) = \theta((pq)(\theta^{-1}(y))) = \theta(q(p(\theta^{-1}(y))))$$
$$= \theta(q)(\theta(p)(y)) = (\theta(p)\theta(q))(y).$$

This shows that $\theta(pq) = \theta(p)\theta(q)$; so, θ is a homomorphism from $\Sigma(X)$ into $\Sigma(Y)$. If $\theta(p)$ is the identity function on Y, we have $\theta(p)(y) = y$, or

$$p(\theta^{-1}(y)) = \theta^{-1}(y)$$

for all $y \in Y$. Thus, p is the identity function on X, and θ is one-to-one. For a permutation q on Y, define the function p on X by the formula

$$p(x) = \theta^{-1}(q(\theta(x))).$$

As before, p is a permutation on X. Clearly we have $\theta(p) = q$; thus, θ is surjective. We have shown that θ is an isomorphism from $\Sigma(X)$ onto $\Sigma(Y)$. □

If the set X is a finite set, the isomorphism class of $\Sigma(X)$ is uniquely determined by the number $|X|$ of elements in X. When $|X| = n$, it is customary to choose $X = \{1, 2, \ldots, n\}$, the set of first n natural numbers, and denote $\Sigma(X) = \Sigma_n$; Σ_n is called *the symmetric group of degree n*.

Definition 7.3. Let X be a set and $\Sigma(X)$ the symmetric group on X. Any subgroup of $\Sigma(X)$ is called a **permutation group** on X.

A permutation group on X is a group whose elements are given as permutations on the set X. It has an advantage in that the group structure can be studied by using properties of permutations.

(7.4) (Cayley) *Let G be a given group. Then there is a set X such that G is isomorphic to a permutation group on X.*

Proof. We choose the set X consisting of all the elements of G. For each element g of G, let π_g be a function on X defined by the formula

$$\pi_g(x) = xg \qquad (x \in X).$$

It follows that π_g is a permutation on X. Furthermore, the associative law proves

$$\pi_{gh} = \pi_g \pi_h \qquad (g, h \in G).$$

Thus, the function π is a homomorphism from G into the symmetric group $\Sigma(X)$. Clearly, π_g is the identity function on X if and only if $g = 1$. This means that π is a monomorphism. Hence G is isomorphic to the image $\pi(G)$ which is a permutation group on X. □

The isomorphism π defined above is called the right *regular representation*. The left regular representation can be defined similarly by

$$\pi'_g(x) = g^{-1}x.$$

In general, a homomorphism from a group into $\Sigma(X)$ of some set X is called a *permutation representation* of G on X. The consideration of permutation representations offers one of the most powerful methods of investigating the properties of a group.

We get the following corollary of the Cayley's theorem.

Corollary. *For any given natural number n, the number of isomorphism classes of groups of order n is finite.*

Proof. The proof of Cayley's theorem shows that a finite group of order n is isomorphic to a subgroup of the symmetric group Σ_n of degree n. The group Σ_n is a finite set which forms a group under a definite law of multiplication. So, any subset of Σ_n becomes a subgroup of Σ_n in at most one way. This implies that the number of isomorphism classes containing subgroups of Σ_n is finite; the number of isomorphism classes of finite groups of order n is, therefore, finite. □

From the remark made after (6.8), we conclude that the minimal number of generators of any finite group G of order n is at most $a = [\log_2 n]$. Any homomorphism from G into Σ_n is determined by assigning a permutation of Σ_n to each member of a generating set of G. Hence, the number of isomorphism classes of finite groups of order n does not exceed $(n!)^a < n^{an}$ (Sims–Gallagher). The number is, in fact, far less. It has been conjectured that the number should be less than n^{kb} for some constant k, where $b = a^2$ and either $a = [\log_2 n]$ or a is the number of prime factors in n. In support of this conjecture, we have the following result. Let the number of isomorphism classes of groups of order p^m, where p is a prime number, be p^l with $l = Am^3$. The Higman–Sims theorem asserts that $A \to 2/27$ as $m \to \infty$ (G. Higman [3], Sims [1]; cf. Chapter 4, §4). If the number of isomorphism classes of *simple* groups of order n is bounded above by n^{cb} for some constant c, then the aforementioned conjecture is true (with $k = c + 1$). This is a theorem of P. Neumann [2]. Among the known simple groups, there are at most two isomorphism classes of nonabelian simple groups of the same order.

Definition 7.5. Let G be a group. A **G-set** is a pair (X, ρ) of a set X and a homomorphism ρ from G into the symmetric group $\Sigma(X)$ on X.

When a G-set (X, ρ) is given, each element g of G determines a permutation $\rho(g)$ on X. For $x \in X$, we write

$$x^g = \rho(g)(x).$$

Since ρ is a homomorphism, we have

(7.6) $x^{gh} = (x^g)^h$ $(g, h \in G)$.

Thus, ρ is a permutation representation of G on the set X. But, if we view this same situation emphasizing the set X, then we call ρ the **action** of G on X. Or, we say that G *acts on X via the action ρ*. A G-set is a set on which an action of G is defined. Sometimes we simply say that a set X is a G-set without explicitly mentioning ρ; but, in this case, the action of G is implicitly given so as to make X a G-set. The same set can be the underlying set of several distinct G-sets.

Definition 7.7. Let X be a G-set. A subset

$$\{x^g \mid g \in G\}$$

is called an **orbit** containing $x \in X$; we write it as O_x (or $O(x)$). The number of elements in O_x, $|O_x|$, is called the **length** of the orbit O_x. A subset Y of X is said to be **G-invariant** if for all g of G

$$y \in Y \Rightarrow y^g \in Y.$$

We often regard geometrically X as a set of points, and the elements of X are called *points*.

EXAMPLE 1. Let X be the set of points in a plane P. Consider the set G consisting of all the rotations about a fixed point A of P. Like §1, Example 5, G forms a group with respect to the composition of motions. Each element of G is considered as a permutation on the set X, and, with this action, X is a G-set. The orbit containing $p \in X$ is a circle with center A; this may be the origin of the term orbit.

The basic properties of orbits are collected in next lemma.

(7.8) *Let X be a G-set and let O_x be the orbit containing a point x of X.*

(i) *If $y \in O_x$, we have $O_y = O_x$.*
(ii) *If O is an orbit different from O_x, then $O \cap O_x = \varnothing$.*
(iii) *A nonempty subset of X is an orbit if and only if it is a minimal G-invariant subset.*
(iv) *Any G-invariant subset of X is a disjoint union of orbits.*

Proof. (i) By definition, there is an element g of G such that $y = x^g$. If h is an element of G, by (7.6), we have

$$y^h = (x^g)^h = x^{gh};$$

thus, $O_y \subset O_x$. Conversely, if $x^k \in O_x$,

$$x^k = x^{gl} = (x^g)^l = y^l$$

with $l = g^{-1}k \in G$; thus, $O_x \subset O_y$. We conclude $O_y = O_x$.

(ii) $O \cap O_x \ni y$ implies that $O = O_y = O_x$ by (i). So, (ii) is a corollary of (i).

(iii) It follows easily from the definition that an orbit is G-invariant. Suppose that O_x contains a nonempty G-invariant subset Y. For an element y of Y, we have $y^g \in Y$ for all g of G; so, $O_y \subset Y$. Since $O_y = O_x$ by (i), an orbit is a minimal nonempty G-invariant subset. Conversely, if Y is a minimal nonempty G-invariant subset, we have $O_y \subset Y$ for any $y \in Y$. By minimality, we get $O_y = Y$.

(iv) This is obvious (cf. (ii) and (iii)). $\square$

Theorem 7.9. *Let X be a G-set, and let $O = O_x$ be an orbit containing a point x of X. Let H be the subset of G defined by*

$$H = \{g \mid g \in G, \, x^g = x\}.$$

(i) *H is a subgroup of G.*

(ii) *There is a one-to-one mapping φ from the set O onto the set of right cosets of H which satisfies*

$$\varphi(x^g) = \varphi(x)g$$

for all $x \in X$ and $g \in G$.

(iii) *If O is a finite set, we have $|O| = |G : H|$. If G is a finite group, the length of an orbit is a divisor of the order $|G|$ of G.*

Proof. (i) Using (7.6), it is easy to verify the two conditions (i) and (ii) of Definition 2.1 for H to be a subgroup of G.

(ii) An element y of O_x is written $y = x^u$ ($u \in G$). We claim that $x^u = x^v$ if and only if $Hu = Hv$. If $x^u = y = x^v$, we have (for $w = v^{-1}$) $x^{uw} = (x^u)^w = y^w = x$. So, $uv^{-1} \in H$ and $Hu = Hv$ by (3.2) (v). Conversely, $Hu = Hv$ gives us $v = hu$ with $h \in H$. Hence we have $x^v = x^{hu} = (x^h)^u = x^u$.

Define a mapping φ from O_x to the set of right cosets of H as follows:

$$\varphi(y) = Hu \quad \text{when} \quad y = x^u.$$

As shown above, this function is one-to-one, surjective, and well defined independent of the choice of an element u of G. If $g \in G$, we get $y^g = (x^u)^g = x^{ug}$. Therefore, we have

$$\varphi(y^g) = Hug = (Hu)g = \varphi(y)g.$$

(iii) Since the mapping φ defined in (ii) is one-to-one, $|O|$ is equal to the index $|G : H|$. The last assertion follows from Theorem 3.3. $\square$

Definition 7.10. The subgroup H defined in Theorem 7.9 is called the **stabilizer** of x; we write $H = S_G(x)$.

(7.11) *Using the same notation as in Theorem 7.8, if a point y is in O_x, the stabilizer of y is conjugate to the stabilizer of x. More precisely, we have*

$$S_G(x^g) = S_G(x)^g.$$

Proof. By definition, $S_G(x^g) = \{h \in G \,|\, (x^g)^h = x^g\}$. So, the following equivalences prove (7.11):

$$(x^g)^h = x^g \Leftrightarrow ghg^{-1} \in S_G(x) \Leftrightarrow h \in S_G(x)^g. \quad \square$$

Theorem 7.12. *Let X be a finite G-set. Then X is a finite disjoint union of orbits $O_1, O_2, \ldots, O_m$:*

$$X = O_1 \cup O_2 \cup \cdots \cup O_m \qquad (O_i \cap O_j = \varnothing \ (i \neq j)).$$

If x_i is an element of O_i for $i = 1, 2, \ldots, m$, we have

$$|X| = \sum_{i=1}^{m} |G : S_G(x_i)|.$$

Proof. The first half follows from (7.8) (iv). Counting the number of elements, we get

$$|X| = |O_1| + |O_2| + \cdots + |O_m|.$$

So, Theorem 7.12 follows from Theorem 7.9 (iii). $\quad \square$

EXAMPLE 2. Let G be a group, and for a natural number k let G_k be the totality of all subsets of G consisting of k elements. For each element g of G and any element A of G_k, the set $Ag = \{ag \,|\, a \in A\}$ is an element of G_k; the mapping $A \to Ag$ is a permutation on G_k. Let $\rho(g)$ be the permutation $A \to Ag$. Then ρ defines an action of G on G_k; we get a G-set (G_k, ρ). If $k = 1$, this is the right regular representation. If we define another action of G on G_k by assigning

$$\rho'(g) \colon A \to g^{-1}A,$$

we obtain a different G-set (G_k, ρ').

EXAMPLE 3. There is still another way to make G_k a G-set. For an element $g \in G$, we define a function $\gamma(g)$ on G_k by

$$\gamma(g)(A) = g^{-1}Ag.$$

Then, $\gamma(g)$ is a permutation on G_k, and γ defines another action of G on G_k. If $k = 1$, the orbits under this action are the conjugacy classes of G (Definition 2.9). The formula in Theorem 7.12 is the so-called **class equation** of the finite group G:

$$(7.13) \qquad |G| = \sum_{i=1}^{m} |G : C_G(x_i)|.$$

In this formula, $\{x_i\}$ is a set of representatives from the conjugacy classes. By the definition of the action $\gamma(g)$, the stabilizer $S_G(x_i)$ coincides with the centralizer $C_G(x_i)$ of x_i. Each term of the right side of the class equation (7.13) is a divisor of the order $|G|$. We have

$$|G : C_G(x)| = 1 \Leftrightarrow x \in Z(G).$$

Hence, exactly $|Z(G)|$ of the terms in the right side of (7.13) are equal to 1. Thus, the class equation may be written

$$(7.13)' \qquad |G| = |Z(G)| + \sum{}' |G : C_G(x_j)|,$$

where each term in the summation $\sum'$ is greater than one.

EXAMPLE 4. Let H be a subgroup of a group G. Any G-set may be viewed as an H-set by restricting the action to H. Consider the G-set (G_1, ρ) of Example 2 as an H-set. In this case, an H-orbit is a left coset gH. Similarly, a right coset can be obtained as an H-orbit. Thus, (3.2) (i, ii, iii) is a special case of (7.8).

EXAMPLE 5. Let K be a subgroup of a group G, and let X be the set of all right cosets of K. If $|K| = k$, X is a subset of G_k of Example 2; in fact, X is an orbit in the G-set (G_k, ρ). Since K is viewed as a coset of K, K is an element of X. It follows from the definition that we have

$$S_G(K) = K.$$

(In this case, $S_G(K) = \{g \,|\, Kg = K\}$.) So, by (7.11), the stabilizer of Kx is the conjugate subgroup K^x.

Let H be a subgroup of G and consider X to be an H-set. From the definition of ρ, it is easy to see that an H-orbit containing a point Kx of X is the set of all the cosets of K which lie in the double coset KxH. Since

$$S_H(Kx) = H \cap S_G(Kx) = H \cap K^x,$$

the results concerning double cosets, (3.8) and Theorem 3.9, are special cases of the theorems in this section.

EXAMPLE 6. Let R be a subset of a group G. The totality of conjugates of R is an orbit in the G-set of Example 3, and the stabilizer $S_G(R)$ with respect to this action coincides with the normalizer $N_G(R)$.

Definition 7.14. A G-set X is said to be a **transitive G-set** if X itself is an orbit. A permutation representation ρ on X of a group G is said to be **transitive** if the G-set (X, ρ) is transitive. In this case, we say that ρ is a transitive action.

Let O be an orbit in a G-set X. We may view O as a G-set by restricting the action to O. In this case, O is a transitive G-set; an orbit is sometimes called *a domain of transitivity.*

(7.15) *Let u and v be two elements of a transitive G-set. Then there is an element g of G such that $v = u^g$.*

Proof. If a G-set X is transitive, $X = O_x$ for some x of X. By (7.8) (i), we have $O_u = O_x = X$. So, for any $v \in X$, there is an element g of G such that $v = u^g$. $\square$

If a subgroup H of a group G is given, the totality of right cosets of H becomes a transitive G-set Γ_H as in Example 5. Conversely, if X is a transitive G-set, then a subgroup H (which is the stabilizer $S_G(x)$ of a point x of X) and a one-to-one mapping φ from X onto Γ_H are determined (Theorem 7.9 (ii)). The function φ satisfies $\varphi(x^g) = \varphi(x)g$ for all $x \in X$ and $g \in G$. This shows that a transitive G-set X is essentially identical with the G-set Γ_H constructed from a subgroup H. (We may say that X and Γ_H are *isomorphic G-sets*.)

(7.16) *Let G be a group. The following three conditions are equivalent.*

   (i)   *There is a transitive permutation representation of G on a set of n elements.*

   (ii)   *There is a homomorphism from G into the symmetric group Σ_n of degree n such that the image of G is transitive.*

   (iii)   *The group G has a subgroup of index n.*

If a subgroup H of index n is given, an action ρ on the set Γ_H of cosets of H is determined (as in Example 5). The kernel of the action ρ is the intersection of all conjugate subgroups of H :

$$\mathrm{Ker}\ \rho = \bigcap H^x \qquad (x \in G).$$

Proof. If there is a transitive permutation representation ρ of G on $X = \{1, 2, \dots, n\}$, the G-set (X, ρ) is transitive. Hence the stabilizer $H = S_G(1)$ is a subgroup of index n. Conversely, if H is a subgroup of index n, the set Γ_H of cosets of H is a G-set by the action $Hx \to Hxg$. Hence, there is a homomorphism ρ from G into $\Sigma(\Gamma_H)$, and this representation is transitive. By (7.2),

$\Sigma(\Gamma_H)$ is isomorphic to Σ_n. So, there is a transitive homomorphism from G into Σ_n. Let K be the kernel of ρ. Then K fixes all the cosets Hx. Hence we have

$$K = \bigcap_x S_G(Hx) = \bigcap_x H^x$$

(cf. Example 5). □

Definition 7.17. Let H be a subgroup of a group G. Define

$$\text{Core}_G\, H = \bigcap_{x \in G} H^x.$$

This subgroup is called the **core** of H.

(7.18) *The core of H is the largest normal subgroup of G which is contained in H.*

Proof. Clearly $\text{Core}_G\, H \lhd G$ (by (4.4)). If K is a normal subgroup of G which is contained in H, we have $K = K^x \subset H^x$ for all x in G. Hence

$$K \subset \bigcap_{x \in G} H^x = \text{Core}_G\, H \subset H.$$

Thus, $\text{Core}_G\, H$ is the largest normal subgroup of G contained in H. □

Exercises

1. Suppose that a finite group G acts on a finite set X and that there are exactly m orbits. Let $\chi(g)$ denote the number of fixed points of the element $g \in G$. Show that the following formula holds:

$$\sum_{g \in G} \chi(g) = m\,|G|.$$

(*Hint.* Let $X = O_1 \cup \cdots \cup O_m$ be the disjoint union of the orbits. Suppose that an element $g \in G$ has exactly $\chi_i(g)$ fixed points in O_i. Count the number of pairs (x, g) such that $x \in O_i, g \in G$, and $x^g = x$, in two different ways to get

$$\sum_{g \in G} \chi_i(g) = |S_G(x)| \cdot |G : S_G(x)| = |G|.)$$

2. Show that the following two conditions on a group G are equivalent:

(1) There is a homomorphism φ from G into Σ_n such that $\varphi(g) \neq 1$ for some $g \in G$.

(2) The group G contains a proper subgroup of index at most n.

3. A group G is said to be *finitely generated* if G is generated by a finite number of elements.

(a) Prove that a finitely generated group contains only a finite number of subgroups of index n for each natural number n.

(b) Show that if a finitely generated group G contains a subgroup H of finite index, then there is a characteristic subgroup K of G such that $K \subset H$ and $|G : K|$ is finite.

(*Hint.* A subgroup of index n determines a homomorphism from G into Σ_n (7.16); for a finitely generated group, the number of such homomorphisms is finite (cf. the proof of (6.8)). So a subgroup of index n contains the kernel of one of these homomorphisms; thus, the first proposition holds. To prove (b), use (a) and Cor. 1 of (3.13).)

§8. Operator Groups; Semidirect Products

Definition 8.1. Let G and H be two groups. If a homomorphism φ from G into Aut H is given, we say that G acts on H via φ and G is an **operator group** on H; the homomorphism φ is called an **action** of G.

The action φ of G is not necessarily an isomorphism. In particular, φ can be the trivial action, that is, $\varphi(g) = 1$ for all g in G. Via the trivial action, any group can act on H.

For any $g \in G$, $\varphi(g)$ is an automorphism of H; we denote the image $\varphi(g)(h)$ of an element h of H simply by h^g. The action on H is given if and only if the functions

$$\varphi(g) : h \rightarrow h^g$$

are defined for each $g \in G$ and satisfy the following formulas for all $u, v \in H$ and $x, y \in G$:

(8.2)
$$(uv)^x = u^x v^x,$$
$$u^{xy} = (u^x)^y,$$
$$u^1 = u,$$

where 1 is the identity of G. In fact, the first formula of (8.2) is equivalent to the statement of $\varphi(x)$ is an endomorphism of H, while the second one is the homomorphism condition for φ. The last condition says $\varphi(1) = 1$, and insures that $\varphi(g)$ is an automorphism of H for each $g \in G$.

EXAMPLE 1. Let G and H be subgroups of a group L such that $G \subset N_L(H)$. The conjugation by an element g of G induces an automorphism θ_g of H (cf. the proof of Theorem 6.11). The function θ is proved to be a homomorphism from G into Aut H; thus, θ is an action of G on H. As remarked earlier, the abbreviation $\theta_g(h) = h^g$ will cause no confusion because $\theta_g(h)$ is the conjugate of h by the element g.

EXAMPLE 2. If G is a subgroup of Aut H, then the injection of G into Aut H defines an action of G on H.

When we consider the action of a group on another group, we take two separate groups G and H in order to avoid loss of generality. But, it is easier to handle the special situation of Example 1 when both groups are contained in a large group and the action is induced by conjugation. So, the following construction of a large group is important.

(8.3) *Let φ be an action of a group G on another group H. Let L be the direct product set of G and H; that is, the totality of pairs (g, h) of elements $g \in G$ and $h \in H$. Define the product of two elements of L by the formula*

$$(g, h)(g', h') = (gg', \varphi(g')(h)h').$$

Then L forms a group with respect to this operation.
    Let 1_G denote the identity of G; similarly, let 1_H denote the identity for H. Define, for $g \in G$ and $h \in H$,

$$\gamma(g) = (g, 1_H), \qquad \eta(h) = (1_G, h).$$

We set

$$\bar{G} = \{\gamma(g) \mid g \in G\}, \qquad \bar{H} = \{\eta(h) \mid h \in H\}.$$

Then, γ is an isomorphism from G onto $\bar{G}$, η is an isomorphism from H onto $\bar{H}$, and we have

$$\bar{H} \lhd L = \bar{G}\bar{H}, \qquad \bar{G} \cap \bar{H} = \{1\}$$

where $1 = (1_G, 1_H)$ is the identity of L. Furthermore, the formula

$$\gamma(g)^{-1}\eta(h)\gamma(g) = \eta(h^g)$$

holds for any $g \in G$ and $h \in H$.

Proof. First, we will verify the associative law.
    Let $g, u, x \in G$ and $h, v, y \in H$. Then,

$$[(g, h)(u, v)](x, y) = (gu, h^u v)(x, y)$$
$$= ((gu)x, (h^u v)^x y) = (g(ux), ((h^u)^x v^x)y)$$
$$= (g(ux), h^{ux}(v^x y)) = (g, h)(ux, v^x y)$$
$$= (g, h)[(u, v)(x, y)].$$

We have used (8.2) and the associative law in G and H.

By definition, $1 = (1_G, 1_H)$ is the identity; the inverse of (g, h) is given by $(g^{-1}, \varphi(g^{-1})(h)^{-1})$. So, L forms a group with respect to the operation defined above.

The definition of the operation also shows that

$$\gamma(gg') = \gamma(g)\gamma(g'), \qquad \eta(hh') = \eta(h)\eta(h');$$

so, both γ and η are homomorphisms and $\bar{G} = \gamma(G)$ and $\bar{H} = \eta(H)$ are subgroups. Furthermore, by definition, we have

$$(g, h) = \gamma(g)\eta(h),$$

$$\gamma(g)^{-1}\eta(h)\gamma(g) = \eta(h^g).$$

Hence $\bar{H} \triangleleft L = \bar{G}\bar{H}$. Clearly, we have $\bar{G} \cap \bar{H} = \{1\}$. $\square$

Definition 8.4. The group constructed in (8.3) is called the **semidirect product** of G and H with respect to the action φ.

Usually, the subgroups $\bar{G}$ and $\bar{H}$ are identified with G and H via injections γ and η, respectively, and consider G and H as subgroups of the semidirect product L using the above identifications of G with $\bar{G}$ and of H with $\bar{H}$.

It is important to note that the conjugation by an element g of G induces an automorphism of H which coincides with the automorphism $\varphi(g)$ given by the action φ of G on H.

When an operator group G on a group H is given, the set of elements of H left invariant by all $g \in G$,

$$\{h \mid h \in H, \varphi(g)(h) = h \quad \text{for all } g \in G\},$$

is a subgroup of H. This subgroup corresponds to the subgroup $H \cap C_L(G)$ in the semidirect product. We shall denote it by

$$C_H(G)$$

by stretching the meaning of the notation. Similarly, the meaning of $C_G(H)$ should be clear; it is the kernel of φ in G. Generally, for any subgroup U and for any subset S of a group G, we use the notation

(8.5) $C_U(S) = U \cap C_G(S).$

When a group G and an operator group on G are involved, we embed both of these groups into their semidirect product, and we use the notations introduced in (8.5) within the semidirect product.

In Example 1, GH is a subgroup of L. But, it should be noted that GH is not necessarily isomorphic to the semidirect product with respect to θ because the condition $G \cap H = \{1\}$ need not hold in L. Thus, the convention (8.5) might give two meanings to the same symbol. But, the action of an element $g \in G$ produces the same effect on H in L and in the semidirect product. So, the two definitions agree and no confusion will arise.

Definition 8.6. A group G is said to be an **internal semidirect product** of two subgroups H and K, if

$$H \lhd G = HK, \qquad H \cap K = \{1\}.$$

The definition of an internal semidirect product is not symmetric with respect to H and K. So, when it is important to distinguish between them, it should be stated clearly which subgroup is normal in G.

The semidirect product of G and H with respect to an action φ of G on H is, according to Definition 8.4, one of the internal semidirect products defined in Definition 8.6. We will prove that any internal semidirect product is isomorphic to the semidirect product with respect to some action. For this reason, the adjective *internal* is often omitted.

(8.7) *Let G be an internal semidirect product of two subgroups H and K such that $H \lhd G = HK$ and $H \cap K = \{1\}$. Let $\theta(u)$ be the automorphism of H induced by the conjugation of $u \in K$. Then θ is an action of K on H, and the semidirect product of K and H with respect to θ is isomorphic to G.*

Proof. In Example 1, we proved that θ is an action of K on H. Let L be the semidirect product of K and H with respect to θ. Then the element of L is a pair (k, h) of elements of $k \in K$ and $h \in H$. Let f be the function defined by

$$(k, h) \rightarrow kh \in G.$$

By assumption, $G = HK = KH$; so, the function f is surjective. For any k, $u \in K$ and $h, v \in H$, the definition of the product in L gives us

$$(k, h)(u, v) = (ku, \theta(u)(h)v).$$

On the other hand, in G, we have

$$(kh)(uv) = (ku)(h^u v).$$

Since $h^u = u^{-1}hu = \theta(u)h$, the above two formulas tell us that the function f is a homomorphism. If the element (k, h) lies in the kernel of f, then we have $kh = 1$, or $k = h^{-1} \in K \cap H = \{1\}$. Thus, the kernel of f is 1; f is an isomorphism. $\square$

The structure of the semidirect product depends not only on the groups H and G, but also on the action φ. If the action φ is trivial, that is, $\varphi(g) = 1$ for all $g \in G$, the semidirect product with respect to the trivial action is called simply the *direct product*. The direct product of any two groups G and H always exists, and its isomorphism class is uniquely determined by G and H.

A semidirect product $H \lhd G = HK$, $H \cap K = \{1\}$ is a direct product if and only if $K \lhd G$. In fact, if G is a direct product, the action of K on H is trivial. So H and K commute elementwise. This implies that K is a normal subgroup of G. On the other hand, if $K \lhd G$, then H and K commute elementwise by (3.16). So, G is a direct product.

(8.8) *Suppose that a group is a product of two subgroups H and K such that $H \cap K = \{1\}$. (For example, suppose G is a semidirect product of H and K.) Then every element g of G can be written as hk ($h \in H$, $k \in K$) in a unique way. In addition, if G is finite, then we have*

$$|G| = |H| \times |K|.$$

Proof. Since $G = HK$, g can be written as a product $g = hk$. If we have

$$hk = h'k' \qquad (h,\ h' \in H,\ k,\ k' \in K),$$

then $(h')^{-1}h = k'k^{-1} \in H \cap K = \{1\}$, and we get $h = h'$ and $k = k'$. Thus, the expression is unique. The last assertion is obvious (from the first part or from (3.11)). $\square$

The next proposition is useful in many applications.

(8.9) *Let G be a semidirect product of subgroups H and K, such that $H \lhd G$. Then, $N_G(K) \cap H$ commutes elementwise with K, and we have*

$$N_G(K) \cap H = C_H(K), \qquad N_G(K) = KC_H(K).$$

Proof. The Second Isomorphism Theorem proves that

$$N_G(K) \cap H \lhd N_G(K).$$

Since $H \cap K = \{1\}$, (3.16) shows that K and $N_G(K) \cap H$ commute elementwise. So, $N_G(K) \cap H \subset C_H(K)$. Clearly we have

$$C_H(K) \subset N_G(K) \cap H,$$

so the first formula holds.

Theorem 3.14, when applied to $N_G(K)$, gives us

$$N_G(K) = G \cap N_G(K) = K(H \cap N_G(K)).$$

So, the first formula implies the second.   □

Definition 8.10. Let φ be an action of a group G on another group H. A subgroup U of H is said to be **G-invariant** if $\varphi(g)U = U$ for all $g \in G$.

If a subgroup U is G-invariant, each $\varphi(g)$ induces an automorphism $\psi(g)$ of U; so, G becomes an operator group on U via the restriction ψ of φ. If a G-invariant subgroup U is normal, then G acts on the factor group H/U. The action is defined by

$$(Ux)^g = Ux^g.$$

In order for the action to be well defined, we must show that $Ux = Uy$ implies $Ux^g = Uy^g$. If $Ux = Uy$, then $xy^{-1} \in U$ by (3.2) (v). Since U is G-invariant,

$$(xy^{-1})^g = x^g(y^{-1})^g = x^g(y^g)^{-1} \in U,$$

and we have $Ux^g = Uy^g$. It is easy to verify the formulas (8.2). For example,

$$(UxUy)^g = (Uxy)^g = U(xy)^g$$
$$= Ux^g y^g = Ux^g Uy^g = (Ux)^g(Uy)^g.$$

Thus, G acts on H/U via $Ux \rightarrow Ux^g$. This action is called the (*canonical*) *induced action*. We always consider G as acting on G-invariant subgroups, or on the factor group by a G-invariant normal subgroup, via the restriction, or the canonically induced action.

(8.11) *Let G be an operator group on H, and let φ be the action. Let L be the semidirect product of G and H with respect of φ. We consider G and H to be subgroups of L.*
  (i) *A subgroup U of H is G-invariant if and only if*

$$G \subset N_L(U).$$

In this case, UG is isomorphic to the semidirect product of U and G with respect to the restriction of φ on U.
  (ii) *If U is a G-invariant normal subgroup of H, then $U \lhd L$, and L/U is isomorphic to the semidirect product of H/U and G with respect to the induced action.*

Proof. The action φ is realized by conjugation of L; so, the first half of (i) follows immediately. Proposition (8.7) proves the rest of (i). The second statement (ii) is proved similarly. Set $\bar{L} = L/U$. Then we have $\bar{H} \triangleleft \bar{L} = \bar{H}\bar{G}$ and $\bar{H} \cap \bar{G} = \{1\}$. Since $\bar{H} \cong H/U$ and $\bar{G} \cong G$, we get (ii) by (8.7). $\square$

EXAMPLE 3. Let p be a prime number. For any natural number n and any positive divisor d of $p^n - 1$, we can construct a nonabelian group of order $p^n d$. For each prime power p^n, there exists a finite field F of p^n elements. The multiplicative group F^* of nonzero elements of F is a cyclic group of order $p^n - 1$. So, by the Cor. of (5.6), F^* contains an element λ of order d. Consider the function $\varphi(\lambda^n)$ defined by the formula

$$\varphi(\lambda^n)(\alpha) = \lambda^n \alpha$$

where $\alpha \in F^+$. Clearly, $\varphi(\lambda^n)$ is an automorphism of the additive group F^+ (since we have $\lambda^n(\alpha + \beta) = \lambda^n\alpha + \lambda^n\beta$). It is easy to see that φ is a homomorphism from the cyclic group $\langle \lambda \rangle$ generated by λ into Aut F^+. Thus, the cyclic group $G = \langle \lambda \rangle$ of order d acts on the group F^+ of order p^n. By (8.8), the semidirect product L with respect to this action has order $p^n d$. The conjugation by λ induces in F^+ the nontrivial automorphism $\alpha \to \lambda\alpha$, so L is nonabelian.

In this example, we have $\lambda^n\alpha \neq \alpha$ unless $\lambda^n = 1$. So, in the G-set $F^+ - \{0\}$, every orbit has length d. Such an action is called *regular*. If a subgroup K of F^+ is G-invariant, then each orbit of G in $K - \{0\}$ has length d. This implies that $|K| \equiv 1 \pmod{d}$. If, in particular, the divisor d satisfies the property that $p^m \equiv 1 \pmod{d}$ implies $m = n$ or $m = 0$, then the only G-invariant subgroups of F^+ are trivial. In this case, we say that the action of G is *irreducible*. For $p^m = 16$ and $d = 5$, we have an irreducible action.

Exercises

1. Let G be the group of 2×2 nonsingular matrices in §1, Example 3. Show that G is a semidirect product of the group of matrices with determinant 1 and the multiplicative group R^*. Describe an action associated with this semidirect product.

(*Hint.* The action is not unique. Why not?)

2. Let G be the group defined in §1, Example 4. Show that G is a semidirect product of the additive group $\mathbb{R}^+$ of the field and the multiplicative group $\mathbb{R}^*$. Describe the action associated with this group.

(*Hint.* The action is unique.)

3. Let L be the group constructed in Example 3. Show that, if φ is irreducible, then G is a maximal subgroup of L.

(*Hint.* Apply Theorem 3.14 to a maximal subgroup of L which contains G, and use (8.11). In general, if H is a normal subgroup, any subgroup K acts on H and leaves both H and $H \cap K$ invariant. There is a one-to-one correspondence between the set of K-invariant subgroups of H which contains $H \cap K$ and the set of those subgroups of HK which contain K. To describe this correspondence using Theorems 4.12 and 3.14 is an exercise similar to Exercise 3.)

4. Prove that the group of Example 3 is isomorphic to the group of matrices

$$\begin{pmatrix} \lambda^m & 0 \\ \alpha & 1 \end{pmatrix} \qquad (\alpha \in F, \quad m = 0, 1, 2, \ldots, d-1).$$

§9. General Linear Groups

A theorem about general groups is of course important in group theory, but the study of a special class of groups is equally significant. In this section, we will study the group of matrices over a general ground field F.

Definition 9.1. Let F be an arbitrary field and let n be a fixed natural number. The group formed of the totality of $n \times n$ nonsingular matrices with respect to the operation of matrix multiplication is called the **general linear group** of degree n over F; it is denoted by $GL(n, F)$. The set of matrices having determinant 1 forms a subgroup which is called the **special linear group** and is written $SL(n, F)$.

If F is the finite field of q elements, we use a special notation, and write $GL(n, q)$ and $SL(n, q)$ for $F = \mathbb{F}_q$.
We will solve the problems of

   (I)   finding a suitable set of generators, and
   (II) determining all the normal subgroups of the general (special) linear group over F.

The problem (I) can be solved using some elementary results from linear algebra. We will fix the ground field F and a natural number n. Let I be the identity matrix, let E_{ij} be the $n \times n$ matrix such that the (i, j) entry is 1, but all the other entries are 0, and for an element λ of F and $i \neq j$, let

$$B_{ij}(\lambda) = I + \lambda E_{ij}.$$

Let A be any $n \times n$ matrix with entries in F. The left multiplication of $B_{ij}(\lambda)$ to A gives an operation of adding λ times the jth row of A to the i-th row of A. Similarly, the right multiplication of $B_{ij}(\lambda)$ produces an operation of adding λ times i-th column to the j-th column of A. These are elementary operations on rows and columns. A theorem in elementary linear algebra asserts that any nonsingular square matrix can be brought to a diagonal matrix

$$D_\delta = \begin{pmatrix} 1 & & & & 0 \\ & 1 & & & \\ & & \ddots & & \\ & & & 1 & \\ 0 & & & & \delta \end{pmatrix}, \qquad \delta = \det A \neq 0$$

by a finite number of elementary (row) operations of the type described above. We have

$$B_{ij}(\lambda)^{-1} = B_{ij}(-\lambda) \qquad (i \neq j).$$

So, the preceding discussion shows that any matrix having determinant 1 can be written as a finite product of matrices of the form $B_{ij}(\lambda)$ $(i \neq j, \ \lambda \in F)$, and that any matrix of determinant δ can be written as a finite product of matrices $B_{ij}(\lambda)$ and the matrix D_δ defined above. As $B_{ij}(\lambda) \in SL(n, F)$, it follows from Definition 2.7 that the following theorem has been proved.

Theorem 9.2. *We have*

$$SL(n, F) = \langle B_{ij}(\lambda), i \neq j, \lambda \in F \rangle$$
$$GL(n, F) = \langle B_{ij}(\lambda), D_\delta, i \neq j, \lambda \in F, \delta \in F^* \rangle.$$

The function $A \rightarrow \det A$ is a homomorphism from $GL(n, F)$ onto the multiplicative group F^* of F (§5, Example 2). By Homomorphism Theorem, we have

(9.3) $SL(n, F) \lhd GL(n, F),$

and the factor group $GL(n, F)/SL(n, F)$ is isomorphic to the multiplicative group F^* of the ground field F.

It is more convenient to use geometric language in order to study the group $GL(n, F)$. Let V be the n-dimensional vector space over the field F. We write the totality of linear transformations from V onto V as $GL(V)$. Elements of $GL(V)$ are, in fact, isomorphisms from V onto V, so the composite of two elements f and g of $GL(V)$ is also an element of $GL(V)$. With this composition as the operation, $GL(V)$ forms a group.

As we learned in elementary linear algebra, the action of any element of $GL(V)$ can be represented by a matrix with entries in F. We will briefly review this connection between linear transformations and matrices. Let $\{v_1, \ldots, v_n\}$ be a basis of the vector space V. For any $f \in GL(V)$, set

$$f(v_i) = \sum_{j=1}^{n} \alpha_{ij} v_j.$$

The coefficients α_{ij} are elements of F, and the $n \times n$ matrix (α_{ij}) is the one representing the linear transformation f; we write $(\alpha_{ij}) = M_f$. If g is another element of $GL(V)$ and if

$$g(v_j) = \sum_{k=1}^{n} \beta_{jk} v_k,$$

then $M_g = (\beta_{jk})$ is the matrix representing g. We have

$$fg(v_i) = g(f(v_i)) = g(\sum \alpha_{ij} v_j)$$
$$= \sum \alpha_{ij} g(v_j) = \sum \alpha_{ij} \beta_{jk} v_k.$$

Thus, the formula $M_{fg} = M_f M_g$ holds for $f, g \in GL(V)$. In this formula, the right side is the matrix multiplication of M_f and M_g. The identity function of V corresponds to the identity matrix I. Hence, if g is the inverse of f, then $M_f M_g = I$ which shows that M_f is nonsingular. If any $n \times n$ nonsingular matrix $M = (\alpha_{ij})$ with entries in F is given, clearly there exists a linear transformation f such that $M = M_f$ (if we define f by the above formula

$$f(v_i) = \sum \alpha_{ij} v_j$$

and extend linearly.) Therefore, we have the following isomorphism:

(9.4) $GL(V) \cong GL(n, F)$.

One of the isomorphisms is given by the correspondence $f \to M_f$. This isomorphism is not unique since it depends on the choice of the bases of V. In fact, if $\{v'_1, v'_2, \ldots, v'_n\}$ is another basis of the vector space V, we have

$$v'_i = \sum_{j=1}^{n} s_{ij} v_j;$$

similarly, $v_k = \sum_{l=1}^{n} t_{kl} v'_l$ where the matrix $T = (t_{kl})$ is the inverse of the matrix (s_{ij}). So, if

$$f(v'_i) = \sum_{j=1}^{n} \alpha'_{ij} v_j,$$

then $f \to M'_f = (\alpha'_{ij})$ is an isomorphism from $GL(V)$ onto $GL(n, F)$, and we have $M'_f = T^{-1} M_f T$. This formula shows that $\det M_f = \det M'_f$. Thus, the determinant of the corresponding matrix does not depend on the choice of the bases of V. We can, therefore, define the special linear group $SL(V)$ as a subgroup of $GL(V)$ which consists of linear transformations corresponding to matrices having determinant 1.

We consider a hyperplane of V. We mean here by a *hyperplane* a subspace of codimension 1, that is, a subspace spanned by $n - 1$ linearly independent vectors. A linear transformation of $SL(V)$ is said to be a *transvection* if it is not the identity, but fixes all the elements of a hyperplane of V. Let $\{v_1, \ldots, v_n\}$ be a basis of V. Then the linear transformation t which is represented by $B_{ij}(\lambda)$ with respect to this basis is a transvection; in fact, if H denotes the hyperplane spanned by $v_1, \ldots, v_{i-1}, v_{i+1}, \ldots, v_n$ (all the elements of the basis except v_i), then t fixes all the elements of H, and t is a transvection, provided $\lambda \neq 0$. Theorem 9.2 can be stated as follows.

(9.5) *The group $SL(V)$ is generated by transvections.*

It follows from the preceding discussion that, for a given hyperplane H in V and for any one-dimensional subspace U of H, there is a transvection t such that t fixes all the elements of H and $U = (t - 1)V$. In fact, if we choose a basis of V in such a way that $v_1 \notin H$, $v_2 \in U - \{0\}$, and the remaining elements $v_3, \ldots, v_n$ are in H, then the linear transformation represented by the matrix $B_{12}(\lambda)$ is a transvection with the required properties.

Conversely, let t be a transvection. We show that there is a basis $\{v_i\}$ of V such that t is represented by the matrix $B_{12}(1)$ with respect to $\{v_i\}$. By definition, t fixes every element of a hyperplane H. Also, there is an element u of V such that $t(u) \neq u$; thus, $V = H + Fu$. Since $t \in SL(V)$, we have $t(u) - u \in H$. So, we can choose a basis of V such that $v_1 = u$, $v_2 = t(u) - u$, and $v_3, \ldots, v_n \in H$. With respect to this basis $\{v_i\}$, t is represented by $B_{12}(1)$.

In general, let f and g be two elements of $GL(V)$ such that the matrix M_f representing f with respect to a basis $\{v_i\}$ of V is identical to the matrix M_g' which represents g with respect to another basis $\{v_i'\}$ of V. Let T be the matrix which transforms the basis $\{v_i'\}$ into the basis $\{v_i\}$. With respect to $\{v_i\}$, the element g is represented by a matrix M_g such that $M_g' = T^{-1} M_g T$. Since M_g' is identical to M_f, we have $T^{-1} M_g T = M_f$. Now, all nonsingular matrices can be written in the form M_x $(x \in GL(V))$. So, let $T = M_x$ and $x^{-1} g x = h$. We have

$$M_f = T^{-1} M_g T = M_x^{-1} M_g M_x = M_h.$$

By (9.4), we get $f = h = x^{-1} g x$. Thus, we have proved the first part of the following proposition.

(9.6) *Two transvections are conjugate in $GL(V)$. If $n = \dim V \geq 3$, then the transvections are conjugate in $SL(n, V)$.*

Proof. If f and g are two transvections, then there exists an element x of $GL(V)$ such that $f = x^{-1} g x$. Choose a basis $\{v_i\}$ of V with respect to which f is represented by $B_{12}(1)$. Let y be the element of $GL(V)$ which is represented by the matrix D_δ ($\delta = \det x$). Since $n \geq 3$, y leaves v_1 and v_2 invariant. So, y commutes with f; we have

$$y^{-1}fy = f = x^{-1}gx.$$

Let $z = xy^{-1}$. We have then $f = z^{-1}gz$; furthermore, $M_z = M_x M_y^{-1}$ and $\det M_z = \delta \cdot \delta^{-1} = 1$. Thus, f and g are conjugate in $SL(V)$. $\square$

(9.7) *A linear transformation which is conjugate to a transvection is also a transvection.*

Proof. By definition, an element t of $SL(V)$ is a transvection if and only if the kernel H of the transformation $t - 1$ is a hyperplane. Assume that t is a transvection. Then for any conjugate t^x, the kernel of $t^x - 1$ is the hyperplane $x(H)$. By (9.3), t^x lies in $SL(V)$; so, t^x is a transvection. $\square$

We note that a transvection t can be characterized as an element of $SL(V)$ such that $\dim (t - 1)V = 1$. Furthermore, for any t and x, we have

$$(t^x - 1)V = x((t - 1)V).$$

Next we will determine the centralizer of $SL(V)$ in $GL(V)$.

(9.8) *The following conditions for an element z of $GL(V)$ are equivalent :*

(1)   *The element z commutes with all transvections;*
(2)   *The element z fixes every one-dimensional subspace of V ;*
(3)   *There exists an element λ of F such that*

$$z(v) = \lambda v \quad \text{for all } v \in V.$$

(4)   *The element z lies in the center of $GL(V)$.*

Proof. (1) $\Rightarrow$ (2) Let U be a one-dimensional subspace. There is a transvection t such that $U = (t - 1)V$. Since z commutes with t by assumption, we have

$$U = (t - 1)V = (t^z - 1)V = z((t - 1)V) = z(U):$$

this proves (2).

(2) $\Rightarrow$ (3) Let $\{v_1, v_2, \ldots, v_n\}$ be a basis of V. By assumption, z fixes the one-dimensional subspaces Fv_i ($i = 1, 2, \ldots, n$), so $z(v_i) = \lambda_i v_i$ for some $\lambda_i \in F$. We will show that $\lambda_1 = \lambda_2 = \cdots = \lambda_n$. For $i \neq j$, z fixes the one-dimensional subspace $F(v_i + v_j)$ as well. Hence, there is an element $\lambda \in F$ such that $z(v_i + v_j) = \lambda(v_i + v_j)$. On the other hand, we have

$$z(v_i + v_j) = z(v_i) + z(v_j),$$

so $\lambda = \lambda_i = \lambda_j$ because v_i and v_j are linearly independent. This proves (3).

$(3) \Rightarrow (4)$ Let z have the form in (3). For any $f \in GL(V)$, we have

$$(zf)(v) = f(z(v)) = f(\lambda v) = \lambda f(v) = z(f(v)) = (fz)(v).$$

Hence z commutes with any $f \in GL(V)$.

$(4) \Rightarrow (1)$ The proof is trivial. $\square$

Corollary 1. *The center Z of $GL(V)$ consists of the linear transformations $v \to \lambda v$ ($\lambda \in F^*$). Z is isomorphic to F^* and coincides with the centralizer of $SL(V)$.*

Corollary 2. *The center of $SL(V)$ is $Z(GL(V)) \cap SL(V)$, and it is isomorphic to the finite cyclic group consisting of all the n-th roots of unity contained in F ($n = \dim V$).*

A solution to the problem (II), which is the aim of this section, is given by the following theorem.

Theorem 9.9. *Normal subgroups of $GL(V)$ either contain the special linear group $SL(V)$ or are contained in the center of $GL(V)$, except when $\dim V = 2$ and the ground field F contains at most 3 elements. Conversely, any subgroup of $GL(V)$ which either contains $SL(V)$ or is contained in the center is a normal subgroup of $GL(V)$.*

First of all, the converse statement is obvious. Any subgroup of the center is always normal. As to the subgroups which contain $SL(V)$, we note that $SL(V) \lhd GL(V)$ and the factor group $GL(V)/SL(V)$ is an abelian group isomorphic to the multiplicative group F^* (§1, Example 2). Since any subgroup of the abelian group F^* is normal, every subgroup which contains $SL(V)$ is normal in $GL(V)$ by the Correspondence Theorem (4.13) (iii).

To survey subgroups contained in F^* is a problem which depends on the specific nature of the ground field F and is rather complicated in general; however, if the field F is finite, then the group F^* is cyclic and all the subgroups can be found by using the corrollary of (5.6).

We will prove the first part of Theorem 9.9 in a slightly more general form.

(9.10) *Let H be a subgroup of $GL(V)$. Suppose that H is not contained in the center of $GL(V)$, but that H is normalized by $SL(V)$. Then H contains $SL(V)$, except when $|F| \leq 3$ and $\dim V = 2$.*

Proof. (a) Suppose that $\dim V \geq 3$. Since H contains an element h not in the center of $GL(V)$, by (9.8), there is a transvection t which does not commute with h. Let g be the commutator of t and h:

$$g = t^{-1}h^{-1}th = (h^{-t})h = t^{-1}t^h.$$

We have $g \neq 1$ since t does not commute with h. By assumption, $SL(V)$ normalizes H; so, $g = (h^{-1})h$ lies in H. On the other hand, both t^{-1} and t^h are transvections; so, $g = t^{-1}t^h$ is a product of two transvections and g belongs to $SL(V)$. Furthermore, g fixes every element in the subspace of codimension ≤ 2 defined by

$$\text{Ker} \, (t^{-1} - 1) \cap \text{Ker} \, (t^h - 1).$$

Let $U = \text{Ker} \, (g - 1)$ and $W = (g - 1)V$. We have codim $U \leq 2$; so dim $W = $ codim $U \leq 2$. Thus, dim W is either one or two.

If dim $W = 1$, g is a transvection. If f is any transvection, f is conjugate to g by (9.6); in fact, there is an element x of $SL(V)$ such that $f = g^x$. Since H contains g and H is normalized by $SL(V)$, f is an element of H. Thus, every transvection lies in H, and by (9.5), we have $SL(V) \subset H$.

Suppose that dim $W = 2$. In this case, there is a hyperplane P which contains W (since dim $V \geq 3$). As $g(P) \subset P + W$, g leaves P invariant. But, g is not a transvection, so there is an element u of P such that $g(u) \neq u$. Choose an element v of $V - P$. Then, there is a transvection s which fixes all the elements of P and moves v onto $v + u$. Set $g' = g^{-1}s^{-1}gs$. We can prove, as before, that g' lies in $H \cap SL(V)$. Clearly, s^g is a transvection which fixes all the elements of $g(P) = P$ and sends $g(v)$ onto $g(v) + g(u)$. If $v \in U = \text{Ker} \, (g - 1)$, we have $s^g \neq s$. Thus, $g' = s^{-g}s$ is a transvection. On the other hand, if we can not choose $v \in U$, then U is contained in P, and the restriction of g on P is not a transvection. So, by (9.8), we can choose an element u of P so that $g(u)$ and u are linearly independent. With this choice of u, the transvection s defined above is certainly different from s^g, so that $g' = s^{-g}s$ is a transvection in $H \cap SL(V)$. Thus, we have reduced the case when dim $W = 2$ to the previously proved case when g is a transvection.

(b) Assume that dim $V = 2$. In this case, we have an element h of H and an element v of V such that v and $h(v)$ form a basis of V. With respect to the basis $\{v, h(v)\}$, the element h is represented by the matrix

$$\begin{pmatrix} 0 & 1 \\ \alpha & \beta \end{pmatrix} \quad (\alpha \neq 0).$$

We consider H as a subgroup of the matrix group $GL(2, F)$ and compute the commutators of h and other suitable matrices to repeat the argument similar to the one in case (a). The commutator

$$\begin{pmatrix} \lambda & 0 \\ 0 & \lambda^{-1} \end{pmatrix}^{-1} \begin{pmatrix} 0 & 1 \\ \alpha & \beta \end{pmatrix}^{-1} \begin{pmatrix} \lambda & 0 \\ 0 & \lambda^{-1} \end{pmatrix} \begin{pmatrix} 0 & 1 \\ \alpha & \beta \end{pmatrix} = \begin{pmatrix} \lambda^{-2} & \gamma \\ 0 & \lambda^2 \end{pmatrix},$$

where $\gamma = \alpha^{-1}\beta(\lambda^{-2} - 1)$, lies in $H \cap SL(V)$. Similarly,

$$\begin{pmatrix} 1 & \mu \\ 0 & 1 \end{pmatrix}^{-1}\begin{pmatrix} \lambda^{-2} & \gamma \\ 0 & \lambda^2 \end{pmatrix}^{-1}\begin{pmatrix} 1 & \mu \\ 0 & 1 \end{pmatrix}\begin{pmatrix} \lambda^{-2} & \gamma \\ 0 & \lambda^2 \end{pmatrix} = \begin{pmatrix} 1 & \mu(\lambda^4 - 1) \\ 0 & 1 \end{pmatrix}$$

also lies in $H \cap SL(V)$. If the ground field F contains a nonzero element λ such that $\lambda^4 \neq 1$, then H contains $B_{12}(\xi)$ for all $\xi \in F$. Since we have

$$\begin{pmatrix} 0 & 1 \\ -1 & 0 \end{pmatrix}^{-1}\begin{pmatrix} 1 & \xi \\ 0 & 1 \end{pmatrix}\begin{pmatrix} 0 & 1 \\ -1 & 0 \end{pmatrix} = \begin{pmatrix} 1 & 0 \\ -\xi & 1 \end{pmatrix},$$

$B_{21}(\xi)$ belongs to H for every $\xi \in F$. Thus, we have $SL(2, F) \subset H$.

We assume that every nonzero element of F satisfies $\lambda^4 = 1$. Since we may assume $|F| \geq 4$, we have $|F| = 5$. In this case, if $\lambda = 2$,

$$\begin{pmatrix} \lambda^{-2} & \gamma \\ 0 & \lambda^2 \end{pmatrix} = \begin{pmatrix} -1 & -2\alpha^{-1}\beta \\ 0 & -1 \end{pmatrix} \in H \cap SL(V).$$

The square of this element is $B_{12}(-\alpha^{-1}\beta)$ and is in $H \cap SL(V)$. Thus, if $\beta \neq 0$, then H contains at least one of $B_{12}(\lambda)$ $(\lambda \neq 0)$. Since $|F| = 5$, H contains all $B_{12}(\xi)$ and, as before H contains $SL(V)$.

If $\beta = 0$, we will replace h by another element h' of H which, with respect to the basis $\{v', h'(v')\}$, is represented by the matrix

$$\begin{pmatrix} 0 & 1 \\ \alpha' & \beta' \end{pmatrix}$$

with $\beta' \neq 0$. Consider the element

$$h' = \begin{pmatrix} 0 & \beta \\ -\beta^{-1} & \delta \end{pmatrix}^{-1}\begin{pmatrix} 0 & 1 \\ \alpha & 0 \end{pmatrix}^{-1}\begin{pmatrix} 0 & \beta \\ -\beta^{-1} & \delta \end{pmatrix}\begin{pmatrix} 0 & 1 \\ \alpha & 0 \end{pmatrix}$$

$$= \begin{pmatrix} \delta^2 - \alpha\beta^2 & -\alpha^{-1}\beta^{-1}\delta \\ \beta^{-1}\delta & -\alpha^{-1}\beta^{-2} \end{pmatrix}$$

which lies in $H \cap SL(V)$ but not in the center of $GL(V)$ provided $\delta \neq 0$. This element is represented in the above form with β' equal to the trace $\delta^2 - \varepsilon - \varepsilon^{-1}$. In the field F_5 of 5 elements, we have $\delta^2 = \pm 1$, and $\varepsilon + \varepsilon^{-1} = 0$ or ± 2. So, $\beta' \neq 0$ and we can replace h by h' to finish the proof. $\square$

Corollary. *If either* dim $V \geq 3$ *or* $|F| > 3$, *a proper normal subgroup of* $SL(V)$ *is contained in the center* Z_0. *Hence, the factor group* $SL(V)/Z_0$ *is simple.*

We write this simple group as $PSL(V)$ and will take up the study of this group again later (Chapter 3, §3).

Early in the twentieth century, the preceding theorem was proved by

Dickson assuming that the ground field F was a finite field. Since then, it has been generalized by many mathematicians to take the form presented here. A simple, elegant proof due to Iwasawa which is based on elementary properties of permutation groups can be found in Nagao [3], [AS] I pp. 172–3, or Dieudonné [1]. The above proof can be modified to treat the case when F is a skew-field (Artin [3]).

If the field F is finite, the group $GL(n, F)$ is a finite group. We will find its order.

Let p be the characteristic of F and let

$$|F| = q = p^m.$$

The order of $GL(n, F)$ is given by

(9.11)
$$|GL(n, F)| = (q^n - 1)(q^n - q) \cdots (q^n - q^{n-1})$$

$$= \prod_{i=0}^{n-1} (q^n - q^i)$$

$$= q^{n(n-1)/2} \prod_{i=1}^{n} (q^i - 1).$$

Furthermore, we have $|GL(n, F)| = (q - 1)|SL(n, F)|$.

Proof. Let V be an n-dimensional vector space over F. We will determine the order $|GL(V)|$. Clearly, we have $|V| = q^n$. Let $\{v_1, \ldots, v_n\}$ be a fixed basis of V over F. For any element f of $GL(V)$, $\{f(v_1), \ldots, f(v_n)\}$ is a basis of V. Conversely, for any basis $\{u_1, \ldots, u_n\}$ of V, there is a unique element f of $GL(V)$ such that $f(v_i) = u_i$ for all $i = 1, 2, \ldots, n$. Thus, $|GL(V)|$ is equal to the number of distinct bases of V over F. The first element v_1 can be any one of the $(q^n - 1)$ nonzero elements of V. If the first i ($i < n$) elements $v_1, \ldots, v_i$ have been chosen, the next element v_{i+1} can be any element of V which is not written as a linear combination of $v_1, \ldots, v_i$. Thus, there are $(q^n - q^i)$ possibilities for v_{i+1}. By (9.4), $GL(V) \cong GL(n, F)$; so we have (9.11). $\square$

(9.12) *Let F be a finite field of characteristic p, and let $|F| = q$. The largest power of p which divides the group order $|GL(V)|$ is q^r where $r = n(n-1)/2$. The group $GL(V)$ contains a subgroup of order exactly equal to q^r.*

Proof. The first part is obvious from (9.11) when we note that all factors of the form $(q^i - 1)$ are prime to p. In order to prove the last assertion, let U be the set of upper-triangular matrices with 1 on the main diagonal:

$$U = \left\{ \begin{pmatrix} 1 & \alpha & \beta & \cdots \\ & 1 & & \\ & & \ddots & \\ 0 & & & 1 \end{pmatrix}, \quad \alpha, \beta, \ldots \in F \right\}.$$

Entries above the main diagonal are arbitrary elements of F, and there are exactly $n(n-1)/2$ places above the diagonal; thus, $|U| = q^r$. It is easy to check that U is a subgroup of G. $\square$

The linear group $GL(n, F)$ satisfies the following property which corresponds to Cayley's theorem for symmetric groups.

(9.13) *Let F be a fixed field. For any given finite group G, the group $GL(n, F)$ contains a subgroup isomorphic to G, provided that n is sufficiently large.*

Proof. Let Γ be the totality of linear combinations

$$\alpha = \sum_{g \in G} \lambda_g g \qquad (\lambda_g \in F)$$

of group elements g with coefficients λ_g in F. For another element $\beta = \sum \mu_g g$ of Γ, we define the sum $\alpha + \beta$ by

$$\alpha + \beta = \sum (\lambda_g + \mu_g)g$$

and $\lambda\alpha = \sum (\lambda\lambda_g)g$ $(\lambda \in F)$; with respect to these operations, Γ forms a finite-dimensional vector space over F. Each element h of G defines a mapping ρ_h on Γ by the formula.

$$\rho_h(\alpha) = \sum \lambda_g(gh).$$

It follows easily from the definitions that

$$\rho_h(\alpha + \beta) = \rho_h(\alpha) + \rho_h(\beta).$$
$$\rho_h(\lambda\alpha) = \lambda\rho_h(\alpha).$$

So, the function ρ_h is a linear transformation on V. We have

$$\rho_{hk} = \rho_h \rho_k,$$
$$\rho_1 = \text{the identity function,}$$

which show that ρ_h is an invertible linear transformation, and that ρ is a homomorphism from G into the general linear group $GL(\Gamma)$. If an element h of g is contained in the kernel of ρ, we have $\rho_h(\alpha) = \alpha$ for all α, so $gh = g$ and $h = 1$. Therefore, ρ is an isomorphism from G into $GL(\Gamma)$. If we set $n = \dim \Gamma$, then $GL(\Gamma) \cong GL(n, F)$ and (9.13) holds. $\square$

(9.14) *Let p be a fixed prime number. If an abelian group E satisfies the property that $x^p = 1$ for any element x of E, then E is isomorphic to a vector space over the field $\mathbf{F}_p$ of p elements. Thus, E has a basis B. In particular, if E is* *finite, then its order* $|E|$ *is a power of* p. *If* $|E| = p^d$, *then we have* Aut $E \cong GL(d, p)$.

Proof. We will consider E as an additive group, and use the additive notation. The assumption says that every element x of E satisfies $px = 0$. For any integers m and n, the formulas (1.8) are written additively as

$$(m + n)x = mx + nx, \qquad (mn)x = m(nx).$$

This proves that the group $\mathbb{Z}$ of integers acts on E. Since $px = 0$ for all $x \in E$, the action of $\mathbb{Z}$ induces an action of the quotient ring $\mathbb{Z}/(p)$ where (p) is the ideal generated by p. We identify $\mathbb{Z}/(p)$ with the field $\mathbf{F}_p$ of p elements. In this way $\mathbf{F}_p$ acts on E, and it is easy to check that E is a vector space over $\mathbf{F}_p$. An automorphism f of E satisfies

$$f(nx) = nf(x) \qquad (n \in \mathbb{Z}),$$

so f is a linear transformation over $\mathbf{F}_p$. If we set dim $E = d$, then $|E| = p^d$ and Aut $E \cong GF(d, p)$. $\square$

Exceptional Cases of Theorem 9.9

We will consider the following cases which were excluded in Theorem 9.9 and (9.10). In both cases $SL(V)$ contains a proper normal subgroup not contained in the center.

$SL(2, 2)$ In this case, the order is 6 by (9.11). The elements

$$\sigma = \begin{pmatrix} 1 & 1 \\ 1 & 0 \end{pmatrix}, \qquad \tau = \begin{pmatrix} 0 & 1 \\ 1 & 0 \end{pmatrix}$$

satisfy $\sigma^3 = \tau^2 = 1$ and $\tau^{-1}\sigma\tau = \sigma^{-1}$. Hence $\langle \sigma \rangle$ is a normal subgroup of order 3 and σ is not in the center.

$GL(2, 3)$ Again by (9.11), the order is 48. If a normal subgroup N of $GL(2, 3)$ contains an element having determinant -1, then we may assume, as in (9.10) (b) that

$$\begin{pmatrix} 0 & 1 \\ 1 & \beta \end{pmatrix} \in N.$$

Simple calculations of commutators show the following:

$$\begin{pmatrix} 1 & 0 \\ 1 & 1 \end{pmatrix} \begin{pmatrix} -\beta & 1 \\ 1 & 0 \end{pmatrix} \begin{pmatrix} 1 & 0 \\ -1 & 1 \end{pmatrix} \begin{pmatrix} 0 & 1 \\ 1 & \beta \end{pmatrix} = \begin{pmatrix} 1 & -1 \\ 1 & 0 \end{pmatrix},$$

$$\begin{pmatrix} 1 & 0 \\ 1 & 1 \end{pmatrix} \begin{pmatrix} 1 & -1 \\ 1 & 0 \end{pmatrix} \begin{pmatrix} 1 & 0 \\ -1 & 1 \end{pmatrix} = \begin{pmatrix} -1 & -1 \\ 0 & -1 \end{pmatrix} \in N.$$

Hence, as shown in (9.10) (b), N contains all transvections and we have $SL(2, 3) \subset N$; so, $N = GL(2, 3)$ in this case.

Let σ, τ, and ρ be the elements of $SL(2, 3)$ defined as follows:

$$\sigma = \begin{pmatrix} 1 & 1 \\ 1 & -1 \end{pmatrix}, \qquad \tau = \begin{pmatrix} 0 & -1 \\ 1 & 0 \end{pmatrix}, \quad \text{and} \quad \rho = \begin{pmatrix} -1 & 1 \\ -1 & 0 \end{pmatrix}.$$

The following relations hold:

$$\sigma^4 = \rho^3 = 1, \qquad \rho^{-1}\sigma\rho = \tau, \qquad \rho^{-1}\tau\rho = \sigma\tau,$$
$$\sigma = \rho^{-1}(\sigma\tau)\rho = \tau\sigma\tau, \quad \text{and} \quad \tau^{-1}\sigma\tau = \sigma^{-1}.$$

Let Q be the subgroup $\langle \sigma, \tau \rangle$ generated by σ and τ. Then, Q is of order 8 and $\rho^{-1}Q\rho = Q$, which proves that $Q \lhd SL(2, 3)$. It is easy to check that Q is a normal subgroup of $GL(2, 3)$.

From the preceding discussion (and (3.5)), we know that the proper normal subgroups of $GL(2, 3)$ are $SL(2, 3)$, Q, and the center of $GL(2, 3)$. The factor group $GL(2, 3)/Q$ is a group of order 6 which is isomorphic to $SL(2, 2)$.

Exercises

1. (a) Let F be an arbitrary field. Show that any transvection of $SL(2, F)$ is conjugate to $B_{12}(\lambda)$ in $SL(2, F)$.

(b) Prove that two elements $B_{12}(\lambda)$ and $B_{12}(\mu)$ are conjugate in $SL(2, F)$ if and only if $\mu = \lambda\delta^2$ for some $\delta \in F^*$.

(c) When F is a finite field, find the number of conjugacy classes of transvections in $SL(2, F)$.

(*Hint.* (c) The number is one if the characteristic of F is 2; otherwise, it is two. In general, the number of conjugacy classes is $|F^* : (F^*)^2|$ where $(F^*)^2$ is the subgroup consisting of the squares of elements F^*.)

2. Let V be a finite-dimensional vector space over a field F. Let f be an element of $GL(V)$ such that $f(u) = v$ and u and v are linearly independent. Show that, if $\dim V \geq 3$, there exists an element w of V such that u, v, and w are linearly independent and

$$f(w) \notin Fv + Fw.$$

3. (a) Suppose that two elements u and v of a finite-dimensional vector space V are linearly independent. Show that there is a transvection which sends v onto u.

(b) Let U be a subspace of V. If the images $\bar{u}$ and $\bar{v}$ of u and v in the factor space $\bar{V} = V/U$ are linearly independent, there is a transvection which sends v onto u and leaves every element of U invariant.

(*Hint.* There is a hyperplane which contains U and $u - v$, but not v.)

4. An element of $GL(V)$ which satisfies the condition (3) of (9.8) is said to be a *similarity transformation*. Let $n = \dim V$, and let m_σ be the minimum number of transvections necessary to express an element σ of $SL(V)$ as a product of transvections.

(a) Show that $m_\sigma \leq n$ if σ is not a similarity transformation.

(b) Show that $m_\sigma \leq n + 1$ if σ is a similarity transformation.

(c) Show that $m_\sigma = n + 1$ if σ is a nonidentity similarity.

(*Hint.* (a) First, prove the statement for $n = 2$. Assume next that $n \geq 3$. If $f \in SL(V)$ is not a similarity, there is an element u of V such that u and $f(u) = v$ are linearly independent (9.8). Choose an element w which satisfies the condition of Exercise 2, and let t be a transvection such that $t(w) = w$ and $t(v) = u$. Then ft fixes u and induces a map on V/Fu which is not a similarity transformation. Continue this process to get transvections $t_1 = t, t_2, \ldots, t_r$ such that the subspace of fixed elements of $ft_1 \cdots t_r$ has dimension at least r. When $r = n - 1$, the resulting element $ft_1 \cdots t_{n-1}$ is either a transvection or the identity. This problem gives an alternate proof of (9.5).

(b) This follows immediately from (a).

(c) Suppose that $f = t_1 \cdots t_n$ where t_i are transvections, and consider the action of t_n on the subspace $\bigcap \mathrm{Ker}(t_i - 1)$ $(i = 1, 2, \ldots, n - 1)$. This problem is a theorem of Dieudonné [2].)

Chapter 2

Fundamental Theorems

§1. Theorems about p-Groups

Throughout this section, the letter p stands for a fixed prime number.

Definition 1.1. A finite group is said to be a **p-group** if its order is a power of p.

The concept of p-groups is very important in the theory of finite groups. Its significance stems from the following two reasons. For one, the class of p-groups has many special properties which make it possible to study this class extensively. For the other, by the theorems of Sylow which will be discussed in the next section, any finite group G contains the so-called Sylow p-subgroups which are p-groups and are closely related to the structure of G. In 1933, P. Hall published a classical paper on the structure of p-groups which laid a foundation of the theory. Many works have followed since then which investigate the general class of p-groups, as well as various special classes. Furthermore, the recent advances in the theory of finite simple groups have thrown a new light on p-groups and have suggested investigations in different directions. The replacement theorem of Thompson and Glauberman is one of the interesting results of this new development and will be discussed in Chapter 4 along with the general theory of P. Hall. In this section, we will discuss the most fundamental theorems of p-groups. (In this section, p-groups are assumed to be finite.)

(1.2) *Let G be a p-group. Any subgroup of G is also a p-group; its index is a power of p. If a subgroup H is normal, the factor group G/H is a p-group.*

This follows easily from Theorem 3.3 of Chapter 1 and Definition 1.1. The next lemma about p-groups is fundamental.

(1.3) *Let G be a p-group, and let X be a nonempty finite G-set. If $|X| \not\equiv 0$ (mod p), then X contains a G-invariant point; that is, there is a fixed point of the action of G on X.*

Proof. By Theorem 7.12 of Chapter 1, the set X is partitioned into a disjoint union of orbits:

$$X = O_1 \cup O_2 \cup \cdots \cup O_m.$$

For each i, the length of the orbit O_i is equal to the index of the stabilizer $S_G(x_i)$ for $x_i \in O_i$. By assumption, G is a p-group, so by (1.2), $|G : S_G(x_i)|$ is a power of p. Since $|X| \not\equiv 0 \,(\mathrm{mod}\ p)$ by assumption,

$$\sum |G : S_G(x_i)| \not\equiv 0 \qquad (\mathrm{mod}\ p)$$

and, for at least one index i, $|G_i : S_G(x_i)|$ is not divisible by p. A power p^n of p is not divisible by p only when $n = 0$; so we have $|O_i| = p^0 = 1$. Thus, the orbit O_i consists of a single G-invariant point. $\square$

Corollary. *Suppose that a p-group Q acts on another p-group G. If $G \neq \{1\}$, then there is a Q-invariant element of G other than the identity.*

Proof. Let $X = G - \{1\}$. By the assumptions, X is a nonempty Q-set such that $|X| \not\equiv 0 \,(\mathrm{mod}\ p)$. So, by (1.3), there is a Q-invariant element of X. $\square$

We will derive several properties of p-groups from the preceding lemma and its corollary.

Theorem 1.4. *Let G be a p-group, and let H be a normal subgroup of G. If $H \neq \{1\}$, then $H \cap Z(G) \neq \{1\}$. In particular, if $G \neq \{1\}$, then the center of the p-group G contains an element different from the identity 1; that is, $G \neq \{1\}$ implies that $Z(G) \neq \{1\}$.*

Proof. By (1.2), H is a p-group. Since $H \lhd G$, the p-group G acts on the p-group $H \neq \{1\}$ via conjugation. Corollary of (1.3) asserts that H contains a G-invariant element z other than 1. By the definition of the action, we have $z = z^g = g^{-1}zg$ for all $g \in G$. Thus, z is contained in the center of G:

$$z \in H \cap Z(G) \neq \{1\}.$$

This proves the first part of Theorem 1.4; the last assertion is obtained by setting $G = H$. $\square$

We proved the commutativity of groups of order p^2 in the Example of Chapter 1, §3. This commutativity may also be proved using Theorem 1.4 and (2.26) of Chapter 1. In fact, (2.26) of Chapter 1 shows that the factor group $G/Z(G)$ by the center is not cyclic unless G is abelian. So, if G is a p-group, we have

$$|G : Z(G)| = 1 \quad \text{or} \quad \geq p^2.$$

(1.5) (Matsuyama [1]) *Let H be a subgroup of a p-group G. Then either $H \lhd G$, or a conjugate subgroup H^x different from H is contained in $N_G(H)$.*

Proof. Suppose that H is not normal, and let X be the totality of conjugates H^x of H different from H. Then, X is a nonempty set on which H acts via conjugation. By Theorem 3.6 of Chapter 1, the number of conjugates of H is equal to the index $|G : N_G(H)|$, so by the assumption and (1.2), we have

$$|X| = |G : N_G(H)| - 1 \not\equiv 0 \qquad (\text{mod } p).$$

Hence, by (1.3), X contains an H-invariant element, say H^y. Since H^y is H-invariant, we get $H \subset N_G(H^y)$. Setting $x = y^{-1}$, we have $H \neq H^x \subset N_G(H)$. $\qquad \square$

The next theorem follows easily from (1.5).

Theorem 1.6. *If H is a proper subgroup of a p-group G, then we have $N_G(H) \neq H$. Thus, the normalizer of a proper subgroup H is strictly larger than H.*

Proof. If $H \lhd G$, then $N_G(H) = G \neq H$, so the theorem holds. If H is not normal, by (1.5), $N_G(H)$ contains a conjugate H^x of H which is different from H. So, $N_G(H) \neq H$ in this case, too. $\qquad \square$

Corollary. *Any maximal subgroup M of a p-group G is normal, and the factor group G/M is the cyclic group of order p. In particular, $|G : M| = p$.*

Proof. The normalizer of M is, by Theorem 1.6, strictly larger than M, so $M \lhd G$. By the Correspondence Theorem, the factor group G/M contains no nontrivial subgroup. So, by Theorem 2.24 of Chapter 1, G/M is a cyclic group of prime order. Thus, we have $|G : M| = p$ by (1.2). $\qquad \square$

Theorems 1.4 and 1.6 are the most basic properties of p-groups, and they lead to many other important properties.

If a p-group G is different from $\{1\}$, then the center $Z(G)$ is not equal to $\{1\}$. So, if $G \neq Z(G)$, the factor group $G/Z(G)$ is, by (1.2), also a p-group different from $\{1\}$. By Theorem 1.4, the center of $G/Z(G)$ is not equal to $\{1\}$. Let $Z_2(G)$ be the subgroup of G corresponding to $Z(G/Z(G))$: that is, (cf. Theorem 4.12 of Chapter 1)

$$Z_2(G)/Z(G) = Z(G/Z(G)).$$

Then, we have $Z_2(G) \neq Z(G)$. If $Z_2(G) \neq G$, then the center of $G/Z_2(G)$ is not equal to $\{1\}$. This leads to the following definition.

Definition 1.7. For any group G, we define the subgroups $Z_i(G)$ for $i = 0, 1, 2,$ ... as follows. (We abbreviate $Z_i(G) = Z_i$.) Define $Z_0 = \{1\}$, and for $i > 0, Z_i$ is the subgroup of G corresponding to $Z(G/Z_{i-1})$ by the Correspondence Theorem:

$$Z_i/Z_{i-1} = Z(G/Z_{i-1}).$$

The sequence of subgroups

$$Z_0 \subset Z_1 \subset Z_2 \subset \cdots$$

is called the **upper central series** of G; its i-th term Z_i is called the **i-th center** of G. A group G is said to be **nilpotent** if $Z_m(G) = G$ for some integer m; in this case, the smallest integer c such that $Z_c(G) = G$ is called the **class** of G.

The trivial group $\{1\}$ consisting of a single element is the only nilpotent group of class 0; abelian groups $\neq \{1\}$ are the nilpotent groups of class 1. Thus, the concept of nilpotent groups is a generalization of the concept of abelian groups.

Repeated applications of Theorem 1.4 prove the following theorem.

Theorem 1.8. *The upper central series of a p-group G reaches G; that is, there is an integer m such that $Z_m(G) = G$. Thus, any p-group is nilpotent.*

Theorem 1.6 is valid for any nilpotent group. The proof is easy. However, since we will study the class of nilpotent groups in detail in a later chapter, we will postpone the proof.

Theorem 1.9. *Let G be a p-group of order p^n. For any subgroup H of G, let*

$$(\mathscr{G}): G_0 = G \supset G_1 \supset \cdots \supset G_r = H$$

be a series of subgroups satisfying the property that G_i is a maximal subgroup of G_{i-1} for all $i = 1, 2, \ldots, r$. Then we have $G_i \lhd G_{i-1}$ and $|G_{i-1} : G_i| = p$. If, in particular, $H = \{1\}$, then we have $r = n$ and the series $(\mathscr{G})$ is a composition series of length n. Thus, all the composition factors of a p-group are cyclic groups of order p.

Proof. By the Corollary of Theorem 1.6, we get $G_i \lhd G_{i-1}$ and $|G_{i-1} : G_i| = p$. The proposition (3.5) of Chapter 1 proves the equality $r = n$; the remaining assertion is easy to prove by Definition 5.8 of Chapter 1. $\square$

Corollary. *Let H be a subgroup of a p-group G. There is a composition series of G through H.*

Definition 1.10. A finite group G is said to be a **solvable group** if all the composition factors of G are cyclic groups of prime orders.

The theory of finite groups appeared for the first time in the history of Mathematics just before the works of Abel and Galois on the theory of algebraic equations; their brilliant success was one of the proud applications of finite group theory. In modern terms, according to Galois theory, the algebraic equation $f(x) = 0$ is solvable by radicals if and only if the Galois group of $f(x)$ satisfies the condition stated in Definition 1.10. Thus, we call a group solvable in Definition 1.10. The term 'nilpotent' came from the other source of group theory, the Lie theory. When G is a Lie group, the corresponding Lie algebra L satisfies the condition $L^m = \{0\}$ for some integer m if and only if $Z_m(G) = G$. This is the origin of the term 'nilpotent'.

We get the following corollary of Theorem 1.9.

(1.9)′ *A p-group is solvable.*

In fact, any finite nilpotent group is solvable. The group constructed in Example 3 of Chapter 1, §8 is solvable, but not nilpotent (because the center consists of only the identity).

A p-group has a series of subgroups which satisfies much stronger conditions than does the series of Theorem 1.9. We will state the following definition.

Definition 1.11. A sequence of subgroups of a group G

$$(\mathscr{H}): H_0 = G \supset H_1 \supset \cdots \supset H_s = \{1\}$$

is said to be a **principal series**, or **chief series** if each H_i is a normal subgroup of G and if for each $i = 1, 2, \ldots, s$, H_i is maximal among normal subgroups of G which are properly contained in H_{i-1}.

The Jordan-Hölder Theorem is also valid for chief series. We will prove this in a more general form in §3. The following theorem holds for p-groups.

Theorem 1.12. *Let G be a p-group of order p^n.*
    (i) *The group G has a chief series of length n :*

$$H_0 = G \supset H_1 \supset H_2 \supset \cdots \supset H_n = \{1\}.$$

We have $H_i \lhd G$ and $|H_{i-1} : H_i| = p$ for $i = 1, 2, \ldots, n$.
    (ii) *Let $K_1 \subset K_2 \subset \cdots \subset K_k$ be a chain of normal subgroups of G. Then there is a chief series $(\mathscr{H})$ of G such that all the given groups K_i are members of $(\mathscr{H})$, and the length of $(\mathscr{H})$ is exactly n.*

We begin with the following lemma.

(1.13) *Let G be a p-group. If N is a minimal normal subgroup of G, then we have $N \subset Z(G)$ and $|N| = p$.*

Proof. By the definition of minimal normal subgroups, $N \neq \{1\}$ and Theorem 1.4 shows that $N \cap Z(G) \neq \{1\}$. On the other hand, $N \cap Z(G)$ is a normal subgroup of G which is contained in N. So, the minimality of N proves that $N \cap Z(G) = N$, or $N \subset Z(G)$. Since all subgroups of the center are normal, N contains only trivial subgroups. By Theorem 2.24 of Chapter 1, we get $|N| = p$. ☐

Proof of Theorem 1.12. We note that (i) is a special case of (ii) when $K_1 = G$ and $k = 1$. We will prove the second proposition by induction on n. We may assume $K_1 \neq \{1\}$ (by removing the trivial subgroups among the $\{K_i\}$), and let N be a minimal normal subgroup of G contained in K_1. The above lemma tells us that $|N| = p$. By the inductive hypothesis, the factor group $\bar{G} = G/N$ has a chief series

$$\bar{H}_0 = \bar{G} \supset \bar{H}_1 \supset \cdots \supset \bar{H}_{n-1} = \{1\}$$

such that all groups $\bar{K}_j = K_j/N$ appear among the groups $\{\bar{H}_i\}$. We have $|\bar{H}_{i-1} : \bar{H}_i| = p$ for all $i = 1, 2, \ldots, n-1$. By the Correspondence Theorem, there are subgroups H_i of G such that $H_i \triangleleft G$, $\bar{H}_i = H_i/N$, and $|H_{i-1} : H_i| = p$ for $i = 1, 2, \ldots, n-1$. So, $H_{n-1} = N$ and we get a chief series of G by adding $H_n = \{1\}$ at the end:

$$H_0 = G \supset H_1 \supset \cdots \supset H_{n-1} \supset H_n = \{1\}.$$

It is clear from the construction that this chief series satisfies the required condition. ☐

Corollary. *Let A be an abelian normal subgroup of a p-group G. If A is maximal among abelian normal subgroups of G, then A satisfies $C_G(A) = A$. In particular, A is maximal among abelian subgroups of G.*

Proof. Set $H = C_G(A)$. Since A is abelian, A is a part of H. Furthermore, as A is normal, Theorem 2.19 of Chapter 1 shows that $H \triangleleft G$. If $A \neq H$, then by Theorem 1.12, there would be a chief series in which both A and H would appear. Hence, there would be a normal subgroup B such that $A \subset B \subset H$, $B \triangleleft G$, and $|B : A| = p$. Since A is a maximal subgroup of B, we would be able to find an element x of B such that $B = \langle A, x \rangle$. By definition, we have $B \subset C_G(A)$, and so $A \subset Z(B)$. Then, (2.26) of Chapter 1 shows that B would be abelian. This contradicts the maximality of A. Thus, we have $A = C_G(A)$. ☐

With a slight change, Theorem 1.12 is also valid for any nilpotent group; this can be proved by using the Refinement Theorem of Schreier which will be proved in §3.

The Basis Theorem of Burnside and the Φ-Subgroup

Definition 1.14. Let $\mathfrak{M}$ be the set of maximal subgroups of a group G. The intersection of all subgroups of $\mathfrak{M}$ is called the **Frattini subgroup** or the **Φ-subgroup** of G; we denote it by $\Phi(G)$.

(1.15) *Let* $\Phi = \Phi(G)$ *be the* Φ-*subgroup of a* p-*group* G. *Then the following propositions hold:*

    (i) Φ char G

    (ii) *The factor group* G/Φ *is an abelian group in which every element satisfies* $x^p = 1$.

    (iii) *For a subset* X *of* G,

$$\langle X, \Phi \rangle = G \Rightarrow \langle X \rangle = G.$$

Proof. (i) Any automorphism of G sends a maximal subgroup into a maximal subgroup. Thus, the set $\mathfrak{M}$ is invariant by any automorphism, and so is Φ.

(ii) By the Cor. of Theorem 1.6, $M \lhd G$ for any $M \in \mathfrak{M}$, and the factor group G/M is cyclic group of order p. Hence, for any elements x and y of G, we have $x^p \in M$ and $x^{-1}y^{-1}xy \in M$ for all $M \in \mathfrak{M}$. Thus, x^p and $x^{-1}y^{-1}xy$ are contained in the intersection of these subgroups; that is, $x^p \in \Phi$ and $x^{-1}y^{-1}xy \in \Phi$. We have

$$(x\Phi)^p = 1 \quad \text{and} \quad x\Phi y\Phi = y\Phi x\Phi.$$

This proves the second assertion.

(iii) If $\langle X \rangle$ were a proper subgroup of G, then there would be a maximal subgroup M which would contain X. Then, we would have $\langle X, \Phi \rangle \subset M$, contrary to the assumption. □

Clearly, (1.15) (i) and (iii) are valid for any finite group G. If G is a p-group, by (1.15) (ii) and (1.14) of Chapter 1, the factor group G/Φ is isomorphic to a vector space over the field $\mathbf{F}_p$ of p elements. Set $V = G/\Phi$. We will consider V as a vector space over $\mathbf{F}_p$.

Theorem 1.16. *Let* Φ *be the Frattini subgroup of a* p-*group* G, *and consider* $V = G/\Phi$ *to be a vector space over* $\mathbf{F}_p$. *Set* $|G : \Phi| = p^d$. *Let* $x_1, \ldots, x_n$ *be elements of* G, *and let* $v_i = \Phi x_i$ *for* $i = 1, 2, \ldots, n$.

    (i) *The dimension of* V *over* $\mathbf{F}_p$ *is* d.

    (ii) *We have* $G = \langle x_1, \ldots, x_n \rangle$ *if and only if* V *coincides with the space spanned by* $v_1, v_2, \ldots, v_n$. *In particular, if* $G = \langle x_1, \ldots, x_n \rangle$, *then we have* $n \geq d$.

    (iii) *The group* G *can be generated by exactly* d *elements. The subset* $\{x_1, x_2, \ldots, x_d\}$ *generates* G *if and only if* $\{v_1, v_2, \ldots, v_d\}$ *is a basis of the vector space* V *over* $\mathbf{F}_p$.

Proof. (i) An m-dimensional vector space over $\mathbf{F}_p$ contains exactly p^m elements. So, the assertion (i) is obvious.

(ii) If $G = \langle x_1, \ldots, x_n \rangle$, then V is generated by $v_1, \ldots, v_n$. An elementary theorem of linear algebra says that a d-dimensional vector space cannot be spanned by less than d elements. Hence, we have $n \geq d$.

Conversely, suppose that $V = \langle v_1, \ldots, v_n \rangle$. Setting

$$X = \{x_1, \ldots, x_n\},$$

we have $\langle X, \Phi \rangle = G$, so by (1.15)(iii) we get $G = \langle X \rangle$.

(iii) If $\{v_1, \ldots, v_d\}$ is a basis of V over $\mathbf{F}_p$, then, by (ii), we have $G = \langle x_1, \ldots, x_d \rangle$. Conversely, if $G = \langle x_1, \ldots, x_d \rangle$, then by (ii), we get $V = \langle v_1, \ldots, v_d \rangle$. In this case, $v_1, \ldots, v_d$ are linearly independent, and $\{v_1, \ldots, v_d\}$ is a basis of V. $\square$

Theorem 1.16 is called the Basis Theorem of Burnside. As a corollary, we obtain the following theorem of P. Hall on the automorphism group of a p-group.

Theorem 1.17. *Let Φ be the Frattini subgroup of a p-group G, and let $|G : \Phi| = p^d$ and $|G| = p^n$. Furthermore, let P be the totality of automorphisms of G which leave every element of G/Φ invariant.*

(i) *The set P is a normal subgroup of* Aut G *and the factor group* (Aut $G)/P$ *is isomorphic to a subgroup of $GL(d, p)$.*

(ii) *The subgroup P is a p-group of order less than $p^{(n-d)d}$. Thus, $|$ Aut $G|$ divides*

$$p^m \prod_{i=1}^{d} (p^i - 1),$$

where $m = nd - d(d+1)/2 \leq n(n-1)/2$.

Proof. Let $V = G/\Phi$. Since Φ is characteristic, every automorphism σ of G induces an automorphism $f(\sigma)$ of V. The function f is clearly a homomorphism from Aut G into Aut V. By definition, we have $P = \mathrm{Ker}\, f$. So, by the Homomorphism Theorem, $P \lhd$ Aut G and (Aut $G)/P$ is isomorphic to a subgroup of Aut V.

By (1.16)(i), we have dim $V = d$, so (9.14) of Chapter 1 shows that Aut $V \cong GL(d, p)$. This proves the part (i).

Choose a fixed basis $\{v_1, v_2, \ldots, v_d\}$ of V over $\mathbf{F}_p$. Let X be the totality of $(y_1, y_2, \ldots, y_d)$ such that $y_i \in G$ and $\Phi y_i = v_i$ for all $i = 1, 2, \ldots, d$. Clearly, we get $|X| = p^{(n-d)d}$. Since an element ρ of P satisfies $v_i^\rho = v_i$ for all i, the group P acts on X via the action defined by

$$(y_1, \ldots, y_d) \rightarrow (y_1^\rho, \ldots, y_d^\rho).$$

Let O be an orbit under this action. By Theorem 7.9 (iii) of Chapter 1, the length of O is the index of the stabilizer $S_G(Y)$ of an element $Y = (y_1, \ldots, y_d)$. If $\rho \in S_G(Y)$, then $y_i^\rho = y_i$ for all $i = 1, \ldots, d$. Since $G = \langle y_1, \ldots, y_d \rangle$ by the Basis Theorem, we get $\rho = 1$. Thus, $S_G(Y) = \{1\}$ and $|O| = |P|$. Since O is an arbitrary orbit, each orbit is of length $|P|$; hence, $|P|$ divides $|X|$. This proves that P is a p-group whose order is a divisor of $p^{(n-d)d}$. The upper bound of Aut G is obtained from (9.11) of Chapter 1. $\square$

Corollary 1. *Let ρ be an automorphism of a p-group G. If the order of ρ is prime to p and if p leaves every element of $G/\Phi(G)$ invariant, then $\rho = 1$.*

Corollary 2. *Let A be an abelian normal subgroup of maximal order of a p-group G. If $|G| = p^n$ and $|A| = p^a$, we have $2n \le a(a + 1)$.*

Proof. Corollary 1 follows immediately from Theorem 1.17 (ii). In Corollary 2, we have $C_G(A) = A$ by the Corollary of Theorem 1.12. Theorem 6.11 of Chapter 1 shows that G/A is a subgroup of Aut A. The inequality of Theorem 1.17 (ii) gives us $n - a \le a(a - 1)/2$, or $2n \le a(a + 1)$. $\square$

Exercises

1. From the class equation ((7.13)' of Chapter 1), deduce that the center of a p-group G is different from $\{1\}$ if $G \ne \{1\}$.

2. Let H be a normal subgroup of a p-group G, and let $|H| = p^i$. Show that H is contained in the i-th center $Z_i(G)$.
   (*Hint.* Use induction on i. Consider $G/Z(G)$.)

3. Let U be the subgroup of $GL(n, p)$ over the field $\mathbf{F}_p$ of p elements consisting of all the upper-triangular matrices with 1 on the main diagonal:

$$ U = \left\{ \begin{pmatrix} 1 & & * \\ & \ddots & \\ 0 & & 1 \end{pmatrix} \right\}. $$

Define the subgroups U_m by the formula

$$ U_m = \left\{ \begin{pmatrix} I_m & A \\ 0 & I_{n-m} \end{pmatrix} \right\} $$

where I_m and I_{n-m} are the identity matrices of size m and $n - m$, respectively, and A ranges over all the $m \times (n - m)$ matrices. Show that U_m is a maximal abelian normal subgroup of U for each m.
   (*Hint.* Find a homomorphism φ from U with Ker $\varphi = U_m$ in order to prove $U_m \lhd U$. Next, verify that $C_U(U_m) = U_m$. We can see from this exercise

that a p-group can contain several nonisomorphic maximal abelian normal subgroups. If $m = 1$, U_1 provides an example in which the equality in Cor. 2 of Theorem 1.17 holds.)

4. Find the center of the p-group U defined in Exercise 3.

(*Hint.* From Exercise 3 (or by direct computation), it is easy to show that the center consists of upper-triangular matrices whose nonzero entries off the main diagonal are at the right upper corner. Exercises 3 and 4 can be restated for any finite field F in place of $\mathbf{F}_p$. When $n = 3$, the group U is an example of a nonabelian group of order p^3 with $|U : Z(U)| = p^2$.)

§2. Theorems of Sylow

In this section we will only consider groups of finite order, except for a short discussion on general π-groups towards end. So, by "groups" we mean finite groups, unless explicitly stated otherwise. The letter p stands for a fixed prime number as in §1. A subgroup is called a **p-subgroup** if its order is a power of p.

Definition 2.1. Let G be a group. We write

$$|G| = p^n m \qquad (p, m) = 1.$$

A subgroup of G is said to be a **Sylow p-subgroup** if its order is exactly p^n. A Sylow p-subgroup is sometimes abbreviated an S_p-subgroup.

A subgroup U of G is an S_p-subgroup if and only if

(i) U is a p-group, and
(ii) the index $|G : U|$ is prime to p.

If a subgroup U is an S_p-subgroup of G, then so is any conjugate of U.

The next theorem of Sylow is the most fundamental theorem in finite group theory.

Theorem 2.2. *Let G be a group.*

(i)   *The group G has an S_p-subgroup.*
(ii)   *Any two S_p-subgroups are conjugate in G.*
(iii) *Any p-subgroup of G is contained in an S_p-subgroup of G.*
(iv)   *The number of S_p-subgroups of G is a divisor of G and is congruent to 1 modulo p.*

We begin with a lemma.

(2.3) *Let L be a group, and let H be an S_p-subgroup of L. For any subgroup K of L, there exists an element x of L such that $K \cap H^x$ is an S_p-subgroup of K.*

Proof. Applying Theorem 3.9 of Chapter 1, we get

$$|L : H| = \sum_x |K : K \cap H^x|,$$

where the sum ranges over a complete system of representatives from double cosets. Since H is an S_p-subgroup of L, the left side of the above equation is prime to p. Hence, at least one of the terms in the sum is prime to p. Let x be an element such that $|K : K \cap H^x|$ is prime to p. On the other hand, $K \cap H^x$ is a p-subgroup since it is a subgroup of a p-group H^x. Thus, $K \cap H^x$ is an S_p-subgroup of K.   □

We now return to the proof of Theorem 2.2. By (9.13) of Chapter 1, the group G is isomorphic to some subgroup of the general linear group $GL(n, F)$ where F is the field of p elements and n is sufficiently large. The group $GL(n, F)$ contains an S_p-subgroup by (9.12) of Chapter 1. Then, (2.3) shows that G has an S_p-subgroup. This proves (i).

Let H be an S_p-subgroup of G, and let K be any p-subgroup of G. Lemma (2.3) shows that there is an element x of G such that $K \cap H^x$ is an S_p-subgroup of K. Since the order $|K|$ is a power of p, K itself is the S_p-subgroup of K. So, $K \cap H^x = K$ or $K \subset H^x$. But, H^x is also an S_p-subgroup of G and (iii) holds. If K is an S_p-subgroup of G, we get $K = H^x$ since they have the same order. Thus, (ii) is proved.

It follows from (ii) and Theorem 3.6 of Chapter 1 that the number of S_p-subgroups divides $|G|$. Before proceeding to the proof of the last assertion, we need another lemma.

Lemma. *Let S be an S_p-subgroup of a group G. Any p-subgroup U of $N_G(S)$ is contained in S.*

Proof. Let $H = N_G(S)$. Then, S is an S_p-subgroup of H. Theorem 2.3 (ii) and (iii) applied to H proves that $U \subset S^x = S$ as $x \in H$.   □

We will prove (iv) in a slightly generalized form.

(2.4) *Let U be a p-subgroup of a group G. The number of S_p-subgroups of G which contain U is congruent to 1 modulo p. (If $U = \{1\}$, (2.4) reduces to (iv).)*

Proof. Let S be an S_p-subgroup of G, and let $H = N_G(S)$. Theorem 3.9 of Chapter 1 applied here to H and U gives us

$$|G : H| = \sum_x |U : U \cap H^x|.$$

Since U is a p-subgroup, each term in the right side is a power of p by (1.2). If $|U : U \cap H^x| = 1$, then

$$U = U \cap H^x \Rightarrow U \subset H^x \Rightarrow xUx^{-1} \subset N_G(S).$$

By the above lemma, we have $xUx^{-1} \subset S$, or $U \subset S^x$. Conversely, if $U \subset S^x$, then retracing the preceding argument, we get $|U : U \cap H^x| = 1$. Let $n(U)$ be the number of S_p-subgroups of G which contain U. Since $S^x = S^y$ implies $HxU = HyU$, exactly $n(U)$ terms among $|U : U \cap H^x|$ are one. Thus, we have

$$|G : H| \equiv n(U) \qquad (\text{mod } p).$$

The left side is independent of U. If $U = S$, obviously we get $n(S) = 1$. So, $|G : H| \equiv 1 \equiv n(U) \,(\text{mod } p)$. $\square$

Corollary (Cauchy). *If the prime number p divides the order of a group G, then G contains an element of order p.*

Proof. Let S be an S_p-subgroup of G. By Sylow's Theorem, $S \neq \{1\}$ and every non-identity element x of S has order a power of p. So, S contains an element of order p. $\square$

Sylow's Theorem is so fundamental to the study of groups that we will present an alternate proof of it based on an entirely different method which was used by Wielandt.

Set $|G| = p^n m$ $(p, m) = 1$, as before. Consider the set G_k consisting of all the subsets of k elements where $k = p^n$. The set G_k becomes a G-set as in Example 2 of Chapter 1, §7. It is important to note that $|G_k| \not\equiv 0$ (mod p). This can be checked by using a property of binomial coefficients. By Theorem 7.12 of Chapter 1, the G-set G_k is partitioned into a disjoint union of orbits. Since $|G_k| \not\equiv 0$ (mod p), there exists an orbit whose length is not divisible by p. Let O be such an orbit and let A be an element of O. By definition $|O| \not\equiv 0$ (mod p). Let S be the stabilizer $S_G(A)$. Then, by Theorem 7.9 (iii) of Chapter 1, we have $|G : S| = |O| \not\equiv 0$ (mod p). Hence, by Theorem 3.3 of Chapter 1, $|S|$ is divisible by p^n; we have $|S| \geq p^n$. On the other hand, for an element a of A, the definition of the action shows that $aS \subset A$. Hence we get

$$|S| = |aS| \leq |A| = p^n.$$

Thus, $|S| = p^n$, and S is an S_p-subgroup of G.

Let U be a p-subgroup of G. The orbit O defined above may be considered as a U-set by restricting the action on U. Since $|O|$ is prime to p, (1.3) shows that U has a fixed point B. Hence $U \subset S_G(B)$. Since O is an orbit of G, by (7.11) of Chapter 1, $S_G(B)$ is conjugate to $S_G(A)$. So, there is an element x such that $U \subset S^x$. This proves (iii) (and (ii)) of Theorem 2.3.

In order to prove (iv), we will slightly modify the above proof by Wielandt (and incidentally, we shall be able to prove (i) at the same time). As before, we set $|G| = p^n m$, but this time we will allow p to be a divisor of m. Consider the

G-set G_k for $k = p^n$, and let O be one of its orbits. Let A be an element of O which contains the identity 1 of G, and let S be the stabilizer of A. By Theorem 7.9 of Chapter 1, we have $|O| = |G : S|$. Since $AS = A$, A is a union of the left cosets of S, and $|S|$ divides $|A| = p^n$. Thus, unless $|S| = p^n$, the length $|O|$ is divisible by pm. If $|S| = p^n$, then we have $A = S$ because $A \ni 1$. In this case, the orbit O coincides with the totality of right cosets of S. Clearly, if U is a subgroup of order p^n, then the totality of right cosets of U forms an orbit of the G-set G_k. Let $N(p^n)$ be the number of subgroups of order p^n in G. Then the preceding discussion proves that $|G_k| \equiv mN(p^n) \pmod{mp}$.

Let C be the cyclic group of order $|G|$. By the Corollary of (5.6), Chapter 1, C contains exactly one subgroup of order p^n. So, if we apply the preceding method to the cyclic group C, we get $|C_k| \equiv m \pmod{mp}$. Since $|G_k| = |C_k|$, we get

$$N(p^n) \equiv 1 \pmod{p}.$$

(This method was discovered by Gallagher and McLaughlin.) We note again that we did not assume $(p, m) = 1$. Namely, if p^n divides $|G|$, the number $N(p^n)$ is of the form $1 + kp$ for some nonnegative integer k. This proves (i) and (iv). $\square$

The next proposition is a useful theorem with frequent applications.

(2.5) *Let P be a p-subgroup of a group G. If P is an S_p-subgroup of $N_G(P)$, then P is an S_p-subgroup of G.*

Proof. Suppose that P were not an S_p-subgroup of G. By Sylow's Theorem (iii), there would be an S_p-subgroup S of G which would contain P as a proper subgroup. A fundamental theorem (Theorem 1.6) on p-groups asserts that $N_S(P) \neq P$. Since $N_S(P)$ is a p-subgroup of $N_G(P)$ with order larger than $|P|$, P could not be an S_p-subgroup of $N_G(P)$. $\square$

Theorem 2.6. *Let H be a normal subgroup of a group G, and let S be an S_p-subgroup of G. The intersections $S \cap H$ is an S_p-subgroup of H; SH/H is an S_p-subgroup of G/H.*

Proof. First of all, $S \cap H$ is a p-subgroup of H. By Theorem 5.7 of Chapter 1, we have $|H : S \cap H| = |SH : S|$. Since S is an S_p-subgroup of G, S is an S_p-subgroup of SH, and $|H : S \cap H|$ is prime to p. Thus, $S \cap H$ is an S_p-subgroup of H.

For the factor group, we have $SH/H \cong S/S \cap H$, so, SH/H is a p-subgroup of G/H. By the Correspondence Theorem, the index of SH/H is equal to $|G : SH|$ which is prime to p. This proves that SH/H is an S_p-subgroup of G/H. $\square$

The next theorem is also quite useful, and its proof illustrates a typical application of Sylow's Theorem.

Theorem 2.7. *Let H be a normal subgroup of a group G. If S is an S_p-subgroup of H, then we have $G = N_G(S)H$.*

Proof. Consider the conjugate S^g by an element g of G. Since $S \subset H \lhd G$, S^g is contained in H and S^g is an S_p-subgroup of H. So, by Sylow's Theorem (ii), S^g is conjugate to S in H. Therefore, there is an element h of G such that $S^g = S^h$. Let $n = gh^{-1}$. Then, $S^n = S$ and $n \in N_G(S)$. Since $g = nh$, we get $g \in N_G(S)H$. So, we have $G = N_G(S)H$. $\square$

As an application, we get the next proposition.

(2.8) *Let S be an S_p-subgroup of a group G. If a subgroup H of G contains $N_G(S)$, then we have $H = N_G(H)$.*

Proof. By assumption, $S \subset N_G(S) \subset H$. So, we apply Theorem 2.7 to $L = N_G(H)$ and get

$$N_G(H) = N_L(S)H \subset \langle N_G(S), H \rangle = H.$$

Clearly, $H \subset N_G(H)$, so we have $H = N_G(H)$. $\square$

In various applications, it is important to generalize Sylow's Theorem to cover groups with operator groups. Assume that an operator group Q on a group G is given. Let L be the semidirect product of Q and G with respect to the given action. As in Chapter 1, §8, we identify G and Q as subgroups of L so that $G \lhd GQ = L$.

Theorem 2.9. *Let q be a prime number. Assume that the operator group Q is a q-group and that q does not divide $|G|$. Then, the following extension of Sylow's Theorem holds.*

    (i)   *There exists a Q-invariant S_p-subgroup.*
    (ii) *Two Q-invariant S_p-subgroups are conjugate by an element of $C_G(Q)$.*
    (iii) *Any Q-invariant p-subgroup is contained in a Q-invariant S_p-subgroup of G.*

Proof. We note that, under the assumptions, the group Q is an S_q-subgroup of L.

    (i) Let S be an S_p-subgroup of G. Since $G \lhd L$, Theorem 2.7 shows that $L = N_L(S)G$. By Theorem 5.7 of Chapter 1, we have $|N_L(S): N_G(S)| = |L:G|$. So, for the prime q, we know from Sylow's Theorem that $N_L(S)$ contains an S_q-subgroup Q_1 of L, which is conjugate to Q in L. Thus, if $Q = Q_1^x$, then $Q = Q_1^x \subset N_L(S)^x = N_L(S^x)$ (see Theorem 2.19 of Chapter 1 for the last equality). By (8.11) of Chapter 1, S^x is Q-invariant, so it is a Q-invariant S_p-subgroup of G. Thus, the proposition (i) holds.

(ii) Let S_1 and S_2 be two Q-invariant S_p-subgroups of G. By Sylow's Theorem, there is an element x of G such that $S_2 = S_1^x$. Since both S_1 and S_2 are Q-invariant, we have $Q \subset N_L(S_i)$ for $i = 1$ and 2 (by (8.11) of Chapter 1). Since

$$Q^x \subset N_L(S_1)^x = N_L(S_1^x) = N_L(S_2),$$

both Q and Q^x are S_q-subgroups of $N_L(S_2)$. So, by applying Sylow's Theorem to the prime q, we conclude that Q and Q^x are conjugate in $N_L(S_2)$. On the other hand, $L = GQ = QG$ and, by Theorem 3.14 of Chapter 1, we get

$$N_L(S_2) = QG \cap N_L(S_2) = Q(G \cap N_L(S_2)).$$

So, by (3.15) of Chapter 1, there is an element y of $G \cap N_L(S_2)$ such that $Q = (Q^x)^y$. Let $z = xy$. Then, $z \in N_L(Q) \cap G$. As $N_L(Q) \cap G = C_G(Q)$ by (8.9) of Chapter 1, we have $z \in C_G(Q)$. By definition, $z = xy$, $S_1^x = S_2$ and $y \in N_L(S_2)$. Therefore, we get

$$S_1^z = S_1^{xy} = S_2^y = S_2.$$

Thus, S_1 and S_2 are conjugate by an element of $C_G(Q)$.

(iii) Let P be a p-subgroup of G which is maximal among Q-invariant p-subgroups of G. We need to show that P is an S_p-subgroup of G. By assumption, P is Q-invariant; so is $N_G(P)$. Hence Q acts on $N_G(P)/P$. As we have proved in (i), $N_G(P)/P$ contains a Q-invariant S_p-subgroup $\bar{S}$. Let S be the subgroup of $N_G(P)$ corresponding to $\bar{S}$. Then, for any $s \in S$ and $x \in Q$, we have $\bar{s}^x \in \bar{S}^x = \bar{S}$, so $s^x \in S$. This proves that S is Q-invariant. Thus, S is a Q-invariant p-subgroup of G which contains P. By the maximality, we get $P = S$. This means that P is an S_p-subgroup of $N_G(P)$. So, by (2.5), P is an S_p-subgroup of G. $\square$

Corollary. *Suppose that, in addition to the assumptions of Theorem 2.9, the action of Q is fixed-point-free; that is, suppose that $C_G(Q) = \{1\}$. Then there is a unique Q-invariant S_p-subgroup of G.*

This corollary follows immediately from Theorem 2.9. The action which satisfies $C_G(Q) = \{1\}$ is a very special one which should impose rather severe restrictions on G. In particular, the famed theorem of Thompson asserts that if the operator group Q is a cyclic group of order q and satisfies $C_G(Q) = \{1\}$, then G is nilpotent. We will prove this theorem later (Chapter 5). It has been conjectured that G should be solvable if there is an operator group Q which satisfies $C_G(Q) = \{1\}$.

Theorem 2.9 can be generalized in the following manner. Let Q be an operator group of G. We will only assume that the order of Q is relatively prime to the order of G. Then, the three statements of Theorem 2.9 hold. (see Theorem 8.11 of this Chapter.) This is a very useful result, but, unfortunately, the only known proof uses a very deep, difficult theorem of Feit–Thompson which asserts the solvability of finite groups of odd order. So, we cannot present the complete proof of the generalization in this book.

There is an alternate proof of Theorem 2.9 (i) based on (1.3). Let X be the totality of S_p-subgroups of G. Then, the group Q acts on X in an obvious way. By Sylow's theorem (iv), $|X|$ divides $|G|$. Thus, $|X|$ is prime to q under the assumptions of Theorem 2.9. We may apply (1.3) for the action of a q-group Q on X and conclude that there is a fixed point. This fixed point is a Q-invariant S_p-subgroup of G.

The first proof has an advantage in that it can be adapted without major changes to the general case mentioned above.

Notion of π-Groups

We will introduce the notion of π-groups mainly for notational convenience. In this subsection, we will consider infinite groups as well as finite ones. The letter π stands for a fixed set of prime numbers; π may be the totality of prime numbers, or π may be just a small collection of prime numbers.

Let π be a fixed set of prime numbers. The set of primes not contained in π will be denoted π'. When π consists of a single prime p, we usually use the notations p and p' in place of the correct notations $\pi = \{p\}$ and $\pi' = \{p\}'$, respectively. A natural number n is said to be a π-*number* if all prime factors of n belong to π. By definition, 1 is a π-number for any π. Any natural number n is represented uniquely as a product of a π-number n_π and a π'-number $n_{\pi'}$. In this case we call n_π the π-*component* of n.

Definition 2.10. Let G be a (not necessarily finite) group. Let g be an element of finite order. We say g is a π-**element** if the order of g is a π-number. A group G is said to be a π-**group** if all of its elements are π-elements. When G is a finite group, let $\pi(G)$ be the set of all prime numbers which divide $|G|$:

$$\pi(G) = \{p \mid \ |G| \equiv 0 \ (\mathrm{mod}\ p)\}.$$

(2.11) *Let G be a finite group. The following three conditions are equivalent :*

  (i)   *The group G is a π-group.*
  (ii) *The order $|G|$ is a π-number.*
  (iii) $\pi(G) \subset \pi$.

Proof. (i) $\Rightarrow$ (ii) If $|G|$ is not a π-number, some prime $p \notin \pi$ divides $|G|$. By the Corollary of Sylow's Theorem, G contains an element of order p; so, G is not a π-group.

(ii) $\Rightarrow$ (iii) Obvious.

(iii) $\Rightarrow$ (i) By Corollary 2 of Theorem 3.3 of Chapter 1, the order of any element is a divisor of $|G|$. Hence, every element of G is a $\pi(G)$-element. Since $\pi(G) \subset \pi$, G is a π-group by definition. $\square$

(2.12) *Let G be a (finite or infinite) group. If G is a π-group, all subgroups or factor groups are again π-groups.*

This is clear from the definitions.

Theorem 2.13. *Let g be an element of finite order. Then we can decompose $g = xy = yx$ such that x is a π-element and y is a π'-element. This decomposition is unique. Furthermore, each factor is a power of g, and the order of x is the π-component of the order of g. Similarly, the order of y is the π'-component of the order of g.*

We call x and y the π-*component* and π'-component of g, respectively.

Proof. Let the π-component of the order of g be n, and let the π'-component be m. Since m and n are relatively prime, we can find a pair of integers a and b such that $am + bn = 1$. Let $x_0 = g^{am}$ and $y_0 = g^{bn}$. By definition, we have

$$g = g^{am + bn} = x_0 y_0 = y_0 x_0.$$

Since the order of g is mn, we get $x_0^n = g^{amn} = 1$; similarly, we get $y_0^m = 1$. So, the decomposition stated in Theorem 2.13 is possible.

Next, we will prove the uniqueness of this decomposition. Suppose that $g = xy = yx$, that the order u of x is a π-number, and that the order v of y is a π'-number. Then, $(u, v) = 1$ and there are integers c and d such that $cu + dv = 1$. We have

$$g^{cu} = (xy)^{cu} = x^{cu} y^{cu} = y^{cu} = y^{1 - dv} = y;$$

similarly, x is also a power of g. Then, $x_0, y_0, x,$ and y are mutually commutative. Since $x_0 y_0 = g = xy$, we get $x^{-1} x_0 = y y_0^{-1}$. In the last formula, (2.22) of Chapter 1 shows that the left side is a π-element and the right side is a π'-element. Since π and π' are complementary, we get $x^{-1} x_0 = y y_0^{-1} = 1$; that is, we have $x = x_0$ and $y = y_0$. This proves the uniqueness of the decomposition. $\square$

Corollary. *Any element of finite order in a general (finite or infinite) group can be written uniquely as a product of mutually commutative elements of prime power orders.*

Various Applications of Sylow's Theorem

Sylow's Theorem is of central importance in finite group theory. Consequently, we encounter applications of Sylow's Theorem throughout our study of finite groups. Here, we will give some examples of its many applications.

EXAMPLE 1. *Let p and q be two prime numbers. A group of order pq is solvable.* We may assume $p > q$ (if $p = q$, the group is abelian: Example of Chapter 1, §3). By Sylow's Theorem, the number of S_p-subgroups is a divisor of group order and of the form $1 + kp$ for some integer $k \geq 0$; the only possible value is 1. Hence, the S_p-subgroup is normal, and the composition factors consist of two cyclic groups of orders p and q. This proves that a group of order pq is solvable. (The normality of the S_p-subgroup can also be proved by using (3.13) of Chapter 1.)

If $p > q$ but $p \not\equiv 1$ (mod q), then an S_q-subgroup is also normal. Hence, by (3.16) of Chapter 1, any element of order p is commutative with any element of order q. The product of these elements is of order pq (by (2.22) of Chapter 1), so the group is cyclic. Thus, if $p > q$ and $p \not\equiv 1$ (mod q), the only group of order pq is the cyclic group. For example, the group of order 15 or 35 is cyclic.

On the other hand, if $p \equiv 1$ (mod q), there is a nonabelian group of order pq (see Example 3 of Chapter 1, §8). We will prove that the number of isomorphism classes of nonabelian groups of order pq is at most one. For the proof, we will need the following result of elementary number theory. The residue classes modulo p which contain integers prime to p form a multiplicative group of order $p - 1$. This group is isomorphic to the multiplicative group $\mathbf{F}_p^*$ of the finite field of p elements which is cyclic. So, if q divides $p - 1$, there is an integer r such that $r \not\equiv 1$ (mod p), but $r^q \equiv 1$ (mod p). In our discussion of Example 1, we will consider this integer r to be fixed. The congruence $x^q \equiv 1$ (mod p) has exactly q solutions and each solution is an integer congruent to a distinct element of the set $\{1, r, r^2, \ldots, r^{q-1}\}$.

Let G be a nonabelian group of order pq where $p \equiv 1$ (mod q). Let P be the S_p-subgroup, and let Q be one of the S_q-subgroups of G. Then $G = PQ$ and every element of G can be written uniquely as a product uv ($u \in P$, $v \in Q$). If $x \in P - \{1\}$ and $y \in Q - \{1\}$, then x^y is an element of $P - \{1\}$. So, $x^y = x^a$ for some positive integer a. Since y does not commute with x, we have $a \not\equiv 1$ (mod p). By (2.16) of Chapter 1, we get $(x^a)^n = (x^y)^n = y^{-1}x^n y$, so

$$y^{-2}xy^2 = y^{-1}(x^y)y = (x^a)^y = x^{a^2}.$$

In general, $y^{-m}xy^m = x^b$ where $b = a^m$. Since $y^q = 1$, y^q commutes with x. This means that $a^q \equiv 1$ (mod p). Thus, a is a solution of the congruence $x^q \equiv 1$ (mod p); we have $a \equiv r^c$ (mod p) for some positive integer c less than q. Then there is an integer d such that $cd \equiv 1$ (mod q) and we have $a^d \equiv r^{cd} \equiv r$ (mod q).

The preceding discussion is summarized as follows. Let $P = \langle x \rangle$ and $Q = \langle y \rangle$. Then $y^{-1}xy = x^a$ with $a \equiv r^c \pmod{q}$ for some c. But, by choosing a suitable generator of Q, we can assume $c = 1$:

$$y^{-1}xy = x^r.$$

Here, r is a fixed integer satisfying $r \not\equiv 1 \pmod{q}$ and $r^q \equiv 1 \pmod{q}$. We will show that the multiplication of two elements of G is uniquely determined by the formula $y^{-1}xy = x^r$. Any element of G can be written as $y^i x^j$, and we have

$$(y^i x^j)(y^k x^l) = y^{i+k} x^m \quad (m = jr^k + l).$$

Since r is a fixed integer, the above formula shows that the operation in G is completely determined by the relations $x^p = y^q = 1$ and $y^{-1}xy = x^r$. These are called the *defining relations* among the generators x and y of G. In a later section (§6) we will consider generators and relations in general groups.

The following simple remark is useful in the discussion of Example 2.

(2.14) *Let H be a subgroup of a finite group G, and let P be an S_p-subgroup of H. Then there is an S_p-subgroup S of G such that $P = S \cap H$. In fact, any S_p-subgroup of G which contains P satisfies the formula $P = S \cap H$. Two distinct S_p-subgroups of H are never contained in the same S_p-subgroup of G.*

Proof. Since P is a p-subgroup of G, there is an S_p-subgroup S of G which contains P. Then, we have $P \subset S \cap H$. As $S \cap H$ is a p-subgroup of H and P is an S_p-subgroup of H, we get $P = S \cap H$. The remaining assertions are obvious. □

EXAMPLE 2. *Let p and q be prime numbers. Then, a finite group G of order $p^n q$ is solvable.*

Proof. Use induction on $|G|$.

(a) Let us consider the case when G contains a nontrivial normal subgroup H. Both the subgroup H and the factor group G/H have orders p^m or $p^m q$ which are less than $|G|$. So, by the inductive hypothesis or (1.9)', both H and G/H are solvable. As we can see from the Correspondence Theorem, the composition factors of G/H are isomorphic to some of the composition factors of a composition series of G through H. Thus, the set of composition factors of G is a union of the composition factors of H and of G/H. Since both H and G/H are solvable, G is solvable.

(b) By (a), we may assume that G is simple. We will derive a contradiction from this assumption.

(c) *The group G contains exactly q S_p-subgroups.*

Proof. Let S be an S_p-subgroup of G. Then, we have $|G : S| = q$. Since q is a prime number and S is not normal, we get $S = N_G(S)$. Theorem 3.6 of Chapter 1 and Sylow's Theorem prove that G contains exactly q S_p-subgroups.

Let D be a subgroup of G which is maximal among the intersections of pairs of S_p-subgroups of G, and let $H = N_G(D)$.

(d) *H has at least 2 S_p-subgroups.*

Proof. By definition, there are two distinct S_p-subgroups S and T of G such that $D = S \cap T$. So, D is a proper subgroup of S and, similarly, of T. A basic theorem on p-groups (Theorem 1.6) shows that $N_S(D) \neq D$ and $N_T(D) \neq D$. If H were to have only one S_p-subgroup P_0, then the S_p-subgroup P of G containing P_0 would contain $N_S(D)$, so $D \subsetneqq S \cap P$. By the definition of D, we would have $S = P$. Similarly, we would have $T = P$, and we would get the contradiction $S = T$. $\square$

(e) *The group H has exactly q S_p-subgroups. All S_p-subgroups of H contain D.*

The proof of the first part is similar to that of (c). The second assertion follows from the proposition that any S_p-subgroup of a group contains a normal p-subgroup. (This is a special case of Theorem 2.6.)

(f) *We have $D = \{1\}$.*

The remark (2.14), (c), and (e) show that every S_p-subgroup of G contains D. Since the intersection of all the S_p-subgroups is a normal subgroup (by Sylow's Theorem (ii)), we get $D = \{1\}$ from (b).

(g) *Two distinct S_p-subgroups of G have only the identity element in common.*

This is clear from (f) and the maximality of D.

We will derive a final contradiction. Each S_p-subgroup contains exactly $p^n - 1$ nonidentity elements, and there are exactly q S_p-subgroups. So, G has exactly $q(p^n - 1)$ nonidentity p-elements. Since $|G| = p^n q$, the number of q-elements is at most q. There is an S_q-subgroup which contains these q elements. So, there are no more q-elements. Thus, G has exactly one S_q-subgroup, which is obviously a nontrivial normal subgroup. This contradicts the statement (b). Hence, we must conclude that our assumption was false and, therefore, that any finite group G of order $p^n q$ is solvable. $\square$

More generally, *for prime numbers p and q, any group of order $p^a q^b$ is solvable* (Theorem of Burnside). This theorem was proved in early years of this century as an application of the character theory of finite groups. Since then, its proof by group theoretical methods not depending on the character theory or representation theory eluded the efforts of many mathematicians for a long time. Recently, such a character-free proof was finally obtained by the combined works of Thompson, Goldschmidt, Bender, and Matsuyama. We will present the proof of Goldschmidt–Matsuyama later (Chapter 5).

EXAMPLE 3. *Let $\sigma_p(G)$ be the number of S_p-subgroups of a finite group G. Then the following formula holds :*

$$\sigma_p(G) = \prod \sigma_p(S_i) \prod m_j,$$

where the first product is over all the composition factors $S_1, \ldots, S_r$ of G, and each m_j in the second product is a prime power which satisfies $m_j \equiv 1$ (mod p).

(This is a theorem of M. Hall [1]. When G is solvable, the formula has been proved by P. Hall [1].)

Proof. If G is simple, the formula is trivially true (with $m_j = 1$). We will proceed by induction on the length of the composition series. It is sufficient to prove that, for a normal subgroup H of G, we get

$$\sigma_p(G) = \sigma_p(G/H)\sigma_p(H) \prod n_i,$$

where n_i are prime powers satisfying $n_i \equiv 1$ (mod p).

Let $\mathscr{S}$ be the totality of pairs of subgroups U and P such that $P \subset H \subset U$, P is an S_p-subgroup of H, and U/H is an S_p-subgroup of G/H.

(a) *If (U, P) and (V, Q) are elements of $\mathscr{S}$, there exists an element g of G such that $V = U^g$ and $Q = P^g$.*

Proof. By definition, U/H and V/H are S_p-subgroups of G/H, so, by Sylow's Theorem (ii), there is an element x of G such that $V = U^x$. Also, P^x and Q are S_p-subgroups of H. Hence, there is an element y of H such that $Q = P^{xy}$. Set $g = xy$. Then, as $y \in H \subset V$, we get $U^g = U^{xy} = (U^x)^y = V$. This proves (a). $\square$

(b) *If S is an S_p-subgroup of G, then $(SH, S \cap H) \in \mathscr{S}$.*
This follows from Theorem 2.6.

(c) *For each $(U, P) \in \mathscr{S}$, the number of those S_p-subgroups S of G which satisfy $SH = U$ and $S \cap H = P$ is a constant not depending on U and P. Let this constant be n. Then we have $\sigma_p(G) = n|\mathscr{S}|$.*
This follows easily from (a) and (b).

Clearly, we have $|\mathscr{S}| = \sigma_p(G/H)\sigma_p(H)$. So, it suffices to prove that the integer n defined in (c) is the product of prime powers n_i such that $n_i \equiv 1$ (mod p). Let (U, P) be a fixed element of $\mathscr{S}$. If an S_p-subgroup S of G satisfies $SH = U$ and $S \cap H = P$, then $S \cap H = P \triangleleft S$ by Theorem 5.7 of Chapter 1. Hence, we have $S \subset N_U(P)$. This proves that $n = \sigma_p(N_U(P)) = \sigma_p(N_U(P)/P)$.

(d) *Let $K = H \cap N_U(P)$. Then we have $K \triangleleft N_U(P) = KS$.*
Theorem 5.7 of Chapter 1 proves $K \triangleleft N_U(P)$, while Theorem 3.14 of Chapter 1 gives us $N_U(P) = KS$.

(e) *Let $\bar{K} = K/P$, $\bar{S} = S/P$, $\bar{L} = N_U(P)/P$, and $\bar{N} = N_{\bar{L}}(\bar{S})$. Then the p-group $\bar{S}$ acts on the p'-group $\bar{K}$. Furthermore, we have $n = |\bar{L} : \bar{N}|$.*

Proof. It follows from the Correspondence Theorem and (d) that $\bar{S}$ acts on $\bar{K}$. Since $n = \sigma_p(\bar{L})$, Sylow's Theorem proves the last formula. $\square$

In the remaining proof, we will consider only the group $\bar{L}$ and its subgroups. So, we will simplify the notation by removing the bars and using L, S, K, and N. Let $n = n_1 \cdots n_s$ be the canonical decomposition of n into prime powers; thus, if $i \neq j$, n_i and n_j are powers of different primes. We need to show that $n_i \equiv 1 \pmod{p}$ for all i.

By (e), L is a semidirect product of S and K. So, by (8.9) of Chapter 1, we have $N \cap K = C_K(S)$. Assume that n_i is a power of a prime number r. An S_r-subgroup R_0 of $C_K(S)$ is clearly an S-invariant r-subgroup of K. Then, by applying Theorem 2.9 (iii) to a p'-group K with an operator group S, we know that there is an S-invariant S_r-subgroup R of K which contains R_0. Since R is S-invariant, RS is a subgroup of L which contains $R_0 S$. By definition, R_0 is an S_r-subgroup of $N \cap K$. So, by (2.14), we have $R_0 = R \cap N$. Hence n_i is equal to $|R : R_0|$. On the other hand, Theorem 3.14 of Chapter 1 gives us $RS \cap N = (R \cap N)S = R_0 S$. So, $R_0 S$ is the normalizer of S in RS and

$$n_i = |R : R_0| = |RS : R_0 S| = \sigma_p(RS) \equiv 1 \pmod{p}.$$

The last congruence follows from Sylow's Theorem (iv). $\square$

The formula for $\sigma_p(G)$ shows that, if G is solvable, $\sigma_p(G)$ is a product of prime powers which are of the form $1 + kp$. (Note that $\sigma_p(S_i) = 1$ if G is solvable.) This is not valid for general groups. Consider $SL(2, 5)$, the special linear group over the field of 5 elements. As we proved in Chapter 1, §9, a proper normal subgroup of $SL(2, 5)$ is contained in the center. So, an S_5-subgroup is not normal. The number of S_5-subgroups is a divisor of $2^3 3$ and of the form $1 + 5k$. So, $\sigma_5(SL(2, 5)) = 6$; it is not a product of prime powers of the form $1 + 5k$. We can prove (Exercise 9 of Chapter 3, §2) that the only simple group G with $\sigma_p(G) = 6$ is isomorphic to $PSL(2, 5)$.

EXAMPLE 4. We have shown that the factor group $PSL(2, 7)$ of the special linear group $SL(2, 7)$ by its center is a simple group of order $168 = 7 \cdot 3 \cdot 8$. *Conversely, a group of order 168 such that no Sylow subgroup is normal must be isomorphic to $PSL(2, 7)$.* The proof of this assertion which will be discussed here is not easy, but it is interesting because it involves several typical methods used in finite group theory: applications of Sylow's Theorem, counting arguments, and the proof of the uniqueness of the structure of a group.

Let G be a group with the said properties, let P be an S_7-subgroup, let $H = N_G(P)$, and let $n_7 = |G : H|$. By Sylow's Theorem and Theorem 3.6 of Chapter 1, we have $n_7 = \sigma_7(G)$. We will prove several propositions.

(a) *The subgroup H is a maximal subgroup of G; we have $n_7 = 8$ and $|H| = 21$. There are exactly $48 = 6 \cdot 8$ elements of order 7.*

Proof. By assumption, $n_7 > 1$. Since n_7 is a divisor of $|G|$ and is of the form $1 + 7k$, we have $n_7 = 8$. If H were not maximal, there would be a proper subgroup G_0 containing H properly. We would have $|G_0 : H| \equiv 1 \pmod{7}$. This would give $|G_0| = |G|$, a contradiction. So, H is maximal. Each element of order 7 is contained in exactly one S_7-subgroup, and each S_7-subgroup contains exactly 6 nonidentity elements; the total number of 7-elements is $8 \cdot 6 = 48$.   □

(b) *Let Q be an S_3-subgroup of H and let $R = N_G(Q)$. Then, $N_H(Q) = Q$ and $|G : R| = n_3 = 28$. The group H is a nonabelian group of order 21, and G contains exactly $2 \cdot 28 = 56$ elements of order 3.*

Proof. Since $n_3 = \sigma_3(G)$, we have $n_3 \equiv 1 \pmod{3}$. If $N_H(Q) \neq Q$, then $N_H(Q)$ would coincide with H and $H \subset R$. Since H is maximal (by (a)) and $|G : H| \not\equiv 1 \pmod{3}$, we would get $R = G$. This contradicts the assumption $n_3 > 1$. Therefore, $N_H(Q) = Q$ and H contains exactly 7 conjugates of Q. If $n_3 = 7$, all S_3-subgroups of G would be contained in H. By the remark after (4.4) of Chapter 1, S_3-subgroups of G generate a normal subgroup. So, $H \lhd G$. This is a contradiction because G has an S_7-subgroup which is not contained in H. Hence $n_3 > 7$; we see that $n_3 = 28$.   □

(c) *The group R is a nonabelian group of order 6.*

Proof. Suppose that R were abelian. Then R would be cyclic group of order 6 (Example 1). By (6.13) (iii) of Chapter 1, Q would be a characteristic subgroup of R. Hence, we would get $N_G(R) \subset N_G(Q) = R$ by (6.14) of Chapter 1. The number of conjugates of R would also be 28. Each conjugate of R would contain exactly two elements of order 6. So, there would be 56 elements of order 6. Thus, the number of elements of order 7, 3, or 6 would be $48 + 56 + 56 = 160$. An S_2-subgroup of G would contain 8 elements. Since $|G| = 168$, G would have exactly one S_2-subgroup. This contradicts the assumption that S_2-subgroups are not normal.   □

The structure of R is uniquely determined by Example 1; the same is true for H. Let t be an element of R of order 2. Then we have $H = PQ, R = \langle Q, t \rangle$, and $H \cap R = Q$.

(d) *Furthermore, we have $H^t \cap P = \{1\}$. The double coset HtP contains $21 \cdot 7$ elements and we have a partition into double cosets : $G = H \cup HtP$.*

Proof. Since $t \notin H$, we get $P^t \neq P$. Clearly, P^t is the only S_7-subgroup of H^t, so $P \cap H^t = \{1\}$. By (3.8) (iv) of Chapter 1, HtP contains $|P : P \cap H^t|$ cosets. Since $|P : P \cap H^t| = 7$, we get $G = H \cup HtP$ by counting the number of elements in the right side.   □

(e) *Each element of HtP is written uniquely as utv where $u \in H$ and $v \in P$. If x is an element of $P - \{1\}$,*

$$txt = f(x)g(x)th(x)$$

where $f(x) \in P$, $g(x) \in Q$, $h(x) \in P$, and the functions f, g, and h are uniquely determined.

Proof. There are at most $|H| \cdot |P|$ elements of the form utv ($u \in H$, $v \in P$). Since $|HtP| = |H| \cdot |P|$, the expression must be unique. If $x \in P - \{1\}$, then $x^t \in P^t - \{1\}$. Since $t^2 = 1$ and $H \cap P^t = \{1\}$ (by (d)), we get $x^t \in HtP$. Thus, the functions f, g, and h are uniquely determined. □

(f) *The product of any two elements of G is uniquely determined by knowing the functions f, g, and h.*

Proof. By (d), all elements of G are contained in either H or HtP. If two elements both lie in H, then their product is known by the (unique) structure of H. Let utv be an element of HtP where $u \in H$ and $v \in P$, and let $w \in H$. Then the product $w(utv)$ is equal to $(wu)tv$, which can be determined by the structure of H. The product $(utv)w$ can be computed similarly. The element vw is an element of H, so $vw = sx$ where $s \in Q$ and $x \in P$. The pair (s, x) of elements s and x are uniquely determined by the structure of H. Since R is nonabelian by (c), we have $ts = s^{-1}t$ and $(utv)w = us^{-1}tx$. This proves that the product is computable by knowing the structure of H. If wty ($w \in H$, $y \in P$) is another element of HtP, then we get

$$(utv)(wty) = us^{-1}txty$$

where, as before, $vw = sx$, $s \in Q$, and $x \in P$. If $x = 1$, then the above product is $us^{-1}y$ which is an element of H. On the other hand, if $x \neq 1$, then the product is

$$(us^{-1}f(x)g(x))t(h(x)y).$$

The element appearing left of t lies in H, while the element $h(x)y$ belongs to P. Since the elements s and x are determined by the structure of H, the proposition (f) is proved. □

The preceding consideration proves that the structure of G is determined by the three functions f, g, and h. We will show that the functions f, g, and h are indeed uniquely determined.

Let x be a generator of P, and let $txt = fgth$ where $f = f(x)$, $g = g(x)$, and $h = h(x)$. Taking the inverses, we get $tx^{-1}t = h^{-1}gtf^{-1}$. So, we have

$f(x^{-1}) = h(x)^{-1}$, $g(x^{-1}) = g(x)$, and $h(x^{-1}) = f(x)^{-1}$. Next, we apply conjugation by an element s of Q. Using $st = ts^{-1}$, we have

$$t(s^{-1}xs)t = stxts^{-1} = sfs^{-1} \cdot s^2g \cdot tshs^{-1}.$$

Since $sfs^{-1} \in P$, $s^2g \in Q$, and $shs^{-1} \in P$, we have

$$f(x^s) = sf(x)s^{-1}, \quad g(x^s) = s^2g(x), \quad h(x^s) = sh(x)s^{-1}.$$

By (b), all elements of $P - \{1\}$ are written either x^s or $(x^{-1})^s$. So, the functions f, g, and h are completely known once the values at the generator x are known.

Let us consider the function g first. Since $g(x^s) = s^2g(x)$ and $|Q| = 3$, there are exactly two elements satisfying $g(u) = 1$. Let u be one of those elements which satisfy $g(u) = 1$. The other element is u^{-1}. We have

$$tut = vtw$$

where both v and w are elements of P (because $g(u) = 1$). Since $t^2 = 1$, we get $tvt = utw^{-1}$. Since u, $w \in P$, this formula implies that $g(v) = 1$. As $|P|$ is odd, we must have $w \neq w^{-1}$. By the uniqueness in (e), u and v are different. So, we have $v = u^{-1}$. Similarly, we have $w = u^{-1}$. Thus,

$$f(u) = u^{-1}, \quad g(u) = 1, \quad h(u) = u^{-1}.$$

As we have shown earlier, these data determine the functions f, g, and h uniquely. For example, if $u^s = u^2$, then

$$f(u^2) = sf(u)s^{-1} = su^{-1}s^{-1} = u^{-4} = u^3, \quad g(u^2) = s^2;$$
$$f(u^4) = s^{-1}f(u)s = s^{-1}u^{-1}s = u^{-2} = u^5, \quad g(u^4) = s.$$

The formula $f = g$ is valid in general. If we choose u^{-1} as a generator of P instead of u, we get exactly the same function because $f(u^{-1}) = u$. Hence, by (f), the structure of G is unique. The simple group $PSL(2, 7)$ has order 168 (see (9.11) of Chapter 1). Thus, the simple group $PSL(2, 7)$ is one of the groups which satisfy our assumptions. Let G be any other group which satisfies these assumptions. We have seen that the structure of G is uniquely determined. Therefore, G must be isomorphic to $PSL(2, 7)$. $\square$

In the simple group $PSL(2, 7)$, the elements which correspond to the elements u and t are the cosets containing

$$\begin{pmatrix} 1 & 0 \\ 1 & 1 \end{pmatrix} \quad \text{and} \quad \begin{pmatrix} 0 & 1 \\ -1 & 0 \end{pmatrix},$$

respectively. The method used in the preceding proof may be applied to the similar problem of characterizing $PSL(2, q)$ with $q \equiv -1 \pmod 4$ (see Exercise 9). Any simple group of order 60 is isomorphic to $PSL(2, 5)$. But, the above procedure does not quite determine the functions (u^{-1} is conjugate to u by an element of Q). It is simpler to use other methods. Either try to establish the isomorphism onto $PSL(2, 4)$ by a method similar to the preceding, or, as suggested in a later exercise (Exercise 9, Chapter 3, §2), try to prove that the group is indeed isomorphic to A_5.

Since $PSL(3, 2)$ has order 168, we have

$$PSL(3, 2) \cong PSL(2, 7).$$

As noted above, $PSL(2, 4)$ is isomorphic to $PSL(2, 5)$. It is known that, among the simple groups in the PSL series, there is exactly one more coincidence of orders (Artin [2]):

$$|PSL(4, 2)| = |PSL(3, 4)|.$$

In the last case, however, the two groups are not isomorphic. This can be seen by noting that the S_2-subgroups of these simple groups are not isomorphic. (The S_2-subgroups are the groups of upper-triangular matrices over the finite fields $\mathbf{F}_2$ and $\mathbf{F}_4$, respectively. The orders of their centers are 2 and 4, respectively, so they are not isomorphic (see Exercise 4 of §1).)

Exercises

1. Suppose that a subgroup H of a finite group G satisfies one of the following two conditions:

(i) For any nonidentity element x of H, we have $C_G(x) \subset H$;

(ii) If K is a subgroup of H and $K \neq \{1\}$, then $N_G(K) \subset H$.

Show that the order $|H|$ of H is relatively prime to its index $|G : H|$.

(*Hint.* The conclusion is equivalent to saying that any S_p-subgroup of H is an S_p-subgroup of G for all prime divisors p of $|H|$.)

2. Suppose that a subgroup H of a finite group G satisfies the following condition:

$$x \notin H \Rightarrow H \cap H^x = \{1\}.$$

Show that the following propositions hold:

(a) The order of H is relatively prime to its index $|G : H|$.

(b) If $|G : H|$ is a power of a prime number p, then an S_p-subgroup of G is normal.

(c) Suppose that another subgroup K of G satisfies the similar condition: $y \notin K \Rightarrow K \cap K^y = \{1\}$. If both K and H are nontrivial, then there is an element u of G such that $H \cap K^u \neq \{1\}$.

(*Hint.* (b) Use the last procedure in Example 2.

(d) Suppose that (c) is false. Count the number of elements of G which lie in all the conjugates of H and K.

Even if the index $n = |G : H|$ is arbitrary, the conclusion of (b) is valid; G *contains a normal subgroup of order* n. This is a famous theorem of Frobenius. All the proofs known so far use character theory (see Chapter 6). In (c), the group K is indeed conjugate to H. However, we apparently need the theorem of Frobenius to prove the conjugacy of K and H.)

3. Let p and q be prime numbers.

(a) Show that a group of order $p^2 q$, or $p^2 q^2$, has at least one normal Sylow subgroup.

(b) Show that if a group G of order $p^3 q$ has no normal Sylow subgroup, then G is isomorphic to the symmetric group Σ_4 on 4 letters.

(*Hint.* (b) Sylow's Theorem and the preceding Exercise 2 (b) give $p = 2$ and $q = 3$. Then, apply (7.16) of Chapter 1.)

4. Let H be a composition subgroup of a finite group G. Show that, for any S_p-subgroup S of G, the intersection $H \cap S$ is an S_p-subgroup of H.

(*Hint.* See Exercise 3 of Chapter 1 §5 for the definition of a composition subgroup. It is conjectured that a subgroup H of a finite group G would be a composition subgroup if $H \cap S$ is an S_p-subgroup of H for any prime p and any S_p-subgroup S of G.)

5. Under the same assumptions and notations as in Theorem 2.9, show that, if S is any Q-invariant S_p-subgroup of G, $S \cap C_G(Q)$ is an S_p-subgroup of $C_G(Q)$.

6. Let F be the field of $32 (= 2^5)$ elements. Let Q be the Galois group of F over the prime field. (By the Galois Theory, Q is a cyclic group of order 5 generated by the automorphism $\alpha \to \alpha^2$ of F.) Let $G = SL(2, F)$. For each element σ of Q, we define the mapping φ_σ of G onto itself as follows:

$$\varphi_\sigma : \begin{pmatrix} \alpha & \beta \\ \gamma & \delta \end{pmatrix} \to \begin{pmatrix} \alpha^\sigma & \beta^\sigma \\ \gamma^\sigma & \delta^\sigma \end{pmatrix}.$$

Verify that φ_σ is an automorphism of G, and that, if we identify σ and φ_σ, the group Q acts on G. Since $|G| = 2^5 \cdot 31 \cdot 33$, G and Q satisfy all the conditions of Theorem 2.9.

Show that the number of Q-invariant S_{31}-subgroups is either 2 or 3. (So, Sylow's Theorem (iv) cannot be extended.)

(*Hint.* By Sylow's Theorem, the number n of S_{31}-subgroups of G is either 32 or $16 \cdot 33$. If the number of Q-invariant S_{31}-subgroups is m, we have $n = 5k + m$. (Consider orbits of Q acting on the set of S_{31}-subgroups of G.)

Since $C_G(Q) = SL(2, 2)$, we get $m \le 6$ by Theorem 2.9 (ii). Thus, $m = 2$ or 3. The group G contains exactly $16 \cdot 33$ S_{31}-subgroups. So, the actual value of m is 3. Thus, a generalization of Sylow's Theorem (iv) fails in the operator group case. However, the number of Q-invariant S_p-subgroups is always a divisor of $|G|$. This follows easily from Theorem 2.9 (ii).)

7. (a) Show that a nonabelian finite simple group contains at least one maximal subgroup which is not abelian.

(b) Show that, if all maximal subgroups of a finite group are abelian, at least one of maximal subgroups is normal.

(*Hint.* (a) Suppose that all maximal subgroups are abelian, and show that they satisfy the condition of Exercise 2. We get a contradiction from Exercise 2 (c). This method may be applied even when all maximal subgroups are assumed to be nilpotent. Choose a pair of maximal subgroups H and K such that $H \ne K$ and their intersection $H \cap K$ has maximal possible order. Try to show that $H \cap K = \{1\}$ by using the generalization of Theorem 1.6 (to a general nilpotent group).

(b) Use induction.)

8. Let n be a natural number. Prove that there is only one isomorphism class of groups of order n if and only if the following two conditions are satisfied:

(a) n is square-free (there is no square factor in n);

(b) if p and q are prime divisors of n, then $p \not\equiv 1 \pmod{q}$.

(*Hint.* The direct product of a group of order a and another group of order b has order ab (see (8.8) of Chapter 1). By Example 1 and Example of Chapter 1, §3, the two conditions (a) and (b) are necessary. The converse proposition may be proved by using induction on n. From Example 3 of Chapter 1, §8, and the remark following Exercise 4, §1, we can write down those conditions on n under which all the finite groups of order n are abelian.)

9. Let p and q be odd prime numbers such that $p = 2q + 1$. Let G be a group of order $p(p^2 - 1)/2$. Show that if Sylow subgroups for the prime numbers p, q, and 2 are not normal, then G is a simple group isomorphic to $PSL(2, p)$.

(*Hint.* Imitate the proof of Example 4. The parts (a) and (b) follow from elementary number theory. For the part (c), there are at least p conjugates of an element of order 2. Another exercise: Consider the case when an S_2-subgroup is normal.

The prime numbers $p = 11, 23$ satisfy the assumption of this exercise. It is known that a simple group of order $p(p^2 - 1)/2$ is isomorphic to $PSL(2, p)$ for any prime number p (Brauer [1]). But, the theory of characters is needed for this proof.)

10. Let G be a (not necessarily finite) π-group. Show that, if the index of a subgroup H is finite, then $|G : H|$ is a π-number.

(*Hint.* Use the remark which follows Corollary 2 of Chapter 1, (3.13).)

§3. Subnormal Series: Schreier's Refinement Theorem

Definition 3.1. A finite sequence of subgroups of G such that

$$(\mathcal{G})\colon G_0 = G \supset G_1 \supset \cdots \supset G_r = \{1\}$$

is said to be a **subnormal series** of length r, if for each $i = 1, 2, \ldots, r, G_i$ is a normal subgroup of G_{i-1}.

The composition series defined in Definition 5.8 of Chapter 1 is an example of a subnormal series. In a subnormal series, we may have repetitions $G_{i-1} = G_i$, and the factor group G_{i-1}/G_i need not be simple.

We will introduce the concept of groups with operator domains and prove the general Refinement Theorem.

Definition 3.2. A pair (Ω, G) consisting of a set Ω and a group G is said to be a **group with an operator domain** Ω if there is a function θ from Ω into End G, the set of endomorphisms of G. The function θ is said to be an **action** of Ω on G; each element of Ω is called an **operator**.

A group with an operator domain Ω is often simply called an Ω-group. If $\sigma \in \Omega$, $\theta(\sigma)$ is an endomorphism of G; we write $\theta(\sigma)x = x^{\sigma}$. Sometimes, the action θ is not explicitly mentioned.

EXAMPLE 1. If a group Q is an operator group on G, then the pair (Q, G) is a group with the operator domain Q.

In general, an operator domain Ω need not be a group. Even if Ω has some structure, the action θ need not transport this structure into End G.

Definition 3.3. Let G be an Ω-group. A subgroup H of G is said to be Ω-**invariant**, or an Ω-**subgroup**, if $H^{\sigma} \subset H$ for all $\sigma \in \Omega$.

If H is an Ω-subgroup, each element $\sigma \in \Omega$ induces an endomorphism of H. Thus, by restricting the domain, H becomes an Ω-group. If both H and K are Ω-subgroups of G, then so are $H \cap K$ and $\langle H, K \rangle$. If an Ω-subgroup H is normal, then each element of Ω defines an endomorphism of the factor group G/H in exactly the same way as in Chapter 1, §8 (see the discussion following Definition 8.10 of Chapter 1). If $\sigma \in \Omega$, the mapping $Hx \to Hx^{\sigma}$ is the endomorphism of G/H. In what follows, the factor group G/H by an Ω-invariant normal subgroup H is always considered to be an Ω-group with the induced action.

EXAMPLE 2. Let $\Omega = $ Inn G. By the natural action, Inn G is an operator domain of G; thus, G is an Ω-group. In this case, the Ω-subgroups are the normal subgroups of G. Similarly, Aut G may be an operator domain of G. If

$\Omega = \text{Aut } G$, the Ω-subgroups are the characteristic subgroups. If $\Omega = \text{End } G$, then the Ω-subgroups are the fully invariant subgroups.

Let G and G' be two Ω-groups. A homomorphism f from G into G' is said to be an Ω-**homomorphism**, or an **operator homomorphism**, if for any $\sigma \in \Omega$ and $x \in G$, f satisfies the formula

$$f(x^\sigma) = f(x)^\sigma.$$

An Ω-isomorphism is defined similarly.

It is significant that the Homomorphism Theorems are valid for all groups with operator domains.

Theorem 3.4. *Let G be an Ω-group, and let H be an Ω-invariant normal subgroup of G. The canonical homomorphism from G to G/H is an Ω-homomorphism.*

Let f be an Ω-homomorphism from G into another Ω-group G'. Then the following hold :

(1) *The kernel K of f is an Ω-invariant normal subgroup of G.*
(2) *There is an Ω-isomorphism g from G/K onto G' such that $f = g \circ \theta$ where θ is the canonical homomorphism of G onto G/K. Thus, the diagram*

is commutative.

Proof. By the definition of the induced action, we have $(Hx)^\sigma = Hx^\sigma$ for any $x \in G$. This means that the canonical homomorphism is an Ω-homomorphism. For (1), we need to show that the kernel K is Ω-invariant. If $f(x) = 1$, then, for any $\sigma \in \Omega$, we get $f(x^\sigma) = f(x)^\sigma = 1$. Thus, K is Ω-invariant. Finally, for (2), we need to show that the isomorphism g is an Ω-isomorphism. By definition, we get

$$g((Hx)^\sigma) = g(Hx^\sigma) = f(x^\sigma) = f(x)^\sigma = (g(Hx))^\sigma.$$

This shows that g is an operator homomorphism. $\square$

By Theorem 3.4, all the theorems in Chapter 1, §5 can be generalized to theorems about Ω-groups; we simply need to replace groups, subgroups, homomorphisms, etc. by Ω-groups, Ω-subgroups, Ω-homomorphisms, etc. In particular, the Correspondence Theorem, and the First and Second Isomorphism Theorems are valid for Ω-groups.

Definition 3.5. Ω-**composition series** and Ω-**subnormal series** may be defined accordingly. If Inn $G = \Omega$ acts naturally on G, the Ω-composition series are the chief series of G of Definition 1.11. If Aut $G = \Omega$ acts naturally on G, an Ω-composition series is called a **characteristic series** of G.

The notions defined above, such as Ω-groups and operator homomorphisms, are also important in the theory of rings and modules, for which readers are referred to text books on algebras ([IK], or [AS]).

We need a lemma and two definitions before we can prove the Refinement Theorem of Schreier.

(3.6) (Lemma of Zassenhaus) *Let G be an Ω-group and let U_1, U_2, V_1, and V_2 be Ω-subgroups of G such that $U_2 \lhd U_1$ and $V_2 \lhd V_1$. Then we have*

$$U_2(U_1 \cap V_2) \lhd U_2(U_1 \cap V_1), \quad V_2(V_1 \cap U_2) \lhd V_2(V_1 \cap U_1),$$

and the corresponding factor groups are Ω-isomorphic :

$$U_2(U_1 \cap V_1)/U_2(U_1 \cap V_2) \cong V_2(V_1 \cap U_1)/V_2(V_1 \cap U_2).$$

Proof. By assumption $U_2 \lhd U_1$, so $U_2(U_1 \cap V_2)$ is an Ω-subgroup of U_1. Similarly, $U_2(U_1 \cap V_1)$ is an Ω-subgroup of U_1. Since $V_2 \lhd V_1$, the conjugation by an element of $U_1 \cap V_1$ leaves $(U_1 \cap V_1) \cap V_2 = U_1 \cap V_2$ invariant. Thus, $U_2(U_1 \cap V_2)$ is a normal subgroup of $U_2(U_1 \cap V_1)$. We apply the Second Isomorphism Theorem for Ω-groups to $H = U_2(U_1 \cap V_2)$ and $K = U_1 \cap V_1$. Then,

$$H \cap K = U_2(U_1 \cap V_2) \cap (U_1 \cap V_1)$$

which is equal to $(U_2 \cap V_1)(U_1 \cap V_2)$ by Theorem 3.14 of Chapter 1, and we get

$$HK/H \cong (U_1 \cap V_1)/(U_2 \cap V_1)(U_1 \cap V_2).$$

This is an Ω-isomorphism, and $HK/H = U_2(U_1 \cap V_1)/U_2(U_1 \cap V_2)$. Since the right side of the above formula is symmetric in U and V, we get a similar isomorphism by exchanging U and V. This proves the isomorphism:

$$U_2(U_1 \cap V_1)/U_2(U_1 \cap V_2) \cong V_2(V_1 \cap U_1)/V_2(V_1 \cap U_2). \quad \square$$

Definition 3.7. Let G be an Ω-group, and let $(\mathcal{G})$ be an Ω-subnormal series of G:

$$(\mathcal{G}): G_0 = G \supset G_1 \supset \cdots \supset G_r = \{1\}.$$

A **refinement** of $(\mathcal{G})$ is an Ω-subnormal series $(\mathcal{G}')$ such that, for every $i = 1, 2, \ldots, r$, the term G_i of $(\mathcal{G})$ appears as a term of $(\mathcal{G}')$.

An Ω-subnormal series $(\mathcal{H})$: $H_0 = G \supset H_1 \supset \cdots \supset H_s = \{1\}$ is said to be **isomorphic** to $(\mathcal{G})$ if there is a one-to-one correspondence between the set of factor groups $\{G_0/G_1, G_1/G_2, \ldots, G_{r-1}/G_r\}$ and the set $\{H_0/H_1, H_1/H_2, \ldots, H_{s-1}/H_s\}$ such that the corresponding factor groups are Ω-isomorphic.

A refinement of $(\mathcal{G})$ is obtained by inserting several terms between neighboring subgroups G_{i-1} and G_i of $(\mathcal{G})$ in such a way that the series remains an Ω-subnormal series.

The following is Schreier's Refinement Theorem.

Theorem 3.8. *Let $(\mathcal{G})$ and $(\mathcal{H})$ be two Ω-subnormal series of an Ω-group G. Then, there are refinements $(\mathcal{G}')$ and $(\mathcal{H}')$ of $(\mathcal{G})$ and $(\mathcal{H})$, respectively, such that $(\mathcal{G}')$ and $(\mathcal{H}')$ are isomorphic.*

Proof. We will use the notation introduced in Definition 3.7. We will define a refinement of $(\mathcal{G})$ by projecting each term of $(\mathcal{H})$ into the factor group G_{i-1}/G_i. Let G_{ij} be the subgroup defined by $G_{ij} = G_i(G_{i-1} \cap H_j)$ for all i and j. By definition, G_{ij} are Ω-subgroups. Since $G_i \triangleleft G_{i-1}$ and $H_j \triangleleft H_{j-1}$, Lemma (3.6) shows that

$$G_{is} \triangleleft G_{is-1} \triangleleft \cdots \triangleleft G_{i0}.$$

Since $H_0 = G$ and $H_s = \{1\}$, we have $G_{is} = G_i$ and $G_{i0} = G_{i-1}$. Thus, $\{G_{ij}\}$ is a refinement of $(\mathcal{G})$; its length is rs if $G_{i0} = G_{i-1} = G_{i-1s}$ is regarded as a single term. Similarly, we define $H_{jk} = H_j(H_{j-1} \cap G_k)$. Then $\{H_{jk}\}$ is a refinement of $(\mathcal{H})$ of length rs. Lemma (3.6) shows that

$$G_{ij-1}/G_{ij} \cong H_{ji-1}/H_{ji}.$$

Thus, the Ω-subnormal series $\{G_{ij}\}$ and $\{H_{jk}\}$ are isomorphic (by the correspondence which assigns the factor group over G_{ij} to the factor group over H_{ji}). $\square$

Corollary 1. *The Jordan–Hölder Theorem holds for Ω-composition series.*

Proof. Any refinement of an Ω-composition series is just a repetition of the terms of the original series; thus, an Ω-composition series does not have any proper refinement. In a refinement between G_{i-1} and G_i, there appears exactly one nontrivial factor group, and this factor group is isomorphic to G_{i-1}/G_i. So, the Refinement Theorem shows that any two Ω-composition series are isomorphic; this is the theorem of Jordan–Hölder. $\square$

Corollary 2. *Suppose that an Ω-group G possesses an Ω-composition series. Then, any Ω-subnormal series can be refined to an Ω-composition series.*

Proof. Let $(\mathscr{G})$ be an Ω-subnormal series. By applying the Refinement Theorem to $(\mathscr{G})$ and a given Ω-composition series, we can show that $(\mathscr{G})$ has a refinement $(\mathscr{G}')$ which is isomorphic to a refinement of the Ω-composition series. Since any refinement of an Ω-composition series is essentially another Ω-composition series (with some repetitions), $(\mathscr{G}')$ becomes an Ω-composition series after removing the repeated terms. $\square$

Corollary 3. *Let H be a normal Ω-subgroup of an Ω-group G which possesses an Ω-composition series. Then, both the Ω-group H and the factor group G/H possess Ω-composition series. Also, there is an Ω-composition series of G through H.*

This follows easily from Corollary 2 and the Correspondence Theorem, and it includes Exercise 3 of Chapter 1, §5 and Theorem 1.12 as special cases. Even if an Ω-group G possesses an Ω-composition series, a nonnormal Ω-subgroup H need not possess an Ω-composition series. This is obvious since there are infinite groups which are nearly simple. (For example, the group $SL(2, \mathbb{R})$ over the field of rational numbers possesses a composition series, but there are subgroups isomorphic to $\mathbb{Z}$.)

Schreier's original proof of Theorem 3.8 depends on a clever use of induction similar to that in the proof of Theorem 5.9 of Chapter 1. This ingenious proof is presented in [IK], pages 238–9.

Solvable Groups

Definition 3.9. A group G is said to be **solvable** if G possesses a subnormal series $G_0 = G \supset G_1 \supset \cdots \supset G_r = \{1\}$ such that each factor group G_{i-1}/G_i is abelian for $i = 1, 2, \ldots, r$.

This definition agrees with Definition 1.10 for finite groups. (This follows from Theorem 3.8.)

(3.10) *Let G be a solvable group. Then any subgroup, as well as any factor group, of G is solvable. Conversely, if a normal subgroup H and the factor group G/H are both solvable, then G is solvable.*

Proof. Let H be a subgroup of a solvable group G. Then G possesses a subnormal series $\{G_i\}$ which satisfies the condition of Definition 3.9. Let

$H_i = G_i \cap H$. Then, clearly we have $H_0 = H$, $H_r = \{1\}$, and $H_i = H_{i-1} \cap G_i$. Since $G_i \triangleleft G_{i-1}$, the Second Isomorphism Theorem proves that $H_i \triangleleft H_{i-1}$ and $H_{i-1}/H_i \cong H_{i-1} G_i/G_i$. The group $H_{i-1} G_i/G_i$ is a subgroup of G_{i-1}/G_i. These statements show that $\{H_i\}$ is a subnormal series of H such that each factor group H_{i-1}/H_i is abelian. So, by definition, H is solvable.

Furthermore, let us suppose that H is normal. Let $\bar{G}_i = HG_i/H$. If f is canonical homomorphism from G onto the factor group G/H, we have $f(G_i) = \bar{G}_i$. So, by (5.2) (vi) of Chapter 1, we have $\bar{G}_i \triangleleft \bar{G}_{i-1}$. The First Isomorphism Theorem proves that

$$\bar{G}_{i-1}/\bar{G}_i \cong HG_{i-1}/HG_i.$$

Since $HG_{i-1} = HG_i G_{i-1}$, we get $HG_{i-1}/HG_i \cong G_{i-1}/(HG_i \cap G_{i-1})$. The last factor group is isomorphic to

$$(G_{i-1}/G_i)/((HG_i \cap G_{i-1})/G_i).$$

Thus, $\{\bar{G}_i\}$ is a subnormal series of G/H such that the factor group $\bar{G}_{i-1}/\bar{G}_i$ is abelian for $i = 1, 2, \ldots, r$. Therefore, G/H is solvable.

In order to prove the converse, we will need the following lemma on subnormal series of a factor group G/H.

Let $\bar{G}_0 = \bar{G} \supset \bar{G}_1 \supset \cdots \supset \bar{G}_s = \{1\}$ be a subnormal series of $\bar{G} = G/H$. Let G_i be the subgroup of G which corresponds to $\bar{G}_i$ in the Correspondence Theorem. It follows that $G_i/H = \bar{G}_i$ ($i = 1, 2, \ldots, s$). Then $G_0 = G \supset G_1 \supset \cdots \supset G_s = H \supset \{1\}$ is a subnormal series of G, and we have $G_{i-1}/G_i \cong \bar{G}_{i-1}/\bar{G}_i$ for $i = 1, 2, \ldots, s$.

This follows immediately from Theorem 5.5 of Chapter 1 and Definition 3.1.

If $\bar{G}$ is solvable, all the factor groups G_{i-1}/G_i can be taken to be abelian. If H is also solvable, we can refine the above series so as to make all the factor groups abelian. Thus, G is solvable if both H and G/H are solvable. ☐

Definition 3.11. Let x and y be two elements of a group G. The element of the form $x^{-1}y^{-1}xy$ is called the **commutator** of x and y; we write $[x, y] = x^{-1}y^{-1}xy$. The subgroup of G generated by all of the commutator defined in G is called the **derived group** or the **commutator subgroup**. We denote it by G', $G^{(1)}$, or $D(G)$. The commutator subgroup of the commutator subgroup, that is, $D(G')$ is called the **second derived group** and is written G'', $G^{(2)}$, or $D^2(G)$. In general, the commutator subgroup of $G^{(i)}$ is written $G^{(i+1)}$, and the series of subgroups

$$G = G^{(0)}, \; G^{(1)}, \; G^{(2)}, \ldots, G^{(i)}, \ldots$$

is called the **derived series**. We write $G^{(\infty)} = \bigcap_{i=0}^{\infty} G^{(i)}$; if G is finite, we have $G^{(\infty)} = G^{(n)}$ for some integer n.

We have already used the concept of commutators back in Chapter 1. There, the following properties of commutators were most prominent:

$$[x, y] = 1 \Leftrightarrow x \text{ and } y \text{ are commutative;}$$
$$[x, y] = x^{-1}x^y = (y^{-1})^x y.$$

Thus, the commutator is a quotient of two conjugates.

Another significant property of the commutators is the following proposition.

(3.12) *A homomorphism carries a commutator to the commutator of the images.*

Proof. Let f be a homomorphism. From the definition of a commutator, we have

$$f([x, y]) = [f(x), f(y)]. \quad \square$$

Corollary 1. *The derived group is fully invariant; in particular, the derived group is characteristic. The derived series is also fully invariant.*

Proof. By (3.12), the set of commutators is mapped into itself by any endomorphism. Hence G' is fully invariant. By induction, we get the last assertion. $\square$

Corollary 2. *Let f be a homomorphism from a group G onto another group H. Then, f maps the derived group of G onto the derived group of H :*

$$f(D(G)) = D(f(G)).$$

Proof. This follows immediately from (3.12). $\square$

Corollary 3. *Let N be a normal subgroup of a group G, and let $\bar{G} = G/N$. Let H be the subgroup of G corresponding to the derived group $D(\bar{G})$, so that $H/N = D(\bar{G})$. Then we have $H = ND(G)$.*

Proof. Apply Corollary 2 to the canonical homomorphism from G to $\bar{G}$. Then, $D(G)$ is mapped onto $D(\bar{G})$. The assertion follows from the Correspondence Theorem. $\square$

The derived group is characterized as follows.

Theorem 3.13. *Let D be the derived group of a group G. Then the factor group G/D is abelian. If the factor group G/H by a normal subgroup H is abelian, then H contains D. Thus, D is the smallest normal subgroup having the property that the factor group is abelian.*

Proof. In the factor group G/D, we always have $[\bar{x}, \bar{y}] = 1$. Thus, $\bar{x}$ and $\bar{y}$ are commutative. If the factor group $\bar{G} = G/H$ is abelian, then again we have $[\bar{x}, \bar{y}] = 1$. This means that $[x, y] \in H$. Since D is generated by all the commutators defined in G, we get $D \subset H$. ☐

Theorem 3.14. *A group G is solvable if and only if the derived series reaches $\{1\}$ in a finite number of steps; thus, G is solvable if and only if $G^{(n)} = \{1\}$ for some integer n.*

Proof. If $G^{(n)} = \{1\}$ for some n, then the derived series is a subnormal series such that all the factor groups are abelian (by Theorem 3.13). So, G is solvable.

In order to prove the converse, we need the following (almost obvious) lemma.

(3.15) *If H is a subgroup of a group G, then, for any positive integer k, we have $H^{(k)} \subset G^{(k)}$.*

Proof. Since $H \subset G$, the set of generators of $H^{(1)}$, or in other words, the set of commutators of the elements of H, is a part of the set of generators of $G^{(1)}$; so, $H^{(1)} \subset G^{(1)}$. By induction, we have $H^{(k)} \subset G^{(k)}$. ☐

We will now prove the second part of Theorem 3.14. Suppose that G is solvable. Then G possesses a subnormal series $\{G_i\}$ such that $G_r = \{1\}$ and all the factor groups G_{i-1}/G_i ($i = 1, 2, \ldots, r$) are abelian. We will show that $G^{(i)} \subset G_i$ by induction on i. The case where $i = 0$ is obvious. Suppose $i > 0$ and $G^{(i-1)} \subset G_{i-1}$. Since the factor group G_{i-1}/G_i is abelian, we have $D(G_{i-1}) \subset G_i$ by Theorem 3.13. On the other hand, from (3.15) and the fact that $G^{(i-1)} \subset G_{i-1}$, we know that $G^{(i)} = D(G^{(i-1)}) \subset D(G_{i-1})$. Hence, we get $G^{(i)} \subset G_i$, thus completing the proof by induction. As $G_r = \{1\}$, we have $G^{(r)} = \{1\}$. This proves Theorem 3.14. ☐

Corollary 1. *If a group G is solvable, then there is a characteristic series $G_0 = G \supset G_1 \supset \cdots \supset G_s = \{1\}$ such that all the factor groups G_{i-1}/G_i are abelian.*

Corollary 2. *If a nontrivial group G is solvable, then G contains a characteristic subgroup which is abelian and different from $\{1\}$.*

Proof. The derived series is one of the characteristic series which satisfies the condition of Corollary 1. For Corollary 2, we may take the last term of the derived series which is different from $\{1\}$. ☐

The assertion of (3.10) concerning the solvability of a subgroup is a trivial consequence of Theorem 3.14 and (3.15). Similarly, the solvability of the factor group follows immediately from Theorem 3.14 and the formula

$$f(G^{(k)}) = f(G)^{(k)}$$

where f is any homomorphism. (The above formula is proved by induction starting from Corollary 2 of (3.12).) However, the method used in the proof of (3.10) can be applied to the following general situation.

Definition 3.16. Let P be a given group theoretical property. A group G is said to be a **poly-P-group** if G possesses a subnormal series $\{G_i\}$ such that $G_0 = G$, $G_r = \{1\}$, and every factor group G_{i-1}/G_i is a P-group for $i = 1, 2, \ldots, r$.

If P is the property of being commutative, then a *poly-P-group* is nothing but the solvable group of Definition 3.9. Clearly, if P is a group theoretical property, then so is *poly-P*.

(3.17) *If P is a property such that any subgroup of a P-group is also a P-group, then any subgroup of a poly-P-group is a poly-P-group. If P is a property such that any factor group of a P-group is also a P-group, then a factor group of a poly-P-group is a poly-P-group.*

The proof is similar to the proof of (3.10).

If P is the property of being cyclic, then we get the class of **poly-cyclic groups**.

Corollary. *Any subgroup, as well as any factor group, of a poly-cyclic group is also poly-cyclic.*

Proof. We have shown ((5.6) of Chapter 1) that any subgroup, as well as any factor group, of a cyclic group is cyclic. $\square$

The class of poly-cyclic groups has many interesting properties. Since the publication of Hirsch's classical work on poly-cyclic groups, many new results have been obtained. We will, however, discuss only a few of the fundamental theorems of Hirsch (see Exercises 5, 6, and 7 at the end of this section).

(3.18) *Any subgroup of a poly-cyclic group is generated by a finite set of elements.*

Proof. We will prove that any poly-cyclic group is generated by a finite set of elements. By definition, a poly-cyclic group G possesses a subnormal series $\{G_i\}$ such that $G_0 = G$, $G_r = \{1\}$, and G_{i-1}/G_i are cyclic for $i = 1, 2, \ldots, r$. We will proceed by induction on r. Since the factor group G_0/G_1 is cyclic, we can take an element x of G such that $\bar{x}$ generates G_0/G_1. Then, we have $G = \langle x, G_1 \rangle$. Since G_1 is a poly-cyclic group, we can conclude from the inductive hypothesis that G_1 is generated by a finite set of elements. Clearly, G is also generated by a finite set of elements.

By the Corollary of (3.17), every subgroup of a poly-cyclic group is poly-cyclic. So, it is generated by a finite set of its elements. $\square$

A group is said to be *finitely generated* if it has a finite set of generators.

(3.19) *A finitely generated abelian group is poly-cyclic.*

Proof. If an abelian group A is finitely generated, then we can write $A = \langle a_1, a_2, \ldots, a_n \rangle$. Let A_i be the subgroup of A which is generated by the first $n - i$ elements $a_1, \ldots, a_{n-i}$:

$$A_i = \langle a_1, \ldots, a_{n-i} \rangle \quad \text{and} \quad A_n = \{1\}.$$

The sequence A_i is a subnormal series of A such that A_{i-1}/A_i are cyclic for all $i = 1, 2, \ldots, n$. Thus, A is poly-cyclic. $\square$

Corollary. *If an abelian group is finitely generated, its subgroups are also finitely generated.*

A poly-cyclic group G possesses a subnormal series $\{G_i\}$ of length r such that all factor groups G_{i-1}/G_i are cyclic. But, if G is infinite, its length r is not bounded; in fact, for any integer k, there is a subnormal series $\{G_i\}$ of length r without repetitions such that $r > k$ and all the factor groups G_{i-1}/G_i are cyclic. The following theorem is valid, however.

Theorem 3.20. *Let G be a poly-cyclic group with a subnormal series $(\mathcal{G}) = \{G_i\}$ such that all the factor groups G_{i-1}/G_i are cyclic. Then, the number of infinite cyclic groups in the set $\{G_{i-1}/G_i\}$ is a constant which depends only on G.*

Proof. Let $(\mathcal{H}) = \{H_j\}$ be another subnormal series with cyclic factors $\{H_{j-1}/H_j\}$. We need to show that $(\mathcal{G})$ and $(\mathcal{H})$ have the same number of infinite cyclic factors. By the Refinement Theorem, there are refinements $(\mathcal{G}')$ of $(\mathcal{G})$ and $(\mathcal{H}')$ of $(\mathcal{H})$ such that $(\mathcal{G}')$ is isomorphic to $(\mathcal{H}')$. Then, obviously, $(\mathcal{G}')$ and $(\mathcal{H}')$ have the same number of infinite cyclic factors. Let us compare $(\mathcal{G})$ and $(\mathcal{G}')$. Since $(\mathcal{G}')$ is a refinement of $(\mathcal{G})$, an infinite factor of $(\mathcal{G}')$ comes from an infinite factor of $(\mathcal{G})$. However, by (2.25) of Chapter 1, every proper subgroup of an infinite cyclic group has finite index. Hence the refinement of an infinite cyclic group involves exactly one infinite factor. Therefore, $(\mathcal{G})$ and its refinement $(\mathcal{G}')$ have the same number of infinite factors. Similarly, $(\mathcal{H})$ has the same number of infinite factors as $(\mathcal{H}')$. So, $(\mathcal{G})$ and $(\mathcal{H})$ has the same number of infinite cyclic factors. $\square$

Subnormal Subgroups

Definition 3.21. A subgroup H of a group G is said to be **subnormal** if there is a sequence of subgroups $\{G_i\}$ such that $G_0 = G \rhd G_1 \rhd G_2 \rhd \cdots \rhd G_r = H$; we write $H \lhd \lhd G$ when H is a subnormal subgroup of G.

(3.22) *Let H be a subnormal subgroup of a group G. Then the following propositions hold :*

- (i) *For any subgroup K, we have* $H \cap K \lhd\lhd K$.
- (ii) *If H is contained in a subgroup K, then* $H \lhd\lhd K$.
- (iii) *If* $K \lhd\lhd H$, *then we have* $K \lhd\lhd G$.
- (iv) *If* $K \lhd\lhd G$, *then* $H \cap K \lhd\lhd G$.
- (v) *For any homomorphism f, we have* $f(H) \lhd\lhd f(G)$.

Proof. By assumption, there is a sequence $\{G_i\}$ of subgroups which satisfies the condition of Definition 3.21. For (i), define the subgroups K_i by $K_i = K \cap G_i$. Then, the Isomorphism Theorem shows that $K_i \lhd K_{i-1}$. Since $K_r = K \cap G_r = K \cap H$ and $K_0 = K \cap G_0 = K$, $K \cap H$ is a subnormal subgroup of K. The proposition (ii) is a special case of (i), and (iii) follows easily from the definitions. The part (iv) is a corollary of (i) and (iii). The last statement follows by induction starting from (5.2) (vi) of Chapter 1. $\quad\square$

(3.23) *Let H and K be subnormal subgroups of a finite group G. Then we have* $\langle H, K \rangle \lhd\lhd G$.

Proof. Let $|G : H| = h$ and $|G : K| = k$. We will use induction on $(\max(h, k), \min(h, k))$ with lexicographical order (following Scott [1]). By assumption, there is a series of subgroups

$$G_0 = G \rhd G_1 \rhd \cdots \rhd G_r \rhd H.$$

We may and will assume that $G_1 \neq G$ and $G_r \neq H$. We will derive a contradiction assuming the result is false: thus, by hypothesis, $\langle H, K \rangle$ is not subnormal. The following propositions are to be proved under this hypothesis.

(a) *We have* $G = \langle G_r, K \rangle$.

Proof. Since $G_r \lhd\lhd G$ and $G_r \neq H$, we may apply the inductive hypothesis to K and G_r to conclude that $\langle K, G_r \rangle \lhd\lhd G$. Set $L = \langle K, G_r \rangle$. Then, by (3.22) (ii), we have $K \lhd\lhd L$ and $H \lhd\lhd L$. If L were a proper subgroup of G, then $\langle H, K \rangle \lhd\lhd L$ by the inductive hypothesis. Since $L \lhd\lhd G$, we would get $\langle H, K \rangle \lhd\lhd G$ by (3.22) (iii). This is a contradiction. So, we have $L = \langle K, G_r \rangle = G$. $\quad\square$

(b) *The subgroup K normalizes H*: $K \subset N_G(H)$.

Proof. Suppose that $H^x \neq H$ for an element x of K. Since $H \subset G_1 \lhd G$, H^x is a subgroup of G_1; in fact, we have $H^x \lhd\lhd G_1$ by (3.22) (v). Since $G_1 \neq G$, we have $\langle H, H^x \rangle \lhd\lhd G_1$. So, by (3.22) (iii), we get $\langle H, H^x \rangle \lhd\lhd G$. Since $\langle H, H^x \rangle \neq H$, again by the inductive hypothesis, $\langle H, H^x, K \rangle \lhd\lhd G$. However, as $x \in K$, we have $\langle H, K \rangle = \langle H, H^x, K \rangle \lhd\lhd G$. This is a contradiction. $\quad\square$

By (a) and (b), we have $G = \langle G_r, K \rangle \subset N_G(H)$; thus, we have $H \lhd G$. By symmetry, we get $K \lhd G$. But, then, $\langle H, K \rangle \lhd G$. This contradicts the hypothesis that $\langle H, K \rangle$ is not subnormal. $\square$

The proposition (3.23) is a theorem of Wielandt [1] adapted for finite groups. The union of two subnormal subgroups need not be subnormal in an infinite group. Robinson [1] gives a comparatively simple counterexample of two subnormal subgroups such that they generate a subgroup which is not subnormal. (The first counterexample of this kind was found by Zassenhaus around 1955.)

Exercises

1. (a) Let $H \lhd \lhd G$. State and prove the refinement theorem concerning the subnormal series between G and H.

(b) Let H be a composition subgroup of a group G (cf. Exercise 3 of Chapter 1, §5). Show that if $K \lhd \lhd G$ and $H \subset K$, then K is a composition subgroup of G.

(Composition subgroups are subnormal, but there are subnormal subgroups which are not composition subgroups. Of course, these two notions are the same for finite groups.)

2. Let H and K be composition subgroups of a group G. Show that $\langle H, K \rangle$ is also a composition subgroup of G.

(*Hint.* Use induction as in the proof of (3.23), replacing the index by the length of the composition series. Exercise 3 of Chapter 1, §5 will be needed.)

3. Show that *poly-(poly-P)*-groups are *poly-P*-groups.

4. A group G is said to satisfy the *maximal condition* for subgroups if any collection of subgroups has at least one maximal element. Show that if G satisfies the maximal condition for subgroups, then any subgroup is finitely generated. Also prove that if every subgroup of G is finitely generated, then there is no infinitely increasing sequence of subgroups: $H_1 \subsetneqq H_2 \subsetneqq \cdots$; that is, show that the group G satisfies the *ascending chain condition* for subgroups.

(If the axiom of choice is assumed, then the ascending chain condition is equivalent to the maximal condition (see [IK] p. 55 or [AS] p. 60). This exercise can be generalized for Ω-groups.)

5. Prove that a solvable group is poly-cyclic if and only if it satisfies the maximal condition for subgroups.

6. Show that a poly-cyclic group G contains a characteristic subgroup C which satisfies two conditions (1) $|G : C|$ is finite, and (2) the identity is the only element of finite order in C.

(*Hint.* By Exercise 3 (b) of Chapter 1, §7, we need only to prove the existence of such a subgroup. Use induction on the length of the subnormal series with cyclic factors, and consider two cases according to whether the first factor group G/G_1 is finite or not.)

7. Let g be a nonidentity element of a poly-cyclic group G. Show that there is a normal subgroup $N(g)$ of finite index which does not contain g. Thus, the intersection of all normal subgroups of finite index is $\{1\}$.

(*Hint.* Use induction as in the proof of Exercise 6, and consider two cases according to whether or not the first term G_1 contains g. By Exercise 3 (b) of Chapter 1, §7, we can take $N(g)$ as a characteristic subgroup of G. This is also a theorem of Hirsch. A group which has the property of this exercise is called *residually finite*. For example, consider the group $SL(n, \mathbb{Z})$ which consists of all the $n \times n$ matrices with integral entries and determinant 1. Then, $SL(n, \mathbb{Z})$ is residually finite because we can define homomorphisms onto $SL(n, p)$ for any prime p by replacing all of the entries of the matrices by their residue classes mod p. A theorem of L. Auslander [1] asserts that any poly-cyclic group is isomorphic to a subgroup of $SL(n, \mathbb{Z})$ with sufficiently large n. The assertion of Exercise 7 becomes obvious since any subgroup of a residually finite group is residually finite.

Let P_0 be the group theoretical property defined as follows: a group G is a P_0-group if and only if G is either cyclic or finite. Then, we can generalize the theorem in this book about poly-cyclic groups to theorems on poly-P_0-groups.)

8. Let p be a prime number, and let F be the field of p^p elements. For any prime divisor q of $p^p - 1$, let λ be an element of F^* whose multiplicative order is q. Let G be the group defined in Exercise 4 of Chapter 1, §8:

$$G = \left\{ \begin{pmatrix} \lambda^m & 0 \\ \alpha & 1 \end{pmatrix} \quad \alpha \in F, \quad m = 0, 1, \ldots, q - 1 \right\}.$$

The mapping $\xi \to \xi^p$ on F generates the Galois group Γ of F over the prime field. As in Exercise 6 of §2, Γ acts on G. With respect to this action, let L be the semidirect product of G and Γ, and let P be the subgroup of G generated by the matrix

$$\begin{pmatrix} 1 & 0 \\ 1 & 1 \end{pmatrix}.$$

(a) Show that P is a subnormal subgroup of L.

(b) Show that $N_L(P)$ is an S_p-subgroup of L, as well as a maximal subgroup, but not a normal subgroup of L.

(Thus, although P is a subnormal subgroup of L, the sequence of subgroups $P, N_L(P), N_L(N_L(P)), \ldots$ need not reach L.)

§4. The Krull–Remak–Schmidt Theorem

The direct product of two groups H and K is defined as the semidirect product with respect to the trivial action (Chapter 1, §8). Thus, the direct product of H and K is the cartesian product set $H \times K$ equipped with the law of composition

$$(u, v)(x, y) = (ux, vy).$$

Definition 4.1. Let $H_1, H_2, \ldots, H_n$ be groups. The **direct product** of these groups is the cartesian product set $H_1 \times H_2 \times \cdots \times H_n$ equipped with the law of composition

$$(x_1, x_2, \ldots, x_n)(y_1, y_2, \ldots, y_n) = (x_1 y_1, x_2 y_2, \ldots, x_n y_n).$$

We denote the direct product $H_1 \times H_2 \times \cdots \times H_n$ by using the same notation as for the set. Each group H_i is called the **direct factor** of the direct product.

If 1_i denotes the identity of the group H_i, then the element $(1_1, 1_2, \ldots, 1_n)$ is the identity of $H_1 \times \cdots \times H_n$; similarly, $(x_1, x_2, \ldots, x_n)^{-1} = (x_1^{-1}, x_2^{-1}, \ldots, x_n^{-1})$. It is easy to verify that the direct product is indeed a group. It is also easily proved that $H_1 \times \cdots \times H_n$ is isomorphic to $(H_1 \times \cdots \times H_{n-1}) \times H_n$; in general, we have

$$H_1 \times H_2 \times \cdots \times H_n \cong (H_1 \times \cdots \times H_m) \times (H_{m+1} \times \cdots \times H_n),$$
$$H \times K \cong K \times H.$$

Let G be the direct product of the groups $H_1, \ldots, H_n$. Let K_i be the set of elements of G such that for all $j \neq i$, the j-th components are the identity 1_j of H_j:

$$K_i = \{(1_1, 1_2, \ldots, 1_{i-1}, x_i, 1_{i+1}, \ldots, 1_n)\} \qquad (x_i \in H_i).$$

Then K_i is a subgroup of G, and we have the following proposition.

(4.2) (i) *The subgroup K_i is isomorphic to H_i under the isomorphism* $(1_1, \ldots, x_i \cdots 1_n) \to x_i$.
  (ii) *The subgroup K_i is normal in G.*
  (iii) *If $i \neq j$, K_i commutes elementwise with K_j.*
  (iv) *We have $G = K_1 K_2 \cdots K_n$ and every element of G can be written uniquely as $x_1 x_2 \cdots x_n$ with $x_i \in K_i$.*

Proof. From definition of the product in G, it is easy to verify (i), (ii), and (iii). Any element of G can be written as a product of elements of K_i:

$$(x_1, x_2, \ldots, x_n) = (x_1, 1, \ldots, 1)(1, x_2, \ldots, 1) \cdots (1, 1, \ldots, x_n).$$

This decomposition is unique, because $(x_1, x_2, \ldots, x_n) = (y_1, y_2, \ldots, y_n)$ if and only if $x_i = y_i$ for all i. □

Theorem 4.3. *Let $H_1, H_2, \ldots, H_n$ be normal subgroups of a group G such that $G = H_1 H_2 \cdots H_n$. Then, the following conditions are equivalent :*
   (1) For each $i = 1, 2, \ldots, n$, we have $H_1 H_2 \cdots H_{i-1} \cap H_i = \{1\}$.
   (2) Every element of G can be written uniquely as $x_1 x_2 \cdots x_n$ where $x_i \in H_i$ for all i.
   (3) There is an isomorphism $G \cong H_1 \times H_2 \times \cdots \times H_n$ such that the subgroup H_i of G corresponds to the subgroup K_i (defined just before (4.2)) of the direct product.

Proof. (1) ⇒ (2) By assumption, $G = H_1 H_2 \cdots H_n$, so every element of G can be written as $x_1 x_2 \cdots x_n$ with $x_i \in H_i$ for all i. Suppose that $x_1 x_2 \cdots x_m = y_1 y_2 \cdots y_m$ where x_i and y_i are elements of H_i for all i. Then, setting $u = y_1 \cdots y_{m-1}$, we get $y_m x_m^{-1} = u^{-1} x_1 \cdots x_{m-1}$. The left side is an element of H_m, while the right side is an element of $H_1 H_2 \cdots H_{m-1}$. The condition (1) gives us $H_1 \cdots H_{m-1} \cap H_m = \{1\}$, so that we have $x_m = y_m$ and $x_1 \cdots x_{m-1} = y_1 \cdots y_{m-1}$. By repeating this argument we can verify the uniqueness of the decomposition. Thus, the condition (1) implies the condition (2).
   (2) ⇒ (3) First, we will show that, if $i < j$, $u \in H_i$ and $v \in H_j$ are commutative. By (2), the product vu can be written uniquely as a product of an element of H_i and another element of H_j. We have $vu = u(u^{-1}vu) = (vuv^{-1})v$. Since $H_i \triangleleft G$, vuv^{-1} lies in H_i, and $u^{-1}vu \in H_j$. So, by the uniqueness of the decomposition, we get $v = v^u$, or $uv = vu$.
   Let g be an element of G. By (2), we can write the element g uniquely as $g = x_1 x_2 \cdots x_n$ where $x_i \in H_i$ for all i. We define the function f from G into the direct product $H_1 \times H_2 \times \cdots \times H_n$ by the formula

$$f(g) = (x_1, x_2, \ldots, x_n).$$

Let $h = y_1 \cdots y_n$ $(y_i \in H_i)$ be the decomposition of another element of G. Since x_i commutes with y_j for $i \neq j$, we have

(∗) $gh = (x_1 y_1)(x_2 y_2) \cdots (x_n y_n).$

Thus, $f(gh) = (x_1 y_1, x_2 y_2, \ldots, x_n y_n) = f(g)f(h)$, and this proves that f is a homomorphism. It is easy to verify the rest of the condition (3).
   (3) ⇒ (1) We only need to verify that $K_1 \cdots K_{i-1} \cap K_i = \{1\}$. But, this is almost obvious; the i-th component of an element belonging to the intersection is 1. □

Corollary. Let $H_1, \ldots, H_n$ be normal subgroups of a group G such that $|H_i|$ is relatively prime to $|H_j|$ for $i \neq j$. Then, the subgroup generated by these normal subgroups $H_1, H_2, \ldots, H_n$ is isomorphic to the direct product $H_1 \times H_2 \times \cdots \times H_n$.

Proof. If $i \neq j$, the order of $H_i \cap H_j$ divides both $|H_i|$ and $|H_j|$. So, we have $H_1 \cap H_2 = \{1\}$, and $|H_1 H_2| = |H_1| \cdot |H_2|$ by (3.11) of Chapter 1. By induction, we get $|H_1 \cdots H_i| = |H_1| \cdots |H_i|$ and $H_1 H_2 \cdots H_{i-1} \cap H_i = \{1\}$. Thus, the condition (1) of Theorem 4.3 is satisfied, and $H_1 \cdots H_n$ is isomorphic to the direct product. $\square$

Definition 4.4. A group G is said to be the **internal direct product** of the normal subgroups $H_1, H_2, \ldots, H_n$ if $G = H_1 \cdots H_n$ and if the three conditions of Theorem 4.3 is satisfied. A group G is called **directly decomposable** if it is an internal direct product of proper normal subgroups; otherwise, it is called (directly) **indecomposable**. A family $\{H_i\}$ of normal subgroups of G is said to be **independent** if it satisfies the condition (1) of Theorem 4.3.

By Theorem 4.3, if the family $\{H_i\}$ of normal subgroups is independent, then the subgroup generated by these normal subgroups is the internal direct product of $H_1, \ldots, H_n$. The converse is also true.

Let $H_1, \ldots, H_n$ be groups, and let K_i be the subgroup of $H_1 \times \cdots \times H_n$ as defined before. By (4.2), the family $\{K_i\}$ is independent, and $H_1 \times \cdots \times H_n$ is the internal direct product of $K_1, \ldots, K_n$. Furthermore, if a group G is the internal direct product of normal subgroups $H_1, \ldots, H_n$, then, by Theorem 4.3, G is isomorphic to the direct product $H_1 \times \cdots \times H_n$, and the isomorphism maps the subgroup H_i of G onto the subgroup K_i of the direct product. Thus, we may identify H_i with K_i, and disregard the distinction between the internal direct product and the (external) direct product. We will abuse the notation by denoting $G = H_1 \times \cdots \times H_n$ where G is the internal direct product of $H_1, \ldots, H_n$; in this case, we also call H_i a *direct factor* of G.

The preceding discussion may be carried over to the case when an operator domain is involved. If both H and K are Ω-groups, their direct product $H \times K$ becomes an Ω-group by defining the action of $\sigma \in \Omega$ as follows:

$$(h, k)^\sigma = (h^\sigma, k^\sigma).$$

It is easy to check that this rule does indeed define an action of Ω on $H \times K$. The Ω-group $H \times K$ is called the **direct product** of the Ω-groups H and K. If an Ω-group G is the internal direct product of the normal subgroups $H_1, \ldots, H_n$ and if every factor H_i is an Ω-subgroup, then G is Ω-isomorphic to the direct product of the Ω-groups $H_1, \ldots, H_n$. In order to prove this assertion, we need to verify that the isomorphism f from G to $H_1 \times \cdots \times H_n$ defined in the proof of Theorem 4.3 satisfies the formula $f(g^\sigma) = f(g)^\sigma$. If $g = x_1 \cdots x_n$ $(x_i \in H_i)$, then $f(g) = (x_1, \ldots, x_n)$. On the other hand, $g^\sigma = x_1^\sigma \cdots x_n^\sigma$ where

$x_i^\sigma \in H_i$. So, we have $f(g^\sigma) = (x_1^\sigma, \ldots, x_n^\sigma)$, and the required formula follows from the definition of the action of σ.

(4.5) *Suppose that an Ω-group G is the (internal) direct product of the Ω-subgroups $H_1, \ldots, H_m$. Let N_i be a normal Ω-subgroup of H_i for each i. Then, the following propositions hold.*

(a) *Each N_i is a normal Ω-subgroup of G.*

(b) *Let $N = N_1 N_2 \cdots N_m$. Then, N is the direct product of the Ω-subgroups $N_1, N_2, \ldots, N_m$; the Ω-group G/N is Ω-isomorphic to the direct product of the Ω-groups $H_1/N_1, \ldots, H_m/N_m$.*

Proof. If $i \neq j$, then H_i and H_j commute elementwise. Thus, N_i commutes elementwise with H_j for $i \neq j$, and $N_G(N_i)$ contains all H_j. Therefore, we have $N_i \lhd G$ and (a) holds.

Since the family $\{N_i\}$ is independent, the first half of (b) follows from (4.3). For an element x_i of H_i, let $\bar{x}_i$ denote the image of x_i by the canonical homomorphism onto H_i/N_i. The function defined by

$$g = x_1 \cdots x_m \to (\bar{x}_1, \ldots, \bar{x}_m)$$

is an Ω-homomorphism from G onto the direct product

$$(H_1/N_1) \times \cdots \times (H_m/N_m).$$

It is easy to see that the kernel of this homomorphism is precisely $N = N_1 \cdots N_m$; the Homomorphism Theorem proves the proposition (b). $\square$

(4.6) *Let G be the direct product of the subgroups $H_1, H_2, \ldots, H_m$. Then the following propositions hold:*

(a) $Z(G) = Z(H_1) \times Z(H_2) \times \cdots \times Z(H_m)$.
(b) $D(G) = D(H_1) \times D(H_2) \times \cdots \times D(H_m)$.

Proof. In the direct product, the rule of composition is given by the formula (∗). So, the element g lies in $Z(G)$ if and only if the component x_i of g lies in $Z(H_i)$ for all i. This proves (a).

By (3.15), $D(H_i)$ is contained in $D(G)$. Let D be the right side of the formula (b): $D = D(H_1) \times \cdots \times D(H_m)$. Then we have $D \subset D(G)$. On the other hand, (4.5)(b) gives us

$$G/D \cong (H_1/D(H_1)) \times (H_2/D(H_2)) \times \cdots \times (H_m/D(H_m)).$$

Since $H_i/D(H_i)$ is abelian by Theorem 3.13, G/D is abelian and $D \supset D(G)$. Thus, we have $D = D(G)$. $\square$

Let G be the direct product of the subgroups $H_1, H_2, \ldots, H_n$. Then, an element g of G can be written uniquely as

$$g = x_1 x_2 \cdots x_n \qquad (x_i \in H_i).$$

The rule of composition (*) shows that the function $\varepsilon_i: g \to x_i$ is a homomorphism from G onto H_i. We will prove the following proposition.

(4.7) *Let G be the direct product of the Ω-subgroups $H_1, \ldots, H_n$. Then, there are endomorphisms ε_i of G such that $\varepsilon_1, \ldots, \varepsilon_n$ are summable normal Ω-endomorphisms which satisfy the following conditions:*

$$\varepsilon_1 + \varepsilon_2 + \cdots + \varepsilon_n = 1, \quad \varepsilon_i^2 = \varepsilon_i, \quad \varepsilon_i \varepsilon_j = 0 \qquad (i \neq j).$$

These Ω-endomorphisms are called the *projections* onto the direct factors. For the endomorphisms 1 and 0, see Chapter 1, §6.

Proof of (4.7). Since $\varepsilon_i(G) = H_i$, it follows from (4.2) (iii) that $\varepsilon_i(G)$ and $\varepsilon_j(G)$ commute elementwise if $i \neq j$. So, ε_i and ε_j are summable, and therefore, we conclude that $\varepsilon_1, \ldots, \varepsilon_n$ are mutually summable endomorphisms.

If $g = x_1 \cdots x_n$ ($x_i \in H_i$), then we have $x_i = \varepsilon_i(g)$. Thus, by definition, $(\varepsilon_1 + \cdots + \varepsilon_n)(g) = \varepsilon_1(g) \cdots \varepsilon_n(g)$ and $\varepsilon_1 + \cdots + \varepsilon_n = 1$. Similarly, we get $\varepsilon_i(\varepsilon_i(g)) = x_i$ and $\varepsilon_j(\varepsilon_i(g)) = 1$ for $i \neq j$. So, we have $\varepsilon_i^2 = \varepsilon_i$ and $\varepsilon_i \varepsilon_j = 0$ if $i \neq j$.

For any $\sigma \in \Omega$, we have $g^\sigma = x_1^\sigma x_2^\sigma \cdots x_n^\sigma$. This implies that $\varepsilon_i(g^\sigma) = x_i^\sigma$. Hence, σ is an Ω-endomorphism.

In particular, we can consider $\Omega_0 = \operatorname{Inn} G$ as a domain of operators. In this case, each factor is an Ω_0-subgroup. So, all ε_i are Ω_0-endomorphisms of G; by Definition (6.17) of Chapter 1, the ε_i are normal. $\square$

The main theorem about direct product decompositions is the Krull–Remak–Schmidt Theorem.

Theorem 4.8. *Suppose that an Ω-group G possesses an Ω-chief series. Then, G is a direct product of a finite number of indecomposable factors. If*

$$G = H_1 \times \cdots \times H_n = K_1 \times \cdots \times K_m,$$

where all H_i and K_j are indecomposable and nontrivial, then we have $n = m$ and any r direct factors $H_{i_1}, \ldots, H_{i_r}$ of the first factorization can be replaced by a suitable set of r factors $K_{j_1}, \ldots, K_{j_r}$ of the second factorization. More precisely, there exists a normal Ω-automorphism σ such that $H_{i_1}^\sigma = K_{j_1}, \ldots, H_{i_r}^\sigma = K_{j_r}$ and such that σ leaves every element of H_i invariant for $i \neq i_1, \ldots, i_r$.

Corollary. *Let G be an Ω-group which has an Ω-chief series. If G satisfies at least one of the following three conditions, then there is essentially a unique way*

*to write G as a direct product of indecomposable factors (if we disregard the
arrangement of the factors) :*

(i) $Z(G) = \{1\}$; (ii) $G = D(G)$; (iii) *There is no nontrivial homomorphism from
G into $Z(G)$.*

Proof. The condition (i), as well as (ii), implies (iii). By (6.18) (ii) of Chapter 1, if
G satisfies (iii), then the only normal automorphism of G is the identity
mapping on G. So, the corollary follows from Theorem 4.8.   □

We need some preparation before beginning the proof of Theorem 4.8.

We remark that we may enlarge the operator domain to $\Omega \cup \mathrm{Inn}\ G$ without affecting the theorem. Thus, we will assume that Ω contains $\mathrm{Inn}\ G$.

(4.9) *Assume that $\mathrm{Inn}\ G \subset \Omega$ and that the Ω-group G possesses an Ω-chief
series. If an Ω-endomorphism σ is given, there exists an integer n such that*

$$\sigma^n(G) = \sigma^{n+1}(G) \quad and \quad G = \sigma^n(G) \times \mathrm{Ker}\ \sigma^n.$$

Proof. Since $\mathrm{Inn}\ G \subset \Omega$, any Ω-homomorphism is normal, and any Ω-
subgroup is a normal subgroup. Let r be the length of an Ω-composition
series of G. Then, the sequence of Ω-subgroups $\{\sigma^i(G)\}$ contains at most r
distinct subgroups; we have $\sigma^n(G) = \sigma^{n+1}(G)$ for some $n \leq r$. Set $H = \sigma^n(G)$.
Since H is a normal Ω-subgroup of G, H possesses an Ω-chief series (Corollary 2 of Theorem 3.8). By the definition of H, σ maps H onto itself. So, σ
induces an Ω-isomorphism from $H/H \cap \mathrm{Ker}\ \sigma$ onto H. By comparing the
lengths of the Ω-composition series, we get $H \cap \mathrm{Ker}\ \sigma = \{1\}$, and σ induces
an Ω-automorphism of H. Set $\rho = \sigma^n$. Then, ρ also induces an Ω-
automorphism of H. So we have $H \cap K = \{1\}$ where $K = \mathrm{Ker}\ \rho$, and ρ
induces an Ω-isomorphism from G/K onto H. Therefore, the length of the
Ω-composition series of G/K is equal to that of H. Since $H \cap K = \{1\}$,
HK/K is Ω-isomorphic to H. Again, by comparing the lengths of the Ω-
composition series, we get $HK = G$. As $\mathrm{Inn}\ G \subset \Omega$ and both H and K are
Ω-subgroups, they are normal and $G = H \times K$.   □

Corollary 1. *Assume that an Ω-group G is indecomposable. If $\mathrm{Inn}\ G \subset \Omega$, any
Ω-endomorphism of G is either nilpotent, or an Ω-automorphism.*

Proof. Let σ be an Ω-endomorphism of G. Under the assumptions of the
Corollary, (4.9) shows that for an integer n, we have either
$\sigma^n(G) = \sigma^{n+1}(G) = G$ and $\mathrm{Ker}\ \sigma^n = \{1\}$, or else $G = \mathrm{Ker}\ \sigma^n$. In the latter case,
σ is nilpotent. In the former case, we have $\mathrm{Ker}\ \sigma = \{1\}$. If $x \in G$, then there is
an element y such that $\sigma^n(x) = \sigma^{n+1}(y)$. Since $\mathrm{Ker}\ \sigma^n = \{1\}$, we get $x = \sigma(y)$.
Thus, σ is surjective, and σ is an Ω-automorphism of G.   □

Corollary 2. *Let G be a nontrivial Ω-group with an Ω-composition series. Suppose that $\mathrm{Inn}\ G \subset \Omega$ and that G is indecomposable. If two Ω-endomorphisms σ*

and ρ are summable and $\sigma + \rho$ is an Ω-automorphism, then either σ or ρ is an Ω-automorphism of G.

Proof. Let $\sigma + \rho = \alpha$. If α is an Ω-automorphism, the inverse α^{-1} exists. Both $\beta = \alpha^{-1}\sigma$ and $\gamma = \alpha^{-1}\rho$ are Ω-endomorphisms of G and they satisfy $\beta + \gamma = 1$. We have $\beta\gamma = \gamma\beta$, because $\beta(\beta + \gamma) = (\beta + \gamma)\beta$.

If Corollary 2 were false, neither β nor γ would be Ω-automorphisms. Then, by Corollary 1, we would have $\beta^n = 0 = \gamma^m$ for some positive integers n and m. Since β and γ are commutative, the binomial theorem gives us

$$1 = (\beta + \gamma)^{n+m} = \sum C_i \beta^i \gamma^{n+m-i}.$$

And, as $\beta^i = 0$ for $i \geq n$ and $\gamma^{n+m-i} = 0$ for $i < n$, we would have $1 = 0$. This is a contradiction since $G \neq \{1\}$. $\square$

Proof of Theorem 4.8. Since Inn $G \subset \Omega$, an Ω-composition series is a chief series, and any Ω-endomorphism is normal. If $G = H_1 \times \cdots \times H_n$, then

$$G \supset H_1 \times \cdots \times H_{n-1} \supset \cdots \supset H_1 \times H_2 \supset H_1 \supset \{1\}$$

is an Ω-subnormal series of G. Since it is refined to an Ω-chief series, the number of factors does not exceed the length of any Ω-composition series. So, it is possible to write G as a direct product of a finite number of indecomposable factors.

We may renumber the factors, if necessary, and assume that $i_1 = 1, \ldots,$ $i_r = r$. Let $H = H_1$ and $L = H_2 \times \cdots \times H_n$. Let $\kappa_1, \ldots, \kappa_m$ be the projections corresponding to the second decomposition of G, and let η and λ be the projections corresponding to $G = H \times L$. By (4.7), we have $1 = \kappa_1 + \cdots + \kappa_m$, so $\eta = \kappa_1 \eta + \cdots + \kappa_m \eta$. Consider the restriction of η on H; this restriction is the identity mapping of H. Each $\kappa_i \eta$ is an Ω-endomorphism of H, and H has an Ω-chief series. Since Inn $G \subset \Omega$, clearly Inn $H \subset \Omega$. Cor. 2 of (4.9) shows that $\kappa_i \eta$ is an Ω-automorphism of H for at least one value of i. If $\kappa_i \eta$ is an Ω-automorphism, then κ_i induces an injection from H to K_i, and η is an isomorphism from $\kappa_i(H)$ onto H.

Let us consider $\eta\kappa_i$ and its restriction σ on K_i. Then, σ is an Ω-endomorphism of K_i such that $\sigma(K_i) = \kappa_i(H)$. Since η is a surjection from $\kappa_i(H)$ onto H and κ_i is an isomorphism from H onto $\kappa_i(H)$, σ is an Ω-automorphism of $\sigma(K_i)$. By Cor. 1 of (4.9), σ is an Ω-automorphism of K_i. Thus, κ_i is an Ω-isomorphism from H onto K_i, and η is an Ω-isomorphism from K_i onto H. This proves that $H \cong K_i$.

We will now prove that $\eta\kappa_i$ and λ are summable. We have to show that K_i and L commute elementwise. Since both K_i and L are normal, it suffices to show that $K_i \cap L = \{1\}$ (see (3.16) of Chapter 1 or Theorem 4.3). If $x \in K_i \cap L$, then $\eta(x) = 1$, so $(\eta\kappa_i)(x) = \kappa_i(\eta(x)) = 1$. On the other hand, σ is an Ω-automorphism of K_i, so $(\eta\kappa_i)(x) = \sigma(x) = 1$ for $x \in K_i$ implies $x = 1$.

This shows that $K_i \cap L = \{1\}$. Thus, K_i commutes elementwise with L, and $\eta \kappa_i$ and λ are summable.

Let $\rho = \eta \kappa_i + \lambda$. Then, ρ is an Ω-endomorphism of G. We claim that ρ is in fact an Ω-automorphism of G. If $\rho(x) = 1$, then we have $\rho \eta(x) = \eta \kappa_i \eta(x) = 1$. Since $\kappa_i \eta$ is an automorphism of H, we get $\eta(x) = 1$. Therefore, $\rho(x) = 1$ implies that $\lambda(x) = 1$. Since $1 = \eta + \lambda$ by (4.7), we get $x = \eta(x)\lambda(x) = 1$, and therefore, Ker $\rho = \{1\}$. By (4.9), ρ is an Ω-automorphism of G. It follows from the definition of ρ that $\rho(H) = K_i$ and that ρ leaves every element of L invariant. Thus, ρ maps $H_1 \times H_2 \times \cdots \times H_n$ onto $K_i \times H_2 \times \cdots \times H_n$.

Similarly, there is an Ω-automorphism which maps $K_i \times H_2 \times \cdots \times H_n$ onto $K_i \times K_j \times H_3 \times \cdots \times H_n$ leaving every element of $K_i \times H_3 \times \cdots \times H_n$ invariant. By repeating this process, we can prove in the end that $n = m$, and hence, the last assertion of Theorem 4.8 holds. $\square$

In the general case, if a group does not possess a chief series, then Theorem 4.8 does not necessarily hold. We will give counterexample later (Example 2 of Chapter 3, §1), but there are simple counterexamples among groups with operators. For example, see Curtis–Reiner, §74.

So far in this book, we have only considered the direct product of a finite number of factors. The direct product of infinitely many factors may be defined similarly.

Definition 4.10. Let $\{H_\lambda\}$ be a family of Ω-groups H_λ indexed by a set Λ. We consider the totality of functions f defined on Λ such that $f(\lambda) \in H_\lambda$ for all $\lambda \in \Lambda$, and define the product of two such functions f and g by the formula

$$(fg)(\lambda) = f(\lambda)g(\lambda) \qquad (\lambda \in \Lambda).$$

The totality of such functions equipped with the operation defined above is called the **complete direct product** of the groups H_λ ($\lambda \in \Lambda$). The totality of functions such that $f(\lambda)$ is the identity of H_λ for all but a finite number of λ's is called the **(restricted) direct product** of the groups H_λ ($\lambda \in \Lambda$).

Clearly, the complete direct product forms a group with respect to the operation defined above. If we define the action of Ω by the formula $f^\sigma(\lambda) = f(\lambda)^\sigma$ for all $\lambda \in \Lambda$, then the complete direct product is also an Ω-group; the restricted direct product is an Ω-subgroup. The set K_λ consisting of functions f such that $f(\mu) = 1$ for all $\mu \neq \lambda$ forms a subgroup which is isomorphic to H_λ, and the restricted direct product is the subgroup generated by all the subgroups K_λ with λ ranging over Λ. In this book, the term "direct product" without a qualifier always refers to the restricted direct product.

Definition 4.11. Let G be an Ω-group, and let $\{H_\lambda\}$ be a family of normal Ω-subgroups indexed by a set Λ. The family $\{H_\lambda\}$ is said to be **independent** if every finite subset $\{H_{i_1}, H_{i_2}, \ldots, H_{i_n}\}$ is independent (in the sense of Definition 4.4).

The following proposition holds. The proof will be omitted since it is similar to that of Theorem 4.3.

(4.12) *Let* $\{H_\lambda\}$ ($\lambda \in \Lambda$) *be a family of normal* Ω-*subgroups of an* Ω-*subgroup G. The following conditions are equivalent.*

(1) *The family* $\{H_\lambda\}$ *is independent.*

(2) *If* $\lambda \neq \mu$, H_λ *and* H_μ *commute elementwise, and the relation* $1 = x_{\lambda_1} \cdots x_{\lambda_n}$, *where* $x_{\lambda_i} \in H$ *and* $\lambda_1, \ldots, \lambda_n$ *are distinct elements of* Λ, *is satisfied only when* $x_{\lambda_i} = 1$ *for all i.*

(3) *For any nonidentity element g of* $\langle H_\lambda(\lambda \in \Lambda)\rangle$, *we can find a finite subset* $\{\lambda_1, \lambda_2, \ldots, \lambda_n\}$ *of distinct elements of* Λ *and a set of elements* x_i *such that* $x_i \in H_{\lambda_i} - \{1\}$ *for each i and* $g = x_1 \cdots x_n$. *This set* $\{\lambda_1, \ldots, \lambda_n\}$ *and the elements* $\{x_i\}$ *are uniquely determined by the element g.*

(4) *There is an isomorphism from* $\langle H_\lambda(\lambda \in \Lambda)\rangle$ *onto the direct product of the groups* H_λ *such that each subgroup* H_λ *is mapped onto the group* K_λ.

We say that a group G is an internal direct product of the subgroups H_λ if the family $\{H_\lambda\}$ satisfies the conditions of (4.12); in this case, we write $G = \prod H_\lambda$.

Completely Decomposable Groups

Definition 4.13. An Ω-group is said to be **completely decomposable** if it is a direct product of simple Ω-groups.

The trivial group $\{1\}$ is completely decomposable, because it is the direct product $\prod S_\lambda$ ($\lambda \in \Lambda$) when Λ is the empty set.

(4.14) *Let G be an* Ω-*group. Then, the following four conditions are equivalent :*

(1) *The* Ω-*group G is completely decomposable.*

(2) *There is a family of normal* Ω-*subgroups* $\{S_\lambda\}$ ($\lambda \in \Lambda$) *such that each member* S_λ *of the family is a simple* Ω-*group and* $G = \langle S_\lambda (\lambda \in \Lambda)\rangle$.

(3) *Any normal* Ω-*subgroup H of G is a direct factor, and there is an* Ω-*subgroup K such that K is completely decomposable and* $G = H \times K$.

(4) *Any normal* Ω-*subgroup H is a direct factor; that is, there is an* Ω-*subgroup K such that* $G = H \times K$.

Proof. (1) $\Rightarrow$ (2) By assumption, we have $G = \prod S_\lambda$, where S_λ are simple Ω-groups. Then, S_λ are Ω-subgroups of G, and the family $\{S_\lambda\}$ satisfies the condition (2).

(2) $\Rightarrow$ (3) Let H be a normal Ω-subgroup of G. If $H = G$, we may choose $K = \{1\}$. So, we will assume that $H \neq G$. Since (2) holds, we have the family $\{S_\lambda\}$ of simple Ω-subgroups. Let $\mathscr{A}$ be the set of all subsets A of Λ such that the family $\{H, S_\alpha (\alpha \in A)\}$ is independent. Since $H \neq G$ and $G = \{S_\lambda\}$, $\mathscr{A}$ is not empty. Since the condition imposed on the subsets A is of finite character,

the set $\mathscr{A}$ satisfies the assumptions of Zorn's lemma ([IK], p. 61); thus, $\mathscr{A}$ has a maximal element, say M. Let $L = \langle H, S_\mu(\mu \in M) \rangle$. We will prove that $L = G$. By definition, L is a normal Ω-subgroup of G. If $L \neq G$, there would be a member S_γ of the family $\{S_\lambda\}$ such that $S_\gamma \subsetneqq L$. Then, $L \cap S_\gamma \neq S_\gamma$ and $L \cap S_\gamma$ would be a normal Ω-subgroup of S_γ. Since S_γ is simple, we would get $L \cap S_\gamma = \{1\}$. Thus, the family $\{H, S_\mu(\mu \in M), S_\gamma\}$ would be independent, contradicting the maximality of M. Hence we have $L = G$. Let $K = \langle S_\mu(\mu \in M) \rangle$. Thus, H and K are independent, and we have $G = H \times K$. By the definition of M, the family $\{S_\mu\}$ $(\mu \in M)$ is independent. Also, K is a direct product of the subgroups S_μ and is completely decomposable. Thus, the condition (2) implies (3).

Since the implication (3) $\Rightarrow$ (4) is trivial, we will prove that (4) implies (1). First, we will show that for any proper normal Ω-subgroup H, there is a normal Ω-subgroup S such that $H \cap S = \{1\}$ and S is simple.

By assumption $H \neq G$. So, we can choose an element u of G not contained in H. The set of normal Ω-subgroups which contain H, but do not contain u forms an ordered set which satisfies the conditions of Zorn's lemma. So, there is a maximal element M in the set. We will show that M is indeed a maximal normal Ω-subgroup of G. Suppose that M were contained in a larger proper normal Ω-subgroup L. By the condition (4), L would be a direct factor and $G = L \times V$ for some Ω-subgroup V different from $\{1\}$. We would have $M \subsetneqq L$ and $M \subsetneqq MV$. The maximality of M would give us $u \in MV \cap L$. On the other hand, Theorem 3.14 of Chapter 1 shows $MV \cap L = M(V \cap L) = M$. This is a contradiction as $u \notin M$. Thus, M is a maximal normal Ω-subgroup of G, and G/M is a simple Ω-group. Again from (4), we get $G = M \times S$ where $S \cong G/M$, and so S is simple. Since $H \cap S \subset M \cap S = \{1\}$, there is a simple Ω-subgroup S which satisfies $H \cap S = \{1\}$.

In order to prove the condition (1), we may assume $G \neq \{1\}$. The preceding argument (applied to the case where $H = \{1\}$) shows that G contains a simple normal Ω-subgroup $S \neq \{1\}$. Let us consider the set of all the family $\{S_\lambda\}$ consisting of simple normal Ω-subgroups S_λ such that $\{S_\lambda\}$ are independent. This set satisfies the assumptions of Zorn's lemma; thus, there is a maximal element. Let it be $\{S_\lambda\}$ and let G_0 be the subgroup of G generated by the members S_λ of this maximal element. By (4.12), G_0 is the direct product of S_λ's and is completely decomposable. If G_0 were a proper subgroup, there would be a simple normal Ω-subgroup S such that $G_0 \cap S = \{1\}$. Then the family $\{S_\lambda, S\}$ would still be independent. But, this contradicts the assumption that $\{S_\lambda\}$ is maximal. Thus, $G_0 = G$ and (1) holds. $\square$

Corollary 1. *A normal Ω-subgroup H of a completely decomposable group G is also completely decomposable.*

Proof. By (4), we have $G = H \times K$. The condition (3) applied to K shows that $G = L \times K$ for some completely decomposable group. Then, the Isomorphism Theorem proves that $H \cong G/K \cong L$. So, H is completely decomposable. ☐

Corollary 2. *Let G be an Ω-group. Assume that any operator of Ω induces an automorphism of G. If G possesses a chief series and if G is a simple Ω-group, then G is a direct product of a finite number of simple groups which are mutually isomorphic.*

Proof. By assumption, G has a minimal normal subgroup N. Since each σ of Ω induces an automorphism of G, N^σ is a minimal normal subgroup of G. Clearly, $\langle N^\sigma(\sigma \in \Omega)\rangle$ is Ω-invariant. So, by its simplicity, we have $G = \langle N^\sigma\rangle$. This means that the family of normal subgroups $\{N^\sigma\}$ satisfies the condition (2) of (4.14) with respect to the operator domain Inn G. As we can see from the proof of (4.14), G is a direct product of groups which are isomorphic to N. Since G possesses a chief series, the number of direct factors is bounded. It follows from (4.5)(a) that N is a simple group. ☐

Corollary 3. *Let G be a group with a composition series. A minimal normal subgroup N of G is a direct product of a finite number of mutually isomorphic, simple groups.*

Proof. Consider Inn $G = \Omega$ as a domain of operators of G. Then, N is a simple Ω-group. Corollary 2 proves the validity of this corollary. ☐

Corollary 4. *Let G be a group with a composition series. Any group in the set of factors of a chief series, as well as of a characteristic series, is a direct product of a finite number of mutually isomorphic simple groups.*

Proof. Each factor is a simple Ω-group for $\Omega =$ Inn G or Aut G. ☐

Central Products

Definition 4.15. A group G is said to be a **central product** of two subgroups H and K if H and K commute elementwise and $G = HK$.

The (internal) direct product $H \times K$ is nothing but a central product with the additional condition that $H \cap K = \{1\}$.

(4.16) *Let G be a group which is a central product of two subgroups H and K. Set $D = H \cap K$.*

  (i)   *Both H and K are normal subgroups.*
  (ii) *We have $D \subset Z(H)$ and $D \subset Z(K)$.*
  (iii) *There is a homomorphism from the direct product $H \times K$ onto G. Let H_1 be a group which is isomorphic to H via an isomorphism $\eta: H \to H_1$. Similarly, let $\kappa: K \to K_1$ be an isomorphism. Set $G_1 =$*

$H_1 \times K_1$. *The function* $x \to (\eta(x), \kappa(x)^{-1})$ *from* D *into* G_1 *defines an isomorphism from* D *into* $Z(G_1)$. *If* Z *denotes its image, then we have* $G \cong G_1/Z$.

Proof. (i) Since H and K commute elementwise, we have $K \subset C_G(H)$. So, both H and K are contained in $N_G(H)$, and since $G = HK$, we get $H \lhd G$. Similarly, we have $K \lhd G$.

(ii) The elements of D are contained in $H \cap C_G(H) = Z(H)$. Similarly, we have $D \subset Z(K)$.

(iii) By (4.6) (a), we have $Z(G_1) = Z(H_1) \times Z(K_1)$. So, the group Z is contained in the center $Z(G_1)$. Since H and K commute elementwise, for any h, $h' \in H$ and $k, k' \in K$, we have

$$(hk)(h'k') = (hh')(kk').$$

We define the function f on G_1 by the formula

$$f((\eta(h), \kappa(k)) = hk \in G.$$

Then, f is a homomorphism from G_1 onto G. If an element $(\eta(h), \kappa(k))$ belongs to the kernel of f, then we have $hk = 1$. So, $h = k^{-1} \in H \cap K$, and the element $h = x$ is an element of D. Therefore, we have $k = x^{-1}$. Conversely, if $x \in D$, then $(\eta(x), \kappa(x^{-1}))$ belongs to the kernel of f. Thus, the kernel of f coincides with the group Z and the Homomorphism Theorem gives us $G \cong G_1/Z$. $\square$

By (4.16) (iii), a central product may be considered to be a direct product with an amalgamated central subgroup. The converse of (4.16) (iii) also holds. Suppose that a subgroup D_1 of $Z(H)$ is isomorphic to a subgroup D_2 of $Z(K)$. Then there is a central product of H and K in which D_1 and D_2 are amalgamated according to the prescribed isomorphism and no additional amalgamation takes place. Precisely, we prove the following proposition.

(4.17) *Let* H_1 *and* K_1 *be two groups, and let* D_1 *be a subgroup of* $Z(H_1)$. *Suppose, in addition, that an isomorphism from* D_1 *into* $Z(K_1)$ *is given. The subset* Z *of the direct product* $H_1 \times K_1$ *consisting of elements of the form* $(x, \varphi(x^{-1}))$, *where* $x \in D_1$, *is a subgroup of* $Z(H_1 \times K_1)$. *The factor group* $(H_1 \times K_1)/Z$ *is a central product of* $H = H_1Z/Z$ *and* $K = K_1Z/Z$. *We have* $H \cong H_1, K \cong K_1$, *and the subgroup* D_1Z/Z *of* H *is identified with the subgroup* $\varphi(D_1)Z/Z$ *of* K *via the isomorphism* $\bar{\varphi}$ *induced from* φ.

Proof. It is easy to check that the set Z is a subgroup of the center. Consider H_1 and K_1 as subgroups of the direct product. Then, we have $H_1 \cap Z = \{1\} = K_1 \cap Z$, and the Second Isomorphism Theorem proves that $H = H_1Z/Z \cong H_1$ and, similarly, that $K \cong K_1$. Set $G = (H_1 \times K_1)/Z$. Then,

H and K commute elementwise and $G = HK$. Thus, G is a central product of H and K. Let $D = H \cap K$. Then, each element of D is a coset containing an element of the form $(x, 1)$ where $x \in H_1 \cap K_1 Z = D_1$. Conversely, if $x \in D_1$, then $(x, 1) = (x, \varphi(x^{-1}))(1, \varphi(x)) \in ZK_1$. Thus, we must have $D = D_1 Z/Z = \varphi(D_1)Z/Z$; furthermore, the elements x and $\varphi(x)$ correspond to the same element of G. $\square$

It is easily seen from several examples (see Exercise 5) that the structure of a central product depends not only on the structures of the factors H and K, but also on the isomorphism φ.

The central product of more than two groups can be defined similarly. Suppose that $H_1, \ldots, H_n$ are subgroups of a group G which satisfy the following conditions: if $i \neq j$, then H_i and H_j commute elementwise, and $G = H_1, \ldots, H_n$. In this case, we call G a *central product* of the subgroups H_1, $\ldots, H_n$.

(4.18) *Let G be a central product of the subgroups $H_1, \ldots, H_n$. Then, the function defined by*

$$(h_1, h_2, \ldots, h_n) \to h_1 h_2 \cdots h_n$$

is a homomorphism from the direct product $H_1 \times H_2 \times \cdots \times H_n$ onto G. The kernel of this homomorphism is a subgroup of the center of the direct product.

The proof is similar to that of (4.16).

The structure of a central product is rather complicated; for example, an analogue of Theorem 4.8 for the central product does not hold.

EXAMPLE. Let D be the group which consists of motions of the plane leaving the square invariant (Example 5 of Chapter 1, § 1). Let ρ be the rotation of $90°$ (counterclockwise) around the center, and let τ be the reflection with respect to the perpendicular bisector of a side. Thus, the elements of D are

$$\rho^i \tau^j \qquad 0 \leq i < 4, \quad j = 0, 1,$$

and we have the rule $(\sigma^i \tau^j)(\rho^k \tau^l) = \rho^a \tau^b$, where

$$a \equiv i + (-1)^j k \pmod 4, \qquad b \equiv j + l \pmod 2.$$

(This follows from the relation $\rho\tau = \tau\rho^{-1}$.) It is easy to see that $Z(D)$ is generated by ρ^2.

Let D_i ($i = 1, 2$) be groups isomorphic to D. We consider the central product $G = D_1 D_2$ by identifying the center. Let ρ_i and τ_i be elements of D_i corresponding to ρ and τ, respectively. We consider D_1 and D_2 to be subgroups of G. In forming the central product, we identified the centers, so we have $\rho_1^2 = \rho_2^2$. Set $\rho_1^2 = \zeta$.

We define two subgroups Q_1 and Q_2 by

$$Q_1 = \langle \rho_1, \tau_1 \rho_2 \rangle, \qquad Q_2 = \langle \rho_1 \tau_2, \rho_2 \rangle.$$

Set $\rho_3 = \tau_1 \rho_2$ and $\rho_4 = \rho_1 \tau_2$. Since every element with suffix 1 commutes with every element with suffix 2, we get the following relations:

$$\rho_3^2 = \zeta = \rho_4^2, \qquad \rho_3^{-1} \rho_1 \rho_3 = \rho_1^{-1}, \qquad \rho_4^{-1} \rho_2 \rho_4 = \rho_2^{-1}$$

and $(\rho_1 \rho_3)^2 = \zeta = (\rho_2 \rho_4)^2$. These relations show that both Q_1 and Q_2 are groups of order 8, and the only element of order 2 in Q_i is ζ. Furthermore, the following relations hold:

$$\rho_1 \rho_4 = \rho_4 \rho_1, \qquad \rho_2 \rho_3 = \rho_3 \rho_2, \qquad \rho_3 \rho_4 = \rho_4 \rho_3.$$

The first two relations follow from the commutativity of elements with suffix 1 and elements with suffix 2. The last one is derived as follows:

$$\rho_3 \rho_4 = \tau_1 \rho_2 \rho_1 \tau_2 = \tau_1 \rho_1 \rho_2 \tau_2 = \rho_1 \tau_1 \zeta \tau_2 \rho_2 \zeta$$
$$= \rho_1 \tau_2 \tau_1 \rho_2 = \rho_4 \rho_3.$$

We used the relations $\zeta^2 = 1$ and $\zeta \in Z(G)$. The above relations show that G is a central product of Q_1 and Q_2. The group Q_1 is isomorphic to Q_2. But, Q_1 is not isomorphic to D. In fact, Q_1 contains only one element of order 2, while D contains more than one element of order 2, ρ^2 and τ. (D contains exactly 5 elements of order 2.)

Since neither D nor Q_i can be written as a central product of proper subgroups, the group $G = D_1 D_2 = Q_1 Q_2$ is an example of a group which has two essentially different decompositions.

The group isomorphic to the group Q_i is called the *quaternion group* (see Exercise 4 of Chapter 1, §4).

Subgroups of Direct Products

Let $G = H \times K$ be the direct product, and let η and κ be the projections onto the factors. As before, we consider H and K to be subgroups of G; for any element of G_1 we have $g = hk$ where $h = \eta(g)$ and $k = \kappa(g)$. If a subgroup U is given, U determines the four subgroups $\eta(U)$, $U \cap H$, $\kappa(U)$, and $U \cap K$. Since the direct factors H and K are normal subgroups of G, $U \cap H$ is normal in U; similarly, we have $U \cap K \lhd U$. Let σ be the restriction of η on U. Then σ is a homomorphism from U onto $\eta(U)$, and its kernel is $U \cap K$. Thus, we have $\sigma(U) = \eta(U) \cong U/U \cap K$. The homomorphism σ maps $U \cap H$ onto $U \cap H$. So, by (5.2) (vi) of Chapter 1, we have $U \cap H \lhd \eta(U)$. If

$\sigma(u) \in U \cap H$, then we know that $\sigma(u) = \eta(u)$, from which we get $\sigma(u)^{-1}u = \kappa(u)$. This implies that $\kappa(u) \in U \cap K$. Conversely, if $\kappa(u) \in U \cap K$, then $\eta(u) = \sigma(u) \in U \cap H$. Thus, we have $\sigma^{-1}(U \cap H) = (U \cap H)(U \cap K)$. The Corollary of Theorem 5.5 of Chapter 1 shows that

$$\eta(U)/(U \cap H) \cong U/(U \cap H)(U \cap K).$$

Similarly, we get $\kappa(U)/(U \cap K) \cong U/(U \cap K)(U \cap H)$. Therefore, we have an isomorphism from $\eta(U)/(U \cap H)$ onto $\kappa(U)/(U \cap K)$; this is given by

$$\varphi : \eta(u)(U \cap H) \to \kappa(u)(U \cap K).$$

In fact, if $\eta(u) = \eta(v)$ for two elements u and v of U, then

$$v^{-1}u = \kappa(v)^{-1}\eta(v)^{-1}\eta(u)\kappa(u) = \kappa(v)^{-1}\kappa(u) \in U \cap K.$$

So, the function defined by $\eta(u) \to \kappa(u)(U \cap K)$ is a homomorphism from $\eta(U)$ onto $\kappa(U)/(U \cap K)$, and its kernel is $U \cap H$. Thus, φ is a well-defined isomorphism which is determined uniquely by U. We write $\varphi = \varphi_U$.

(4.19) *Let η and κ be the projections corresponding to the direct product $G = H \times K$. A subgroup U of G determines four subgroups $U \cap H$, $\eta(U)$, $U \cap K$, and $\kappa(U)$, and the isomorphism φ_U from $\eta(U)/(U \cap H)$ onto $\kappa(U)/(U \cap K)$. Conversely, suppose that we have subgroups H_1, H_2 of H and subgroups K_1, K_2 of K, such that $H_2 \lhd H_1$ and $K_2 \lhd K_1$, and an isomorphism θ from H_1/H_2 onto K_1/K_2. Then, there is a subgroup U of G such that $H_1 = \eta(U)$, $H_2 = U \cap H$, $K_1 = \kappa(U)$, $K_2 = U \cap K$, and $\theta = \varphi_U$. The subgroup U is uniquely determined by H_1, H_2, K_1, K_2, and θ.*

Proof. The first part has been proved. In order to prove the converse, we define the subset U of $H \times K$ as follows:

$$U = \{(h, k) \mid h \in H_1, k \in K_1, \theta(hH_2) = kK_2\}.$$

Since H_1 and K_1 are subgroups and θ is an isomorphism, we can easily show that U is a subgroup of $G = H \times K$. Clearly, we have $\eta(U) = H_1$ and $\kappa(U) = K_1$. An element of $U \cap H$ is of the form $(h, 1)$ where $\theta(hH_2) = K_2$. Since θ is an isomorphism from H_1/H_2 onto K_1/K_2, the element h must belong to H_2. Conversely, if $h \in H_2$, then $(h, 1) \in U$. Thus, we have $U \cap H = H_2$; similarly, $U \cap K = K_2$. Then, the definitions give us $\varphi_U = \theta$. It is clear that the subgroup U is uniquely determined by H_1, H_2, K_1, K_2, and θ. $\square$

Corollary. *Let H and K be finite groups of relatively prime orders. Any subgroup of $H \times K$ has the form $H_1 \times K_1$ where $H_1 \subset H$ and $K_1 \subset K$.*

Proof. If U is a subgroup of $H \times K$, the isomorphism φ_U maps $\eta(U)/(U \cap H)$ onto $\kappa(U)/(U \cap K)$. Since these groups have relatively prime orders, both groups must be of order 1. Hence, we get $\eta(U) = U \cap H$, $\kappa(U) = U \cap K$, and $U = \eta(U) \times \kappa(U)$. $\square$

(4.20) *Let H and K be normal subgroups of a group G. Then, $G/H \cap K$ is isomorphic to some subgroup of the direct product $(G/H) \times (G/K)$.*

Proof. Consider the function which maps an element x of G to the element (Hx, Kx) of $(G/H) \times (G/K)$. This is clearly a homomorphism from G into the direct product. The kernel consists of those elements x which satisfy $Hx = H$ and $Kx = K$; thus, the kernel is $H \cap K$. By the Isomorphism Theorem, we have (4.20). $\square$

Corollary. *Let P be a group theoretical property which satisfies the following two conditions :*

(a) *Any subgroup of a P-group is a P-group;*
(b) *The direct product of two P-groups is also a P-group.*
   If both factor groups G/H and G/K are P-groups, so is $G/H \cap K$.

Definition 4.21. Let G be a group, and let P be a group theoretical property. If the set of those normal subgroups H of G for which G/H are P-groups has a smallest element, we denote this element by $O^P(G)$. Thus, $O^P(G)$ is the smallest normal subgroup of G having the property that $G/O^P(G)$ is a P-group.

If P is the commutativity of groups, then $O^P(G)$ is the derived group $D(G)$. If the property P satisfies the conditions (a) and (b) of the preceding corollary, then for any finite group G, the group $O^P(G)$ exists, even though this need not be true for an infinite group G. The group $O^P(G)$ need not exist at all for some properties P, even for finite groups. But, if $O^P(G)$ exists, then it is certainly a characteristic subgroup. It will be clear that, even if $O^P(G)$ exists for all groups, the property P need not satisfy the conditions (a) and (b) of Cor. of (4.20).

Let p be a prime number, and let P be the property of being a p-group. Then, P satisfies the conditions (a) and (b) of Cor. of (4.20). In this case, we write $O^p(G)$ instead of $O^P(G)$. The group $O^p(G)$ is the smallest normal subgroup of G such that the factor group $G/O^p(G)$ is a p-group. In particular, suppose that $G = \mathbb{Z}$. Then, the index of the subgroup H_k generated by k is p^n for $k = p^n$, so $\mathbb{Z}/H_k$ is a p-group. However, the order of $\mathbb{Z}/H_k$ increases to infinity as $n \to \infty$, and $\bigcap H_k = \{0\}$. Thus, $\mathbb{Z}$ does not contain any smallest normal subgroup such that $\mathbb{Z}/H$ is a p-group; $O^p(\mathbb{Z})$ does not exist.

Exercises

1. Let $G = H \times K$ be the direct product of two groups H and K. Show that if there is a homomorphism f from K into $Z(H)$ such that $f(K) \neq \{1\}$, then there is a subgroup K_1 of G such that $K_1 \neq K$ and $G = H \times K_1$.

(*Hint.* Use (6.18) (ii) of Chapter 1 and Theorem 4.8.)

2. Let H be a nonabelian group, and let A be an abelian subgroup of H which is not contained in $Z(H)$. Set $G = H \times K$ where K is isomorphic to A. Show that G has an endomorphism which does not map $Z(G)$ into $Z(G)$.

(*Hint.* The isomorphism $K \to A$ can be extended to an endomorphism of G. There are many groups H which contain an abelian subgroup A satisfying the condition of this exercise. Give some examples.)

3. Prove that the converse of (4.7) is valid.

4. Let H be a simple group, and set $G = H \times H$. Thus, G is the totality of (x, y) $(x, y \in H)$. Show that, for any $\sigma \in \operatorname{Aut} H$, the set $M_\sigma = \{(x, x^\sigma)\ x \in H\}$ is a maximal subgroup of G.

(*Hint.* The product of M_σ and a direct factor of G coincides with G.)

5. Let H be the direct product of the cyclic group C_1 of order p and the cyclic group C_2 of order p^2. Let D be the set of elements of H of order at most p. Thus, D is the subgroup of H which is the direct product of C_1 and the subgroup E of order p contained in C_2. Let K be a group isomorphic to H, and let F be the subgroup of K which corresponds to D. For any ismorphism φ from D to F, we define the central product of H and K which is obtained by identifying D and F via φ. Show that this central product is isomorphic to $C_2 \times C_1 \times C_1$ or $C_2 \times C_2$ according to whether or not the generator of $\varphi(E)$ is a p-th power of some element of K.

6. Let G be a central product of two subgroups H and K, and let $D = H \cap K$. For a subgroup U of G, we write

$$H_U = \{h \,|\, h \in H \quad \text{such that} \quad hk \in U \quad \text{for some} \quad k \in K\},$$

and we define K_U similarly. Show that the following generalization of (4.19) holds.

(a) The set H_U is a subgroup of H which contains D. Similarly, K_U is a subgroup of K which contains D. We have $U \cap H \lhd H_U$ and $U \cap K \lhd K_U$.

(b) Let $h \in H_U$. Choose an element k of K such that $hk \in U$. Then, the mapping φ_U defined by $h(U \cap H) \to k(U \cap K)$ is an isomorphism from $H_U/(U \cap H)$ onto $K_U/(U \cap K)$ and satisfies $\varphi_U(d(U \cap H)) = d^{-1}(U \cap K)$ for all $d \in D$.

(c) Conversely, let H_1, H_2, K_1, and K_2 be four subgroups such that $H_i \subset H$, $K_i \subset K$, $D \subset H_1 \cap K_1$, $H_2 \lhd H_1$, and $K_2 \lhd K_1$. Suppose that there is an isomorphism θ from H_1/H_2 onto K_1/K_2 which satisfies $\theta(dH_2) = d^{-1}K_2$ for all $d \in D$. Then, there is a unique subgroup U of the central product HK such that

$$H_1 = H_U, \qquad H_2 = H \cap U, \qquad K_1 = K_U, \qquad K_2 = K \cap U, \qquad \theta = \varphi_U.$$

(In this case, $U/(U \cap K)$ is isomorphic to H_U/D, but the subgroup H_U need not be a homomorphic image of the subgroup U.)

7. Let $\Phi(G)$ be the Frattini subgroup of a group G. Prove that if G and H are finite groups, then we have

$$\Phi(G \times H) = \Phi(G) \times \Phi(H).$$

(*Hint.* Use (4.19) and Exercise 4. For a group G which has no maximal subgroup at all, we define $\Phi(G) = G$. If H is simple, Exercise 4 proves that $\Phi(H \times H) = \{1\}$. So, if there is a simple group H with no maximal subgroup, then we have

$$\Phi(H \times H) \neq \Phi(H) \times \Phi(H).$$

If both G and H are finitely generated, the equality still holds (see Exercise 5 of Chapter 1, §2).)

8. Let R_1 be the additive group of rational numbers $\mathbb{R}^+$, and let R_2 be the multiplicative group of nonzero rational numbers $\mathbb{R}^*$. Each element r of R_2 induces an automorphism $s \to sr$ of R_1; thus, R_2 acts on R_1. Let G be the semidirect product of R_1 and R_2 with respect to the above action, and consider R_1 as a subgroup of G. Show that R_1 is a minimal normal subgroup of G.

(*Hint.* See Example 5 of Chapter 1, §6. This exercise shows that Corollary 3 of (4.14) need not hold for a group without a composition series.)

§5. Fundamental Theorems on Abelian Groups

We will consider abelian groups in this section and will write them additively except for the operator groups which will be written as multiplicative groups. We will use additive terms, such as *direct sums* in place of direct products, *direct summands* for direct factors, etc. (Since the direct sum means the restricted direct sum, it is the coproduct in the category of abelian groups.)

First, we will prove the fundamental theorem on finitely generated abelian groups following Rado [1]. Let us begin with a lemma.

(5.1) *Let G be a finitely generated abelian group, and let $\{x_1, \ldots, x_n\}$ be a set of generators of G. Let*

$$y_1 = a_1 x_1 + a_2 x_2 + \cdots + a_n x_n$$

be an element of G where $a_i \in \mathbb{Z}$ and the greatest common divisor $(a_1, \ldots, a_n)$ is 1. Then there are $n - 1$ elements $y_2, y_3, \ldots, y_n$ of G such that

$$G = \langle y_1, y_2, \ldots, y_n \rangle.$$

Proof. Let $|a_i|$ denote the absolute value of the integer a_i, and let $m = |a_1| + \cdots + |a_n|$. We will prove (5.1) by induction on m. If $m = 1$, then $y_1 = \pm x_i$ and the lemma holds trivially. Assume $m > 1$. By assumption, we have $(a_1, \ldots, a_n) = 1$, so at least two coefficients a_i and a_j are nonzero. We assume that $|a_i| \geq |a_j| > 0$. Then we can find $e = \pm 1$ such that $|a_i - ea_j| < |a_i|$. The given element y_1 can be written

$$y_1 = a_1 x_1 + \cdots + (a_i - ea_j)x_i + \cdots + a_j(x_j + ex_i) + \cdots.$$

Thus, for $b_i = a_i - ea_j$, $b_k = a_k(k \neq i)$, $z_l = x_l(l \neq j)$, and $z_j = x_j + ex_i$, we have $G = \langle z_1, \ldots, z_n \rangle$, $y_1 = b_1 z_1 + \cdots + b_n z_n$, $(b_1, \ldots, b_n) = 1$, and

$$|b_1| + |b_2| + \cdots + |b_n| < m.$$

So, by the inductive hypothesis, we can find elements $y_2, \ldots, y_n$ of G such that $G = \langle y_1, y_2, \ldots, y_n \rangle$. $\square$

Theorem 5.2. *Any finitely generated abelian group G can be expressed as a direct sum of cyclic groups :*

$$G = E_1 + \cdots + E_m + I_1 + \cdots + I_r,$$

where $I_1, I_2, \ldots, I_r$ are infinite cyclic groups and E_i is a finite cyclic group of order $e_i > 1$ with the additional condition that e_i divides e_{i+1} for each $i = 1, 2, \ldots, m - 1$. Furthermore, the set of integers $(e_1, e_2, \ldots, e_m; r)$ is uniquely determined by G.

Proof. Choose a set of elements $\{x_1, \ldots, x_n\}$ of G satisfying the following conditions. First of all, we have $G = \langle x_1, x_2, \ldots, x_n \rangle$, and the number of elements, n, is minimal. Furthermore, if

$$G = \langle x_1, x_2, \ldots, x_{i-1}, y_i, y_{i+1}, \ldots, y_n \rangle$$

for some i $(1 \leq i \leq n)$ and some set of elements $y_i, y_{i+1}, \ldots, y_n$ of G, then the order of the element x_i is at most the order of y_i. It will be clear that we can choose such a set $\{x_i\}$; pick the element x_i as an element of the smallest order among all possible elements y_i.

Set $\langle x_i \rangle = E_i$ and $e_i = |E_i|$. (If some of the E_i's are infinite cyclic, then set $e_i = \infty$.) We will prove that $G = E_1 + \cdots + E_n$ and that e_i divides e_{i+1} for each i. Suppose that we have the following relation among $x_1, \ldots, x_n$:

$$a_i x_i + \cdots + a_j x_j = 0.$$

We may assume that the relation is written with the indices in increasing order and that for each index k appearing in the relation, we have

$$0 < |a_k| \le e_k.$$

Let d be the greatest common divisor of $a_i, \ldots, a_j$, and let $a_k = db_k$ for each k. Then, we get $(b_i, \ldots, b_j) = 1$. So, by (5.1), there are elements

$$y_i = b_i x_i + \cdots + b_j x_j, \ y_{i+1}, \ldots, y_n$$

such that

$$\langle x_i, x_{i+1}, \ldots, x_n \rangle = \langle y_i, y_{i+1}, \ldots, y_n \rangle.$$

The definition of y_i gives us $dy_i = 0$. Thus, the order of the element y_i is at most d. On the other hand, by the condition imposed on x_i, the order e_i of x_i is at most the order of y_i. Therefore, we have $e_i \le d \le |a_i| \le e_i$, so the equalities $e_i = d = |a_i|$ hold.

Since $|a_i| = e_i$, the relation $a_i x_i + \cdots + a_j x_j = 0$ implies that $a_i x_i = 0$. Thus, the condition (1) of Theorem 4.3 is satisfied (in the reverse order of indices); we have $G = E_1 + \cdots + E_n$.

Similarly, from the relation $e_i x_i + e_{i+1} x_{i+1} = 0$, we arrive at the equality $e_i = d = (e_i, e_{i+1})$ which gives us the additional condition that e_i divides e_{i+1}. Supposing e_m to be the last finite order, we write

$$G = E_1 + \cdots + E_m + I_1 + \cdots + I_r.$$

It remains to show that the set of integers $(e_1, \ldots, e_m; r)$ is unique.

The sequence of subgroups of G which is obtained by removing one direct summand at a time is a subnormal series for which the set of factors consists of cyclic groups. So, by applying Theorem 3.20 to the poly-cyclic group G, we get the invariance of r, the number of infinite cyclic factors.

Clearly, $E_1 + \cdots + E_m = T(G)$ is the set of elements of finite order. Suppose that $T(G) = F_1 + \cdots + F_l$ is a second decomposition into a direct sum of cyclic groups F_i where $f_i = |F_i|$ and f_i divides f_{i+1}. We need to show that $l = m$ and $f_i = e_i$ for all $i = 1, 2, \ldots, m$.

First of all, we have $f_l = e_m$, because they are the maximal values of the orders of the elements in G. Let u be a generator of the group F_l, and let $u = v_1 + \cdots + v_m$ be the decomposition in $E_1 + \cdots + E_m$, where $v_i \in E_i$. Then, at least one of the elements has order e_m. After changing the ordering of the factors if necessary, we may assume that v_m is of order e_m. Then, we have

$$\langle v_m \rangle = E_m \quad \text{and} \quad T(G) = \langle E_1 + \cdots + E_{m-1}, F_l \rangle.$$

Since $|F_l| = e_m$, from (3.11) of Chapter 1, we get

$$(E_1 + \cdots + E_{m-1}) \cap F_l = \{0\}.$$

Thus, $T(G) = (E_1 + \cdots + E_{m-1}) + F_l$. This shows that

$$T(G)/F_l \cong F_1 + \cdots + F_{l-1} \cong E_1 + \cdots + E_{m-1}.$$

By induction, we can prove that $m = l$ and $e_i = f_i$ for all i.   □

Let G be a finitely generated abelian group. The set of integers

$$(e_1, \ldots, e_m; r)$$

which is uniquely determined by G is called the *invariant* of G, and we write it as $\mathrm{inv}(G)$. The integer r is called the *rank* of G.

Corollary 1. *Let G and G' be two finitely generated abelian groups. Then, G is isomorphic to G' if and only if $\mathrm{inv}(G) = \mathrm{inv}(G')$.*

Corollary 2. *Let G be a finitely generated abelian group, and let $T(G)$ be the set of elements of finite order. Then $T(G)$ is a finite group and is a direct summand of G.*

Corollary 3. *Let p be a prime number. The number of isomorphism classes of abelian groups of order p^n is equal to the number of partitions $p(n)$, that is, the number of isomorphism classes is equal to the number of ways of expressing n as a sum of the following form:*

$$n = n_1 + n_2 + \cdots + n_m \qquad (1 \le n_1 \le n_2 \le \cdots \le n_m).$$

Proof. The first two corollaries follow easily from Theorem 5.2. We will prove Corollary 3. Let G be a finite abelian group of order p^n, and let $\mathrm{inv}(G) = (e_1, \ldots, e_m; r)$. Then, we have $r = 0$ and $e_1 e_2, \ldots, e_m = p^n$. Thus, each natural number e_i is a power of p; we write $e_i = p^{n_i}$. Clearly, we have

$$1 \le n_1 \le \cdots \le n_m \quad \text{and} \quad n = n_1 + \cdots + n_m.$$

By Corollary 1, each isomorphism class corresponds to a partition of n; different isomorphism classes correspond to different partitions.   □

(If the integer n becomes large, then the number of partitions $p(n)$ is very large; for example, $p(100) = 190, 569, 292$. If n is large, then $\log p(n)$ is approximately $\pi\sqrt{2n/3}$.)

(5.3) *Let $T(G)$ be the set of elements of finite order in an abelian group G. Then, $T(G)$ is a fully invariant subgroup of G, and the factor group $G/T(G)$ contains no element of finite order except the identity.*

Proof. If the orders of two elements x and y are finite, then, by (2.22) of Chapter 1, the order of $x + y$ is finite. Obviously, the inverse of x is also of finite order. So, $T(G)$ is a subgroup. If $nx = 0$, then, for any endomorphism φ of G, we have $n\varphi(x) = \varphi(nx) = 0$. Thus, $T(G)$ is fully invariant.

Consider the factor group $\bar{G} = G/T(G)$. If an element $\bar{x}$ of $\bar{G}$ is of finite order, then, for any element x of G contained in the coset $\bar{x}$, there is an integer n such that $nx \in T(G)$. Thus, the order of the element x is finite. We have $x \in T(G)$; so, $\bar{x} = 0$. $\square$

Definition 5.4. The subgroup $T(G)$ defined in (5.3) is called **the torsion subgroup** of G. An abelian group G is said to be **torsion free** if $T(G) = \{0\}$; G is called a **torsion group** if $T(G) = G$.

If G is a finitely generated abelian group, then $T(G)$ is a direct summand. But, $T(G)$ need not be a direct summand in general.

EXAMPLE. Let p_n be the n-th prime number (counting from $p_1 = 2$ in increasing order). Let C_n be the cyclic group of order p_n, and let G be the complete direct sum of the groups C_n. In other words, G is the totality of functions defined on natural numbers such that $f(n) \in C_n$. The operation in G is defined by the formula $(f + g)(n) = f(n) + g(n)$.

(a) $f \in T(G) \Leftrightarrow f(n) = 0$ *except for a finite number of values of n.*

Proof. Suppose that the order of f is m. Then, we have $mf = 0$; so, $mf(n) = 0$ for all n. But, $f(n) \in C_n$ and, for sufficiently large n, we have $p_n > m$. So, $f(n) = 0$ for sufficiently large values of n. Thus, there are only finitely many values of n satisfying $f(n) \neq 0$. The converse is easy to prove. $\square$

(b) *The factor group $G/T(G)$ is a **divisible** group; that is, for any element $\bar{a}$ and for any prime p, there is an element $\bar{x}$ such that $\bar{a} = p\bar{x}$.*

Proof. Let a be an element of G which lies in the coset $\bar{a}$. If $p_n \neq p$, then there is an element x_n of C_n such that $px_n = a(n)$. (This is easy to see. Since p_n is relatively prime to p, there are integers u and v such that $up_n + vp = 1$. Then, $a(n) = p(va(n))$ because $p_n a(n) = 0$.) Let x be the function defined by the formula $x(n) = x_n$ if $p_n \neq p$, and by $x(n) = 0$ if $p_n = p$. Then, px and a take the same value for all n such that $p_n \neq p$. Thus, $px - a \in T(G)$; this means $p\bar{x} = \bar{a}$ in $G/T(G)$. $\square$

(c) *The only subgroup of G which is divisible is $\{0\}$.*

Proof. Let H be a divisible subgroup of G, and let f be an element of H. Let n be an arbitrary integer. Since H is divisible, there is an element h of H such that $p_n h = f$. Then $f(n) = p_n h(n) = 0$ because $h(n) \in C_n$. Thus, we have $f = 0$, and $H = \{0\}$. $\square$

(d) *We have $G \neq T(G)$, and $T(G)$ is not a direct summand of G.*

Proof. By (a), $T(G)$ is the direct sum of the groups C_n. Since the index set is infinite, $T(G)$ is a proper subgroup. If $T(G) + H = G$, then H is isomorphic to $G/T(G)$. By (b), $G/T(G)$ is divisible; so, the subgroup H is also divisible. But, by (c), G contains no nontrivial divisible subgroup. This contradicts $G \neq T(G)$. $\square$

Theorem 5.5. *Let G be an abelian torsion group. Let p be a prime number, and let G_p be the set of elements whose orders are powers of p.*

(i) *G_p is a fully invariant subgroup of G.*
(ii) *The group G is the direct sum of G_p: $G = \sum G_p$ (as p ranges over all prime numbers).*
(iii) *If G is finite, G_p is the S_p-subgroup of G.*

Proof. (i) By (2.22) of Chapter 1, G_p is a subgroup. The proof of its full invariance is similar to the proof of (5.3).

(ii) The Cor. of Theorem 2.13 proves that every element is a product of elements whose orders are powers of primes. So, we have $G = \langle G_p \rangle$ (as p ranges over all prime numbers). If π is any set of prime numbers, then the set $G_\pi = \langle G_q \ (q \in \pi) \rangle$ contains only π-elements. So, if $p \notin \pi$, then we have $G_\pi \cap G_p = \{0\}$. This shows that the family $\{G_p\}$ is independent. By (4.12), G is the direct sum of the subgroups G_p.

(iii) If G is finite, each G_p is a p-group by (2.11). As shown above, we have $G = \sum G_p$. Then, $|G| = \prod |G_p|$ and $|G_p|$ is a p-component of $|G|$. Thus, G_p is the S_p-subgroup of G. $\square$

The subgroup G_p is called *the p-primary component* of the torsion abelian group G.

Free Abelian Groups

Definition 5.6. Let X be a set. An abelian group F is said to be **free on X** if the following two conditions are satisfied.

(1) There is a mapping i from X into F.
(2) If g is any mapping from X into an abelian group G, then there is a unique homomorphism f from F into G such that $g = f \circ i$.

The condition (2) means that the following diagram is commutative.

An abelian group F is called a **free abelian group** if F is free on some set X. In this case, we say that F is *generated freely* by iX; the set iX is a *set of free generators*. The number $|X|$ is called the **rank** of a free group F.

(5.7) *Let F be a free abelian group on a set X. Then the following propositions hold.*

   (i) $F = \langle iX \rangle$.
   (ii) *If an abelian group F' is also free on X, then F' is isomorphic to the group F.*
   (iii) *The mapping i is injective.*

Proof. (i) Let $H = \langle iX \rangle$ and apply the condition (2) of Definition 5.6 to the following diagram:

By (2), there is a homomorphism f from F into H such that $i = if$. Let the composite of f and the inclusion of H into F be f_0. Then, f_0 is a homomorphism from F into F such that $if_0 = i$. Clearly, the identity function 1 on F satisfies $i1 = i$. By the uniqueness, we have $f_0 = 1$. So, we have $H = F$.

   (ii) Since both F and F' are free on X, we have the following commutative diagram.

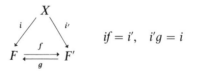

$$if = i', \quad i'g = i$$

Then, fg is the identity function on iX. So, by (i), $fg = 1$ on F. Similarly, we have $gf = 1$ on F'. Thus, f is the inverse of g, and vice versa; we have $F \cong F'$.

   (iii) If x and y are two distinct elements of X, then there is a function g from X into $\mathbb{Z}$ such that $g(x) = 1$ but $g(y) = 0$. The condition (2) gives us a homomorphism f from F into $\mathbb{Z}$ such that $g = if$. So, we get $i(x) \neq i(y)$, and the mapping i is injective. $\square$

Theorem 5.8. *For any given set X, there is an abelian group F which is free on X.*

Proof. Let F_X be the totality of functions φ defined on X such that $\varphi(x) \in \mathbb{Z}$ for all $x \in X$ and $\varphi(x) = 0$ except for finitely many elements of X. If f and g are two functions of F_X, we define the sum $f + g$ as usual: that is, $(f + g)(x) = f(x) + g(x)$. Clearly, F_X becomes an abelian group. For each x of X, let $i(x)$ be the function on X defined as follows:

$$i(x)(x) = 1, \qquad i(x)(y) = 0 \quad \text{if} \quad y \neq x.$$

Then, i is a mapping from X into F_X. We will show that, with the mapping i, F_X is free on X. We only need to verify the condition (2). So, let g be a mapping from the set X into an abelian group G. If $f \in F_X$, we have $f(x) = 0$ for all except a finite number of elements x of X. So, $\sum f(x)g(x)$ is defined as the summation ranges over all $x \in X$, and the function φ defined by

$$\varphi(f) = \sum f(x)g(x)$$

is a homomorphism from F_X into G such that $\varphi(i(x)) = g(x)$ for all x of X. Since any element f of F_X can be written in the form $f = \sum f(x)i(x)$, the above function φ is the unique one which satisfies $\varphi \circ i = g$. Thus, F_X is an abelian group which is free on X. $\square$

Theorem 5.9. *Let i be a mapping from a set X to an abelian group F. Then, F is free on X if and only if F is isomorphic to the direct sum of the groups C_x where C_x is the infinite cyclic group generated by $i(x)$.*

Proof. The proof of Theorem 5.8 shows that the direct sum of the groups C_x is free on X.

Conversely, suppose that F is free on X. For each $x \in X$, we take an infinite cyclic group C'_x, and we define their direct sum $F' = \sum C'_x$. If $i(x)$ is one of the generators of C'_x, then we have a mapping i' from X into F'. We have seen that F' is free on X. So, by (5.7) (ii) we get $F \cong F'$. Since $i'(x) \to i(x)$ can be extended to an isomorphism from F' onto F, the family $\{C_x\}$ is independent. Hence, we have $F \cong \sum C_x$. $\square$

Corollary 1. *An abelian group is free if and only if it is a direct sum of infinite cyclic groups.*

Corollary 2. *Let F be an abelian group which is free on X. Then, for any subset Y of X, $\langle iY \rangle$ is free on Y.*

Theorem 5.10. *Any abelian group is a homomorphic image of a free abelian group.*

Proof. Let G be an arbitrary abelian group. Choose a subset X of G such that $G = \langle X \rangle$. By Theorem 5.8, there is a free abelian group F which is free on X. Then, by the condition (2) of Definition 5.6, there is a homomorphism f from F into G such that $f(ix) = x$ for any element x of X. Since $G = \langle X \rangle$, the mapping f is surjective, and we have $f(F) = G \cong F/\mathrm{Ker}\, f$. $\quad\square$

Theorem 5.11. *Any subgroup of a free abelian group is free.*

Proof. Let F be the free abelian group on a set X. We may identify X with the subset $i(X)$ of F, and we will assume that $X \subset F$. Let U be a subgroup of F. We will prove that U is free.

   Let $\mathscr{A}$ be the totality of pairs (A, B) such that A is a subset of X and B is a set of free generators of $U \cap \langle A \rangle$. Since $\{0\}$ is free on the empty set, the set $\mathscr{A}$ contains $(\varnothing, \varnothing)$; A is not empty. We define an order in $\mathscr{A}$ by

$$(A, B) < (A', B') \Leftrightarrow A \subset A' \quad \text{and} \quad B \subset B'.$$

It is easy to verify that $\mathscr{A}$ satisfies the conditions of Zorn's lemma, so $\mathscr{A}$ has a maximal element. Let (A_0, B_0) be a maximal element of $\mathscr{A}$. If $A_0 = X$, then $U = U \cap \langle X \rangle$ is freely generated by B_0. So, U is free.

   Suppose that $A_0 \neq X$. Then, there is an element x of X not contained in A_0. Set $A_1 = \{A_0, x\}$. Since (A_0, B_0) is maximal, we have $U \cap \langle A_0 \rangle \neq U \cap \langle A_1 \rangle$. (If $U \cap \langle A_0 \rangle = U \cap \langle A_1 \rangle$, then (A_1, B_0) is an element of $\mathscr{A}$ and $(A_0, B_0) < (A_1, B_0)$, contradicting the maximality of (A_0, B_0).) Any element of $U \cap \langle A_1 \rangle$ is of the form $mx + c$ where $c \in \langle A_0 \rangle$. Choose an element b of $U \cap \langle A_1 \rangle$ with the smallest positive n: $b = nx + c$. Then, we have

$$\langle b \rangle \cap (U \cap \langle A_0 \rangle) = \{0\}.$$

Also, we have $U \cap \langle A_1 \rangle = \langle b, U \cap \langle A_0 \rangle \rangle$. This proves that

$$U \cap \langle A_1 \rangle = \langle b \rangle + (U \cap \langle A_0 \rangle),$$

and $(A_1, B_1) \in \mathscr{A}$ if $B_1 = \{b, B_0\}$. This contradicts the maximality of (A_0, B_0). $\quad\square$

Modules over a Ring

The concept of modules over a ring is a generalization of the concept of abelian groups.

Definition 5.12. Let R be a ring. A **module** M over R, or more simply, an **R-module**, is an abelian group M with the operator domain R such that, if the action of $r \in R$ on $m \in M$ is written mr, we have the formulas

$$m(r + s) = mr + ms, \qquad m(rs) = (mr)s$$

for any r and s of R.

An abelian group A is a module over the ring $\mathbb{Z}$; also, A is a module over the endomorphism ring $E(A)$. A vector space over a field K is a K-module such that $m1 = m$ for the identity 1 of multiplication.

Homomorphisms between R-modules are assumed to be operator homomorphisms. The following concept is essential in the subsequent development of the theory. Given two R-modules A and B, we consider the set of all R-homomorphisms from A to B, and we write it as $\mathrm{Hom}_R(A, B)$. If f and g are two R-homomorphisms, we define their sum as before:

$$(f + g)(a) = f(a) + g(a).$$

Then, $f + g$ is also an R-homomorphism from A to B. With respect to this operation, $\mathrm{Hom}_R(A, B)$ forms an abelian group. This is easily verified. The identity 0 is the homomorphism which sends every element to the zero of B; $-f$ is the R-homomorphism such that $(-f)(a) = -(f(a))$.

(5.13) *Let α be an R-homomorphism from an R-module A into another R-module A'. Then, α induces a homomorphism α^* from $\mathrm{Hom}_R(A', B)$ into $\mathrm{Hom}_R(A, B)$ as follows. If $f \in \mathrm{Hom}_R(A', B)$, then $\alpha^*(f)$ is the element of $\mathrm{Hom}_R(A, B)$ which maps an element a of R into*

$$\alpha^*(f)(a) = f(\alpha(a)).$$

If α and β are two R-homomorphisms from A into A', then we have $(\alpha + \beta)^* = \alpha^* + \beta^*$. If γ is an R-homomorphism from A' into another R-module A'', then we have $(\alpha\gamma)^* = \gamma^*\alpha^*$.

Proof. The function $\alpha^*(f)$ is defined as the composite of the two R-homomorphisms α and f. So, $\alpha^*(f)$ is an R-homomorphism from A into B. Therefore, $\alpha^*(f)$ is an element of $\mathrm{Hom}_R(A, B)$. The rest of (5.13) is also easy to verify. $\square$

If β is an R-homomorphism from B into another R-module B', then the R-homomorphism β_* from $\mathrm{Hom}_R(A, B)$ into $\mathrm{Hom}_R(A, B')$ is defined by the formula $\beta_*(f)(a) = \beta(f(a))$. We also have the rules $\beta_* + \beta'_* = (\beta + \beta')_*$ and $(\beta\gamma)_* = \beta_* \gamma_*$.

(5.14) *If an R-module A is the direct sum of the R-modules A_i ($i = 1, 2, \ldots, n$), then $\mathrm{Hom}_R(A, B)$ is the direct sum of $\mathrm{Hom}_R(A_i, B)$.*

Proof. We consider each factor A_i to be a submodule of A. If $f \in \mathrm{Hom}_R(A, B)$, the restriction f_i of f on A_i is an element of $\mathrm{Hom}_R(A_i, B)$. The function defined by $f \to (f_1, \ldots, f_n)$ is easily shown to be an isomorphism from $\mathrm{Hom}_R(A, B)$ onto the direct sum. $\square$

If the number of summands is infinite, $\mathrm{Hom}_R(A, B)$ is isomorphic to the complete direct sum of the family $\{\mathrm{Hom}_R(A_i, B)\}$.

Many theorems about abelian groups can be and have been generalized in various forms to theorems about R-modules. For example, the fundamental theorem (Theorem 5.2) has been generalized to the theorems about the structure of finitely generated modules over a principal ideal domain. This is a far more natural way to consider the problem, but we will refer the reader to standard textbooks on algebra for these and other generalizations. ([IK], Chapter 7, §§3, 4; [AS], II pp. 5–8). For the functor Hom, we will be content to state only the following results and will leave the rest to books on Homological Algebra (cf. [AS] Chapter 5).

Even when the image group is the multiplicative group $\mathbb{C}^* = \mathbb{C} - \{0\}$, we will still use additive notation for $\mathrm{Hom}_{\mathbb{Z}}(A, \mathbb{C}^*)$; thus, the operation in Hom is defined by $(f + g)(a) = f(a)g(a)$. Often, we simplify the notation by writing $A^* = \mathrm{Hom}_{\mathbb{Z}}(A, \mathbb{C}^*)$. If B is a subgroup of an additive group A, we define $B^\perp$ as the subgroup of A^* consisting of the elements which are trivial on B: $B^\perp = \{\varphi \in A^* \mid \varphi(B) = 1\}$.

Theorem 5.15. *Let B be a subgroup of an additive group A. Then, the following propositions hold.*

- (i) $B^\perp \cong (A/B)^*$.
- (ii) $A^*/B^\perp \cong B^*$.
- (iii) *If A is infinite cyclic, we have $A^* = \mathbb{C}^*$. If A is a finite cyclic group of order n, then $A^* \cong A$.*
- (iv) *If A^* is a torsion π-group, then so is A.*
- (v)   *If A is finite, then $A^* \cong A$.*

Proof. (i) The canonical homomorphism θ from A into A/B induces a homomorphism θ^* from $(A/B)^*$ into A^* (5.13). Clearly, Im θ^* is contained in $B^\perp$. If $\varphi \in B^\perp$, then Ker φ contains B. Then, φ induces a homomorphism $\bar{\varphi}$ from A/B into $\mathbb{C}^*$. Since $\theta^*(\bar{\varphi}) = \varphi$, Im θ^* coincides with $B^\perp$. This proves (i).

(ii) We will need a special property of $\mathbb{C}^*$, its injectivity, which will be stated in the next lemma. We will leave the general properties of injective modules to Exercise 6 at the end of this section.

Lemma. *Any homomorphism from B into $\mathbb{C}^*$ can be extended to an element of A^*; that is, for any $\varphi \in B^*$, there is an element ψ of A^* such that $\psi(b) = \varphi(b)$ for all $b \in B$.*

Proof. Let φ be an element of B^*. Consider the totality of pairs (H, η) such that H is a subgroup of A which contains B and η is an element of H^* which is an extension of φ. Let $\mathcal{H}$ be the set of all such pairs (H, η) equipped with the order defined by

$$(H, \eta) < (H', \eta') \Leftrightarrow H \subset H', \quad \eta' \text{ is an extension of } \eta.$$

Then, the set $\mathcal{H}$ satisfies the conditions of Zorn's lemma and has a maximal element (H_0, η_0). We will prove that $H_0 = A$. This will finish the proof of the lemma.

We will use a proof by contradiction, so suppose that $H_0 \neq A$. Choose an element $a \notin H_0$ and set $H_1 = \langle H_0, a \rangle$. If $H_0 \cap \langle a \rangle = \{0\}$, then the function η_1 defined by $\eta_1(h + na) = \eta_0(h)$ for any $h \in H_0$ and $n \in \mathbb{Z}$ would be an extension of η_0; this contradicts the maximality of (H_0, η_0). Therefore, we have $H_0 \cap \langle a \rangle \neq \{0\}$. Let m be the smallest positive integer such that $ma \in H_0$. For this integer m, let γ be one of the m-th roots of $\eta_0(ma)$. Then, $\eta_1(h + na) = \eta_0(h)\gamma^n$ defines a homomorphism from H_1 into $\mathbb{C}^*$, and η_1 is an extension of η_0. Again, we have a contradiction to the maximality of (H_0, η_0). So, we have $H_0 = A$. $\square$

Clearly, the restriction of the elements of A^* on B induces a homomorphism from A^* into B^*. The above lemma shows that the homomorphism is surjective. Since the kernel is $B^\perp$, we get the isomorphism $B^* \cong A^*/B^\perp$.

(iii) If $A = \mathbb{Z}$, then, for any $c \in \mathbb{C}^*$, the function f defined by $f(n) = c^n$ is an element of A^*. If $g \in A^*$, $g(1) = d$ is a nonzero complex number, and we have $(f + g)(n) = (cd)^n$. Thus, the mapping $f \to f(1)$ is an isomorphism from A^* onto $\mathbb{C}^*$. On the other hand, if A is the cyclic group $\mathbb{Z}/(n)$, then $f(\bar{1}) = c$ must satisfy $c^n = 1$. Thus, A^* is isomorphic to the group of n-th roots of unity in $\mathbb{C}$. Since the latter is clearly a cyclic group of order n, we have $A^* \cong A$.

(iv) If $b \in A$ is either of infinite order or a π'-element, then, for $B = \langle b \rangle$, (ii) shows that $B^* \cong A^*/B^\perp$. Thus, A^* contains either an element of infinite order, or a π'-element. This is not the case. So, A is a torsion π-group. (Even if A is torsion, A^* need not be a torsion group. For example, let A be the direct sum of the cyclic groups C_n of the Example following Definition 5.4. Then, A^* is the group G of the Example, and G is not a torsion group (cf. the remark following (5.14)).)

(v) Since a finite abelian group is a direct sum of cyclic groups, (5.14) and (iii) show that $A^* \cong A$. $\square$

Abelian Groups and Operator Groups

We will continue to use the additive notation for abelian groups, but we will write operator groups multiplicatively (even if the operator group is abelian). Furthermore, the action of an endomorphism σ is denoted by juxtaposition on the right, so the image of an element a is written as $a\sigma$. Thus, for $a, b \in A$ and $\sigma, \tau \in E(A)$, we have

$$(a + b)\sigma = a\sigma + b\sigma, \qquad a(\sigma + \tau) = a\sigma + a\tau$$

$$a(\sigma\tau) = (a\sigma)\tau.$$

(5.16) *Let m be a natural number. The following conditions on an additive group A are equivalent.*

(1) *The mapping $a \to ma$ is an automorphism of A.*
(2) *For any $a \in A$, there is a unique solution x of $mx = a$.*
(3) *There is an automorphism θ of A such that $(ma)\theta = a$ for all $a \in A$ and $\theta\sigma = \sigma\theta$ for all $\sigma \in E(A)$.*

Proof. The implications (1) $\Rightarrow$ (2) and (3) $\Rightarrow$ (1) are obvious. We will prove the remaining one: (2) $\Rightarrow$ (3). For each a of A, let $a\theta$ be the unique solution of $mx = a$. We have to show that $\theta \in \operatorname{Aut} A$. Let $b \in A$ and $my = b$. Then, by definition, we have $b\theta = y$, and $m(x + y) = mx + my = a + b$. This means that $(a + b)\theta = x + y = a\theta + b\theta$, so θ is a homomorphism. Clearly, θ is a bijection. This proves that θ is an automorphism of A. If σ is any endomorphism, then we have $m(x\sigma) = (mx)\sigma = a\sigma$, so $(a\sigma)\theta = x\sigma = (a\theta)\sigma$ for all $a \in A$. Thus, $\sigma\theta = \theta\sigma$, and (3) is proved. $\square$

An additive group A is said to be **m-regular** if A satisfies the conditions (1), (2), and (3).

Corollary. *An abelian torsion group A is m-regular if and only if the order of any element of A is relatively prime to m.*

Proof. Suppose that the order d of an element a of A is divisible by a prime divisor p of m. We can write $d = pn$. Then, we have $m(na) = 0$, so there is a nonzero solution x of $mx = 0$. Since (5.16) (2) is not satisfied, A is not m-regular.

Conversely, suppose that the order of any element of A is relatively prime to m. Then, for each $a \in A$, the mapping $a \to ma$ is an automorphism of the cyclic group $\langle a \rangle$ generated by a (see (6.10) of Chapter 1). Hence we have at least one solution x of $mx = a$. Suppose that we also have $my = a$. Then, we get $m(x - y) = 0$. Since m is relatively prime to the order of the element $x - y$, we have $x = y$. Thus, (5.16) (2) holds, and A is m-regular. $\square$

(5.17) *Suppose that a finite group Q of order m acts on an m-regular abelian group A. Then, $C_A(Q)$ is a direct summand of A.*

Proof. We denote by α_x the automorphism of A which is induced by the action of the element x of Q. We will prove that $A = C_A(Q) + A_0$ where

$A_0 = \langle a\alpha_x - a \rangle$ (as x ranges over Q and a ranges over A). Since A is m-regular, there is an automorphism θ satisfying the condition (3) of (5.16). Set $e = (\sum \alpha_x)\theta$ where the summation is over all $x \in Q$. Then, e lies in $E(A)$. For every $y \in Q$, we have

$$e\alpha_y = (\sum \alpha_x)\theta\alpha_y = (\sum \alpha_x \alpha_y)\theta = (\sum \alpha_{xy})\theta = e.$$

It follows that, for any $a \in A$, we get

$$ae^2 = ae(\sum \alpha_x)\theta = a(\sum e\alpha_x)\theta = a(me)\theta = ae.$$

Thus, we have $e^2 = e$. We will prove that $A = Ae + A(1 - e)$ (Exercise 3, §4). We have $a = ae + a(1 - e)$. So, it suffices to prove that

$$Ae \cap A(1 - e) = \{0\}.$$

If $a \in Ae \cap A(1 - e)$, then $a = ae$ because $a \in Ae$ and $e^2 = e$. On the other hand, $a \in A(1 - e)$ gives us $ae = 0$. So, we have $a = ae = 0$. This proves that $A = Ae + A(1 - e)$.

We will show that $C_G(Q) = Ae$ and $A_0 = A(1 - e)$. If $a \in C_G(Q)$, then $a\alpha_x = a$ for all $x \in Q$. Hence, we have

$$ae = a(\sum \alpha_x)\theta = (\sum a\alpha_x)\theta = (ma)\theta = a.$$

Conversely, suppose that $ae = a$. Then, for any $x \in Q$, we have

$$a\alpha_x = (ae)\alpha_x = a(e\alpha_x) = ae = a.$$

Thus, the element a belongs to $C_A(Q)$. We get $Ae = C_A(Q)$.
It follows from the definitions that

$$1 - e = \sum (1 - \alpha_x)\theta = \theta \sum (1 - \alpha_x).$$

This implies that $A(1 - e) \subset A_0$. Since $\alpha_x e = e$ for all x, we get

$$\alpha_x - 1 = \alpha_x(1 - e) - (1 - e),$$

which gives us $A_0 \subset A(1 - e)$. Thus, we have $A_0 = A(1 - e)$ and A is the direct sum of $C_G(Q)$ and A_0. $\square$

(5.18) *Suppose that a finite group Q of order m acts on an m-regular abelian group A. Suppose that a direct summand B of A is Q-invariant. Then, there is a Q-invariant subgroup C such that $A = B + C$.*

Proof. Since B is a direct summand of A, there is a subgroup U such that A is the direct sum of B and U: $A = B + U$. The subgroup U is not unique, so we will show that it is possible to choose a subgroup U which is Q-invariant and satisfies the condition $A = B + U$. As before, let α_x be the automorphism of A induced by the element x of Q. For any $u \in U$, we write

$$u\alpha_x = v + w, \qquad (v \in B, \quad w \in U).$$

If $x \in Q$ is fixed, the mapping $u \to v$ is a homomorphism from U into B; we write it as β_x. Similarly, let γ_x be the automorphism $u \to w$ of U. As $\alpha_{xy} = \alpha_x \alpha_y$, we get

$$u\alpha_{xy} = (u\alpha_x)\alpha_y = (u\beta_x + u\gamma_x)\alpha_y = u(\beta_x\alpha_y + \gamma_x\beta_y) + u(\gamma_x\gamma_y).$$

Thus, we have $\beta_{xy} = \beta_x\alpha_y + \gamma_x\beta_y$ and $\gamma_{xy} = \gamma_x\gamma_y$. Since γ_x is invertible, we get

(*) $$\beta_y = \gamma_x^{-1}\beta_{xy} - \gamma_x^{-1}\beta_x\alpha_y = \gamma_y(\gamma_{xy}^{-1}\beta_{xy}) - (\gamma_x^{-1}\beta_x)\alpha_y.$$

Since A is m-regular, we have the automorphism θ which satisfies the condition (3) of (5.16). The elements xy, as well as x, range over Q, so we can sum the above formula (*) over all $x \in Q$ and apply θ to get

$$\beta_y = \gamma_y\delta - \delta\alpha_y,$$

where $y \in Q$, $\delta = \theta \sum \gamma_x^{-1}\beta_x$, and the sum is over all $x \in Q$. The automorphism θ is the inverse of the mapping $x \to mx$, so θ leaves every subgroup invariant. Thus, δ is a homomorphism from U into B. Let $C = U(1 + \delta)$. Then, we have $U\delta \subset B$, $\langle B, C \rangle = A$, and $B \cap C = \{0\}$; thus, $A = B + C$. We will prove that C is Q-invariant. Since $\alpha_x = \beta_x + \gamma_x$ on U, we have

$$C\alpha_x = U(1 + \delta)\alpha_x = U(\alpha_x + \delta\alpha_x) = U(\beta_x + \gamma_x + \delta\alpha_x)$$
$$= U(\gamma_x + \gamma_x\delta) = U\gamma_x(1 + \delta) = U(1 + \delta) = C.$$

This shows that C is Q-invariant.   □

Definition 5.19. Suppose that a group Q acts on an additive group A. This action is said to be **reducible** if A contains a nontrivial Q-invariant subgroup; otherwise, the action is called **irreducible**. When the action is an isomorphism from Q into $E(A)$, the action is called **faithful**.

An operator group Q is faithful if and only if $C_Q(A) = \{1\}$; in general, the factor group $Q/C_Q(A)$ may be considered to be a faithful operator group of A.

The most important result concerning irreducible abelian groups is the following lemma of Schur.

Schur's Lemma. *Let an operator group Q act irreducibly on an additive group A. Then, the set of all Q-homomorphisms of A forms a skew-subfield of $E(A)$.*

Proof. Let E_Q be the totality of Q-endomorphisms of A. Then, E_Q is a subring of $E(A)$. Let σ be an element of E_Q. If α_x is the automorphism of A induced by the action of x in Q, we have $\alpha_x \sigma = \sigma \alpha_x$ for any $x \in Q$. Then, $A\sigma$ is Q-invariant. If $\sigma \neq 0$, then $A\sigma$ is a nonzero Q-subgroup of A. Since A is irreducible, we get $A = A\sigma$. Clearly, the kernel of σ is also Q-invariant. So, if $\sigma \neq 0$, Ker σ is $\{0\}$. This proves that σ is an automorphism of A. Since σ^{-1} is also a Q-automorphism, E_Q is a skew-field. □

(5.20) *Suppose that an additive group A has an operator group Q which acts irreducibly on A. Then, there is a field F such that A is isomorphic to a direct sum of groups, each of which is isomorphic to the additive group F^+ of the field F.*

Proof. Let D be the totality of Q-endomorphisms of A. By Schur's lemma, D is a skew-field. Let F be the center of D. It is easy to show that F is a subfield of D. Since A is a D-module, it is also an F-module. So, A has a basis $\{a_\lambda\}$ over $F: A = \sum a_\lambda F$. As an additive group, $a_\lambda F$ is a subgroup of A which is isomorphic to the additive group F^+ of the field F. This proves (5.20). □

Corollary. *Let A be an additive group which does not contain any nontrivial characteristic subgroup. Then, there is a field F such that A is isomorphic to a direct sum of groups, each of which is isomorphic to the additive group F^+ of the field F.*

Proof. The group Aut A acts on A irreducibly, so the conclusion follows from (5.20). □

(5.21) *Suppose that a finite abelian group Q acts irreducibly and faithfully on an additive group A. Then Q is cyclic.*

Proof. Since the action of Q is faithful, Q is contained in $E(A)$. Let E_Q be the set of all Q-endomorphisms of A. By Schur's lemma, E_Q is a skew-field. Since Q is abelian, every element of Q induces a Q-automorphism. Thus, Q is contained in the center of the skew-field E_Q. Since the center is a field, Q is a finite subgroup of the multiplicative group of the field. Hence, we can conclude from the remark following the Corollary of (5.6), Chapter 1, that Q is cyclic. □

Definition 5.22. Let p be a prime number. An abelian group E is said to be an **elementary abelian p-group** if every element x of E satisfies $x^p = 1$.

For any abelian p-group G, let $\Omega_1(G)$ be the set of all elements of G which satisfy $x^p = 1$. Clearly, $\Omega_1(G)$ is the set consisting of the elements of order p and the identity. Therefore, $\Omega_1(G)$ is elementary abelian and is a characteristic subgroup of G.

One of the most significant properties of an elementary abelian p-group E is (9.14) of Chapter 1. This theorem states that the group E is isomorphic to a direct sum of cyclic groups of order p. So, E is completely decomposable and satisfies the conditions of (4.14). Another important property of elementary abelian p-groups follows from (5.21).

(5.23) *Let E be a finite elementary abelian p-group which is not cyclic, and let $\mathfrak{M}$ be the set of all maximal subgroups of E: $\mathfrak{M} = \{E_0 || E: E_0| = p\}$. If E acts on a finite additive group A and if p is prime to the order $|A|$ of A, then we have $A = \langle C_A(E_0) \rangle$ where E_0 ranges over $\mathfrak{M}$.*

Proof. Since A is finite, A is a direct sum of directly indecomposable E-subgroups. So, it is sufficient to show that each summand is centralized by some member of $\mathfrak{M}$. Let B be an indecomposable summand. Then $\bar{E} = E/C_E(B)$ acts faithfully on B. By assumption, B is p-regular. So, by (5.17), $C_B(x)$ is a direct summand of B for each $x \in \bar{E}$. Since E is abelian, $C_B(x)$ is not only $\langle x \rangle$-invariant, but also E-invariant. It follows from (5.18) that $C_B(x)$ is a direct summand of B when it is considered to be an E-group. But, B is indecomposable, so $C_B(x) = \{0\}$ for all $x \neq 1$. This means that $\bar{E}$ acts faithfully on a minimal E-invariant subgroup B_0 of B. Since the action of $\bar{E}$ on B_0 is irreducible and faithful, $\bar{E}$ is cyclic by (5.21). This implies that $C_E(B)$ contains a maximal subgroup E_0 of E. Thus, we have $B \subset C_A(E_0)$ for some $E_0 \in \mathfrak{M}$. $\square$

Exercises

1. A subset $\{a_1, \ldots, a_n\}$ of an additive group A is said to be *free* if $\sum n_i a_i = 0$ with $n_i \in \mathbb{Z}$ implies that $n_i = 0$ for all i. A subset $\{a_\lambda\}$ is called free if any finite subset of $\{a_\lambda\}$ is free as defined above.

(a) Show that $\{a_\lambda\}$ ($\lambda \in \Lambda$) is free if and only if the subgroup $\langle a_\lambda \rangle$ (λ ranges over Λ) is freely generated by $\{a_\lambda\}$.

(b) Let $\mathscr{F}$ be the totality of free subsets of A. Show that $\mathscr{F}$ satisfies the conditions in Zorn's lemma (with respect to the natural inclusion). Also show that $\mathscr{F} = \varnothing$ if and only if A is a torsion group.

(c) Let M be a maximal free subset of A. Show that $F = \langle M \rangle$ is a free abelian group and that A/F is a torsion group.

(d) Let N be another maximal free subset of A. Show that the cardinality of N is equal to the cardinality of M.

(e) Let F be a free abelian group freely generated by a set $\{u_1, u_2, \ldots, u_n\}$. By using the theory of linear equations, prove that any set of $n + 1$ elements $a_1, a_2, \ldots, a_{n+1}$ of F is not free.

(*Hint.* (d) Consider two cases according to whether or not $|M|$ is finite. If $|M|$ is finite, use (c) and Theorem 5.2 to compare ranks of $\langle M \rangle$ and $\langle M \rangle \cap \langle N \rangle$. If the cardinality of M is infinite, it is equal to the cardinality of $\langle M \rangle$. (We will need some knowledge on cardinal numbers. See [IK], Chapter 1, §10; Theorem 23.) The cardinality of M is called the *rank* of A. By (e), the invariance of the rank in Theorem 5.2 can be proved without using the theory of poly-cyclic groups.)

2. Let F be a finitely generated free abelian group. Prove the following propositions. If a nontrivial subgroup U of F is given, there are elements $u_1,$ $\ldots, u_n$ and positive integers $e_1, e_2, \ldots, e_m$ ($m \leq n$) such that $\{u_1, \ldots, u_n\}$ is a set of free generators of F, $U = \langle e_1 u_1, \ldots, e_m u_m \rangle$, and e_i divides e_{i+1} for each $i = 1, 2, \ldots, m - 1$. Furthermore, the set of integers $(m; e_1, \ldots, e_m)$ is uniquely determined by the subgroup U.

(*Hint.* Imitate the proof of Theorem 5.2. Choose a set of free generators $\{u_1, u_2, \ldots, u_n\}$ and integers $m, e_1, \ldots, e_m$ which satisfy the following conditions. For each $k = 1, 2, \ldots, m$, $e_k u_k$ is an element of U. If there is a set of elements $v_i, v_{i+1}, \ldots, v_n$ such that $f_i v_i \in U$ for some positive integer f_i and

$$\langle u_1, \ldots, u_n \rangle = \langle u_1, \ldots, u_{i-1}, v_i, \ldots, v_n \rangle,$$

then $i \leq m$ and $e_i \leq f_i$.

Theorem 5.10 and Exercise 2 give an alternate proof of Theorem 5.2. We note that if a free group F is not finitely generated, then there is a subgroup for which the conclusion of Exercise 2 does not hold.)

3. Let $\{u_1, \ldots, u_n\}$ be a set of free generators of a free abelian group F. Take $a_{ij} \in \mathbb{Z}$, and let $v_i = \sum_j a_{ij} u_j$. Show that $\{v_1, \ldots, v_n\}$ is a set of free generators of F if and only if the determinant $|a_{ij}|$ is ± 1.

(From this exercise, we see that Aut F is isomorphic to the group of $n \times n$ matrices over $\mathbb{Z}$ having determinant ± 1.)

4. Let $A, A',$ and A'' be R-modules. A sequence

$$0 \to A' \xrightarrow{\alpha} A \xrightarrow{\beta} A'' \to 0$$

is said to be *exact* if the kernel of the outgoing arrow coincides with the image of the incoming arrow; that is, the above sequence is exact if α is injective, Im $\alpha = $ Ker β, and β is surjective. Show that if the above sequence is exact, then

$$0 \to \mathrm{Hom}_R(A'', B) \to \mathrm{Hom}_R(A, R) \to \mathrm{Hom}_R(A', R)$$

is exact. The second arrow is β^*, while the last one is α^*.

(*Hint.* Compare with Theorem 15.5. In general, the homomorphism α^* need not be surjective.)

5. Let an R-module A be the direct sum of a family of R-modules $\{A_\lambda\}$. Show that $\mathrm{Hom}_R(A, B)$ is the complete direct sum of the groups $\mathrm{Hom}_R(A_\lambda, B)$. Also show that

$$\mathrm{Hom}_R(A, B + B') \cong \mathrm{Hom}_R(A, B) + \mathrm{Hom}_R(A, B').$$

6. An abelian group J is said to be *injective* if the following condition is satisfied: for any exact sequence $0 \to M \to N$ and any homomorphism $\varphi: M \to J$, there is a homomorphism ψ from N into J such that $\varphi = \mu\psi$ where $\mu: M \to N$. Prove that the following propositions hold.

   (a)   An abelian group is injective if and only if it is divisible.
   (b)   If an injective group J is a subgroup of an abelian group A, then J is a direct summand.
   (c)   Let J be an injective abelian group. For any subgroup H of J, the factor group J/H is injective.
   (d)   Let A be a direct sum of subgroups, each of which is isomorphic to the additive group Q of rational numbers. Then, A is injective.
   (e)   Let F be a free abelian group. Then, there is an injective abelian group J which contains F.
   (f)   Let A be an abelian group. There is an injective abelian group J such that A is a subgroup of J.

(*Hint.* (a) The proof that a divisible group is injective is similar to the proof of the Lemma just after Theorem 5.15. For the converse, use the injection $n\mathbb{Z} \to \mathbb{Z}$ and the freeness of the subgroup $n\mathbb{Z}$. (c) Use (a). (f) Use Theorem 5.10 and (e).

Similarly, we can define an injective R-module, and we can prove the corresponding theorems ([IK], pp. 468–71; [AS], Chapter 6, §5). A direct sum of injective R-modules is also injective.)

7. For any additive group A, we write $nA = \{na \,|\, a \in A\}$. A subgroup P of the group A is said to be *pure* if $nP = P \cap nA$ holds for every positive integer n.

   (a) Show that any direct summand is a pure subgroup.

   (b) Show that the torsion subgroup $T(A)$ is pure. Thus, a pure subgroup need not be a direct summand.

   (c) Let P be a pure subgroup of an additive group A such that A/P is a direct sum of cyclic groups. Show that P is a direct summand.

   (d) Let a be an element of maximal order in a finite abelian p-group A. Show that $\langle a \rangle$ is pure and is a direct summand of A.

   (*Hint.* (c) Let C_λ/P be a cyclic direct summand of A/P. Show that there is a cyclic subgroup D_λ such that $C_\lambda = P + D_\lambda$. Furthermore, show that the family $\{P, D_\lambda(\lambda \in \Lambda)\}$ is independent. For a finite abelian group, Part (d) may be used to prove Theorem 5.2.)

8. Let B be a subgroup of an additive group A. Show that the set of subgroups C of A such that $B \cap C = \{0\}$ forms a partially ordered set which satisfies the conditions of Zorn's lemma. Let C be a maximal element of this set. We may write $D = \langle B, C \rangle$ and $G = A/D$.

(a) Show that $D = B + C$ and that G is a torsion group.

(b) For a prime number p, we define $\Omega_1(G_p) = \{\bar{x} \in G \mid p\bar{x} = 0\}$. Show that $\Omega_1(G_p)$ is isomorphic to a subgroup of B/pB.

(*Hint.* (b) Choose an element x of A such that $\bar{x} \in \Omega_1(G_p)$. Then, $px = b + c$ where $b \in B$ and $c \in C$. The mapping $\bar{x} \rightarrow b + pB$ is an isomorphism from $\Omega_1(G_p)$ into B/pB. Part (b) may be used to give an alternate proof of Exercise 6(b).)

9. Let p be a prime number, and let n be a natural number. Suppose that a subgroup B of an additive group A is a direct sum of cyclic groups of order p^n. Show that B is a direct summand of A if and only if $p^n A \cap B = \{0\}$.

(*Hint.* Suppose that $p^n A \cap B = \{0\}$. In the notation of the previous exercise, we may assume $p^n A \subset C$. If q is a prime number different from p, then we have $B = qB$. So, the group G of Exercise 8 is a p-group. If $G \neq \{0\}$, we get a contradiction to the isomorphism of Exercise 8(b).)

10. Let A be a finite abelian p-group. Define

$$A^i = \langle p^i a \mid a \in A \rangle, \quad E^i = \Omega_1(A^i) \quad \text{and} \quad \dim(E^{i-1}/E^i) = d_i$$

for each i.

(a) Show that, in any decomposition $A = \sum A_k$ where A_k are cyclic, there are exactly d_i cyclic groups of order p^i among the direct summands A_k.

(b) Prove the uniqueness of the invariants for finite p-groups using Part (a).

11. Let A be an abelian p-group. Then, A is a direct sum of cyclic groups if and only if there is a sequence $\{A_i\}$ $(i = 1, 2, \ldots,)$ of subgroups such that

(1) $\quad A_i \cap p^i A = \{0\}, \quad$ (2) $\quad A_i \subset A_{i+1}, \quad$ and $\quad$ (3) $\quad A = \bigcup A_i$.

Prove this theorem of Kulikov [1] following the steps (a) through (f).

(a) Suppose that A is the direct sum of the cyclic groups C_n: $A = \sum C_n$. Let A_i be the subgroup generated by all the summands C_k of order at most p^i. Show that the sequence $\{A_i\}$ satisfies the three conditions (1)–(3).

(b) Suppose that there is a sequence $\{A_i\}$ satisfying the three conditions. Let $E_{ij} = \Omega_1(A_i) \cap p^j(A)$. Show that we can choose a basis B_{ij} of E_{ij} which satisfies the property that

$$B_{ij} \cap E_{kl} = B_{kl} \quad \text{for all} \quad E_{kl} \subset E_{ij}.$$

(c) Let B be the union of all B_{ij}. Show that B is a basis of $\Omega_1(A)$.

(d) Each element of B is contained in E_{ij} for some values of i and j. For $b \in B$, let $i(b) = i$ and $j(b) = j$ be the indices of the minimal subgroup E_{ij}

which contains b (where E_{ij} is minimal in the sense of the inclusion relation). Show that there is an element c of A such that $b = p^j c$.

(e) For each $b \in B$, choose an element c of A which is determined as in (d). Show that the subgroup C generated by all those elements c is a direct sum of the cyclic groups $\langle c \rangle$.

(f) We will get a contradiction by supposing that $A \neq C$. (Pick an element x of minimal order in $A - C$. If the order of x is p^{k+1}, then $p^k x$ is a linear combination of the elements in the basis B, and each $b \in B$ which appears in the expression of $p^k x$ satisfies $j(b) \geq k$. So, there is an element y of C such that $p^k x = p^k y$. The element $x - y$ is an element of $A - C$ having order at most p^k. This contradicts the fact that we choose an element x of minimal order.)

12. (a) Suppose that an abelian p-group A is a direct sum of cyclic groups. Show that any subgroup B of A is also a direct sum of cyclic groups.

(b) Suppose that an abelian group A is a direct sum of cyclic groups. Show that any subgroup of A is a direct sum of cyclic groups.

(c) Suppose that the orders of the elements of the torsion subgroup $T(A)$ of an abelian group A are bounded. Show that $T(A)$ is a direct sum of cyclic groups and also that $T(A)$ is a direct summand of A.

(*Hint.* (a) Use Exercise 11. If $\{A_i\}$ is the sequence of subgroups satisfying the three conditions of Exercise 11, then the sequence $\{B_i\}$ defined by $B_i = B \cap A_i$ will satisfy the corresponding properties for B.

(b) If an abelian group A is a direct sum of cyclic groups, then the p-primary components of $T(A)$ are direct sums of cyclic groups, and so is the factor group $A/T(A)$. If U is any subgroup of A, then U satisfies the same conditions by Part (a) and Theorem 5.11. From this observation, we can see that U is a direct sum of cyclic subgroups (see Exercise 7(c)).

(c) That $T(A)$ is a direct sum of cyclic groups follows from Exercise 11. In order to prove that $T(A)$ is a direct summand of A, use Exercise 9 and use induction on the bound of the orders of the elements in $T(A)$. The propositions proved in this exercise are also theorems of Kulikov [1].)

§6. Generators and Relations

Suppose that a set X of generators of a group G is given (Definition 2.7 of Chapter 1). When a finite product $u_1 \cdots u_n$ of elements of X or their inverses ($u_i \in X$ or $u_i^{-1} \in X$) is equal to the identity of G, we say that the expression $u_1 \cdots u_n = 1$ is a **relation** among elements of X. To simplify the notation, we set $r = u_1 \cdots u_n$ and say that we have a relation $r = 1$. If there are several relations $r_1 = 1$, $r_2 = 1$, ..., it is convenient to express these relations collectively as R, and we say that we have the relations R.

When we deal with a relation such as $r = 1$, we should note the following point. The left side of the relation is a product of elements, each of which is an element of X or the inverse of an element of X. It is possible for the k-th element $u_k = a$ to belong to X, while at the same time, $b = a^{-1}$ also belongs to X. In this case, if we replace u_k by b^{-1}, we get the same element of G; thus, we get a relation which is essentially the same relation as the old one in G. But we should consider these two as different relations. A simple example, $xx = 1$, will illustrate the point. This relation implies that $x^{-1} = x$ in G. But, if we replace x by x^{-1}, we get the trivial relation which is certainly different from the original relation. Thus, in the relation $u_1 \cdots u_n = 1$, *we must have a definite rule for each k about whether the k-th element u_k is an element of X or the inverse of an element of X.*

Let X be a set of generators of a group G, and let R be a set of relations among elements of X. Suppose that we have a group H and a function φ from X into H. We extend φ on the set X^{-1} by defining $\varphi(x^{-1}) = \varphi(x)^{-1}$ for all $x \in X$. If

$$r = u_1 \cdots u_n = 1$$

is a relation in R, we have a definite rule as to whether u_k is an element of X or an element of X^{-1}. Thus, $\varphi(u_k)$ is determined accordingly, and so is $\varphi(r) = \varphi(u_1) \cdots \varphi(u_n)$. We will say that the function φ **preserves the relations** R if $\varphi(r) = 1$ for all relations $r = 1$ in R.

Definition 6.1. Let G be a group, let X be a set of generators, and let R be a set of relations among elements of X. The group G is said to be **defined** by the subset X and the relations R if the following condition is satisfied.

If a function φ from X into any group H preserves the relations R, then φ can be extended to a homomorphism from G into H; that is, there is a homomorphism f from G into H such that $f(x) = \varphi(x)$ for all $x \in X$.

Since X generates G, the extension f is clearly unique.

EXAMPLE 1. As we have seen in Example 1, §2, a nonabelian group of order pq (where p and q are prime) is defined by a suitable set of two elements x and y and the relations

$$x^p = y^q = 1, \qquad y^{-1}xy = x^r \qquad (r^q \equiv 1 \not\equiv r(\mathrm{mod}\ p)).$$

When a group G is defined by the subset X and the relations R, we write $G = \langle X \,|\, R \rangle$, and call it a **presentation** of G; R is called the **defining relations**.

Among the most fundamental problems in this field, we will single out two for discussion. Does any group G have a presentation, and, for any given set X and the relations R among elements of X, is there a group G for which $\langle X \,|\, R \rangle$ becomes a presentation? We will solve these problems in this section.

Definition 6.2. Let X be a set. A group F is said to be **free on** X, if the following conditions are satisfied:

(1) There is a function i from X into F.
(2) For any function g from X into an arbitrary group G, there is a unique homomorphism f from F into G such that $if = g$.

The function i of Condition (1) is called the *canonical mapping*. A group F is called a **free group** if it is free on a set X; the cardinality $|X|$ is called the *rank* of F.

The free group F which is free on X or, to put it more simply, the free group on X, is a group in which a set of free generators are determined by the canonical mapping. Let F' be the derived group of F (cf. Definition 3.11). Then *the factor group* $\bar{F} = F/F'$ *is an abelian group which is free on* X. This can be seen as follows. Let g be a mapping from X into an abelian group A. Then, as F is free on X, g can be extended to a homomorphism φ from F into A. But, A is abelian, so by Theorem 3.13, Ker φ contains the derived group F'. By the Cor. of Theorem 5.3, Chapter 1, φ induces a homomorphism f from $\bar{F}$ into A. It is clear that f is unique, so $\bar{F}$ is free on X.

Proposition (5.7) can be generalized to nonabelian free groups, and the proof is almost identical to the one given there. Any two free groups with the same rank are isomorphic. Conversely, isomorphic free groups have the same rank. This may be proved by using the fact that $\bar{F}$ is the free abelian group of rank $|X|$. An alternate proof may be given as follows. Let p be a fixed prime number, and consider the vector spaces over the field of p elements. A homomorphism φ from a free group F into a vector space V is completely described by the vectors $\varphi(iX)$. Thus, if dim $V > |X|$, then φ cannot be surjective, while if dim $V \leq |X|$, there is a surjective homomorphism from F onto V. The property of having or not having a surjective homomorphism is a group theoretical property. So, the isomorphic free groups have the same rank.

The canonical mapping i is one-to-one. This, together with Definition 6.1, proves the following proposition.

(6.3) *Let F be the free group on X, and let i be the canonical mapping. Then, F is defined by the subset iX and the empty set of relations. Conversely, a group defined by a set X and the empty set of relations is free on X.*

Theorem 6.4. *For any set X, there is a free group which is free on X.*

Proof. We will follow the method of [AN] (pp. 24–25), adapting it for free groups. Let Y be a set such that its cardinality is $|X|$ and $X \cap Y = \varnothing$. Let $X = \{x_\alpha \,|\, \alpha \in A\}$. Since there is a bijection between X and Y, let y_α denote the element of Y corresponding to an element x_α of X. Consider the totality of finite sequences $(u_1, u_2, \ldots, u_n)$ of elements of $X \cup Y$. If

$$P = (u_1, \ldots, u_n) \quad \text{and} \quad Q = (v_1, \ldots, v_m)$$

are two such sequences, the combination $(u_1, \ldots, u_n, v_1, \ldots, v_m)$ is another sequence of elements of $X \cup Y$; we write

$$(P, Q) = (u_1, \ldots, u_n, v_1, \ldots, v_m).$$

If $n = 0$, we still consider ( ) as a sequence, and write $1 = ( )$.

We will define the reduction of sequences. If $P = (u_1, \ldots, u_n)$ contains x_α and y_α as adjacent terms (that is, if P contains x_α, y_α or y_α, x_α), then we may remove these two terms to get a shorter sequence. This process is called a *reduction*. A sequence which admits no reduction is called *irreducible*. A reduction makes the number of terms smaller, so any sequence can be reduced to an irreducible one after a finite number of reductions. The important fact here is that a given sequence P is reduced to a unique irreducible sequence $\rho(P)$ no matter how the reductions are performed. We will prove the uniqueness by induction on the length n of the sequence $P = (u_1, \ldots, u_n)$.

Suppose that we get an irreducible sequence $\rho_1(P)$ following a series of reductions and get another irreducible one $\rho_2(P)$ after another series of reductions. We will show that $\rho_1(P) = \rho_2(P)$. Let P_k be the sequence obtained from P by removing u_k and u_{k+1} ($k = 1, 2, \ldots, n - 1$). Suppose that the first step of the first series of reductions reduces P to P_i and that the first step of the second series reduces P to P_j. If $i = j$, then we get the same sequence P_i after the first step. By the inductive hypothesis, we have $\rho_1(P) = \rho_2(P)$. If $i \neq j$, we may assume that $i < j$. If $j = i + 1$, then the three consecutive terms (u_i, u_{i+1}, u_{i+2}) are either $(x_\alpha, y_\alpha, x_\alpha)$ or $(y_\alpha, x_\alpha, y_\alpha)$. In this case, we clearly have $P_i = P_{i+1}$. Thus, we have $\rho_1(P) = \rho_2(P)$ as before. Suppose $j > i + 1$. Let P_{ij} be the sequence obtained from P by removing u_i, u_{i+1}, u_j, and u_{j+1}. Then, P_{ij} can be obtained from P_i by removing u_j and u_{j+1}. By the inductive hypothesis, the reduction of P_i gives a unique irreducible sequence; thus, we get $\rho_1(P) = \rho_1(P_i) = \rho(P_{ij})$. Similarly, P_j is reduced to P_{ij}, and we have $\rho_2(P) = \rho_2(P_j) = \rho(P_{ij})$. This proves that $\rho_1(P) = \rho_2(P)$ in all cases.

Let F be the totality of irreducible sequences. For two elements P and Q of F, we define the product of P and Q as the reduced sequence $\rho(P, Q)$; we denote the product by juxtaposing P and Q: $PQ = \rho(P, Q)$. We will prove that the set F forms a group with respect to this operation.

First of all, the operation satisfies the associative law. By definition, we have

$$(PQ)R = \rho(\rho(P, Q), R) \quad \text{and} \quad P(QR) = \rho(P, \rho(Q, R)).$$

Both are obtained by reducing the sequence (P, Q, R). As we have seen, the reduction process gives a unique irreducible sequence $\rho(P, Q, R)$. Thus, we have $(PQ)R = P(QR)$. The empty sequence $1 = ( )$ is the identity. If the sequence (x_α) consists of a single term, its inverse is (y_α). In general, the inverse of a sequence P is obtained by reversing the order and exchanging x and y. For example, the inverse of (x_1, y_2, x_1, x_2) is (y_2, y_1, x_2, y_1).

We will prove that F is free on X. For each $x_\alpha \in X$, let $i(x_\alpha)$ be the sequence (x_α). (Since (x_α) is irreducible, it is an element of F.) Suppose that G is a group and that g is a function from X into G. We extend g on Y by defining $g(y_\alpha) = g(x_\alpha)^{-1}$. For a finite sequence $P = (u_1, \ldots, u_n)$ of elements $u_i \in X \cup Y$, we let

$$f(P) = g(u_1)g(u_2) \cdots g(u_n).$$

We will show that $f(P) = f(\rho(P))$ where $\rho(P)$ is the irreducible sequence obtained by a finite number of reductions. Each reduction removes an adjacent pair (x_α, y_α) or (y_α, x_α). Since $g(y_\alpha) = g(x_\alpha)^{-1}$, the removal of the pair does not affect f as an element of G, and we get $f(P) = f(\rho(P))$. The restriction of f on F defines a function from F into G. By the definition of f, the formula

$$f(P)f(Q) = f((P, Q))$$

holds. So, if P and Q are elements of F, we have

$$f(P)f(Q) = f((P, Q)) = f(\rho(P, Q)) = f(PQ).$$

Therefore, f is a homomorphism from F into G which satisfies $if = g$ on X. Since

$$P = (u_1, \ldots, u_n) = (u_1) \cdots (u_n),$$

f is uniquely determined by the formula $if = g$. Thus, F is free on X. $\square$

The group which is free on X is uniquely determined; that is, all groups which are free on X are isomorphic to each other ((5.7) (ii)). The group F constructed above is the group which is free on X. By definition, an element v of F is an irreducible sequence $(u_1, \ldots, u_n)$ where $u_i \in X$ or $u_i^{-1} \in X$. Thus, we have

$$v = (u_1, \ldots, u_n) = (u_1) \cdots (u_n) = i(u_1) \cdots i(u_n).$$

(If $u_i^{-1} \in X$, then we define $i(u_i) = i(u_i^{-1})^{-1}$.) In the above formula, the element v is written as a product of elements of the form ix_α or $(ix_\alpha)^{-1}$. Furthermore, no adjacent pair is ix and $(ix)^{-1}$ for any $x \in X$. The expression of v satisfying the above properties is called the **standard form** of v. For any choices of $u_1, \ldots, u_n$, the product $(iu_1) \cdots (iu_n)$ is an element of F. And, the standard form of this product is uniquely determined; this is the essence of the preceding proof of Theorem 6.4.

We often identify the elements of X with their images under the canonical mapping. So, we consider X to be a subset of the free group F. Then, the standard form of an element v of F is $v = u_1 \cdots u_n$. The integer n is called the *length* of v, and we denote it by $l(v)$. We have $l(v) \geq 0$, and we note that $l(v) = 0$ if and only if $v = 1$. Furthermore, the formula

$$l(vv') \leq l(v) + l(v')$$

holds for any elements v and v' of F.

The standard form of an element of the free group F is uniquely determined only after the canonical mapping i designates the set of free generators of F. So, the lengths of the elements are also defined only after fixing the set of free generators.

Theorem 6.5. *Any group is a homomorphic image of a free group.*

Proof. Let G be a group, and let X be a set of generators of G. By Theorem 6.4, there is a free group F which is free on X. By applying the condition (2) of Definition 6.2 for the injection of X into G, we see that there is a homomorphism f from F into G such that $f(ix) = x$. Since $\langle X \rangle = G$, the homomorphism f is surjective. Hence, we have $G = f(F)$. $\square$

The notion of a group which is free on a given set can be used to restate the concept of relations among generators in terms of the free group. Let G be a group. Suppose that a set X of generators of G is given. We will consider the set X to be fixed during the following discussion. Let F be the free group on the set X, and let i be the canonical mapping from X into F. Then, there is a unique homomorphism from F onto G such that $f(ix) = x$ for all x in X.

Suppose that $r = 1$ is a relation among the generators in G. Then, r is expressed as a product $u_1 u_2 \cdots u_n$ of elements u_i of $X \cup X^{-1}$, and, for each k, it is definitely determined whether u_k is an element of X or an element of X^{-1}. So far, the canonical mapping i is defined only on X. We extend the definition of i by setting $i(x^{-1}) = (ix)^{-1}$ for all $x \in X$. Then, since we know whether the k-th element u_k is an element of X or the inverse of an element of X, the element iu_k is defined without ambiguity. So, the element ir defined by $(iu_1)(iu_2) \cdots (iu_n)$ is uniquely determined as an element of F. The homomorphism f from F onto G satisfies $f(ix) = x$ and $f(ix^{-1}) = x^{-1}$ for all $x \in X$.

Therefore, we have $f(ir) = u_1 u_2 \cdots u_n = r = 1$, and the element ir lies in the kernel of f.

Conversely, let v be an element of $\mathrm{Ker}\, f$, and let

$$v = (iu_i) \cdots (iu_n)$$

be the standard form of v with respect to the set of free generators iX. Since $f(v) = 1$, we get $f(v) = u_1 u_2 \cdots u_n = 1$ in G. The uniqueness of the standard form of v shows that, for each k, it is definitely known whether u_k is an element of X or the inverse of an element of X. Then, $r = u_1 u_2 \cdots u_n = 1$ is a relation among the generators in X. This relation is reduced in the sense that no neighboring pair consists of an element and its inverse. The preceding discussion shows that the reduced relations among the elements of X and the elements of the free group F which lie in $\mathrm{Ker}\, f$ are in a one-to-one correspondence.

So far in our discussion of relations, we have assumed that the set X of generators is a subset of a group G. Therefore, any relation among the generators is actually presented as a product in G. It is convenient to remove this restriction so that we can consider any given set and the relations among its elements. Let F be the free group on X. Any element of F is a product of elements in $iX \cup iX^{-1}$. Thus, if $v \in F$, we can write

$$v = (iu_1)(iu_2) \cdots (iu_n)$$

where $u_i \in X \cup X^{-1}$. In this expression, we can replace ix and ix^{-1} by x and x^{-1}, respectively, by simply dropping the i from each element. The expressions obtained in this way are called formal products of elements in X; they are also called the **words** with respect to a given set X. A *relation* among the elements of X is a formula $w = 1$, where w is a word with respect to X.

We will now consider the problems posed early in this section. Suppose that a set X and a collection R of relations among the elements of X are given. The problem is to determine whether or not there is a group G which has the presentation $\langle X \mid R \rangle$. Strictly speaking, if $G = \langle X \mid R \rangle$, then G is the most general among those groups which are generated by the set X such that the elements of X satisfy the given relations R. In this strict sense, the group $G = \langle X \mid R \rangle$ need not exist. For example, let $X = \{x, y\}$ be a set of two distinct elements, and let R be the set consisting of one relation: $xy^{-1} = 1$. In any group which contains the set X, the relation R does not hold. However, the following theorem does hold.

Before we state the theorem, we need to introduce some new notations. Suppose that a set X and a collection R of relations $r_\alpha = 1$ ($\alpha \in A$) are given. Let F be the free group on X, and let i be the canonical mapping from X into F. By definition, each word r_α comes from an element ir_α of F by dropping the mapping i. Let K be the smallest normal subgroup of F which contains all the elements $ir_\alpha (\alpha \in A)$. Furthermore, let

$$G = F/K, \qquad \theta X = \{\theta x \mid x \in X\}, \qquad \theta R = \{\theta r_\alpha = 1 \ (\alpha \in A)\}$$

where $\theta x = K(ix)$ is the element of G containing ix and θr_α are obtained from r_α by replacing every $x \in X$ by θx and every x^{-1} ($x \in X$) by $(\theta x)^{-1}$, and by designating the latter term as an element of $(\theta X)^{-1}$.

Theorem 6.6. *Under the notations introduced in the preceding paragraph, θR is a set of relations among the generators in θX, and we have $G = \langle \theta X \mid \theta R \rangle$.*

Proof. Since $F = \langle iX \rangle$ and since θ is the composite of the canonical mapping i and the canonical homomorphism from F onto G, we have $G = \langle \theta X \rangle$. According to the definition, the normal subgroup K contains all the elements ir_α. Hence, we have $K(ir_\alpha) = K$. So, by definition, $\theta r_\alpha = 1$ in G. Each θr_α is written as a product of elements of the form θx or $(\theta x)^{-1}$, and the latter is designated to lie in $(\theta X)^{-1}$. Thus, $\theta r_\alpha = 1$ is a relation for each $\alpha \in A$.

We will show that $G = \langle \theta X \,|\, \theta R \rangle$. Let φ be a function from θX into a group H which preserves the relations θR. Since F is free on X, we can apply the condition (2) of Definition 6.2 to the function $\theta \varphi$ from X into H to show that there is a homomorphism f from F into H such that $f(ix) = \varphi(\theta x)$ for all $x \in X$. Since f is a homomorphism, we have

$$f((ix)^{-1}) = f(ix)^{-1}.$$

On the other hand, φ is extended on θX^{-1} by the formula

$$\varphi(\theta x^{-1}) = \varphi(\theta x)^{-1}.$$

Thus, $f(ix^{-1}) = \varphi(\theta x)^{-1} = \varphi(\theta x^{-1})$, and, for any relation $\theta r_\alpha = 1$, we have $f(ir_\alpha) = \varphi(\theta r_\alpha)$. By assumption, φ preserves the relations θR. Hence, we get $\varphi(\theta r_\alpha) = 1$, which gives us the relation $ir_\alpha \in \mathrm{Ker}\, f$. The definition of K shows that $K \subset \mathrm{Ker}\, f$. Hence, f induces a homomorphism g from $G = F/K$ into H (see the Cor. of Theorem 5.3, Chapter 1) such that

$$g(\theta x) = f(ix) = \varphi(\theta x).$$

Thus, φ can be extended to a homomorphism from G into H. We conclude that $G = \langle \theta X \,|\, \theta R \rangle$. $\square$

Corollary 1. *Let X be a set of generators of a group G. Then, there is a set R of relations among elements of X such that $G = \langle X \,|\, R \rangle$.*

Corollary 2. *Let X be a set of generators of a group G. Let F be the free group on X, and let i be the canonical mapping from X to F. Let f be the homomorphism from F onto G such that $f(ix) = x$. Let R be a set of relations $r_\alpha = 1$ $(\alpha \in A)$. Then, $G = \langle X \,|\, R \rangle$ if and only if $\mathrm{Ker}\, f$ is the smallest normal subgroup of F which contains all the elements ir_α of F.*

Proof. The homomorphism f induces an isomorphism from $F/\mathrm{Ker}\, f$ onto G. This isomorphism maps $(\mathrm{Ker}\, f)ix$ onto x. If $R = \{ir_\alpha\}$ is a set of generators of $\mathrm{Ker}\, f$, $\mathrm{Ker}\, f$ coincides with the smallest normal subgroup K which contains all the elements of the form ir_α $(\alpha \in A)$, and Theorem 6.6 gives us $G = \langle X \,|\, R \rangle$ (as $\theta x \to x$ and $\theta r_\alpha \to r_\alpha$ under the isomorphism). This proves Corollary 1.

Conversely, suppose that $G = \langle X \,|\, R \rangle$. Since R is a set of relations in G, we have $K \subset \mathrm{Ker}\, f$. Let φ be the function from X into the factor group $H = F/K$

defined by $\varphi(x) = Kix$. For any relation $r_\alpha = 1$ of R, we have $\varphi(r_\alpha) = Kir_\alpha = K$. Thus, φ preserves the relations R. By the definition of $G = \langle X \mid R \rangle$, φ can be extended to a homomorphism g from G into H. If $v \in \mathrm{Ker}\, f$, then we can write $v = ir$ where $r = 1$ is a relation in G. Let $r = u_1 u_2 \cdots u_n$ where $u_i \in X \cup X^{-1}$. Then, we have

$$K = g(r) = g(u_1)g(u_2) \cdots g(u_n)$$
$$= \varphi(u_1)\varphi(u_2) \cdots \varphi(u_n) = Kir.$$

This is equivalent to saying that $ir = v \in K$. So, $\mathrm{Ker}\, f \subset K$ and, combined with the previous containment $K \subset \mathrm{Ker}\, f$, we have $K = \mathrm{Ker}\, f$. $\square$

The mapping θ which appears in Theorem 6.6 is not necessarily an injection. It is known that, for arbitrarily given X and R, there is no algorithm to decide whether the mapping θ is an injection or not. This is the word problem for groups. We refer the reader to the book by Takeuti [1] on this subject. For the problem of how to find the defining relations of a concrete group, we refer the reader to Chapter 3, §2 where the cases of the symmetric and alternating groups are treated.

EXAMPLE 2. Example 4, §2 may be regarded as dealing with the problem of finding generators and the defining relations of the group $G = PSL(2, 7)$. The parts (a) and (b) of Example 4, §2 prove that there are elements x and y which satisfy the relations $x^7 = y^3 = 1$ and $y^{-1}xy = x^2$, while part (c) determines the existence of an element t satisfying $t^2 = 1$ and $t^{-1}yt = y^{-1}$. Part (d) proves that $G = \langle x, y, t \rangle$ and that if x is suitably chosen, the relation $txt = x^{-1}tx^{-1}$ holds. The last part (f) shows that the above relations are the defining relations of G. The last relation is equivalent to $(xt)^3 = 1$.

When a set X and a collection R of relations are given, it is customary to write $G = \langle X \mid R \rangle$ when G is the group discussed in Theorem 6.6. The group G is, in fact, generated by θX, and the mapping θ is not necessarily an injection. So, a misuse of the notation is involved here. But, it is much more convenient to use this notation, although a little caution must be exercised.

(6.7) *Let $\langle X \mid R \rangle$ be a presentation of a group G. If R' is another set of relations, then the group $H = \langle X \mid R \cup R' \rangle$ is a homomorphic image of G. Thus, if we add more relations, we get a presentation of a factor group.*

Proof. Let θ be the function from X into G given by Theorem 6.6. Similarly, let θ' be the function from X into H. Then, the function θ' preserves the relations R. Hence, by the Definition of $\langle X \mid R \rangle$, there is a homomorphism f from G into H such that $f(\theta(x)) = \theta'(x)$ for all $x \in X$. This homomorphism, is clearly surjective, and so $f(G) = H$. $\square$

Generally speaking, it is difficult to study the structure of a group G from its presentation $\langle X \mid R \rangle$. There is no algorithm to decide when $\langle X \mid R \rangle$ is nontrivial. Only a few scattered results have been obtained for various specific presentations. The following theorem of Coxeter is one of the most remarkable results in this direction.

Let l, m, and n be three natural numbers greater than 1. Define $G(l, m, n)$ to be the group given by the following presentation:

$$X = \{x, y, z\} \quad \text{and} \quad R = \{x^l = y^m = z^n = xyz = 1\}.$$

Then $G(l, m, n)$ is finite if and only if

$$2/g = (1/l) + (1/m) + (1/n) - 1 > 0,$$

where g is the order of $G(l, m, n)$. If $1 < l \leq m \leq n$, then the above condition is satisfied only in the following cases:

(i) $l = m = 2$ and n is arbitrary integer ≥ 2.
(ii) $l = 2, m = 3$, and $3 \leq n \leq 5$.

For the proof of this theorem, and for many more results about the presentations of specific groups, the readers are referred to Coxeter-Moser [1]. G. Higman [1] contains an outline of a geometric proof of the above theorem (see also Exercise 2, Chapter 3 §4). By the way, in Case ii, not only the orders of the groups ($g = 12$, 24, and 60) but also their structures are determined (see Example 3, Exercise 1, and Example 4 in this section). The structure of the group in Case i is completely described by the following theorem which has many applications in group theory; the last assertion is particularly important.

Theorem 6.8. *Let D be a group generated by two elements of order 2. Then, D has the following presentation:*

$$D = \langle x, y \mid x^2 = y^2 = (xy)^n = 1 \rangle,$$

where n is either a natural number, or $n = \infty$. In the latter case, the relation $(xy)^n = 1$ will be omitted. If n is finite, the group D is isomorphic to the dihedral group of order $2n$. Let $z = xy$ and $D_0 = \langle z \rangle$. Then, we have

$$x^{-1}zx = y^{-1}zy = z^{-1},$$

and D_0 is a normal subgroup of index 2. Furthermore, if n is an odd integer, then x and y are conjugate in D.

Proof. Since $z = xy$, $x^{-1} = x$, and $y^{-1} = y$, we get

$$x^{-1}zx = x^{-1}(xy)x = yx = y^{-1}x^{-1} = (xy)^{-1} = z^{-1}.$$

Therefore, the group D_0 is a normal subgroup. Clearly, we have

$$D = \langle D_0, x \rangle = D_0\langle x \rangle,$$

and D is a semidirect product of D_0 and $\langle x \rangle$. So, the structure of D is uniquely determined. The last assertion follows from Sylow's theorem. It can also be proved by computation as follows. The formula

$$x^{-1}z^m x = z^{-m}$$

holds for every $m \in \mathbb{Z}$. If the order n of z is odd, we set $n = 2k - 1$. Then,

$$z^{-k}xz^k = z^{-2k}x = z^{-1}x = y. \quad \square$$

The Coset Enumeration Method

When a presentation of a finite group G is given, there is a general procedure to determine the order of G from the given presentation. The method, which is called the coset enumeration method of Coxeter and Todd, is mechanical and has been adapted for the computer. Because of the recent advances in computer technology, it has become possible to handle groups of fairly large order. In the past 15 years, several new sporadic simple groups have been "constructed" by computers using variants of this method. We will only discuss the method in a most primitive form. Detailed discussion of the actual computations by the computer can be found in Leech [1].

EXAMPLE 3. As a simple example, we will consider the group G given by the following presentation:

$$\langle x, y \mid x^3 = y^3 = (xy)^2 = 1 \rangle.$$

Setting $xy = z^{-1}$, we see that this is the group $G(3, 3, 2)$ in the theorem of Coxeter. We will verify that its order is 12 by the coset enumeration method.

The first step is to choose a known group H which is contained in G as a subgroup. Our aim is to determine the index $|G : H|$. Of course, we may take $H = \{1\}$, but the choice of a larger subgroup will reduce the amount of laborious computation. So, we will choose $H = \langle x \rangle$, and we note that the relation $x^3 = 1$ makes $|H| = 3$.

The next step is to record the relations $r = 1$ on the top of a sheet of paper. We write each relation in the form $u_1 \cdots u_n = 1$ where u_i are elements of X or X^{-1}. On top of the sheet, we write $u_1, u_2, \ldots, u_n$ in this order, place a comma after each relation, and continue with the next relation. In our case, we have three relations and the completed top line is as follows:

$$x \quad x \quad x, \qquad y \quad y \quad y, \qquad x \quad y \quad x \quad y$$

Each of these elements will be the head of a column which goes downward. We will enumerate the right cosets of H. Let the numerals 1, 2, 3, ..., represent the various cosets. We will begin by designating the coset H by 1, and each apparently new coset will be identified with the succeeding numeral. We will be inserting numerals in between the columns on our sheet of paper according to the following rule. At the k-th line, suppose we have just written the numeral i immediately to the left of the column headed by an element u. Then, we write the numeral j which represents the coset iu just to the right of this same column. We repeat the above process for each column, and we will continue to fill in lines until the multiplication table between the cosets and the generators is complete.

The first line of our example will be written as follows:

$$x \quad x \quad x \quad , \quad y \quad y \quad y \quad , \quad x \quad y \quad x \quad y$$
$$1 \quad 1 \quad 1 \quad 1 \quad \quad 2 \quad 3 \quad 1 \quad \quad 1 \quad 2 \quad 4 \quad 1$$

We begin by writing 1 at the left end of the line. Since $H = \langle x \rangle$, we have four 1's at the beginning. Then, we have a new coset $1y = 2$; the next coset $2y$ is also new. But, by the relation $y^3 = 1$, $3y$ must be the coset 1. In any line, the numerals under the commas, as well as that at the right end of the line, coincide with the numeral at the left end of the line. The first line continues with $1x = 1$, $1y = 2$, and an apparently new coset $2x = 4$. The last numeral must be 1. Thus, we get $4y = 1$. On the other hand, we already have the relation $3y = 1$. So, we get the *coincidence* $4 = 1y^{-1} = 3$. (A coincidence occurs when two distinct numerals actually represent the same coset.) Such coincidences can also happen in a more elaborate fashion. When the coincidence of i and j ($j > i$) occurs, we replace all the j's which appear in our table by the numeral i. This can lead to new coincidences, and when they occur, we always eliminate the larger number. After completely filling in a line, we proceed to the next line, beginning with the next coset.

The second line of our example begins with the coset 2. Since the numeral 4 has been replaced by 3, the second line continues $2x = 3$, $3x = 5$, and $5x = 2$. The first 3 lines are as follows:

$$x \quad x \quad x \quad , \quad y \quad y \quad y \quad , \quad x \quad y \quad x \quad y$$
$$1 \quad 1 \quad 1 \quad 1 \quad \quad 2 \quad 3 \quad 1 \quad \quad 1 \quad 2 \quad 3 \quad 1$$
$$2 \quad 3 \quad 5 \quad 2 \quad \quad 3 \quad 1 \quad 2 \quad \quad 3 \quad 1 \quad 1 \quad 2$$
$$3 \quad 5 \quad 2 \quad 3 \quad \quad 1 \quad 2 \quad 3 \quad \quad 5 \quad 6 \quad 7 \quad 3$$

At the end, we get a coincidence $7 = 3y^{-1} = 2$ which, in turn, leads to the coincidence $6 = 2x^{-1} = 5$. These coincidences give the relation $5y = 5$. Thus, the multiplication table between cosets and generators has been completed:

	1	2	3	5
x	1	3	5	2
y	2	3	1	5

(In this table, the x row has the coset ix under i; the same is true for the y row.) Let K be the union of the cosets 1, 2, 3, and 5. Then, the table shows that $Kx = K = Ky$. This means that $K\langle x, y \rangle = KG = K$. Since the subset K contains the identity of G, we obtain $K = KG = G$.

The preceding computation shows not only that $|G : H| = 4$, but also how the generators act on the cosets. Thus, we get a permutation representation of the group G, and the structure of the group G can be described explicitly as a permutation group. In this manner, several new sporadic simple groups have recently been constructed by computer. In each case, the computer method has produced a permutation representation of the group which is given by a set of generators and relations. Sims [2] has constructed a permutation representation of Lyons' group, a group in which the number of distinct cosets is close to 100,000. Truely ingenious ideas are required in order to handle problems of this magnitude within reasonable limits of computer time and space.

EXAMPLE 4. Let us consider the group G generated by the two elements x and y and the relations $x^5 = y^3 = (xy)^4 = 1$ and a further condition that the element $(xy)^2$ belongs to the center of G. In Example 3, we followed the general procedure closely, just as in a computer program. But, if we perform our calculations on paper, we may modify the method depending on the particular features of the problem at hand. Set $u = xy$, $z = u^2$, and $H = \langle x, z \rangle$. Then, we have $z \in Z(G)$, $H = \langle x \rangle \times \langle z \rangle$, and $|H| = 10$. Furthermore, for any coset of H, we have $Haz = Hza = Ha$ and $Ha(xyu) = Hau^2 = Ha$.

In the following table, an underlined numeral i indicates that the coset i is defined there, and a bar, $i - j$, shows that a relation $iw = j$ is appearing for the first time. The action of u is involutory. Thus, $iu = j$ and $ju = i$ are equivalent, and so we will only designate whichever one comes first. We arrange the rows not in order of appearance, but in such a way as to avoid coincidences. In the part under the relation $xyu \in H$, we proceeded from both ends and tried to capture the possible relation in the middle. Under the relations $x^5 = 1$ and $y^3 = 1$, we omitted the cyclical arrangements and also used rearrangement of the rows.

x	y	u	,	x	x	x	x	x	,	y	y	y
1 -	1 -	2 -		1	1	1	1	1		2 -	3 -	1
2 -	3	1		2	3 -	4 -	5 -	6 -		2		
6	2	3 -		6								
3	4 -	6		3						4	6 -	7 - 4
5	6	7 -		5								
7 -	8 -	5		7	8 -	9 -	10 -	11 -		7		
11	7	4 -		11								
4	5 -	11		4						8	5	11 - 8
10	11	8 -		10								
8	9 -	10		8						9	10 -	12 - 9
9	10	12 -		9								
12 -	12	9		12								

The table gives us the following multiplication table of the cosets and the generators:

	1	2	3	4	5	6	7	8	9	10	11	12
x	1	3	4	5	6	2	8	9	10	11	7	12
y	2	3	1	6	11	7	4	5	10	12	8	9

Thus, we get $|G| = 120$. In this example, the permutation representation obtained for G is not faithful; it is the representation of the factor group $G/\langle z \rangle$. If we want to determine the structure of the group, we must either begin the Coxeter-Todd method starting from the group $\langle x \rangle$, or else we can use the following argument.

Let X and Y be the elements of $SL(2, 5)$ defined by

$$X = \begin{pmatrix} 1 & 1 \\ 0 & 1 \end{pmatrix}, \qquad Y = \begin{pmatrix} -1 & -1 \\ 1 & 0 \end{pmatrix}.$$

Then, we have $X^5 = Y^3 = 1$ and

$$XY = \begin{pmatrix} 0 & -1 \\ 1 & 0 \end{pmatrix}.$$

So, $(XY)^4 = 1$ and $(XY)^2$ belongs to the center of $SL(2, 5)$. Thus, by (6.7), there is a homomorphism f from G into $SL(2, 5)$. But, as $SL(2, 5)$ has order 120 (Chapter 1, (9.11)), we get $G \cong SL(2, 5)$.

Applying (6.7) again, we see that the group G_1 defined by

$$\langle x, y \,|\, x^5 = y^3 = (xy)^2 = 1 \rangle$$

is isomorphic to a factor group of $SL(2, 5)$. In this case, it is clear that the group is isomorphic to the factor group of $SL(2, 5)$ by its center, $G_1 \cong PSL(2, 5)$.

If we omit the relation $(xy)^4 = 1$ from the presentation of the group G at the beginning of Example 4, we still get a finite group (provided we keep the relation that $(xy)^2$ lies in the center). In this case, if we start from the subgroup $H = \langle x \rangle$, the total number of cosets is 360. It would be difficult to finish the computation by pencil and eraser within a day; this is certainly a job for a computer. We will determine the structure of this group later (§9, Example 2) using Schur's theory of central extensions.

G. Higman [5] gives the following general description of the Coxeter-Todd method and proves a theorem of Mendelsohn [1] which states that the method will succeed in finding the permutation representation of a group G provided that the index $|G : H|$ is finite.

Let $\langle X \,|\, R \rangle$ be a presentation of a group G. We will assume that X is finite and that R consists of a finite number of relations $r_1 = r_2 = \cdots = r_s = 1$. Let $X = \{x_1, \ldots, x_m\}$. Then, each r_i is a word which is written as a product of elements in X or X^{-1}. For the sake of convenience, we will consider R to include the trivial relations $x_i x_i^{-1} = 1$ for all generators x_i. Let H be a subgroup of G. We will assume that a set of generators $\{u_k\}$ of H is given and that each u_k is explicitly given as a word in $X \cup X^{-1}$.

Let Ω_n be the set of words which are written as products of at most n elements of $X \cup X^{-1}$. The empty product 1 is contained in all Ω_n.

We consider the weakest equivalence relation on Ω_n which satisfies the following three conditions, and we write $w_1 \equiv w_2$ if and only if w_1 and w_2 are equivalent under this equivalence relation.

(1) If two elements w_1 and w_2 of Ω_{n-1} are equivalent, then for any $y \in X \cup X^{-1}$, we have $w_1 y \equiv w_2 y$.
(2) If the word wr_i is defined for $w \in \Omega_n$ and some i, then we have $wr_i \equiv w$.
(3) If $1 \cdot u_k$ is defined for some k, then we have $1 \cdot u_k \equiv 1$.

Let E_n be the set of equivalence classes. For each word w of Ω_n, let $[w]$ be the class of E_n which contains w. If $w \in \Omega_n$ and $y_1, y_2, \ldots, y_t \in X \cup X^{-1}$, then $wy_1 y_2 \cdots y_t$ is a word of Ω_{n+t}. But, by (1), its class depends only on the class $[w]$. Thus, the formula

$$[w]y_1 \cdots y_t = [wy_1, \ldots, y_t]$$

defines an action of the word $y_1 y_2 \cdots y_t$ on E_n.

The equivalence relation $\equiv$ and the set E_n can be defined mechanically. In fact, the method we used in the previous examples is essentially a practical way to find the set E_n. Once the set E_n is defined, we are in one of the following two situations.

(A) Some class of E_n contains only words having length n.
(B) Every class of E_n contains at least one word of length less than n.

If we have the situation (B), we have reached the terminal stage. In this case, every class $[w]$ contains a word w of length less than n. Hence, for any $y \in X \cup X^{-1}$, we can find the class $[wy]$ in E_n. Since $[w]y = [wy]$, the free group F on X acts on E_n. The condition (2) proves that the kernel of this action contains all the elements ir_α. So, by Corollary 2 of Theorem 6.6, the action of F induces the action of G. The condition (3) shows that the stabilizer of $[1]$ contains the given subgroup H. Thus, $Hw_1 = Hw_2$ implies that $[w_1] = [w_2]$. We will show that, conversely, $[w_1] = [w_2]$ implies $Hw_1 = Hw_2$. Let the relation $w_1 \sim w_2$ be defined by the condition $Hw_1 = Hw_2$. This induces an equivalence relation on Ω_n which satisfies the three conditions (1)–(3). As we have seen, $Hw_1 = Hw_2$ implies $[w_1] = [w_2]$. So, the relation $w_1 \sim w_2$ is not stronger than the relation $w_1 \equiv w_2$. Since the relation $w_1 \equiv w_2$ is the weakest equivalence relation, it coincides with the relation $w_1 \sim w_2$. Thus, we have $[w_1] = [w_2]$ if and only if $Hw_1 = Hw_2$. So, E_n is the totality of the cosets of H, and $|E_n| = |G : H|$.

Since the set X is finite, the set Ω_n is also finite. Thus, $|E_n|$ is finite. So, unless $|G : H|$ is finite, we never reach the situation (B). However, if $|G : H|$ is finite, E_n satisfies the conditions of the situation (B) for a sufficiently large n. This may be obvious; at any rate, it can be easily proved as follows.

Let T be a transversal of the subgroup H such that $H \cap T = \{1\}$. We assume that each element of T is explicitly given as a word in X. For any $t \in T$ and $y \in X \cup X^{-1}$, the product ty is an element of G. So, there is an element s of T such that $tys^{-1} \in H$. The word tys^{-1} can be written as a product of the elements $r_1, \ldots, r_s$, their conjugates in F, $u_1, \ldots, u_k$, and the inverses of these elements. So, if n is large enough, the word tys^{-1} can actually be written in the above manner in Ω_n. By the conditions (2) and (3), we get $[1]tys^{-1} = [1]$. Hence, we have $[ty] = [s]$ in Ω_n. We have assumed that $|T| = |G : H|$ is finite. So, if n is sufficiently large, for all $t \in T$ and all $y \in X \cup X^{-1}$, there is an element s of T such that $[ty] = [s]$ in E_n. Furthermore, we can assume that, for each $s \in T$, s can be written as a word of length less than n.

We will prove that, for any w in Ω_n, there is an element s of T such that $[w] = [s]$ by using induction on the length of w. If the length is 0, then $w = 1$ and the assertion is obvious. Assume that the length is positive. Let y be the last letter of the word w. Set $w = w'y$. Then, w' is shorter than w. By the inductive hypothesis, there is an element t of T such that $[w'] = [t]$. Then, we have

$$[w] = [w'y] = [w']y = [t]y = [ty].$$

As shown earlier, there is an element s of T such that $[ty] = [s]$. This proves that, for sufficiently large n, E_n satisfies the conditions of the situation (B). Thus, if $|G : H|$ is finite, the coset enumeration will reach the terminal stage after a finite number of steps, and the index $|G : H|$ and the permutation representation of the group G on the cosets of H are obtained (Mendelsohn [1]). In this case, however, a practical upper bound for n in terms of the presentation $\langle X \,|\, R \rangle$ and the set $\{u_1, \ldots, u_k\}$ of generators of H has not been found, so we cannot predict whether or not the coset enumeration of a particular group will be finished within the capacity of the available computer.

The Generating Set of a Subgroup

We will discuss the method of Reidemeister of finding a generating set of elements for a subgroup H when a set of generators of a group G is given.

Assume that a right transversal T of a subgroup H is given. We write $\bar{g} = T \cap Hg$, where $\bar{g}$ is the representative of the right coset Hg. Sometimes, we will use the functional notation and write $\bar{g} = \tau(g)$. The fundamental properties of this function τ which assigns the representative of each coset Hg to any element g of G are:

$$g\tau(g)^{-1} \in H, \qquad \tau(\tau(g)h) = \tau(gh),$$

and $\tau(t) = t$ for any $t \in T$. These properties are easy to verify from the formula $Hg = H\tau(g)$. For any t and x of G, let

$$(t, x) = \tau(t)x\tau(tx)^{-1}.$$

Then, (t, x) is an element of H. It is proved that the following formulas hold: $(\tau(t), x) = (t, x)$ and $(t, x^{-1}) = (tx^{-1}, x)^{-1}$.

With these notations, the following theorem holds.

Theorem 6.9. *Let G be a group. Suppose that a set X of generators for G and a transversal T of a subgroup H satisfying $T \cap H = \{1\}$ are given. Then, the subgroup H is generated by the elements of the form*

$$(t, \cdot x) \qquad (t \in T, \quad x \in X).$$

Proof. An arbitrary element h of H can be written as a word in X, so we have $h = u_1 u_2 \cdots u_n$ where $u_i \in X \cup X^{-1}$ for all i. Set $v_0 = 1$, $v_i = v_{i-1}u_i$ and $t_i = \tau(v_i)$ for $i = 0, 1, \ldots, n$. Then, $t_0 = t_n = 1$, and we have

$$h = t_0 u_1 t_1^{-1} \cdot t_1 u_2 t_2^{-1} \cdot t_2 u_3 t_3^{-1} \cdots t_{n-1} u_n t_n^{-1}.$$

Since $t_i = \tau(v_i) = \tau(v_{i-1}u_i) = \tau(\tau(v_{i-1})u_i) = \tau(t_{i-1}u_i)$, we see that $t_{i-1}u_i t_i^{-1} = (t_{i-1}, u_i)$ and

(6.10)
$$h = (t_0, u_1)(t_1, u_2) \cdots (t_{n-1}, u_n).$$

The elements u_i are contained in $X \cup X^{-1}$, and if $u_i \in X^{-1}$, then $u_i = x^{-1}$ ($x \in X$) and $(t, u_i) = (tu_i, x)^{-1} = (\tau(tu_i), x)^{-1}$. Thus, the formula (6.10) shows that h is written as a product of elements of the form (t, x) or their inverses. $\square$

Corollary 1. *Let G be a finitely generated group. If a subgroup H of G has finite index, then H is also finitely generated.*

Proof. The set $\{(t, x)\}$ of generators of H is finite because both T and X are finite. $\square$

Corollary 2. *Any finite group G has a finite presentation; that is, $G = \langle X \mid R \rangle$ where X is finite and R is also a finite set of relations.*

Proof. We can choose a finite set X of generators. Let F be the free group on X. Then, F is finitely generated. There is a homomorphism f from F onto G and $G \cong F/\text{Ker } f$. Hence, $\text{Ker } f$ is finitely generated by Corollary 1. Let $\{ir_\alpha\}$ be a finite set which generates $\text{Ker } f$. Then, $R = \{r_\alpha = 1\}$ is a set of relations and Corollary 2 of Theorem 6.6 proves that $\langle X \mid R \rangle$ is a presentation of the group G. $\square$

Theorem of Nielsen–Schreier. *Any subgroup of a free group is also free.*

We need some preparation before we can prove the Nielsen–Schreier Theorem.

Let F be a free group, let G be a subgroup of F, and let T be a transversal of G in F. Let X be a set of free generators of F. Then, F is free on X, and we can consider X to be a subset of F.

A transversal T is said to be a **Schreier transversal** if $H \cap T = \{1\}$ and the following condition is satisfied. Let $t = u_1 \cdots u_n$ be the standard form of an element t of T. For any $i \leq n$, the partial product $u_1 u_2 \cdots u_i$ belongs to T.

Lemma. *Any subgroup has a Schreier transversal.*

Proof. Let Gx be a right coset and let $l(Gx)$ be the length of the shortest element in Gx. We will show that we can choose a Schreier transversal such that every element t satisfies $l(t) = l(Gt)$, using induction on $n = l(Gt)$. If $n = 0$, then the coset contains the identity. We merely need to choose $\{1\} = T \cap H$. Assume that $n > 0$. Let us consider a coset Gx of length n.

Choose an element x_0 of minimal length, and let $x_0 = u_1 \cdots u_n$ be its standard form. Since $l(Gu_1 \cdots u_{n-1}) \leq n - 1$, the representative of the coset $Gu_1 \cdots u_{n-1}$ has been chosen by the inductive hypothesis; let it be s. If $s = v_1 \cdots v_m$ is the standard form of s, then $m = l(Gs)$ and the partial products $v_1 \cdots v_j$ are in T. Set $t = su_n$. Then, we have $Gt = Gsu_n = Gx$ and $l(t) \leq l(s) + 1$. Since $l(s) \leq n - 1$, we get

$$l(t) \leq l(s) + 1 \leq n.$$

On the other hand, t is an element of Gx, and $l(Gx) = n$. Hence, we get $l(t) = n$ and $m = n - 1$. By definition,

$$t = su_n = v_1 \cdots v_m u_n,$$

and the right side is the standard form of t (u_n cannot be cancelled). We take the element t as a representative of Gx. It is now clear that we have a Schreier transversal. $\square$

The following proposition will prove the freeness of a subgroup of a free group.

(6.11) *Let X be a set of free generators of F, and let T be a Schreier transversal of a subgroup G. Then, G is freely generated by the nonidentity elements of the form (t, x) where $t \in T$ and $x \in X$.*

Proof. Let Y be the set consisting of all the pairs (C, x) where C is a (right) coset of G and x is an element of X. Let E be the group free on Y. We may assume that Y is a subset of E in the following proof. For each coset C, $\tau(C) = T \cap C$ is the representative of C; in particular, we have $\tau(G) = 1$. Since $\tau(C)x\tau(Cx)^{-1}$ is an element of G, the mapping

$$(C, x) \rightarrow \tau(C)x\tau(Cx)^{-1}$$

can be extended to a homomorphism α from the free group E into G. The image $\alpha((C, x))$ is the element $(\tau(C), x)$ of G. So, by Theorem 6.9, α is surjective.

We will prove that we can define an element $h(C, u)$ of E which satisfies the following formulas for any coset C of G and any elements u and v of F:

(6.12) $h(C, 1) = 1, \qquad h(C, uv) = h(C, u)h(Cu, v).$

For each coset C and an element x of X, we will define

$$h(C, x) = (C, x) \quad \text{and} \quad h(C, x^{-1}) = (Cx^{-1}, x)^{-1}.$$

where the right sides are generators of E or their inverses. For an element u of F, let $u = u_1 \cdots u_n$ be the standard form of u. Set $v = u_1 \cdots u_{n-1}$. We define $h(C, u)$ by induction on $l(u)$ as follows:

$$h(C, 1) = 1, \qquad h(C, u) = h(C, v)h(Cv, u_n)$$

We will prove that this function h satisfies the formulas (6.12). The first formula is satisfied by definition, so we will proceed to verify the second one by induction on $l(v)$. Suppose that $l(v) = 1$. Let $u = u_1 \cdots u_n$ be the standard form of u, and consider the two cases according to whether $u_n = v^{-1}$ or $u_n \neq v^{-1}$. If $u_n = v^{-1}$, then $w = u_1 \cdots u_{n-1}$ is the standard form of uv and, by definition, we have

$$h(C, u) = h(C, w)h(Cw, u_n).$$

Since $h(Cw, v^{-1}) = h(Cwv^{-1}, v)^{-1} = h(Cu, v)^{-1}$, we get

$$h(C, u)h(Cu, v) = h(C, uv).$$

On the other hand, if $u_n \neq v^{-1}$, $uv = u_1 \cdots u_n v$ is the standard form of uv. Hence, by definition, we have the second formula of (6.12). This finishes the case of $l(v) = 1$. If $l(v) > 1$, we write $v = yz$ where $l(y) = l(v) - 1$ and $l(z) = 1$. Hence, by the inductive hypothesis and the case of $l(z) = 1$, we get

$$\begin{aligned} h(C, uv) &= h(C, uy)h(Cuy, z) \\ &= h(C, u)h(Cu, y)h(Cuy, z) \\ &= h(C, u)h(Cu, yz). \end{aligned}$$

This proves that the formulas (6.12) are satisfied.

If $x \in X$, then we have

$$\alpha(h(C, x)) = (\tau(C), x) \in G,$$

where α is the homomorphism from E onto G defined earlier. Using induction on $l(u)$ and the formula (6.12), we can prove that the following formula (6.13) holds for all $u \in F$:

(6.13) $\alpha(h(C, u)) = (\tau(C), u).$

We define the mapping β from G into E by the formula

$$\beta(u) = h(G, u).$$

Then, by (6.12), β is a homomorphism from G into E, and the formula (6.13) gives us $\alpha(\beta(u)) = u$ for all $u \in G$. Note that $\tau(G) = 1$ and $(1, u) = u$ if $u \in G$. This shows that β is an isomorphism from G into E. Set $\omega = \alpha\beta$. Then, ω is an endomorphism of E such that $\omega^2 = \omega$. Since α is surjective, we get

$$\omega(E) = \beta(G) \cong G.$$

Set $K = \text{Ker } \omega$. By the First Isomorphism Theorem, we have

$$E/K \cong \omega(E) \cong G.$$

By definition, E is freely generated by $Y = \{(C, x)\}$. We will show that K is generated by all $\omega(y)^{-1}y$ ($y \in Y$) and their conjugates. An element e of E can be written in the standard form, so we have $e = uv$ where $u \in Y \cup Y^{-1}$ and the length of v is smaller than that of e. Then, we have

$$\omega(e)^{-1}e = \omega(v)^{-1}(\omega(u)^{-1}u)\omega(v) \cdot \omega(v)^{-1}v.$$

If $u = y^{-1}$ ($y \in Y$), then $\omega(u)^{-1}u = y(\omega(y)^{-1}y)^{-1}y^{-1}$. Thus, by using induction on the length, we see that $\omega(e)^{-1}e$ can be written as a product of elements of the form $\omega(y)^{-1}y$ for $y \in Y$ or their conjugates. In particular, if $e \in \text{Ker } \omega$, then e can be written in such a form. Since $\omega^2 = \omega$, K is generated by the elements of the form $\omega((C, x))^{-1}(C, x)$ and their conjugates.

We define two normal subgroups K_0 and K_1 of E as follows. Let K_0 be the subgroup generated by those pairs (C, x) which satisfy $(\tau(C), x) = 1$ and by their conjugates; let K_1 be the subgroup generated by all $h(G, t)$ and their conjugates where $t \in T$. The definition of α is $\alpha(C, x) = (\tau(C), x)$. Hence, every generator of K_0 is contained in the kernel of ω, and so we have $K_0 \subset K$. From the definition of K_0, we know that K_0 is the smallest normal subgroup which contains all the pairs (C, x) such that $(\tau(C), x) = 1$. So, K_0 is the smallest normal subgroup containing a part of the free generating set of E. It is clear that the factor group E/K_0 is free on the remaining set of generators, the set of those (C, x) which satisfy $(\tau(C), x) \neq 1$ (see Theorem 6.6). Since $E/K \cong G$, it suffices to show $K = K_0$ in order to prove the theorem.

We will compute $\omega((C, x))$ for a generator (C, x) of E. By the definition and (6.12), we get

$$\omega((C, x)) = \beta(\alpha(C, x)) = \beta(\tau(C)x\tau(Cx)^{-1})$$
$$= h(G, \tau(C))h(C, x)h(Cx, \tau(Cx)^{-1}).$$

On the other hand, we have $h(C, x) = (C, x)$, and by (6.12),

$$h(Cx, \tau(Cx)^{-1}) = h(G, \tau(Cx))^{-1}.$$

Thus, for $y = (C, x)$, we get

$$\omega(y)^{-1}y = h(G, \tau(Cx))h(G, \tau(C))^{-y}.$$

The right side is a product of some of the generators of K_1, while the left side is a typical generator of K. Thus, we have $K \subset K_1$.

Finally, we will show the containment $K_1 \subset K_0$, which will force the equality $K = K_1 = K_0$. Take an element t of T, and let $t = u_1 \cdots u_n$ be its standard form. Set $s = u_1 \cdots u_{n-1}$ and $u = u_n$. Since T is a Schreier transversal, s is another element of T. By the formulas (6.12), we have

$$h(G, t) = h(G, s)h(Gs, u).$$

If $u \in X$, then $h(Gs, u) = (Gs, u)$. Since $\tau(Gs) = s$ and $\tau(Gsu) = t$, we get

$$(\tau(Gs), u) = \tau(Gs)u\tau(Gsu)^{-1} = sut^{-1} = 1.$$

This gives us $(Gs, u) \in K_0$. On the other hand, if $u^{-1} = x \in X$, then $h(Gs, u) = (Gsu, x)^{-1}$ and we get $(Gsu, x) \in K_0$ as before. In all cases, the term $h(Gs, u)$ is an element of K_0. So, by induction, all the elements of the form $h(G, t)$ with $t \in T$ belong to K_0. Hence, we have $K_1 \subset K_0$. This proves the Nielsen–Schreier Theorem. □

If F is a free group of finite rank, we can compute the rank of any subgroup of finite index from (6.11).

(6.14) *Let F be a free group of finite rank n, and let G be a subgroup of finite index j in F. Then, G is a free group of rank $nj - j + 1$.*

Proof. Let X be a set of free generators of F, and let T be a Schreier transversal of G. Let Y be the set of pairs (t, x) such that $t \in T$, $x \in X$, and $tx = \tau(tx)$. By (6.11), G is freely generated by the elements of the form $tx\tau(tx)^{-1}$ such that $(t, x) \notin Y$, so the rank of G is $nj - |Y|$.

Let $t = u_1 \cdots u_n$ be the standard form of an element t of T. Since T is a Schreier transversal, $s = u_1 \cdots u_{n-1}$ also belongs to T. Hence, if $u_n \in X$, then we have $(s, u_n) \in Y$, while if $u_n = x^{-1}$ $(x \in X)$, then $(t, x) \in Y$. Thus, each nonidentity element $t \in T$ determines a pair (s, x) in Y. We will show that the element t is determined by the pair (s, x). If the last term of the standard form of s is not equal to x^{-1}, then we have $t = sx$; otherwise, we have $t = s$. This proves that $|Y| = j - 1$. □

There are many known proofs of the Nielsen–Schreier Theorem based on a variety of different principles. The preceding proof is the algebraic proof of Weir [1] which is an improved version based on the original idea of Kuhn [1]. The proof of Nielsen deals with the standard forms of the elements of G.

For a very readable account of this proof, the readers are referred to Nielsen [1]. The proof we have given here can be adapted to the subgroup theorem of Kurosh on free products. A streamlined proof of this generalization is given by MacLane [1]. There are many other proofs of the Subgroup Theorem. Baer–Levi [1] uses a topological argument, while Higgins [1] gives a proof using graph theory. Among the proofs based on the cancellation argument, Takahashi's proof [1] is one of the simplest.

Free Products with Amalgamation

When a family of groups G_λ ($\lambda \in \Lambda$) is given and we consider the presentations $\langle X_\lambda | R_\lambda \rangle$ of these groups, it is usually assumed that the sets X_λ are mutually disjoint. We will assume that this is the case, so we have $X_\lambda \cap X_\mu = \emptyset$ if $\lambda \neq \mu$. Let X be the union of these sets X_λ, and let $R = \bigcup R_\lambda$. The group G with the presentation $\langle X | R \rangle$ is called the **free product** of the groups G_λ. If F is a free group on X, then F is the free product of the infinite cyclic groups $\langle x \rangle$ generated by $x \in X$ (see (6.3)).

Suppose that we have a more general situation: we are given a group H and an injection φ_λ from H into G_λ for each $\lambda \in \Lambda$. We define a group G generated by all G_λ's in which the element $\varphi_\lambda(h)$ of G_λ is identified with the element $\varphi_\mu(h)$ of G_μ for each $h \in H$ and $\lambda, \mu \in \Lambda$. Then, we have

$$G = \langle X | R, \varphi_\lambda(h) = \varphi_\mu(h) \text{ for all } h \in H, \lambda, \mu \in \Lambda \rangle.$$

The group G is called the **free product with amalgamation**. If $H = \{1\}$, we get the free product.

The free product of the groups G_λ contains a subgroup isomorphic to G_λ. This is obvious. For the free product with amalgamation, the corresponding statement that there is a subgroup isomorphic to G_λ holds, but it is not obvious (because the added relations $\varphi_\lambda(h) = \varphi_\mu(h)$ might impose further restrictions on the set of generators X_λ).

We have the following theorem. Let G be the free product of the groups G_λ with amalgamation. Then, G contains each of the groups G_λ (or a subgroup isomorphic to it), and the intersection $G_\lambda \cap G_\mu$ is the subgroup obtained by identifying the subgroup $\varphi_\lambda(H)$ of G_λ with the subgroup $\varphi_\mu(H)$ of G_μ according to the isomorphism $\varphi_\lambda(h) \to \varphi_\mu(h)$ between them. Thus, in G, we have

$$G_\lambda \cap G_\mu = \varphi_\lambda(H) = \varphi_\mu(H).$$

This can be proved by adapting the method used to prove Theorem 6.4 to this more general situation.

Consider the totality of finite sequences $(a_1, \ldots, a_n)$ of elements a_i of G_{λ_i}. Among sequences, we will consider the following operations.

(1)   If some member a_i is the identity, then we remove a_i to make a shorter sequence.

(2)   If neighboring terms belong to the same group (that is, if $\lambda_i = \lambda_{i+1}$ for some i), then we replace the neighboring terms (a_i, a_{i+1}) by their product $a_i a_{i+1}$ in order to shorten the sequence.

(3)   When $a_{i+1} = \varphi_{\lambda_{i+1}}(h)b_{i+1}$ for some $h \in H$, we replace the pair (a_i, a_{i+1}) by $(a_i \varphi_{\lambda_i}(h), b_{i+1})$.

We call two sequences equivalent if one is obtained from the other by applying a finite number of the operations described above or the inverses of these operations. Let G be the set of equivalence classes of this relation. We define the product of two sequences $(a_1, \ldots, a_n)$ and $(b_1, \ldots, b_m)$ as the sequence $(a_1, \ldots, a_n, b_1, \ldots, b_m)$. Then, it is easy to verify that this induces an operation in G.

This will make G a group. In fact, G is the free product with amalgamation. For each λ, the totality of the classes which contain one term sequences (a_λ) with $a_\lambda \in G_\lambda$ forms a subgroup isomorphic to G_λ, and this subgroup will be identified with G_λ. If $\lambda \neq \mu$, then $G_\lambda \cap G_\mu$ is a subgroup isomorphic to H, and the isomorphism is given by φ_λ which coincides with φ_μ on $G_\lambda \cap G_\mu$. We will identify $\varphi_\lambda(H)$ with H and write $G_\lambda \cap G_\mu = H$. Let T_λ be a transversal of H in G_λ such that $T_\lambda \cap H = \{1\}$. Then, any element g of G can be written uniquely in the form $h t_1 t_2 \cdots t_n$ for some integer $n \geq 0$ where $h \in H$, $t_i \in T_{\lambda_i} - \{1\}$, and $\lambda_i \neq \lambda_{i+1}$ for all i. This is called the *standard form* of the element g of G.

The preceding statements are theorems of Schreier. The associative law and the uniqueness of the standard form can be verified by methods similar to those used in the proof of Theorem 6.4, but these proofs are long and tedious. We will present the outline of a simplified proof based on an idea of van der Waerden [1].

Let W be the totality of the standard forms. Thus, W is a set of elements of the form $w = h t_1 \cdots t_n$ where $h \in H$, $t_i \in T_{\lambda_i} - \{1\}$ and $\lambda_i \neq \lambda_{i+1}$. If $w' = h' s_1 \cdots s_m$ is another element of W, we define $w = w'$ if and only if $h = h'$, $n = m$, and $t_i = s_i$ for all $i = 1, 2, \ldots, n$. For each λ, we will define an action of the group G_λ on W. Let g be an element of G_λ. Then, for $w = h t_1 \cdots t_n$, we define wg by induction on n.

(1)   If $\lambda_n = \lambda$, then $wg = (h t_1 \cdots t_{n-1})(t_n g)$.
(2)   If $\lambda_n \neq \lambda$ but $g = \varphi_\lambda(u)$ for some $u \in H$, then $wg = hu$ if $n = 0$, and $wg = w \varphi_{\lambda_n}(u)$ if $n > 0$.
(3)   If $\lambda_n \neq \lambda$ and $g = ht$ ($h \in \varphi_\lambda(H)$ and $t \in T_\lambda - \{1\}$), then $wg = (wh) \cdot t$ where the right side is the element of W obtained by annexing t to the end of the standard form of wh.

From the definition, it follows that this does indeed define an action of G_λ; that is, we have $w(gg') = (wg)g'$ for any elements g and g' of G_λ. Among the elements of W, we have the unique identity element 1 for which we have $n = 0$ and $h = 1$. If an element g of G_λ has the form $\varphi_\lambda(h)t$ where $t \in T_\lambda$, then $1 \cdot g = h$ if $t = 1$, and otherwise, $1 \cdot g = ht$. Thus, the permutation representation of G_λ on W is faithful. Let G be the permutation group on W generated by all G_λ ($\lambda \in \Lambda$). Then, we have

$$G_\lambda \cap G_\mu = \varphi_\lambda(H) = \varphi_\mu(H).$$

This is obvious if we look at the image of the special element 1 of W. So, it follows from (6.7) that G is a homomorphic image of the free product with amalgamation. But, an element g of G can be written in the standard form and this is identical to the element $1 \cdot g$ of W. Thus, the standard form of g is unique, and G is isomorphic to the free product with amalgamation.

The concept of free products with amalgamation is useful in constructing examples of groups with rather curious properties (see Exercises 7–11).

Exercises

1. Show that the group $G(4, 3, 2)$ is isomorphic to the symmetric group Σ_4.

(*Hint.* This is similar to Example 3. The group $G = \langle x, y \rangle$ which satisfies the relations $x^4 = y^3 = 1$ and $(xy)^2 \in Z(G)$ is also a finite group of order 288.)

2. Show that the group $G(3, 3, 3)$ contains an abelian normal subgroup of index 3.

(*Hint.* Let $G = \langle x, y \mid x^3 = y^3 = (xy)^3 = 1 \rangle$. For a primitive cubic root ω of unity, the mapping $x, y \to \omega$ will preserve the defining relations of G. So, it can be extended to a homomorphism from G onto $\langle \omega \rangle$. Let A be the kernel of this homomorphism. Find the generators of A by using Theorem 6.9. The set $\{1, x, x^2\}$ is a transversal of A in G, and the subgroup A is generated by yx^2, xyx, and x^2y. Note that $xy^{-1} \in A$. The relation $(xy)^3 = 1$ implies that the first two elements are commutative. Furthermore, we have $yx^2 \cdot xyx \cdot x^2y = 1$. Thus, A is abelian. In fact, A is a free abelian group freely generated by two elements, and $G(3, 3, 3)$ is infinite. Similarly, $G(4, 4, 2)$ contains an abelian normal subgroup of index 4, while $G(6, 3, 2)$ has an abelian normal subgroup of index 6. The result of Exercise 2 has some applications in finite group theory (Feit–Thompson [2]).)

3. Let F be a free group of rank n. Show that if $n > 1$, the derived group F' is not finitely generated.

4. Let H be a subgroup of a group $G = \langle X \mid R \rangle$. Show that a presentation of the group H can be obtained in the following manner.

Consider X to be a subset of G. Let φ be a homomorphism from the free group F on X into G such that $\varphi(ix) = x$. Set $K = \text{Ker } \varphi$ and $L = \varphi^{-1}(H)$. Choose a Schreier transversal T of L in F. Define (t, ix) as in (6.11), and let Y be the set of those (t, ix) which are not the identity. An element u of L is written as a word in X. By using (6.10), we can write u as a word in Y. Let the standard form of u as a word in Y be denoted by u^*. Let $R^* = \{(t^{-1}ir_\alpha t)^* = 1\}$ where α ranges over all relations $r_\alpha = 1$ in R and $t \in T$. Then $H = \langle Y \mid R^* \rangle$.

(*Hint*. By (6.11), L is free on Y. So, by Corollary 2 of Theorem 6.6, it suffices to show that K is the smallest normal subgroup of L which contains all the elements $(t^{-1}ir_\alpha t)^*$. As a corollary of this exercise, we have the following proposition. Let H be a subgroup of a finitely presented group G. If $|G:H|$ is finite, then H is also finitely presented.)

5. (a) Let X be a set of generators of a group G, and let w be an element of G. Then, w can be written as a word in X. Choose one such expression, and consider it to be fixed throughout this exercise. Let $x \in X$. We define $e_x(w)$ as follows. In the expression of w, suppose that x appears exactly a times and x^{-1} exactly b times. Then, we let $e_x(w) = a - b$. Show that w is contained in the derived group G' if $e_x(w) = 0$ for all $x \in X$.

(b) Suppose that the group G is free on X. Show that $w \in G'$ if and only if $e_x(w) = 0$ for all $x \in X$.

(c) Let Y be a set with the same cardinality as X, and let $x_\lambda \leftrightarrow y_\lambda$ be a bijection between them. Let A be the free abelian group freely generated by Y. Consider both A and G to be subgroups of the direct product $G \times A$, and set $H = \langle x_\lambda y_\lambda \rangle$ ($\lambda \in \Lambda$). Show that we have $G' = H' = G \cap H$.

(*Hint*. The equality $G' = H'$ is easy to verify. Part (a) gives us $G \cap H \subset G'$. The converse of Part (a) does not hold in general, but if $w \in G'$, there is an expression of w as a word in X such that $e_x(w) = 0$ for every $x \in X$.)

6. Suppose that x, y, z, and w are four elements of a finite group which satisfy the relations

$$x^{-1}yx = y^2, \qquad y^{-1}zy = z^2, \qquad z^{-1}wz = w^2, \qquad w^{-1}xw = x^2.$$

Show that $x = y = z = w = 1$.

(*Hint*. Let k and l be the orders of x and y, respectively. Then $x^k = 1$ implies that $2^k \equiv 1 \pmod{l}$. If k and l are greater than 1, then the smallest prime divisor of k is smaller than the smallest prime divisor of l. In fact, if a prime number q divides l, then we can find a divisor a and a prime divisor p of k such that $2^a \not\equiv 1 \pmod{q}$ but $2^{ap} \equiv 1 \pmod{q}$. Then, p divides $q - 1$ by Fermat's theorem.)

7. Let G be the group generated by four elements x, y, z, and w satisfying the defining relations stated in Exercise 6. Show that $G \neq \{1\}$. Exercise 6 shows that G has no nontrivial finite homomorphic image. But, since G is finitely generated, G possesses a maximal normal subgroup N. So, prove that G/N is a finitely generated simple group of infinite order (G. Higman [1]).

(*Hint*. Let $H_i = \langle x_i, y_i \mid x_i^{-1} y_i x_i = y_i^2 \rangle$. Then, the mapping

$$x_i \to \begin{pmatrix} \frac{1}{2} & 0 \\ 0 & 1 \end{pmatrix}, \qquad y_i \to \begin{pmatrix} 1 & 1 \\ 0 & 1 \end{pmatrix}$$

is extended to a homomorphism from G into an infinite group of matrices. So, both $\langle x_i \rangle$ and $\langle y_i \rangle$ are infinite, and there are the free products with amalgamations:

$$K_1 = \langle H_1, H_2 \,|\, x_2 = y_1 \rangle \quad \text{and} \quad K_2 = \langle H_3, H_4 \,|\, x_4 = y_3 \rangle.$$

Considering the standard forms of the elements of K_1, we see that the subgroup $\langle x_1, y_2 \rangle$ is the free group on $\{x_1, y_2\}$. Similarly, $\langle x_3, y_4 \rangle$ is the free group on $\{x_3, y_4\}$. Therefore, there is the generalized free product

$$\langle K_1, K_2 \,|\, x_1 = y_4, x_3 = y_2 \rangle,$$

which is the group G in Exercise 7. Hence, we have $G \neq \{1\}$.

The simple factor group of G was the first example of a finitely generated simple group of infinite order. Actually, the cardinality of the isomorphism classes of such infinite simple groups is the cardinality of continuum (Camm [1]). We have seen in §9 of Chapter 1 that the factor group of $SL(n, F)$ by its center Z is almost always simple. In this case, $SL(n, K)/Z$ is finitely generated if and only if the field F is finite (so the group $SL(n, F)$ is also finite).)

8. Let G be the free product of two groups A and B: $G = A * B$. Let $g = g_1 \cdots g_n$ be the standard form of a nonidentity element g of G; that is, $g_i \neq 1$, g_i is an element of A or B, and no neighboring pair of terms belongs to the same group.

(a) Show the uniqueness of the standard form following the procedure of Theorem 6.4. Conversely, show that if every element of $G_0 = \langle A, B \rangle$ has a unique standard form, then G_0 is isomorphic to $A * B$.

(b) Let $a \in A - \{1\}$ and $b \in B - \{1\}$. Show that a is not conjugate to b in G. Also show that the order of ab is infinite.

(c) Let U and V be subgroups of A, and let $b \in B - \{1\}$. Show that $\langle U^b, V \rangle$ is isomorphic to $U^b * V$.

(*Hint.* (b) This follows almost immediately from the uniqueness of the standard form.

(c) It suffices to show that, for $u_i \in U$ and $v_i \in V$,

$$u_1^b v_1 u_2^b v_2 \cdots u_n^b v_n = 1$$

implies that $u_1 = \cdots = u_n = v_1 = \cdots = v_n = 1$. Consider the standard form of the left side in G.)

9. Let G be a free product of G_λ with amalgamation. Let φ_λ be the injection from H into G_λ, let T_λ be a transversal of $\varphi_\lambda(H)$ in G_λ such that $T_\lambda \cap \varphi_\lambda(H) = \{1\}$, and let

$$g = h t_1 \cdots t_n \qquad (h \in H, \quad t_i \in T_{\lambda_i} - \{1\}, \quad \lambda_i \neq \lambda_{i+1})$$

be the standard form of g. The integer n is the *length* of g. By using the uniqueness of the standard form and induction on the length, show that any element of finite order is conjugate to an element of some factor G_λ. Thus, if each G_λ is torsion-free, so is G_λ. (A group is called *torsion-free* if the identity is the only element of finite order.)

(*Hint.* If g has finite order, then $\lambda_1 = \lambda_n$ and the length of $t_n g t_n^{-1} = (t_n h t_1) t_2 \cdots t_{n-1}$ is shorter.)

10. Let p be a prime number, and let $\mathcal{P}$ be the set of all p-subgroups of a group G. Then $\mathcal{P}$ is a partially ordered set ordered by the containment relation. The maximal elements of $\mathcal{P}$ are called S_p-*subgroups* of G.

(a) Show that $\mathcal{P}$ is inductive and that G has at least one S_p-subgroup.

(b) Let $G = A * B$. Show that any S_p-subgroup of G is conjugate to an S_p-subgroup of one of the factors, A or B. Show that any S_p-subgroup of A (or B) is an S_p-subgroup of G. Finally, show that two S_p-subgroups of an infinite group are not necessarily isomorphic, and that even if they are isomorphic, they need not be conjugate.

(c) If $\mathcal{P}$ has only a finite number of maximal elements, then the extended Sylow's theorem holds; that is, two S_p-subgroups are conjugate and the number of S_p-subgroups is of the form $1 + kp$ for some integer k.

(*Hint.* (b) Let P be an S_p-subgroup of G. If $x \neq 1$ is an element of P, the order of x is finite. So, x is conjugate to an element of one of the factors by Exercise 9. If $x^t \in A$, then $A \cap P^t \neq \{1\}$. So, it suffices to show that if $A \cap Q \neq \{1\}$ for some $Q \in \mathcal{P}$, then $Q \subset A$. Suppose that this proposition is false, and choose elements a and u such that $1 \neq a \in A \cap Q$ and $u \in Q - A$. Since both u and au are of finite order, they are conjugate to elements of A or B. Let $u = g^{-1} v g$ for $v \in A$ (or B). Then, either the standard form of au has even length, or the first term of the standard form does not cancel the last term. This is a contradiction. The last part follows from Exercise 8(b).

(c) Let X be the set of all maximal elements of $\mathcal{P}$. The group G acts on X via conjugation. Let P be an S_p-subgroup of G. Then, the length of a P-orbit is finite and equal to the index of some subgroup of P. So, the length of a P-orbit is a power of p (Exercise 10, §2). If P normalizes an S_p-subgroup Q, then PQ is a p-subgroup, and we get $P \subset Q$. Let Y be a G-orbit in X. If $P \in Y$, then Y is a disjoint union of P-orbits and only one of the P-orbits has length 1. Thus, we get $|Y| \equiv 1 \pmod{p}$. If $X \neq Y$, we take an S_p-subgroup R not contained in Y. Then, Y is a disjoint union of R-orbits and none of the R-orbits has length 1. Thus, we get the contradiction that $|Y| \equiv 0 \pmod{p}$. Therefore, we have $X = Y$ and the Sylow's theorem holds.)

11. (a) Suppose that two subgroups A and B of a group G are isomorphic via $f: a \to b$. Let

$$G_1 = \langle u \rangle * G \quad \text{and} \quad G_2 = \langle v \rangle * G$$

where $\langle u \rangle$ and $\langle v \rangle$ are infinite cyclic. Show that the subgroups $\langle A^u, G \rangle$ and $\langle B^v, G \rangle$ are isomorphic via an isomorphism which extends the mapping $a^u \to b^v$ and $g \to g$ $(g \in G)$. Let G_3 be the free product of G_1 and G_2 with amalgamation such that the subgroups $\langle A^u, G \rangle$ and $\langle B^v, G \rangle$ are identified using isomorphism defined above. Show that G may be considered to be a subgroup of G_3 and that G_3 contains an element w such that $w^{-1}aw = f(a)$ for all $a \in A$. Thus, any isomorphism between two subgroups can be realized by conjugation in a larger group.

(b) Let A be a subgroup of a group G. Suppose that there is a family $\{f_\lambda\}$ of isomorphisms from A onto subgroups B_λ $(\lambda \in \Lambda)$ of G. Show that there is a group G_1 and a family $\{t_\lambda\}$ of elements of G_1 such that G_1 contains G and, for all $a \in A$ and $\lambda \in \Lambda$, we have $t_\lambda^{-1}at_\lambda = f_\lambda(a)$.

(c) Suppose that a group G is torsion-free. Show that there is a group G_1 which contains G and satisfies the following two properties: (1) G_1 is torsion-free, and (2) any two elements of $G - \{1\}$ are conjugate in G_1.

(d) Suppose that a group G is torsion-free. Show that there is a group G^* which contains G and satisfies the property that any two elements of $G^* - \{1\}$ are conjugate.

(*Hint.* (c) Set $G - \{1\} = \{a_\lambda\}$. If a and a_λ are two elements of $G - \{1\}$, the mapping $a \to a_\lambda$ induces an isomorphism from $\langle a \rangle$ to $\langle a_\lambda \rangle$. By Part (b), there is a group G_1 such that $G_1 \supset G$ and, for all $\lambda \in \Lambda$, a is conjugate to a_λ in G_1. It follows from the construction and Exercise 9 that G_1 is torsion-free.

(d) Given a group G, there is a group G_1, such that $G \subset G_1$ and the elements of $G - \{1\}$ are conjugate to each other in G_1. Again, use Part (c) to get a group G_2 such that $G_1 \subset G_2$ and the elements of $G_1 - \{1\}$ are conjugate to each other in G_2. Thus, we have an increasing sequence $G \subset G_1 \subset \cdots$ of groups G_n. Let $G^* = \bigcup G_n$. Then, G^* satisfies the required properties. This exercise is due to G. Higman–B. H. Neumann–H. Neumann [1]. As Part (d) shows, there are (simple) infinite groups in which all nonidentity elements form a single conjugacy class. On the other hand, the only finite group which has such a property is the cyclic group of order 2.)

§7. Extensions of Groups and Cohomology Theory

A normal subgroup H of a group G determines the factor group G/H. We write $F = G/H$ and call G an **extension** of H by F. Extension theory concerns the problem of studying the structure of an extension of H by F using knowledge about the properties of the two given groups H and F. Thus, the main problem is to find all the possible extensions of H by F.

Let H and F be two given groups. We will write the elements of F as $1, \sigma, \tau, \rho, \ldots$ (mostly using Greek letters). Let G be an extension of H by F. Then, an element σ of F is a left coset of H in G. We choose a representative from each coset and let the representative of an element σ of F be $t(\sigma)$. In order to simplify the notation, we will assume that the representative of the coset H is the identity: $t(1) = 1$. The product $t(\sigma)t(\tau)$ belongs to the coset $\sigma\tau$, so we can find an element $f(\sigma, \tau)$ of H such that

(7.1) $t(\sigma)t(\tau) = t(\sigma\tau)f(\sigma, \tau)$.

Since $H \lhd G$, an automorphism $T(\sigma)$ of H is induced by conjugation by each element $t(\sigma)$. In general, the automorphism $T(\sigma)$ depends on the choice of the representative $t(\sigma)$. By definition, we have

(7.2) $t(\sigma)^{-1}ht(\sigma) = T(\sigma)(h)$ $(h \in H)$.

Since $t(1) = 1$, the following formulas hold for any $\sigma, \tau \in F$:

(7.3) $T(1) = 1, \quad f(\sigma, 1) = f(1, \tau) = 1$.

From (7.1), we now get

(7.4) $T(\sigma)T(\tau) = T(\sigma\tau)if(\sigma, \tau)$,

where $if(\sigma, \tau)$ is the inner automorphism of H induced by the element $f(\sigma, \tau)$. From the formula (7.1), we also get

$$(t(\sigma)t(\tau))t(\rho) = t(\sigma\tau)f(\sigma, \tau)t(\rho)$$
$$= t((\sigma\tau)\rho)f(\sigma\tau, \rho)T(\rho)(f(\sigma, \tau)).$$

On the other hand,

$$t(\sigma)(t(\tau)t(\rho)) = t(\sigma(\tau\rho))f(\sigma, \tau\rho)f(\tau, \rho).$$

So, the associative law implies the following formula:

(7.5) $f(\sigma, \tau\rho)f(\tau, \rho) = f(\sigma\tau, \rho)T(\rho)(f(\sigma, \tau))$.

If a family $\{T(\sigma)\}$ of automorphisms of the group H and a family $\{f(\sigma, \tau)\}$ of elements of H satisfy the conditions (7.3), (7.4), and (7.5), we call f a **factor set** belonging to the family T. If a factor set f is obtained from the extension G as in the preceding discussion, we call f a *factor set associated with the extension G*. A factor set associated with the extension G depends not only on the extension G, but also the choice of a transversal t of H. (The conditions (7.3) are not essential for the subsequent discussions, but they help simplify some of the computations.)

The following theorem holds.

Theorem 7.6. *Let $\{f(\sigma, \tau)\}$ be a factor set belonging to a family $\{T(\sigma)\}$ of automorphisms of H. Then, there is an extension G of H by F such that, for a* *suitable choice of a transversal of H, f is a factor set associated with the extension G.*

Proof. We define an operation in the direct product set $F \times H$ as follows:

(7.7) $(\sigma, x)(\tau, y) = (\sigma\tau, f(\sigma, \tau)T(\tau)(x)y)$.

We will show that the set $F \times H$ forms a group under this operation. First, we will prove the associative law. By definition, we have

$$((\sigma, x)(\tau, y))(\rho, z) = ((\sigma\tau)\rho, f(\sigma\tau, \rho)T(\rho)(f(\sigma, \tau)T(\tau)(x)y)z).$$

From (7.4) and (7.5), we know that the second term of the right side is equal to

$$f(\sigma\tau, \rho)T(\rho)(f(\sigma, \tau))T(\tau\rho)if(\tau, \rho)xT(\rho)(y)z$$

$$= f(\sigma, \tau\rho)T(\tau\rho)(x)f(\tau, \rho)T(\rho)(y)z.$$

On the other hand, the definitions give us

$$(\sigma, x)((\tau, y)(\rho, z)) = (\sigma, x)(\tau\rho, f(\tau, \rho)T(\rho)(y)z)$$

$$= (\sigma(\tau\rho), f(\sigma, \tau\rho)T(\tau\rho)(x)f(\tau, \rho)T(\rho)(y)z).$$

This proves that the operation defined in $F \times H$ satisfies the associative law. The identity is the pair $(1, 1)$. The inverse of (σ, x) is (σ^{-1}, y) where

$$y^{-1} = f(\sigma, \sigma^{-1})T(\sigma^{-1})(x).$$

From the definition of y, we have

$$(\sigma, x)(\sigma^{-1}, y) = (1, 1).$$

This formula is already sufficient to prove that (σ^{-1}, y) is the inverse of (σ, x) (see Exercise 1 of Chapter 1, §1), but the other formula

$$(\sigma^{-1}, y)(\sigma, x) = (1, 1)$$

can also be proved by using (7.4) $(\sigma \to \sigma^{-1}, \tau \to \sigma)$, (7.3) and (7.5) $(\tau = \sigma^{-1}, \rho = \sigma)$.

Let G be the group equipped with the operation (7.7) on the set $F \times H$. Then, we call G the group constructed from the factor set $\{T(\rho), f(\sigma, \tau)\}$. As can be seen from (7.7), the mapping defined by sending (σ, x) to σ is a homomorphism from G onto F, and its kernel K consists of all pairs $(1, x)$:

$$K = \{(1, x)\}.$$

Also from (7.7), we can see that the mapping $(1, x) \to x$ is an isomorphism from K onto H. Thus, the group G is an extension by F of a group isomorphic to H. We can choose a representative $(\sigma, 1)$ from each coset of K. We then have the following formulas:

$$(\sigma, 1)(\tau, 1) = (\sigma\tau, f(\sigma, \tau)) = (\sigma\tau, 1)(1, f(\sigma, \tau))$$
$$(\sigma, 1)^{-1}(1, x)(\sigma, 1) = (1, T(\sigma)(x)).$$

Comparing these with (7.1) and (7.2), we see that the factor set associated with the extension G is precisely $\{T(\rho), f(\sigma, \tau)\}$. ☐

By Theorem 7.6, the problem of studying the extension of a given group H by a factor group F is reduced to a problem about factor sets. But, for a given group G, the corresponding factor set is not unique. If we choose another transversal $\{t'(\sigma)\}$ of H such that $t'(1) = 1$, we have a family of elements $\{h(\sigma)\}$ of H such that

$$t'(\sigma) = t(\sigma)h(\sigma) \quad \text{for all} \quad \sigma \in F.$$

Thus, the corresponding family of automorphisms $\{T'(\sigma)\}$ satisfies

(7.8) $$T'(\sigma) = T(\sigma)ih(\sigma) \qquad (\sigma \in F),$$

where $ih(\sigma)$ denotes the inner automorphism by the element $h(\sigma)$. Furthermore, we have

(7.9) $$f'(\sigma, \tau) = h(\sigma\tau)^{-1}f(\sigma, \tau)T(\tau)(h(\sigma))h(\tau).$$

We call two factor sets

$$\{T(\rho), f(\sigma, \tau)\} \quad \text{and} \quad \{T'(\rho), f'(\sigma, \tau)\}$$

associated or *equivalent* if there is a family $\{h(\sigma)\}$ of elements of H which satisfy the conditions (7.8) and (7.9), as well as the formula $h(1) = 1$. The next proposition supplements Theorem 7.6, and it will be restated later in a more precise form (7.16).

(7.10) *Two extensions constructed from associated factor sets are isomorphic.*

Proof. Let the extension constructed from the factor set

$$\{T(\rho), f(\sigma, \tau)\}$$

be the group G on the set $F \times H$ in which the operation is defined by the formula (7.7). Similarly, let G' be the extension constructed from the factor set

$$\{T'(\rho), f'(\sigma, \tau)\}.$$

If two factor sets are associated, then there is a family $\{h(\sigma)\}$ of elements of H which satisfies (7.8) and (7.9). We define a mapping φ from G into G' by the formula

$$\varphi: (\sigma, x) \to (\sigma, h(\sigma)^{-1}x).$$

It follows from the definition and the formulas (7.8) and (7.9) that φ is an isomorphism from G into G'. In fact, the following formula holds in G':

$$(\sigma, h(\sigma)^{-1}x)(\tau, h(\tau)^{-1}y) = (\sigma\tau, f'(\sigma, \tau)T'(\tau)(h(\sigma)^{-1}x)h(\tau)^{-1}y).$$

By (7.8) and (7.9), the second term of the right side is equal to

$$h(\sigma\tau)^{-1}f(\sigma, \tau)T(\tau)(x)y.$$

Since we have the formula (7.7) in G, we see that φ is indeed a homomorphism from G into G'. Clearly, φ is an isomorphism. We note that $\varphi((1, x)) = (1, x)$. $\square$

We can see from the formula (7.8) that the class containing $T(\sigma)$ in the group Out $H = \text{Aut } H/\text{Inn } H$ does not depend on the choice of representatives $t(\sigma)$. We have a function ψ defined by

$$\psi(\sigma) = T(\sigma) \text{ Inn } H.$$

The formula (7.4) shows that ψ is a homomorphism from F into Out H. This homomorphism depends only on G; we write $\psi = \psi_G$. There is a case in which ψ_G uniquely determines the extension G.

(7.11) *Let H and F be two groups, and let G be an extension of H by F. Suppose that $Z(H) = \{1\}$. Then, the structure of the group G is uniquely determined by the homomorphism ψ_G. Furthermore, if ψ is a homomorphism from F into Out H, then there is an extension G of H by F such that $\psi_G = \psi$. In this case, the isomorphism class of G is uniquely determined. All the extensions are realized in the direct product $F \times \text{Aut } H$. Thus, an extension G of H by F is isomorphic to a subgroup U of $F \times \text{Aut } H$ which satisfies the two conditions $U \cap \text{Aut } H = \text{Inn } H$ and $\pi(U) = F$, where π is the projection from $F \times \text{Aut } H$ onto the first factor F.*

Proof. Let G be an extension of H by F. Since $Z(H) = \{1\}$, we have $H \cap C_G(H) = \{1\}$. As $H \lhd G$, the subgroup $C_G(H)$ is also normal. Therefore, by Theorem 6.11 of Chapter 1, the factor group $G/C_G(H)$ is isomorphic to a subgroup of Aut H. Let θ be this isomorphism from $G/C_G(H)$ onto a subgroup L of Aut H. We see that θ is determined as follows. Let θ_g be the automorphism of H induced by the conjugation of the element g. Then, θ is induced by the mapping $g \to \theta_g$. Clearly, we have

$$\text{Inn } H \subset L \subset \text{Aut } H.$$

Since $H \cap C_G(H) = \{1\}$, Proposition (4.20) shows that G is isomorphic to a subgroup of $(G/H) \times (G/C_G(H))$.

So, G is isomorphic to a subgroup of $F \times \operatorname{Aut} H$. We will describe this embedding of G in more detail. The proof of (4.20) indicates that the element g of G is mapped to (gH, θ_g) of $F \times \operatorname{Aut} H$. If $\sigma = gH$, then θ_g and $T(\sigma)$ differ only by an inner automorphism of H. From the definition of ψ_G, we get $\operatorname{Im} \psi_G = L/\operatorname{Inn} H$. For an element σ of F, $\psi_G(\sigma)$ is a coset of $\operatorname{Inn} H$. If ρ is an element of L contained in $\psi_G(\sigma)$, then, as $\theta(G) \supset \operatorname{Inn} H$, there is an element g of G such that $\sigma = gH$ and $\rho = \theta_g$. Thus, the image of G in $F \times \operatorname{Aut} H$ is the subgroup determined by F, $\operatorname{Ker} \psi_G$, L, $\operatorname{Inn} H$, and the isomorphism from $F/\operatorname{Ker} \psi_G$ onto $L/\operatorname{Inn} H$ induced by ψ_G (see (4.19)). These subgroups and the isomorphism are determined by ψ_G alone. Hence, the structure of G is uniquely determined by ψ_G.

Suppose that ψ is a homomorphism from F into $\operatorname{Out} H$. Let $L/\operatorname{Inn} H$ be the image of F under ψ. By (4.19), the subgroups F, $\operatorname{Ker} \psi$, L, $\operatorname{Inn} H$, and the isomorphism from $F/\operatorname{Ker} \psi$ onto $L/\operatorname{Inn} H$ induced by ψ determine a unique subgroup G of $F \times \operatorname{Aut} H$. Again by (4.19), we have

$$G \cap \operatorname{Aut} H = \operatorname{Inn} H \quad \text{and} \quad G/\operatorname{Inn} H \cong F = \pi(G),$$

where π is the projection into F. By assumption $Z(H) = \{1\}$, so we have $\operatorname{Inn} H \cong H$. Thus, G is an extension of H by F. We can easily check that $\psi_G = \psi$ (see the formula (6.7) of Chapter 1). □

If $Z(H) \neq \{1\}$, a homomorphism ψ from F into $\operatorname{Out} H$ need not have the form ψ_G associated with an extension (see Exercise 3). The preceding proof shows that the structure of the factor group $G/Z(H)$ is uniquely determined by the homomorphism ψ_G (Nagao [1]).

As we can immediately see from the definition of the group G constructed from the factor set $\{T(\rho), f(\sigma, \tau)\}$, G is the semidirect product with respect to the action T if $f(\sigma, \tau) = 1$ for all σ and τ of F. In general, if the factor set associated with the extension G is equivalent to the *trivial* factor set satisfying $f(\sigma, \tau) = 1$ for all $\sigma, \tau \in F$, then by (7.10), the extension G is isomorphic to some semidirect product of H and F. Conversely, the factor set associated with a semidirect product is equivalent to the trivial factor set.

Sometimes, it may be more convenient to study extensions of groups using the notion of exact sequences.

Suppose that we have a sequence $\{G_n\}$ of groups and a sequence of homomorphisms f_i from G_i into G_{i+1}. We will express these homomorphisms by arrows between the groups:

(7.12) $\cdots \longrightarrow G_{n-1} \xrightarrow{f_{n-1}} G_n \xrightarrow{f_n} G_{n+1} \longrightarrow \cdots .$

The letters above the arrows denote the homomorphisms, and (7.12) expresses that f_n is a homomorphism from G_n into G_{n+1}, etc. We will sometimes omit the letters and denote the homomorphisms simply by arrows. The set of suffixes may be finite or infinite.

Definition 7.13. The above sequence (7.12) is said to be **exact** if we have $\operatorname{Im} f_{n-1} = \operatorname{Ker} f_n$ for each n.

With this concept we can say that a group G is an extension of H by F if and only if the following sequence is exact:

$$1 \to H \to G \to F \to 1,$$

where $H \to G$ is the injection of H in G, and $G \to F$ is the canonical homomorphism from G onto F. Any short exact sequence

$$1 \to A \xrightarrow{\kappa} B \xrightarrow{\lambda} C \to 1$$

is equivalent to saying that the middle group B is an extension of $\kappa(A)$ by C. In the following, we may use short exact sequences instead of extensions of groups.

In general, a diagram is a collection of groups and a collection of arrows between some of these groups. For example,

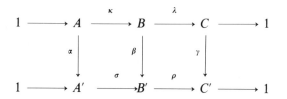

is a diagram. Letters above or to the sides of the arrows designate homomorphisms. In the above diagram, there are two mappings $\kappa\beta$ and $\alpha\sigma$ from A to B'. If $\alpha\sigma = \kappa\beta$, then we say that the left square is a **commutative diagram**. If we have $\lambda\gamma = \beta\rho$ in addition to $\alpha\sigma = \kappa\beta$, we say that the above diagram is commutative.

(7.14) *Assume that the above diagram is commutative and that both the upper and lower rows are exact. If α and γ are monomorphisms (epimorphisms or isomorphisms), then so is β.*

Proof. Assume that both α and γ are monomorphisms. We will show that β is a monomorphism. Suppose that $\beta(b) = 1$ for some $b \in B$. Then, we have

$$1 = \beta\rho(b) = \lambda\gamma(b) = \gamma(\lambda(b)),$$

from which we get $\lambda(b) = 1$ since γ is a monomorphism. By assumption, the upper row is exact. So, we have Ker $\lambda = $ Im κ. Therefore, there is an element a of A such that $b = \kappa(a)$. Then, we have

$$1 = \beta(b) = \kappa\beta(a).$$

The commutativity of the diagram gives us

$$\kappa\beta = a\sigma \quad \text{and} \quad \alpha\sigma(a) = 1.$$

But, by assumption, both α and σ are injective. So, we get

$$a = 1 \quad \text{and} \quad b = \kappa(a) = 1.$$

This proves that β is a monomorphism.

Assume that both α and γ are surjective. Let $b' \in B'$. We will show that $b' \in \beta(B)$. Since γ is surjective, there is an element c of C such that $\rho(b') = \gamma(c)$. By assumption, the upper row is exact. We conclude that $c = \lambda(b)$ for some element b of B. So,

$$\rho(b') = \lambda\gamma(b) = \rho(\beta(b))$$

by the commutativity of the diagram. Set $b' = \beta(b)b''$. Then, we have $\rho(b'') = 1$. Since the lower row is exact, we have Ker $\rho = $ Im σ. Hence, there is an element a' of A' such that $\sigma(a') = b''$. As α is also surjective, we have an element a of A satisfying $\alpha(a) = a'$. Thus,

$$b'' = \alpha\sigma(a) = \beta(\kappa(a)), \quad \text{and we have} \quad b' = \beta(b\kappa(a)) \in \beta(B).$$

This shows that β is an epimorphism. $\square$

We will represent an extension by a short exact sequence.

Definition 7.15. Let H and F be two fixed groups. Two extensions G and G' of H by F are called **equivalent** if there is a homomorphism $\varphi: G \to G'$ such that the following diagram is commutative.

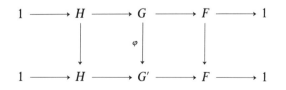

In the above diagram, the two rows are the short exact sequences representing the extensions G and G', and the homomorphisms $H \to H$ and

$F \rightarrow F$ are the identity mappings. By (7.14), if G and G' are equivalent by a homomorphism φ, then φ is an isomorphism. So, G' is equivalent to G by the inverse isomorphism φ^{-1}. Thus, Definition 7.15 defines an equivalence relation among the extensions of H by F. Also, equivalent extensions are isomorphic, but isomorphic extensions need not be equivalent (see Example 2 at the end of this section).

(7.16) *Let G and G' be two extensions. Then, G is equivalent to G' if and only if the factor set associated with G is equivalent to the factor set associated with G'.*

Proof. If the two extensions G and G' are equivalent, we have the commutative diagram of Definition 7.15. Suppose that we choose a transversal $\{t(\sigma)\}$ of H in G and get a factor set $\{T(\rho), f(\sigma, \tau)\}$. From the commutativity of the diagram, $\varphi(t(\sigma))$ is an element of G' contained in the coset corresponding to σ. So, if we choose the transversal $\{\varphi(t(\sigma))\}$ in G', then by applying φ to the formulas (7.1) and (7.2), we can see that the factor set associated with G' is exactly the same as $\{T(\rho), f(\sigma, \tau)\}$. Note that φ leaves every element of H invariant (by the commutativity of the left square). Thus, equivalent extensions give rise to equivalent factor sets.

The converse is also true. If two factor sets are equivalent, then we have the isomorphism φ from G onto G' which we constructed in the proof of (7.10). It is easy to see that φ makes the diagram of Definition 7.15 commutative. □

Cohomology of Groups

In the following discussion, we will consider the equivalence classes of the extensions of an abelian group. We will assume that H is abelian although F may be an arbitrary group. Since this is the case, we have Inn $H = \{1\}$. So, the automorphism $T(\sigma)$ does not depend on the choice of transversals, and it satisfies

$$T(\sigma)T(\tau) = T(\sigma\tau).$$

This means that F acts on H via the action T, and we write $T(\sigma)x = x^\sigma$ as before. The condition for $\{f(\sigma, \tau)\}$ to be a factor set is

$$f(\sigma, \tau\rho)f(\tau, \rho) = f(\sigma\tau, \rho)f(\sigma, \tau)^\rho.$$

We will deal only with factor sets with respect to the given action T. Thus, we will consider the action T, as well as the two groups H and F to be fixed.

The assumption that H is abelian gives rise to another essential difference: we can define an operation among the factor sets so as to make the set of equivalence classes an abelian group. If f and g are two factor sets, the sum $f + g$ is defined by the formula

$$(f + g)(\sigma, \tau) = f(\sigma, \tau)g(\sigma, \tau).$$

It is easy to verify that $f + g$ is a factor set. Clearly, the totality of factor sets forms an abelian group. The identity is the factor set f_0 such that $f_0(\sigma, \tau) = 1$ for all σ and τ in F. The inverse of f is the factor set g satisfying

$$g(\sigma, \tau) = f(\sigma, \tau)^{-1}.$$

Let $Z^2(F, H)$ be this abelian group of factor sets.

Suppose that a function h from F to H is given and satisfies $h(1) = 1$. The formula

$$f(\sigma, \tau) = h(\sigma)^\tau h(\tau)h(\sigma\tau)^{-1}$$

defines a function f from $F \times F$ to H. It is easy to show that f is a factor set; let $f = \delta h$. The totality of factor sets of the form δh will be denoted by $B^2(F, H)$. Since the formula

$$\delta(h + k) = \delta h + \delta k$$

holds, the subset $B^2(F, H)$ forms a subgroup of $Z^2(F, H)$. The factor group

$$H^2(F, H) = Z^2(F, H)/B^2(F, H)$$

is called **the 2-dimensional cohomology group**. By (7.9), two factor sets are equivalent if and only if they belong to the same cosets of $B^2(F, H)$. Thus, by (7.16), the equivalence classes of the extensions of H by F are in a one-to-one correspondence with the elements of the cohomology group.

For any positive integer n, the n-dimensional cohomology group is defined. We have to refer to a standard textbook (for example, Nakayama–Hattori [1]) for more comprehensive discussion of cohomology theory. We will only present some of the more elementary aspects of cohomology theory, and we will limit our discussions to those essential definitions and theorems which pertain to our treatment of group theory.

We begin with the concept of group rings and modules over various group rings. Let $\mathbb{Z}$ be the ring of integers. The **group ring** of a group G over $\mathbb{Z}$ is the totality of finite sums of the form $\sum n_g g$ ($n_g \in \mathbb{Z}$, $g \in G$) equipped with termwise addition and the multiplication induced naturally by the group operation. Thus, we have

$$\sum n_g g + \sum m_g g = \sum (n_g + m_g)g, \qquad (\sum n_g g)(\sum m_g g) = \sum \left(\sum_x n_x m_{x^{-1}g} \right)g.$$

Let Γ be the group ring of G over $\mathbb{Z}$. We will study Γ-modules (see Definition 5.12). Homomorphisms between Γ-modules are always assumed to be Γ-homomorphisms unless it is explicitly stated otherwise.

Free Γ-modules are defined as in Definition 5.6 and have the properties corresponding to (5.7), Theorems 5.8, 5.9, and 5.10 for free abelian groups. We have collected these essential properties in the following omnibus lemma.

(7.17) *The following properties of a subset X of a Γ-module F are equivalent.*

(i) *A mapping from X into a Γ-module A is uniquely extended to a homomorphism of F into A.*

(ii) *For each $x \in X$, the correspondence $x\gamma \to \gamma$ is a Γ-isomorphism between $x\Gamma$ and Γ, and F is the direct sum of all submodules $x\Gamma$.*

A Γ-module which satisfies these conditions is the **free Γ-module** freely generated by X.

Corollary. *Any Γ-module is a homomorphic image of a free Γ-module.*

(7.18) *In the following diagram of Γ-modules, suppose that the row is exact, $f\beta = 0$, and F is free. Then, there exists a homomorphism g from F to A such that $g\alpha = f$.*

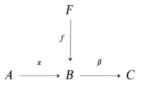

Proof. Let X be a set of free generators of F. If $x \in X$, then $\beta(f(x)) = 0$. Thus, $f(x)$ belongs to Ker β. Since the row is exact, we have Ker $\beta = \mathrm{Im}\ \alpha$, and there is an element a_x of A such that $\alpha(a_x) = f(x)$. By (7.17), the mapping $x \to a_x$ can be extended to a homomorphism g from F into A. Since $g(x) = a_x$, we have $g\alpha(x) = f(x)$ for every $x \in X$. Again by (7.17), F is a direct sum of the submodules $x\Gamma$. So, we have $g\alpha = f$. $\square$

Definition 7.19. We can consider $\mathbb{Z}$ as a Γ-module by setting $n(\sum n_g g) = (\sum n_g)n$. (Note that this action does indeed make $\mathbb{Z}$ a Γ-module; it is called the **trivial Γ-module** $\mathbb{Z}$.) A sequence X consisting of free Γ-modules X_i ($i = 0, 1, 2, \ldots$) and homomorphisms d_i is called a **free resolution** of $\mathbb{Z}$ if the sequence

$$\mathscr{X}: \cdots \longrightarrow X_n \xrightarrow{\ d_n\ } X_{n-1} \longrightarrow \cdots$$

$$\longrightarrow X_1 \xrightarrow{\ d_1\ } X_0 \xrightarrow{\ d\ } \mathbb{Z} \longrightarrow 0$$

is exact.

If A is any Γ-module and $\mathscr{X}$ is a free resolution of $\mathbb{Z}$, we have

$$\cdots \leftarrow \operatorname{Hom}_\Gamma(X_1, A) \leftarrow \operatorname{Hom}_\Gamma(X_0, A) \leftarrow \operatorname{Hom}_\Gamma(\mathbb{Z}, A) \leftarrow 0,$$

where the homomorphism

$$\operatorname{Hom}_\Gamma(X_{i-1}, A) \to \operatorname{Hom}_\Gamma(X_i, A)$$

is defined by d_i^* (see (5.13)). By (5.13), we get $d_i^* d_{i+1}^* = 0$ which implies that $\operatorname{Im} d_i^* \subset \operatorname{Ker} d_{i+1}^*$. The factor group

$$H^n(G, A) = \operatorname{Ker} d_{n+1}^* / \operatorname{Im} d_n^*$$

is called the n-dimensional **cohomology group** of A with respect to Γ (or the action of G). Cohomology groups are usually written additively.

It is important that the cohomology group is uniquely determined independent of the choice of free resolutions of $\mathbb{Z}$. In order to prove this assertion, we need the following lemma.

(7.20) *Let $\mathscr{X}$ and $\mathscr{Y}$ be two free resolutions of the trivial Γ-module $\mathbb{Z}$.*

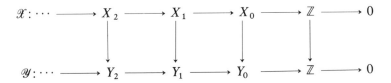

Then, beginning with the identity mapping $\mathbb{Z} \to \mathbb{Z}$, we have a family of homomorphisms f_i from X_i to Y_i such that the above diagram is commutative. If $\{f_i\}$ and $\{f_i'\}$ are two families of homomorphisms which make the diagram commutative, then $\{f_i\}$ and $\{f_i'\}$ are homotopic; that is, there is a family of homomorphisms h_i from X_i into Y_{i+1} such that $f_i - f_i' = h_i d_{i+1}' + d_i h_{i-1}$, where $\{d_i\}$ is the set of homomorphisms of $\mathscr{X}$ and $\{d_i'\}$ is the set of homomorphisms of $\mathscr{Y}$.

Proof. Suppose that we have already defined the homomorphisms $f_0, f_1, \ldots, f_i$ for some i so as to make part of the above diagram commutative. Look at the following diagram:

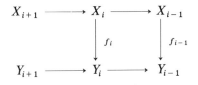

Set $f = d_{i+1} f_i$. Then $f d_i' = d_{i+1} f_i d_i' = d_{i+1} d_i f_{i-1} = 0$ by the commutativity and the exactness of the upper row. Since the lower row is also exact, (7.18) produces a homomorphism f_{i+1} from X_{i+1} into Y_{i+1} such that $f_{i+1} d_{i+1}' = f$.

Thus, with f_{i+1} at the left end, the above diagram is commutative. By induction, we conclude that there is a family of homomorphisms $\{f_i\}$ which satisfies the commutativity conditions.

The existence of a family $\{h_i\}$ of homomorphisms is also proved by induction on i. Suppose that $h_0, \ldots, h_{i-1}$ have been defined to satisfy the relation

$$f_{i-1} - f'_{i-1} = h_{i-1} d'_i + d_{i-1} h_{i-2}.$$

Set $f = f_i - f'_i - d_i h_{i-1}$. Then,

$$fd'_i = (f_i - f'_i)d'_i - d_i h_{i-1} d'_i$$
$$= d_i(f_{i-1} - f'_{i-1}) - d_i h_{i-1} d'_i = d_i(d_{i-1} h_{i-2}) = 0$$

by the commutativity and the exactness. By (7.18) again, there is a homomorphism h_i from X_i into Y_{i+1} such that $f = h_i d'_{i+1}$. $\square$

Theorem 7.21. *The cohomology group $H^n(G, A)$ is uniquely determined and is independent of the choice of the free resolution of $\mathbb{Z}$.*

Proof. Let $\mathscr{X}$ and $\mathscr{Y}$ be two free resolutions. Then, by (7.20), there is a family of homomorphisms $\{f_i\}$ from $\mathscr{X}$ to $\mathscr{Y}$ and a family $\{g_i\}$ from $\mathscr{Y}$ to $\mathscr{X}$. Then, $\{f_i g_i\}$ is a family of homomorphisms from $\mathscr{X}$ to itself. Since both the family $\{f_i g_i\}$ and the family of identity mappings make the diagram $\mathscr{X} \to \mathscr{X}$ commutative, (7.20) shows that they are homotopic, that is, there is a family of homomorphisms $\{h_i\}$ which satisfies the relations

$$f_i g_i - 1 = h_i d_{i+1} + d_i h_{i-1}$$

(we set $h_{-1} = 0$ so that the relation holds for $i = 0$).

By (5.13), g_i^* maps $\mathrm{Hom}_\Gamma(X_i, A)$ into $\mathrm{Hom}_\Gamma(Y_i, A)$, and we have $g_i^*(d'_{i+1})^* = d^*_{i+1} g^*_{i+1}$. Therefore,

$$g_i^*(\mathrm{Ker}\ d^*_{i+1}) \subset \mathrm{Ker}(d'_{i+1})^*$$

and

$$g_i^*(\mathrm{Im}\ d_i^*) \subset \mathrm{Im}(d'_i)^*.$$

This shows that g_i^* maps

$$\mathrm{Ker}\ d^*_{i+1}/\mathrm{Im}\ d_i^* \quad \text{into} \quad \mathrm{Ker}(d'_{i+1})^*/\mathrm{Im}(d'_i)^*.$$

So, g_i^* maps the cohomology group determined by $\mathscr{X}$ into the one which is determined by $\mathscr{Y}$. Similarly, f_i^* gives a homomorphism of the reverse direction.

As $\{f_i g_i\}$ and 1 are homotopic, we have

$$g_i^* f_i^* - 1^* = d_{i+1}^* h_i^* + h_{i-1}^* d_i^*.$$

Hence, if $x \in \operatorname{Ker} d_{i+1}^*$, we have $g_i^* f_i^* x - x = d_i^*(h_{i-1}^* x)$. This shows that $g_i^* f_i^*$ is the identity on the cohomology group defined by $\mathscr{X}$. Similarly, $f_i^* g_i^*$ is also the identity mapping on the cohomology group defined by $\mathscr{Y}$. Therefore, the cohomology group $H^n(G, A)$ does not depend on the choice of the free resolution of $\mathbb{Z}$. $\quad\square$

Standard Bar Resolution

It may be fairly obvious that a free resolution of the trivial Γ-module $\mathbb{Z}$ exists. By the Corollary of (7.17), the Γ-module $\mathbb{Z}$ is a homomorphic image of a free Γ-module X_0. Let $X_0/K_1 \cong \mathbb{Z}$. We express the kernel K_1 as a quotient module X_1/K_2 of another free Γ-module X_1. By repeating this process, we get a free resolution of $\mathbb{Z}$:

$$\to X_n \to X_{n-1} \to \cdots \to X_0 \to \mathbb{Z} \to 0.$$

In order to compute the cohomology groups, a special resolution is useful. We define the **bar resolution** (standard resolution) of $\mathbb{Z}$ as follows.

Let n be any nonnegative integer. For a fixed n, we will consider the symbols $[x_1 | x_2 | \cdots | x_n]$, where $x_1, x_2, \ldots, x_n$ are nonidentity elements of G, and let B_n be the free Γ-module generated freely by the totality of such symbols. We will extend the meaning of the symbols and set $[x_1 | x_2 | \cdots | x_n] = 0$ if any of the elements $x_1, x_2, \ldots, x_n$ is the identity of G. If $n = 0$, then B_0 is a free Γ-module generated freely by a single symbol $[\ ]$ with no element inside the bracket; we have $B_0 \cong \Gamma$. Therefore, there is a Γ-homomorphism ∂_0 from B_0 into $\mathbb{Z}$ which corresponds to the mapping ε from Γ into $\mathbb{Z}$ defined by $\varepsilon(\sum n_g g) = \sum n_g$. We define a Γ-homomorphism ∂_n from B_n into B_{n-1} for $n > 0$ as follows. Since B_n is free on the set of elements $[x_1 | x_2 | \cdots | x_n]$, we need only to define $\partial = \partial_n$ on these symbols:

$$\partial[x_1 | x_2 | \cdots | x_n] = [x_2 | x_3 | \cdots | x_n]$$
$$+ \sum (-1)^i [x_1 | \cdots | x_i x_{i+1} | \cdots | x_n]$$
$$+ (-1)^n [x_1 | \cdots | x_{n-1}] x_n.$$

We will show that $\{B_n\}$ and $\{\partial_n\}$ define a free resolution of $\mathbb{Z}$. We will define a set of homomorphisms s_n from B_n into B_{n+1} (which are just homomorphisms of abelian groups). Each B_n is a free Γ-module. Thus, it is a free abelian group freely generated by $[x_1 | \cdots | x_n]x$ for some $x \in G$. Define s_n as follows. If

$n = -1$, s_{-1} is the mapping from $\mathbb{Z}$ into B_0 defined by $n \to n[\ ]$. If $n \geq 0$, then s_n maps

$$[x_1 | \cdots | x_n]x \quad \text{into} \quad (-1)^{n+1}[x_1 | \cdots | x_n | x].$$

It is easy to verify that the following formulas hold:

$$s_{-1} \partial_0 = 1, \qquad s_i \partial_{i+1} + \partial_i s_{i-1} = 1 \qquad (i \geq 0).$$

The second formula gives us $\text{Ker } \partial_i \subset \text{Im } \partial_{i+1}$ for $i \geq 0$. We have

$$\partial_i = s_i \partial_{i+1} \partial_i + \partial_i s_{i-1} \partial_i.$$

If $i = 0$, then we get

$$s_0 \partial_1 \partial_0 = 0 \quad \text{as} \quad s_{-1} \partial_0 = 1.$$

By the definition of s_0, $s_0 B_0$ contains all the free generators of B_1. So, we get $\partial_1 \partial_0 = 0$. If $i > 0$, we get $\partial_i s_{i-1} \partial_i + \partial_i \partial_{i-1} s_{i-2} = \partial_i$, and

$$\partial_i \partial_{i-1} s_{i-2} = s_i \partial_{i+1} \partial_i.$$

So, induction on i gives us $\partial_{i+1} \partial_i = 0$. This implies that $\text{Im } \partial_{i+1} \subset \text{Ker } \partial_i$. We have shown before that the opposite containment holds, so $\text{Im } \partial_{i+1} = \text{Ker } \partial_i$, and the sequence $\{B_n\}$, together with the homomorphisms $\{\partial_n\}$, is a free resolution of the Γ-module $\mathbb{Z}$.

If B_n is the free Γ-module defined above, then, for any Γ-module A, $\text{Hom}_{\mathbb{Z}}(B_n, A)$ is isomorphic to the additive group $C^n(G, A)$ consisting of all the functions $f(x_1, \ldots, x_n)$ of n variables $x_1, \ldots, x_n \in G$ such that

$$f(x_1, \ldots, x_n) \in A \quad \text{and} \quad f(x_1, \ldots, x_n) = 0$$

whenever any of the elements $x_1, \ldots, x_n$ is the identity of G. The elements of $C^n(G, A)$ are called the **normalized cochains**. The homomorphism $\delta = \partial^*$ is given by the following formula:

(7.22) $\delta f(x_1, \ldots, x_n, x_{n+1}) = f(x_2, \ldots, x_{n+1})$
$$+ \sum (-1)^i f(x_1, \ldots, x_i x_{i+1}, \ldots, x_{n+1})$$
$$+ (-1)^{n+1} f(x_1, \ldots, x_n) x_{n+1}.$$

We write $\text{Ker } \delta_{n+1} = Z^n(G, A)$ and $\text{Im } \delta_n = B^n(G, A)$. The elements of these groups are called *cocycles* and *coboundaries*, respectively. By definition, we get

$$H^n(G, A) = Z^n(G, A)/B^n(G, A).$$

Cohomology groups of low dimensions are easily described using co-chains.

The 1-dimensional cohomology group $H^1(G, A)$.
We find

$$\text{Ker } \delta_2 = \{f \mid f(xy) = f(x)y + f(y)\},$$
$$\text{Im } \delta_1 = \{g \mid g(x) = a - ax \quad (a \in A)\}.$$

An element of Ker δ_2 is called a *derivation* or a *crossed homomorphism*. The elements of Im δ_1 are called *inner derivations* or *principal crossed homomorphisms*.

The 2-dimensional cohomology group $H^2(G, A)$.
An element f of Ker δ_3 satisfies the identity

$$f(x_2, x_3) - f(x_1 x_2, x_3) + f(x_1, x_2 x_3) - f(x_1, x_2)x_3 = 0.$$

Hence, f is nothing but a factor set with respect to the action of G on A. Im δ_2 is the set of functions of the form

$$f(x, y) = g(y) - g(xy) + g(x)y.$$

When A is an abelian group on which the group G acts, the group $H^2(G, A)$ is the group of equivalence classes of the extensions of A by G.

Cohomology of Subgroups

We will study the relationship between the cohomology groups of a group G and its subgroup K. Let $\Gamma = \mathbb{Z}(G)$ be the group ring of G over $\mathbb{Z}$, and let $\Delta = \mathbb{Z}(K)$ be the group ring of K. With these notations we have the following lemma.

(7.23) *The Δ-module Γ is a free Δ-module. In general, any free Γ-module is a free Δ-module.*

Proof. Clearly, any Γ-module becomes a Δ-module by restricting the action to the ring Δ. Let $\{t_\lambda\}$ be a left transversal of K. Then, G is a union of the cosets $t_\lambda K$, so we have $\Gamma = \sum t_\lambda \Delta$. Clearly, each $t_\lambda \Delta$ is Δ-isomorphic to Δ. From (7.17), we conclude that Γ is free. Since any free Γ-module F is a direct sum of submodules isomorphic to Γ, F is a direct sum of free Δ-modules and is free by the first part of this proposition. $\square$

Let $\mathscr{X} : \to X_2 \to X_1 \to X_0 \to \mathbb{Z} \to 0$ be a free resolution of the trivial Γ-module $\mathbb{Z}$. Then, each group X_i is a free Γ-module, so X_i is a free Δ-module by (7.23). Furthermore, since each homomorphism d_i is a Γ-homomorphism, d_i is trivially a Δ-homomorphism. Hence, $\mathscr{X}$ is a free resolution of the trivial Δ-module $\mathbb{Z}$. Thus, we can study simultaneously, cohomology groups of G and K using a single resolution $\mathscr{X}$. (This is one of the effects of Theorem 7.21.) We will consider $\mathscr{X}$ to be a fixed free resolution of the Γ-module $\mathbb{Z}$.

We will define the restriction mapping ρ. Let A be a Γ-module. By restricting the action to Δ, A becomes a Δ-module. So, any element f of $\mathrm{Hom}_\Gamma(X_n, A)$ is a Δ-homomorphism from the Δ-module X_n into the Δ-module A. The element f becomes an element of $\mathrm{Hom}_\Delta(X_n, A)$. Let ρf be the element f viewed as a Δ-homomorphism. We call ρf the restriction of f. The following proposition is fairly obvious.

(7.24) (i) *The restriction mapping ρ gives a homomorphism from* $\mathrm{Hom}_\Gamma(X_n, A)$ *into* $\mathrm{Hom}_\Delta(X_n, A)$.
(ii) *We have $\delta\rho = \rho\delta$.*

The first part asserts that $\rho(f + g) = \rho f + \rho g$. In the second part, the symbol δ represents collectively the homomorphisms induced in the Hom sequences by d_n. The equality indicates the commutativity of the following diagram:

$$
\begin{array}{ccc}
\mathrm{Hom}_\Gamma(X_n, A) & \xleftarrow{\;d_n*\;} & \mathrm{Hom}_\Gamma(X_{n-1}, A) \\
\Big\downarrow{\scriptstyle \rho_n} & & \Big\downarrow{\scriptstyle \rho_{n-1}} \\
\mathrm{Hom}_\Delta(X_n, A) & \xleftarrow{\;d_n*\;} & \mathrm{Hom}_\Delta(X_{n-1}, A)
\end{array}
$$

The proofs of these statements are relatively easy.

It follows from (ii) that the restriction mapping ρ induces a homomorphism from $H^n(G, A)$ into $H^n(K, A)$. We denote this homomorphism by $\rho* = \rho*(G, K)$.

If the index $|G : K|$ is finite, we can define a homomorphism, called the **corestriction**, going in the opposite direction. Let T be a left transversal of K in G. For any element f of $\mathrm{Hom}_\Delta(X_n, A)$, let τf be the function from X_n into A defined by

$$
\tau f(x) = \sum_{t \in T} f(xt)t^{-1}.
$$

We have the following proposition.

(7.25) (i) *The mapping τ does not depend on the choice of the transversal T.*
(ii) *The function τf lies in* $\mathrm{Hom}_\Gamma(X_n, A)$.

(iii) *The mapping τ is a homomorphism from $\operatorname{Hom}_\Delta(X_n, A)$ into $\operatorname{Hom}_\Gamma(X_n, A)$.*

(iv) *We have $\delta\tau = \tau\delta$.*

Proof. (i) Let $T = \{t_\lambda\}$. If we choose another representative t'_λ from $t_\lambda K$, then $t'_\lambda = t_\lambda k$ for some $k \in K$. Since f is a Δ-homomorphism, we get

$$f(xt'_\lambda)t'^{-1}_\lambda = f(xt_\lambda k)(t_\lambda k)^{-1}$$
$$= f(xt_\lambda)kk^{-1}t^{-1}_\lambda = f(xt_\lambda)t^{-1}_\lambda.$$

Thus, τf does not depend on the choice of T.

(ii) Let g be an element of G. It suffices to show that $\tau f(xg) = (\tau f(x))g$. If $T = \{t_\lambda\}$ is a transversal, so is $\{gt_\lambda\}$. Hence, by (i), we have

$$\tau f(xg) = \sum_t f(xgt)t^{-1} = \sum_{gt} f(xgt)(gt)^{-1}g = (\tau f(x))g.$$

(iii) This simply means that $\tau(f + g) = \tau f + \tau g$. The proof is trivial.

(iv) Since d_n is a Γ-homomorphism, $d_n(xt) = (d_n x)t$ holds for all $x \in X_n$ and $t \in T$. Thus, we get

$$\delta(\tau f)(x) = (\tau f)(dx) = \sum_t f((dx)t)t^{-1}$$
$$= \sum_t f(d(xt))t^{-1} = \sum_t \delta f(xt)t^{-1} = \tau(\delta f)(x).$$

This proves that $\tau\delta = \delta\tau$. $\square$

By (7.25) (iii) and (iv), the corestriction mapping τ induces a homomorphism from $H^n(K, A)$ into $H^n(G, A)$ which we denote by $\tau^* = \tau^*(K, G)$.

With these notations, the following theorem of Gaschütz–Eckmann holds.

Theorem 7.26. *Let K be a subgroup of finite index j in a group G. Then, for any element c of $H^n(G, A)$, we have*

$$\tau^*(\rho^* c) = jc.$$

Proof. Let f be an element of $\operatorname{Hom}_\Gamma(X_n, A)$ which represents the cohomology class c. Then, we get $\rho f(xt) = f(x)t$ for any element t of G. The definition of τ gives us

$$\tau(\rho f)(x) = \sum_t \rho f(xt)t^{-1} = \sum f(x) = jf(x).$$

This proves Theorem 7.26. $\square$

Corollary 1. *If the index $|G : K| = j$ is finite, the elements c in the kernel of the restriction mapping $\rho^*(G, K)$ satisfy $jc = 0$.*

Corollary 2. *Let G be a finite group of order g. Then, any element c of $H^n(G, A)$ satisfies $gc = 0$. Thus, if G is a finite π-group, then $H^n(G, A)$ is a π-group.*

Corollary 3. *Let K be a subgroup of a finite group G such that, for a prime number p, K contains at least one S_p-subgroup of G. Then, the p-primary component of $H^n(G, A)$ is isomorphic to a subgroup of the p-primary component of $H^n(K, A)$, and the restriction mapping $\rho^*(G, K)$ gives the isomorphism between them. For any $p \in \pi(G)$ (see the definition just before (2.11)), let S be an S_p-subgroup of G. Then, $H^n(S, A)$ is a p-group, and the p-primary component of $H^n(G, A)$ is isomorphic to a subgroup of $H^n(S, A)$. If $H^n(S, A) = \{0\}$ for all $p \in \pi(G)$, then we have $H^n(G, A) = \{0\}$.*

Corollary 4. *Let G be a finite group. If A is a finitely generated Γ-module, then $H^n(G, A)$ is finite.*

Proof. If $c \in \text{Ker } \rho^*$, then Theorem 7.26 shows that

$$jc = \tau^*(\rho^*c) = 0.$$

This proves Corollary 1. If $K = \{1\}$, each term of the bar resolution is trivial. Hence, we have $H^n(K, A) = \{0\}$ for $K = \{1\}$. So, Corollary 1 implies Corollary 2.

Under the assumptions of Corollary 3, the index $j = |G : K|$ is prime to p. So, by Corollary 1, Ker ρ^* contains no p-element except the identity. Thus, ρ^* maps the p-primary component of $H^n(G, A)$ isomorphically into the p-primary component of $H^n(K, A)$. The remaining parts of Corollary 3 follow from Corollary 2 and the first part of this Corollary.

For the proof of Corollary 4, observe that each term X_n of the bar resolution is a free Γ-module with a finite set of generators because G is a finite group. Furthermore, a finitely generated Γ-module A is also a finitely generated abelian group. We will show that $\text{Hom}_\Gamma(X_n, A)$ is finitely generated. If X_n is free on a finite set $\{u_i\}$ and if A (when considered as an abelian group) is generated by a finite set $\{v_j\}$, then any Γ-homomorphism f from X_n into A is completely determined by the set of integers n_{ij} such that $f(u_i) = \sum n_{ij} v_j$. Let f_{ij} be the Γ-homomorphism from X_n into A defined by $f_{ij}(u_k) = \delta_{ik} v_j$, where $\delta_{ik} = 1$ if $i = k$ and otherwise $\delta_{ik} = 0$. Then, the set $\{f_{ij}\}$ generates $\text{Hom}_\Gamma(X_n, A)$. In fact, we have $f = \sum n_{ij} f_{ij}$. Thus, $\text{Hom}_\Gamma(X_n, A)$ is a finitely generated abelian group. By the Corollary of (3.19), Ker d_{n+1}^* is also finitely generated, and so is $H^n(G, A)$. By Corollary 2, $H^n(G, A)$ is a torsion group. Theorem 5.2 shows that $H^n(G, A)$ is finite. $\square$

A conjugate subgroup K^σ is isomorphic to K, so the cohomology group $H^n(K^\sigma, A)$ is, of course, isomorphic to $H^n(K, A)$. We will write this isomorphism explicitly. For any $f \in \text{Hom}_A(X_n, A)$, let

$$f^\sigma(x) = f(x\sigma^{-1})\sigma.$$

Then, for $k \in K$, we have

$$f^\sigma(xk^\sigma) = f(xk^\sigma \sigma^{-1})\sigma = f(x\sigma^{-1})k\sigma = f^\sigma(x)k^\sigma.$$

Thus, f^σ is a $\mathbb{Z}K^\sigma$-homomorphism. Since $\delta(f^\sigma) = (\delta f)^\sigma$ holds, the mapping $f \to f^\sigma$ induces a homomorphism from $H^n(K, A)$ into $H^n(K^\sigma, A)$. Let I_σ be this homomorphism. Clearly, we have

$$I_1 = 1 \quad \text{and} \quad I_{\sigma\tau} = I_\sigma I_\tau.$$

So, I_σ is an isomorphism for all $\sigma \in G$.

Theorem 7.27. *Let K and L be two subgroups of a group G, and let $S = \{\sigma\}$ be a set of representatives from the double cosets with respect to K and L. Suppose that $|G : K|$ is finite. Then, for any Γ-module A, we have the following formula:*

$$\tau^*(K, G)\rho^*(G, L) = \sum_\sigma I_\sigma \rho^*(K^\sigma, K^\sigma \cap L)\tau^*(K^\sigma \cap L, L).$$

Proof. By (3.8) (iv) of Chapter 1, the double coset $K\sigma L$ contains exactly $|L : L \cap K^\sigma|$ right cosets of K. Let $R^{(\sigma)} = \{l_i^{(\sigma)}\}$ be a right transversal of $L \cap K^\sigma$ in L. Then, $\{\sigma l_i^{(\sigma)}\}$ is a set of representatives of the right cosets of K which are contained in $K\sigma L$. Therefore,

$$T = \{(l_i^{(\sigma)})^{-1}\sigma^{-1}\}$$

is a left transversal of K, where σ range over S. By definition, we have $\tau f(x) = \sum f(xt)t^{-1}$ $(t \in T)$, and

$$\sum_{t \in T} f(xt)t^{-1} = \sum_\sigma \sum_l f(xl^{-1}\sigma^{-1})\sigma l = \sum_\sigma \sum_l f^\sigma(xl^{-1})l.$$

The right side is equal to

$$\sum_\sigma \tau(L \cap K^\sigma, L)(\rho(K^\sigma, L \cap K^\sigma)(I_\sigma f)). \quad \square$$

(7.28) *Suppose that an S_p-subgroup S of a finite group G satisfies the following condition: $S \cap S^x \neq \{1\} \Rightarrow S = S^x$. Then, the p-primary component of $H^n(G, A)$ is isomorphic to the p-primary component of $H^n(N_G(S), A)$.*

Proof. Apply Theorem 7.27 with $K = N_G(S)$ and $L = S$. Then, K^σ contains exactly one S_p-subgroup S^σ. If $S \cap S^\sigma \neq \{1\}$, then $S = S^\sigma$ and $\sigma \in K$. Since $\rho^*(K^\sigma, \{1\}) = 0$, the right side of the formula in Theorem 7.27 vanishes except for the term with $\sigma = 1$, so we get

$$\tau^*(K, G)\rho^*(G, S) = \rho^*(K, S).$$

Corollary 3 of Theorem 7.26 says that both $\rho^*(G, S)$ and $\rho^*(K, S)$ are injective on the p-primary component. So, $\tau^*(K, G)$ is also injective. On the other hand, Theorem 7.26 shows that $\tau^*(K, G)$ maps the p-primary component of $H^n(K, A)$ onto the p-primary component of $H^n(G, A)$. Hence, $\tau^*(K, G)$ gives the isomorphism between them.   □

In the preceding discussions, we were given a fixed free resolution $\mathscr{X}$ of the trivial Γ-module $\mathbb{Z}$, and we defined the restriction and corestriction mappings between the cohomology groups based on $\mathscr{X}$. If we choose a different free resolution, we get another restriction mapping. We will discuss briefly how the change of free resolutions affects the mappings.

Let $\mathscr{Y} = \{Y_n\}$ be another free resolution of the trivial Γ-module $\mathbb{Z}$. Then, by (7.20), there are Γ-homomorphisms $\{f_n\}$ which satisfy the commutative conditions. We will write those homomorphisms collectively as $f = \{f_n\}$, and we call f a Γ-morphism from $\mathscr{X}$ to $\mathscr{Y}$. Since f is a Γ-morphism, f induces a collection of homomorphisms

$$f^* = \{f_n^*\} \quad \text{from} \quad \{\mathrm{Hom}_\Gamma(Y_n, A)\} \quad \text{into} \quad \{\mathrm{Hom}_\Gamma(X_n, A)\}.$$

The proof of Theorem 7.21 shows that f^* induces isomorphisms from the cohomology groups defined by $\mathscr{Y}$ onto the cohomology groups defined by $\mathscr{X}$.

The restriction mapping ρ is defined by looking at the elements of $\mathrm{Hom}_\Gamma(X_n, A)$ as Δ-homomorphisms. The two free resolutions $\mathscr{X}$ and $\mathscr{Y}$ give rise to the following diagram:

$$
\begin{array}{ccc}
H^n(G, A) & \longrightarrow & H^n(K, A) \\
\uparrow & & \uparrow \\
H^n(G, A) & \longrightarrow & H^n(K, A)
\end{array}
$$

The first row is the restriction mapping ρ^* defined by $\mathscr{X}$, while the second row is that defined by $\mathscr{Y}$. The left upward arrow is induced by the Γ-morphism f from $\mathscr{X}$ to $\mathscr{Y}$, while the right arrow is defined by f when regarded as the Δ-morphism. It is obvious that the above is a commutative diagram. Since the upward arrows are isomorphisms, the above commutative diagram shows the uniqueness of the restriction mapping.

If we take the bar resolution $\mathscr{B}(G)$, then the cohomology groups are defined in terms of cochains. In this case, if $f(x_1, \ldots, x_n)$ is a normalized cocycle for G, then we get a cocycle for K by restricting all elements x_i in K. This restriction of the variables induces a homomorphism from $H^n(G, A)$ into $H^n(K, A)$. We will show that this coincides with the restriction mapping. For each n, there is an obvious inclusion mapping $i_n: B_n(K) \to B_n(G)$. This is a Δ-homomorphism, and defines a Δ-morphism i from $\mathscr{B}(K)$ into $\mathscr{B}(G)$. The proof of this amounts to verifying the commutativity of the following diagram; this is trivial.

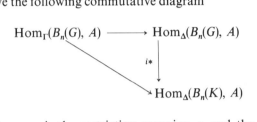

So, the Δ-morphism i induces isomorphisms between the cohomology groups. We have the following commutative diagram

$$\mathrm{Hom}_\Gamma(B_n(G), A) \longrightarrow \mathrm{Hom}_\Delta(B_n(G), A)$$

with a slanted arrow and $i*$ mapping to

$$\mathrm{Hom}_\Delta(B_n(K), A)$$

The horizontal arrow is the restriction mapping ρ, and the slanted arrow coincides with the mapping obtained by restricting the variables to K. So, if the cohomology groups are identified using the isomorphism induced by $i*$, then these two restriction mappings coincide.

Similarly, the corestriction can be expressed in terms of cochains (see Exercise 5).

If the subgroup K is normal, the cohomology groups of G, K, and G/K are connected by the fundamental exact sequences of Hochschild–Serre [1]. We will prove only a few special cases.

Let K be an operator group acting on an additive group A. It is customary to write $C_A(K) = A^K$. If $K \triangleleft G$, the elements of a coset of K induce the same action on A^K. Thus, the group G/K acts on A^K. We will show that the following sequence is exact:

(7.29)

$$0 \longrightarrow H^1(G/K, A^K) \overset{\alpha}{\longrightarrow} H^1(G, A) \overset{\beta}{\longrightarrow}$$

$$\longrightarrow H^1(K, A)^G \overset{\gamma}{\longrightarrow} H^2(G/K, A^K).$$

We begin with the definitions of the mappings and the groups involved in (7.29).

The mapping α is called the *inflation*. It is defined as follows.
If $f \in C^n(G/K, A^K)$, then f defines a cochain g such that

$$g(\sigma_1, \ldots, \sigma_n) = f(K\sigma_1, \ldots, K\sigma_n).$$

Then, we see that δg corresponds to δf. This is obvious from the formula (7.22) if we note that

$$f(K\sigma_1, \ldots, K\sigma_{n-1})\sigma_n = f(K\sigma_1, \ldots, K\sigma_{n-1})K\sigma_n$$

because the value of f is in A^K. Then, the mapping $f \to g$ induces a homomorphism from $H^n(G/K, A^K)$ into $H^n(G, A)$. This is the inflation α.

If $f \in C^1(G/K, A^K)$ corresponds to a coboundary, then there is an element a of A such that $f(K\sigma) = a - a\sigma$ for all $\sigma \in G$. But, if $\sigma \in K$, then $f(K\sigma) = f(K) = 0$. This shows that $a\sigma = a$ for all $\sigma \in K$. So, we have $a \in A^K$ and $f \in B^1(G/K, A^K)$. Thus, α is injective, and the sequence (7.29) is exact up to the second term.

Since K is normal, the mapping I_σ is an automorphism of $H^1(K, A)$. Thus, the group G acts on $H^1(K, A)$ via the mapping $\sigma \to I_\sigma$. The fourth term is the set of fixed elements of this action. The mapping β is induced by the restriction mapping $\rho^*(G, K)$. Note that this may be identified with the mapping on the cochains obtained by restricting the variables. It is clear that the image of ρ^* is contained in $H^1(K, A)^G$. Since the image of the inflation is represented by a cochain which is 0 on K, we have $\alpha\beta = 0$. We will show that Ker $\beta \subset$ Im α. Let f be a crossed homomorphism representing an element of Ker β. Denoting the coboundary mapping (7.22) by δ, we have $\rho f = \delta a$ for some $a \in A$. Set $g = f - \delta a$. Then, g is a crossed homomorphism such that $g(\sigma) = 0$ for all $\sigma \in K$. As g is a crossed homomorphism, we have

$$g(\sigma\tau) = g(\sigma)\tau + g(\tau).$$

If $\sigma \in K$, then we get $g(\sigma\tau) = g(\tau)$ for all $\tau \in G$. On the other hand, if $\tau \in K$, then we have

$$g(\sigma) = g(\sigma\tau) = g(\sigma)\tau.$$

This shows that the value of g is always in A^G and is constant for all elements σ of the right coset $K\tau$. Thus, g represents an element in the image of α. So, Im $\alpha =$ Ker β, and the sequence (7.29) is exact up to the third term.

Before we define the mapping γ, we will express the isomorphism I_σ in terms of cocycles. We will work out the one-dimensional case here and leave the general case to Exercise 6(a). We do not need to assume that K is normal. Consider the bar resolution of the trivial Γ-module $\mathbb{Z}$ where Γ is the group ring of G. If an element f of $\mathrm{Hom}_\Delta(B_1, A)$ satisfies $\delta f = 0$, then we have

$$f([\sigma]\rho) - f([\sigma\rho]) + f([\rho]) = 0$$

for all $\sigma, \rho \in G$. If $\rho \in K$, the formula gives us

$$f^\sigma([\rho^\sigma]) = f([\rho^\sigma]\sigma^{-1})\sigma = f([\sigma^{-1}\rho])\sigma - f([\sigma^{-1}])\sigma$$
$$= f([\rho])\sigma + a - a\rho^\sigma \qquad (a = -f([\sigma^{-1}])\sigma).$$

This shows that if an element α of $H^1(K, A)$ is represented by a crossed homomorphism φ, then the element $I_\sigma\alpha$ of $H^1(K^\sigma, A)$ is represented by the crossed homomorphism φ^σ such that $\varphi^\sigma(\rho^\sigma) = \varphi(\rho)\sigma$.

The mapping γ is defined as follows. Let T be a transversal of K such that $T \cap K = \{1\}$. If a crossed homomorphism φ represents an element of $H^1(K, A)^G$, then, for any $\tau \in T$, φ^τ is in the same cohomology class as φ. So, there is an element a_τ of A such that

$$\varphi^\tau(\rho) = \varphi(\rho) + a_\tau\rho - a_\tau \quad \text{for all} \quad \rho \text{ in } K.$$

If $\tau = 1$, we may choose $a_1 = 0$. We will consider these elements $\{a_\tau\}$ to be fixed, and since any element σ of G can be written uniquely as $\sigma = \kappa\tau$ where $\kappa \in K$ and $\tau \in T$, we define

$$a_\sigma = a_\tau + \varphi(\kappa)\tau$$

for any σ of G. Then, the family $\{a_\sigma\}$ of elements satisfies the following two identities for any $\lambda \in K$ and $\sigma \in G$:

$$a_{\lambda\sigma} = a_\sigma + \varphi(\lambda)\sigma, \quad \varphi^\sigma(\lambda) = \varphi(\lambda) + a_\sigma\lambda - a_\sigma.$$

These formulas can easily be verified. Write $\sigma = \kappa\tau$ with $\kappa \in K$ and $\tau \in T$. We have $\lambda\sigma = (\lambda\kappa)\tau$. So, by definition,

$$a_{\lambda\sigma} = a_\tau + \varphi(\lambda\kappa)\tau.$$

As φ is a crossed homomorphism, $\varphi(\lambda\kappa) = \varphi(\lambda)\kappa + \varphi(\kappa)$ holds. This gives us $a_{\lambda\sigma} = a_\sigma + \varphi(\lambda)\sigma$. The second formula is proved by expanding the formula $\varphi^\sigma(\lambda) = \varphi(\sigma\lambda\sigma^{-1})\sigma$. Since φ is a crossed homomorphism, we have

$$\varphi^\sigma(\lambda) = \varphi(\kappa\tau\lambda\tau^{-1}\kappa^{-1})\sigma$$
$$= \varphi(\kappa)\tau\lambda + \varphi(\tau\lambda\tau^{-1})\tau + \varphi(\kappa^{-1})\kappa\tau$$
$$= a_\sigma\lambda - a_\tau\lambda + \varphi^\tau(\lambda) - \varphi(\kappa)\tau$$
$$= \varphi(\lambda) + a_\sigma\lambda - a_\sigma.$$

We have used the identity $\varphi(\kappa^{-1})\kappa = -\varphi(\kappa)$ which holds for any crossed homomorphism φ.

Set $\alpha(\sigma, \rho) = a_{\sigma\rho} - a_\sigma\rho - a_\rho$. Clearly, α is an element of $Z^2(G, A)$. If λ and μ are elements of K, then we have $\alpha(\lambda\sigma, \mu\rho) = \alpha(\sigma, \rho)$. This is easily proved by using the formulas for $a_{\lambda\sigma}$, $a_{\mu\rho}$, and $\varphi^\sigma(\mu)$. By expressing $\varphi^{\sigma\rho}(\kappa)$ in two different ways, we get $\alpha(\sigma, \rho) \in A^K$. Thus, α may be considered to be an element of $Z^2(G/K, A^K)$. Suppose that we choose another crossed homomorphism $\tilde{\varphi}$ from the cohomology class of φ. Then, there is an element b of A such that

$$\tilde{\varphi}(\kappa) = \varphi(\kappa) + b - b\kappa$$

for all $\kappa \in K$. Let $\{\tilde{a}_\sigma\}$ be the family of elements associated with $\tilde{\varphi}$, just as $\{a_\sigma\}$ is associated with φ. Then, we have

$$\tilde{\varphi}^\sigma(\kappa) = \tilde{\varphi}(\kappa) + \tilde{a}_\sigma \kappa - \tilde{a}_\sigma.$$

An easy computation shows that $\tilde{a}_\sigma = a_\sigma + b - b\sigma + c_\sigma$ for some $c_\sigma \in A^K$. Thus, the element $\tilde{\alpha}$ constructed from $\{\tilde{a}_\sigma\}$ satisfies

$$\tilde{\alpha}(\sigma, \rho) = \alpha(\sigma, \rho) + c_{\sigma\rho} - c_\sigma \rho - c_\rho.$$

So, α and $\tilde{\alpha}$ belong to the same cohomology class. The mapping γ is defined by sending the class $\{\varphi\}$ to the class $\{\alpha\}$. Clearly, γ is a homomorphism from $H^1(K, A)^G$ into $H^2(G/K, A^K)$.

If the class $\{\varphi\}$ lies in the image of β, then φ can be chosen so that it is the restriction of some crossed homomorphism $\tilde{\varphi}$ which is defined on G. We have

$$\tilde{\varphi}^\sigma(\kappa) = \tilde{\varphi}(\sigma\kappa\sigma^{-1})\sigma = \tilde{\varphi}(\sigma)\kappa + \tilde{\varphi}(\kappa) + \tilde{\varphi}(\sigma^{-1})\sigma.$$

Thus, we may choose $\tilde{a}_\sigma = \tilde{\varphi}(\sigma)$ for all $\sigma \in G$. In this case, $\tilde{\alpha}(\sigma, \rho) = 0$ and we get $\beta\gamma = 0$. Conversely, suppose $\gamma(\{\varphi\}) = \{\alpha\} = \{0\}$. Then, we can write $\alpha = \delta c$ for some $c \in C^1(G/K, A^K)$. Let $\{c_\sigma\}$ be the image of c by the inflation mapping. Then, we have

$$\alpha(\sigma, \rho) = c_{\sigma\rho} - c_\sigma \rho - c_\rho.$$

Also, $c_\rho = 0$ for all $\rho \in K$. Let $\{a_\sigma\}$ be the family of elements associated with φ as before, and set $\theta(\sigma) = a_\sigma - c_\sigma$. Since

$$a_{\sigma\rho} - a_\sigma \rho - a_\rho = \alpha(\sigma, \rho) = c_{\sigma\rho} - c_\sigma \rho - c_\rho,$$

θ is a crossed homomorphism defined on G. By the definition of a_σ, we get $a_\rho = \varphi(\rho)$ if $\rho \in K$. Thus, if $\rho \in K$, we have

$$\theta(\rho) = a_\rho - c_\rho = \varphi(\rho).$$

This means that φ is the restriction of the crossed homomorphism θ defined on G. Hence, we have Ker $\gamma \subset$ Im β. This proves that the sequence (7.29) is exact.

There is a similar exact sequence which concerns higher dimensional co-homology groups (Hochschild–Serre [1], Lyndon [1]). Hattori [1] gave an ingeneous proof of this general theorem for which the reader is referred to Nakayama–Hattori [1] §22 or Iyanaga [1], Chapter 1. We will be content with proving the following theorem on 2-dimensional cohomology groups.

(7.30) *Let K be a normal subgroup of G. If $H^1(K, A) = \{0\}$, then we have the exact sequence :*

$$0 \xrightarrow{\quad\quad} H^2(G/K, A^K) \xrightarrow{\quad\alpha\quad} H^2(G, A) \xrightarrow{\quad\beta\quad} H^2(K, A).$$

The mapping α is the inflation, and β is the restriction defined in the proof of the previous exact sequence.

Proof. Let $f \in C^2(G/K, A^K)$ such that the image g of f under the inflation mapping is a coboundary. Denoting the coboundary mapping (7.22) by δ, we have $g = \delta h$ for some function h. If σ and τ are elements of K, then $f(K\sigma, K\tau) = 0$. So, we have

$$0 = g(\sigma, \tau) = h(\tau) - h(\sigma\tau) + h(\sigma)\tau.$$

This formula shows that h defines a crossed homomorphism on K. Since $H^1(K, A) = \{0\}$, there is an element a of A such that

$$h(\sigma) = a - a\sigma \quad \text{for all} \quad \sigma \in K. \qquad \text{Set } k = h - \delta a.$$

Then, we have $\delta k = \delta h = g$ and $k(\sigma) = 0$ for any $\sigma \in K$. We will show that k is the image of an element of $C^1(G/K, A^K)$ under the inflation mapping. Since f is normalized, we have $g(\sigma, \tau) = 0$ if either σ or τ is in K. If $\sigma \in K$, then $g = \delta k$ gives us $k(\tau) = k(\sigma\tau)$. Thus, k is the image of the inflation mapping. If $\tau \in K$, then we get $k(\sigma\tau) = k(\sigma)\tau$. Since $k(\sigma\tau) = k(\sigma)$ for $\tau \in K$, we get $k(\sigma) \in A^K$ for all $\sigma \in G$. Thus, f is a coboundary, and we conclude that the mapping α is injective.

As in the proof of (7.29), we clearly have $\alpha\beta = 0$. Suppose that the restriction of a factor set f lies in $B^2(K, A)$. Then, there is a function g such that $f = \delta g$ on K. We can extend g to a function on G by arbitrarily defining its values for all elements of $G - K$, and we consider $f - \delta g$. This is a factor set in the cohomology class of f. So, we may assume from the beginning that the restriction of f on K is zero. Choose a transversal $T = \{\tau_i\}$ with $T \cap K = \{1\}$. For an element

$$\rho = \kappa\tau_i \quad (\kappa \in K, \tau_i \in T), \quad \text{set} \quad g(\rho) = f(\kappa, \tau_i).$$

If $\lambda \in K$, we have $g(\lambda\rho) = f(\lambda\kappa, \tau_i)$ and $g(\lambda) = 0$. So,

$$\delta g(\lambda, \rho) + f(\lambda, \rho) = f(\kappa, \tau_i) - f(\lambda\kappa, \tau_i) + f(\lambda, \kappa\tau_i)$$
$$= f(\lambda, \kappa)\tau_i = 0.$$

Let $h = f + \delta g$. Then, we get $h(\lambda, \rho) = 0$ if $\lambda \in K$. Since h is a factor set, we have

$$h(\sigma, \rho) - h(\lambda\sigma, \rho) + h(\lambda, \sigma\rho) - h(\lambda, \sigma)\rho = 0.$$

This formula reduces to $h(\sigma, \rho) = h(\lambda\sigma, \rho)$ if $\lambda \in K$. Also, if $\sigma \in K$, then $\lambda\sigma = (\lambda\sigma\lambda^{-1})\lambda$, and it follows from the above formula that

$$h(\lambda, \sigma\rho) = h(\lambda, \sigma)\rho + h(\lambda, \rho).$$

Thus, for any $\lambda \in G$, $h(\lambda, \sigma)$ is a crossed homomorphism as a function of $\sigma \in K$. Since $H^1(K, A) = \{0\}$, there is an element $a(\sigma)$ of A such that

$$h(\sigma, \tau) = a(\sigma) - a(\sigma)\tau \quad \text{for all} \quad \tau \in K.$$

If $\lambda \in K$, then $h(\lambda, \tau) = 0$ and $h(\lambda\sigma, \tau) = h(\sigma, \tau)$. So, we may choose an element of A such that $a(\lambda) = 0$ and $a(\lambda\sigma) = a(\sigma)$ for any $\lambda \in K$ and $\sigma \in G$. Set $k = h + \delta a$. We will show that k is in the image of the inflation mapping. First, we will prove that $k(\sigma, \tau) = 0$ if either σ or τ belongs to K. If $\sigma \in K$, then

$$k(\sigma, \tau) = h(\sigma, \tau) + a(\tau) - a(\sigma\tau) + a(\sigma)\tau = 0.$$

On the other hand, if $\tau \in K$, we also get $k(\sigma, \tau) = 0$ because

$$a(\sigma\tau) = a(\sigma\tau\sigma^{-1}\sigma) = a(\sigma) \quad \text{and} \quad h(\sigma, \tau) = a(\sigma) - a(\sigma)\tau.$$

Since k is a cocycle, we get the following formula for $\lambda, \mu \in K$:

$$k(\lambda\sigma, \mu\tau) = k(\sigma, \tau) = k(\sigma, \tau)\lambda.$$

Thus, k is the image of an element of $C^1(G/K, A^K)$ by the inflation mapping. This completes the proof of the exactness. $\square$

The exact sequence (7.30) is equivalent to an important theorem on algebras over a Galois extension (see [AS] II p. 435).

EXAMPLE 1. Let F be the free group on a set X. We may consider X as a subset of F. We will determine the cohomology groups of F. Let $\Gamma = \mathbb{Z}F$ be the group ring of F over $\mathbb{Z}$. An element of Γ has the form $\sum n_w w$ ($n_w \in \mathbb{Z}$). The mapping ε which sends $\sum n_w w$ to $\sum n_w$ is a Γ-homomorphism from Γ onto the trivial Γ-module $\mathbb{Z}$. Let I_F be the kernel of ε. Clearly, I_F is an abelian group freely generated by $w - 1$ for $w \in F - \{1\}$. The formulas

$$xy - 1 = (x - 1)y + (y - 1), \qquad x^{-1} - 1 = -(x - 1)x^{-1}$$

show that I_F is also a Γ-module generated by $x - 1$ with $x \in X$. The following proposition is important.

The ideal I_F is a free Γ-module generated freely by $x - 1$ $(x \in X)$.

In order to prove this proposition, consider a derivation from F into Γ; that is, consider a mapping D from F into Γ which satisfies the following formula for all u and v of F:

$$D(uv) = D(u)v + D(v).$$

It follows immediately from the above formula that $D(1) = 0$ and that $D(u^{-1}) = -D(u)u^{-1}$. So, the derivation D is uniquely determined by the values $D(x)$ for $x \in X$. Conversely, if $d(x)$ are arbitrarily given elements of Γ, then there exists a derivation D defined over F such that $D(x) = d(x)$ for all $x \in X$. Let $(u_1, \ldots, u_n)$ be a sequence of elements where $u_i \in X \cup X^{-1}$. We will define $D(u_1, \ldots, u_n)$ by induction on n. If $n = 0$, we let $D(1) = 0$. When $n = 1$, $D(u) = D(x)$ if $u = x \in X$, and $D(u) = -D(x)x^{-1}$ if $u = x^{-1}$. For $n > 1$, we let

$$D(u_1, \ldots, u_n) = D(u_1, \ldots, u_{n-1})u_n + D(u_n).$$

We will show that $D(u_1, \ldots, u_n)$ depends only on the element $u_1 \cdots u_n$ of F. It suffices to show that the reduction at the end of the sequence does not change the value. This is easily done. Furthermore, we can show that D is indeed a derivation.

For any $x \in X$, define the derivation D_x by $D_x(x) = 1$ and $D_x(y) = 0$ if y is an element of X different from x. For each derivation D, $w - 1 \to D(w)$ defines an element of $\mathrm{Hom}_{\mathbb{Z}}(I_F, \Gamma)$. Let φ_x be the element defined by D_x. Then, for any $y \in X$ and $\gamma \in \Gamma$, we get

$$\varphi_x((y - 1)\gamma) = \gamma \quad (\text{if } y = x), \qquad = 0 \quad (\text{if } y \neq x).$$

This is proved as follows. Set $\gamma = \sum n_w w$ $(n_w \in \mathbb{Z})$. Then, we have

$$\varphi_x((y - 1)\gamma) = \varphi_x(\sum n_w((yw - 1) - (w - 1)))$$
$$= \sum n_w(D_x(yw) - D_x(w)) = \sum n_w D_x(y)w$$
$$= D_x(y)\gamma.$$

We can now show that the set $\{x - 1\}$ $(x \in X)$ is free. Suppose that $\sum (y - 1)\lambda_y = 0$ for $\lambda_y \in \Gamma$. Then, by applying φ_x, we get $\lambda_x = 0$. Thus, I_F is a free Γ-module on $\{x - 1\}$ $(x \in X)$. $\square$

From the preceding proposition, we can see that the sequence

$$\to 0 \to 0 \to I_F \to \Gamma \xrightarrow{\varepsilon} \mathbb{Z} \to 0$$

is a free resolution of the trivial Γ-module $\mathbb{Z}$. So, by Theorem 7.21, we get $H^n(F, A) = \{0\}$ whenever $n \geq 2$. In general, if there is a free resolution X such that $X_{n+1} = \{0\}$, then we say that the cohomology dimension of the group G is finite and at most n. The smallest value of such an n is called the *cohomology dimension* of G, and we denote it by $cd(G)$. If $cd(G) = n$, then there is a free resolution such that $X_k = \{0\}$ for all $k \geq n + 1$. So, we get $H^k(G, A) = \{0\}$ for all $k > n$. If $cd(G) = 0$, then Γ is isomorphic to $\mathbb{Z}$. So, the group G is the trivial group $\{1\}$. We have shown that a free group is of cohomology dimension ≤ 1. The converse of this theorem is very difficult to prove. Here, we will merely state that it is the Theorem of Stallings–Swan (Stallings [1], Swan [1]) which asserts that if $cd(G) \leq 1$ for a group G, then G is free.

EXAMPLE 2. We will study the cohomology groups of a finite cyclic group G of order m. Let σ be a generator of G. In the group ring $\Gamma = \mathbb{Z}G$, we define two elements

$$\tau = 1 - \sigma \quad \text{and} \quad \mu = 1 + \sigma + \sigma^2 + \cdots + \sigma^{m-1}.$$

The multiplication of these elements induces Γ-endomorphisms of Γ which we will denote by the same letters as τ and μ. If

$$\gamma = \sum a_i \sigma^i \in \Gamma \qquad (a_i \in \mathbb{Z}),$$

we define $\varepsilon(\gamma) = \sum a_i$. Then, we have

$$\gamma\mu = \varepsilon(\gamma)\mu \quad \text{and} \quad \gamma\tau = \sum (a_i - a_{i-1})\sigma^i,$$

where we set $a_{-1} = a_{m-1}$. So, we get

$$\text{Ker } \mu = \text{Ker } \varepsilon = \text{Im } \tau, \qquad \text{Ker } \tau = \text{Im } \mu = \mathbb{Z}\mu.$$

Let $X_n = \Gamma$ for all $n \geq 0$, $d_0 = \varepsilon$, $d_{2i-1} = \tau$, and $d_{2i} = \mu$ for all $i > 0$. Then, the sequence

$$\cdots \to \Gamma \xrightarrow{\mu} \Gamma \xrightarrow{\tau} \Gamma \xrightarrow{\mu} \Gamma \xrightarrow{\tau} \Gamma \xrightarrow{\varepsilon} \mathbb{Z} \to 0$$

is a free resolution of the trivial Γ-module $\mathbb{Z}$. Thus, the cohomology groups of a finite cyclic group are periodic (with period 2). Since this is the case, the groups $\text{Hom}_\Gamma(X_n, A)$ are all isomorphic to A (by the isomorphism $f \to f(1)$), and the mappings μ^* and τ^* are right multiplications by the elements μ and τ of Γ, respectively. By definition, for $i > 0$, we have

$$H^{2i-1}(G, A) = \text{Ker } \mu^*/\text{Im } \tau^*$$
$$H^{2i}(G, A) = \text{Ker } \tau^*/\text{Im } \mu^*.$$

For example, if G acts trivially on $A = \mathbb{Z}$, then we have $n\mu = nm$ and $n\tau = 0$ for any $n \in \mathbb{Z}$. Thus,

$$H^{2i-1}(G, \mathbb{Z}) = \{0\}, \qquad H^{2i}(G, \mathbb{Z}) = \mathbb{Z}/m\mathbb{Z}.$$

On the other hand, if A is a finite abelian group,

$$|H^{2i-1}(G, A)| = |H^{2i}(G, A)|$$

holds. The above equality follows from the lemma of Herbrand which asserts that if μ^* and τ^* are two endomorphisms of a finite abelian group A such that $\mu^*\tau^* = \tau^*\mu^* = 0$, we have

$$|\text{Ker } \mu^* : \text{Im } \tau^*| = |\text{Ker } \tau^* : \text{Im } \mu^*|.$$

This formula can be proved as follows. The assumption $\mu^*\tau^* = \tau^*\mu^* = 0$ gives us two sequences of subgroups:

$$A \supset \text{Ker } \mu^* \supset \text{Im } \tau^* \supset \{0\}$$
$$A \supset \text{Ker } \tau^* \supset \text{Im } \mu^* \supset \{0\}.$$

By the Isomorphism Theorem, we have

$$A/\text{Ker } \mu^* \cong \text{Im } \mu^* \quad \text{and} \quad A/\text{Ker } \tau^* \cong \text{Im } \tau^*.$$

Thus, by comparing the indices, we get

$$|\text{Ker } \mu^* : \text{Im } \tau^*| = |\text{Ker } \tau^* : \text{Im } \mu^*|.$$

Let p be a prime number. If the cyclic group Z_p of order p acts trivially on Z_p, then both μ^* and τ^* are the zero mapping. Hence, we have $H^k(Z_p, Z_p) = Z_p$ for $k > 0$. There are exactly two isomorphism classes of groups of order p^2. So, the cyclic group of order p^2 gives $p - 1$ nonequivalent extensions of Z_p by Z_p (see the remark after Definition 7.15).

EXAMPLE 3. *Let G be a finite p-group which acts on a finite abelian group A. If $H^1(G, A) = \{0\}$, then we have*

$$H^1(K, A) = H^2(K, A) = \{0\}$$

for any subgroup K of G.

Proof. We will prove this proposition by induction on the order $|G|$. Let M be a maximal subgroup of G. By the Corollary of Theorem 1.6, $M \lhd G$ and G/M is a cyclic group of order p. Consider the exact sequence (7.29):

$$0 \to H^1(G/M, A^M) \to H^1(G, A) \to H^1(M, A)^G \to H^2(G/M, A^M)$$

By assumption, we have $H^1(G, A) = \{0\}$. So, the exactness gives us $H^1(G/M, A^M) = \{0\}$. Since G/M is cyclic and A is finite, Herbrand's lemma shows that $H^2(G/M, A^M) = \{0\}$, too. Again, the exactness implies that $H^1(M, A)^G = \{0\}$. Since M is a p-group, Corollaries 2 and 4 of Theorem 7.26 prove that $H^1(M, A)$ is a finite p-group. Thus, a p-group G acts on a p-group $H^1(M, A)$ without fixed points. By the Corollary of (1.3), we have $H^1(M, A) = \{0\}$. Let K be a proper subgroup of G. Then, K is contained in the maximal subgroup M of G. By the inductive hypothesis, we get $H^1(K, A) = H^2(K, A) = \{0\}$. It remains to prove that $H^2(G, A) = \{0\}$. Apply the exact sequence (7.30) for M. Then, the groups at the end are $\{0\}$. The exactness of the sequence proves that $H^2(G, A) = \{0\}$. $\square$

In this case, $H^n(K, A) = \{0\}$ for all n and all subgroups K of G. Such a module A is called *cohomologically trivial*. This is an important class of modules in applications to number theory. The work of Nakayama [1] is quite prominent in this field.

Exercises

1. Assume that the two rows of the following diagram are exact and that the diagram is commutative. Let f_i be the homomorphism $A_i \to B_i$. Prove the following propositions.

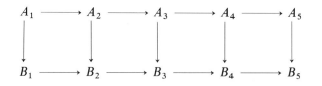

(a) If f_1 is surjective and both f_2 and f_4 are injective, then f_3 is also injective.
(b) If f_5 is injective and both f_2 and f_4 are surjective, then f_3 is surjective.

(*Hint.* This is a generalization of (7.14) and is called the *Five Lemma.*)

2. Show that for any $\mathbb{Z}G$-modules A and B, we have

$$H^n(G, A + B) \cong H^n(G, A) + H^n(G, B).$$

(*Hint.* Let $\{X_n\}$ be a free resolution of the trivial Γ-module $\mathbb{Z}$. Then, we have

$$\text{Hom}_\Gamma(X_n, A + B) \cong \text{Hom}_\Gamma(X_n, A) + \text{Hom}_\Gamma(X_n, B)$$

(Exercise 5, §5). Verify that d^* commutes with the projections.)

3. Let F and H be two given groups. Also suppose that a homomorphism ψ from F to Out H is given. For each $\sigma \in F$, choose an automorphism $T(\sigma)$ of H in the class $\psi(\sigma)$. Then,

(*) $$T(\sigma)T(\tau) = T(\sigma\tau)if(\sigma, \tau),$$

where $if(\sigma, \tau)$ is the inner automorphism induced by an element $f(\sigma, \tau)$ of H. We will choose $T(1) = 1$ from the class $\psi(1)$ and $f(1, \tau) = f(\sigma, 1) = 1$, and we define $k(\sigma, \tau, \rho)$ by the formula

$$f(\sigma, \sigma\rho)f(\tau, \rho) = k(\sigma, \tau, \rho)f(\sigma\tau, \rho)T(\rho)(f(\sigma, \tau)).$$

Prove the following three propositions about k.

(a) The element $k(\sigma, \tau, \rho)$ belongs to the center $Z(H)$, and the cochain k is a cocycle: $k \in Z^3(F, Z(H))$.

(b) The cocycle k depends on the choices of $T(\sigma)$ and $f(\sigma, \tau)$, but the cohomology class of k does not depend on these choices.

(c) The homomorphism ψ coincides with the homomorphism ψ_G of an extension G of H by F if and only if k is a coboundary.

(*Hint.* (a) The elements $f(\sigma, \tau)$ are uniquely determined, mod $Z(H)$. The first half is proved in much the same way that (7.5) is proved using (7.4). For the second half, take the image by $T(\eta)$ of the defining formula of k. In the right side, use the formula (*) to change $T(\rho)T(\eta)$ to $T(\rho\eta)$. Then, eliminate all the terms with $T(\eta)$ and $T(\rho\eta)$ using the definition of k.

(b) If the automorphisms $T'(\sigma) = T(\sigma)id(\sigma)$ $(d(\sigma) \in H)$ are chosen, we can choose

$$f'(\sigma, \tau) = d(\sigma\tau)^{-1}f(\sigma, \tau)(T(\tau)d(\sigma))d(\tau)$$

and get a formula analogous to (*). From this $f'(\sigma, \tau)$, we define k' similarly. Then, $k' = k$. If f' is replaced by f'', then we get

$$f''(\sigma, \tau) = f'(\sigma, \tau)z(\sigma, \tau) \quad \text{for some} \quad z(\sigma, \tau) \in Z(H).$$

In this case, the cocycle k'' defined by f'' is cohomologous to k.

(c) The proof is similar to the one which shows that the extension associated with the factor set in the zero class is isomorphic to a semidirect product.

The cohomology class of the cocycle k defined in this exercise is called the *obstruction* of ψ; the cocycle k prevents ψ from satisfying the formula $\psi = \psi_G$ unless k is cohomologous to 0.)

4. Let F be a group acting on an abelian group A. We consider two extensions G_1 and G_2 of A by F. In order to avoid any possible confusion, let the normal subgroup A in G_i be A_i. Let θ_i be the natural homomorphism from G_i/A_i onto F for $i = 1, 2$, and let f_i be a factor set associated with the extension G_i. We define the subgroups U and D of $G_1 \times G_2$ as follows:

$$U = \{(g_1, g_2)\,|\,\theta_1(g_1) = \theta_2(g_2)\},$$
$$D = \{(a, a^{-1})\,|\,a \in A\}.$$

Show that the following propositions hold:

(a) $D \lhd U$.
(b) Set $\bar{A} = A_2\,D/D$ and $\bar{G} = U/D$. Then, $\bar{A} \cong A$, $\bar{A} \lhd \bar{G}$, and $\bar{G}$ is an extension of $\bar{A}$ by F.
(c) The factor set associated with $\bar{G}$ is $f_1 + f_2$.

(*Hint.* We only need to choose a suitable transversal of $\bar{A}$. As we can see from this exercise, the sum of two equivalence classes of extensions can be defined directly, without referring to the factor sets. This construction is called *the Baer sum of extensions*.)

5. Let K be a subgroup of finite index in G. We can determine the corestriction in terms of cocycles (see (7.25)). Let $\{B_n\}$ be the standard bar resolution for G, and let $\{B_n(K)\}$ be the same for K. The elements of B_n are $\mathbb{Z}$-linear combinations of $[\sigma_1|\sigma_2|\cdots|\sigma_n]\sigma$. We will define $\theta_n: B_n \to B_n(K)$. Let T be a fixed left transversal of K such that $T \cap K = \{1\}$. We can write

$$\sigma = tk \quad (t \in T, k \in K) \quad \text{for all} \quad \sigma \in G,$$

and this decomposition is unique. Using this, set

$$\sigma_n t = t_n k_n \quad (t_n \in T, k_n \in K).$$

Define $t_1, t_2, \ldots, t_{n-1} \in T$ and $k_1, k_2, \ldots, k_{n-1} \in K$ inductively downwards by the formula $\sigma_i t_{i+1} = t_i k_i$, and set

$$\theta([\sigma_1|\sigma_2|\cdots|\sigma_n]\sigma) = [k_1|k_2|\cdots|k_n]k.$$

Let i be the injection of $B_n(K)$ into B_n.

(a) Show that θ is a $\mathbb{Z}K$-homomorphism.
(b) Show that θ defines a $\mathbb{Z}K$-morphism from $\{B_n\}$ to $\{B_n(K)\}$.
(c) Show that θ induces an isomorphism on the cohomology groups $H^n(K, A)$ which is the inverse of the isomorphism induced by the morphism i.

(d) Let f be a cocycle of $Z^n(K, A)$. Show that the corestriction τ maps the cohomology class of f to the class of the cocycle $\tau_c\, f$ defined as follows:

$$\tau_c\, f(\sigma_1, \sigma_2, \ldots, \sigma_n) = \sum f(k_1^{(t)}, k_2^{(t)}, \ldots, k_n^{(t)}) t^{-1},$$

where $[k_1^{(t)} | k_2^{(t)} | \cdots | k_n^{(t)}] = \theta([\sigma_1 | \sigma_2 | \cdots | \sigma_n] t)$ for all $t \in T$ and the summation is over $t \in T$.

(Hint. These are proved by easy computations. The mapping θ depends on the choice of the transversal T, but the isomorphism induced on the cohomology groups does not.

The definition of τ_c in (d) shows that $\tau_c = \theta^* \tau$ where τ is the corestriction of (7.25). We have seen that the mapping $\rho_c = \rho i^*$ is obtained by restricting the variables of cocycles (see p. 213). By (c), θ^* is the inverse of i^*, and we have $\rho_c \tau_c = \rho \tau$. Thus, the corestriction which was determined in terms of the cocycles can be identified with the corestriction τ of (7.25) in the same way that the restriction mapping defined by the cocycles can be identified with the restriction ρ.)

6. (a) Let K be a subgroup of a group G. For $f \in Z^n(K, A)$ and $\sigma \in G$, let g be the element of $C^n(K^\sigma, A)$ defined by

$$g(\sigma_1^\sigma, \sigma_2^\sigma, \ldots, \sigma_n^\sigma) = f(\sigma_1, \ldots, \sigma_n)\sigma.$$

Show that g is a cocycle and that it represents the cohomology class $I_\sigma(\alpha)$ where α is the class of f.

(b) Show that $I_\rho(\eta) = \eta$ for any $\rho \in G$ and $\eta \in H^n(G, A)$.

(Hint. Consider f as an element of $\mathrm{Hom}_A(B_n(K), A)$. Then, θf belongs to $\mathrm{Hom}_A(B_n, A)$ and $\delta(\theta f) = 0$. Set

$$\Delta_k = \delta(\theta f)([\sigma_1^\sigma | \cdots | \sigma_k^\sigma | \sigma^{-1} | \sigma_{k+1} | \cdots | \sigma_n])\sigma.$$

In the expansion of Δ_k in (7.22), let P_k be the sum of first k terms, let the $(k+1)$-th term be $(-1)^k A_k$, and let Q_{k+1} be the sum of the terms after the $(k+3)$-th one. Then, the $(k+2)$-th term is $(-1)^{k+1} A_{k+1}$. Set

$$h_k([\sigma_1^\sigma | \cdots | \sigma_{n-1}^\sigma] f^\sigma) = \theta f([\sigma_1^\sigma | \cdots | \sigma_{k-1}^\sigma | \sigma^{-1} | \sigma_k | \cdots | \sigma_{n-1}]\rho)\sigma.$$

Then, h_k is a $\mathbb{Z}K^\sigma$-homomorphism and satisfies $\delta h_k = P_k - Q_k$. If $\sigma_1, \ldots, \sigma_n$ are elements of K, we see that

$$A_0 = f([\sigma_1 | \cdots | \sigma_n]) \quad \text{and} \quad A_{n+1} = f([\sigma_1^\sigma | \cdots | \sigma_n^\sigma]\sigma^{-1})\sigma.$$

Since $\sum (-1)^k \Delta_k = 0$, we get

$$A_0 = A_{n+1} + \sum (-1)^k \delta h_k.$$

This proves (a). Part (b) is obvious from the definition.)

7. Let F be the free group on the set $\{x_i\}$, and let Γ be the group ring of F over $\mathbb{Z}$. Let D_i be the derivation from F to Γ defined by $D_i(x_i) = 1$ and $D_i(x_j) = 0$ $(j \neq i)$ (see Example 1). Show that for an arbitrary element $w \in F$, we have

$$w - 1 = \sum (x_i - 1)D_i w.$$

(*Hint.* Both sides are derivations as functions of w, and agree on the set of generators.)

8. Define F and Γ as in the preceding exercise, and let I_F be the ideal of Γ as defined in Example 1. Let N be a normal subgroup of F. Set $G = F/N$, and let Δ be the group ring of G over $\mathbb{Z}$. Moreover, let I_N be the ideal of $\mathbb{Z}N$ generated by $n - 1$ $(n \in N)$.

(a) Let η be the homomorphism from Γ to Δ induced by the natural homomorphism from F to G. Show that $\operatorname{Ker} \eta = I_N \Gamma = \Gamma I_N$.

(b) Show that $X = I_F/I_F I_N$ has the structure of a Δ-module and that X is a free Δ-module freely generated by the images $\{e_i\}$ of $\{x_i - 1\}$.

(c) Let N' be the commutator subgroup of N. Show that G acts on N/N' via the following action: for each $v \in N/N'$ and $\xi \in G$, choose $n \in N$ and $x \in G$ which represent the corresponding cosets, and define v^ξ to be the coset of n^x.

(d) Show that the following sequence of Δ-modules is exact:

$$0 \to N/N' \xrightarrow{\theta} X \xrightarrow{\partial} \Delta \xrightarrow{\varepsilon} \mathbb{Z} \to 0,$$

where $\mathbb{Z}$ is the trivial Δ-module and the homomorphisms ε, ∂, and θ are defined as follows:

$$\varepsilon(\sum n_\sigma \sigma) = \sum n_\sigma,$$
$$\partial(\sum e_i \gamma_i) = \sum \eta(x_i - 1)\gamma_i$$

for $\gamma_i \in \Delta$, and

$$\theta(v) = \sum e_i \eta(D_i n)$$

for a representative n of v and the derivations D_i of Exercise 7.

(*Hint.* (a) Choose a transversal T of N. An element γ of Γ can be written as $\sum a_{t,n} tn$ $(t \in T, n \in N)$. Then, $\gamma \in \operatorname{Ker} \eta$ if and only if

$$\sum_n a_{t,n} = 0$$

for all $t \in T$. So, $\gamma \in \operatorname{Ker} \eta$ means that

$$\gamma = \sum_{n \neq 1, t} a_{t,n} t(n - 1) \in \Gamma I_N.$$

Similarly, we have $\operatorname{Ker} \eta = I_N \Gamma$.

(b) If $\lambda \in I_F$ and $y^{-1}x \in N$, then we have

$$\lambda x - \lambda y = \lambda y(y^{-1}x - 1) \in I_F I_N.$$

Thus, X has the structure of a Δ-module. The right ideal I_F is freely generated by $x_i - 1$. Let M be the free right Δ-module freely generated by $\{v_i\}$. Then, the mapping

$$\varphi: \sum (x_i - 1)\gamma_i \to \sum v_i \eta(\gamma_i)$$

is a homomorphism from I_F onto M, and the kernel of φ is precisely $I_F \operatorname{Ker} \eta = I_F I_N$. Clearly, φ induces a Δ-isomorphism from $I_F/I_F I_N$ onto M. Thus, the Δ-module X is free on $\{e_i\}$.

(c) In general, if A is an abelian normal subgroup of a group H, then H acts on A via conjugation. This induces an action of H/A on A.

(d) The composite of the canonical mapping from I_F onto X and the mapping ∂ coincides with the homomorphism η on I_F. Thus, we have

$$\operatorname{Im} \partial = \operatorname{Ker} \varepsilon, \qquad \operatorname{Ker} \partial = (\operatorname{Ker} \eta)/I_F I_N.$$

By Theorem 6.11, the subgroup N is free on a set Y. So, the factor group N/N' is the free abelian group on the set $\{yN'\}$ where $y \in Y$. Also, I_N is the free left $\mathbb{Z}N$-module on the set $\{y - 1\}$ where $y \in Y$. (The proposition in Example 1 is also true when we consider I_N as a left ideal.) We will show that $\operatorname{Ker} \eta = \Gamma I_N$ is the free left Γ-module on $\{y - 1\}$ where $y \in Y$. Suppose that

$$\sum \gamma_y(y - 1) = 0.$$

We can write $\gamma_y = \sum t\lambda_{t, y}$ ($t \in T$, $\lambda_{t, y} \in \mathbb{Z}N$). (See the hint (a) for the definition of T.) Then, we have $\sum \lambda_{t, y}(y - 1) = 0$ for all t. Since I_N is free on $\{y - 1\}$, we get $\lambda_{t, y} = 0$ for all t and y. Thus, ΓI_N is free. (Similarly, the right Γ-module $I_N \Gamma$ is free.)

Next, we will show that N/N' is isomorphic to $(\operatorname{Ker} \eta)/I_F I_N$. Set $c_i = y_i N'$. Then, N/N' is free on $\{c_i\}$. Let ψ be the function such that

$$\psi(\sum \gamma_i(y_i - 1)) = \sum \delta(\gamma_i)c_i$$

where $\gamma_i \in \Gamma$ and δ is the mapping from Γ to $\mathbb{Z}$ such that $\delta(\sigma) = 1$ for all $\sigma \in F$. Since $I_F = \operatorname{Ker} \delta$, we have $\operatorname{Ker} \psi = I_F I_N$. Thus, ψ induces an isomorphism from $(\operatorname{Ker} \eta)/I_F I_N$ onto N/N'. The inverse isomorphism sends the element yN' onto the coset containing $y - 1$. We have the following formulas

$$rs - 1 \equiv (r - 1) + (s - 1)$$
$$r^{-1} - 1 \equiv -(r - 1) \qquad (\operatorname{mod}(I_N)^2)$$

for any elements r and s of N. Since $(I_N)^2 \subset I_F I_N$, the mapping

$$nN' \to n - 1 + I_F I_N$$

is an isomorphism from N/N' onto $(\mathrm{Ker}\,\eta)/I_F I_N$. By the Exercise 7, we get

$$n - 1 = \sum (x_i - 1)D_i(n).$$

Thus, the mapping θ is an isomorphism from N/N' onto $\mathrm{Ker}\,\partial$. It remains to prove that θ is an operator homomorphism. The proof is easy, since

$$(x^{-1}nx - 1) + I_F I_N = (n - 1)x + I_F I_N.$$

The exact sequence of (d) is due to Lyndon [2]. Let $G = F/N$ be a presentation of G where F is free group on a set X. Set $|X| = n$. Suppose that N is the smallest normal subgroup of F which contains the words $r_1, \ldots, r_m$. By the Cor. 2 of Theorem 6.6, we have $G = \langle X \mid R \rangle$ where the set R consists of the relations $r_1 = 1, \ldots,$ and $r_m = 1$. In this case, the Δ-module N/N' is generated by at most m elements. Thus, there is a free resolution

$$\to X_2 \to X_1 \to \mathbb{Z}G \to \mathbb{Z} \to 0$$

of the trivial $\mathbb{Z}G$-module $\mathbb{Z}$ such that X_1 is freely generated by n elements and that X_2 is freely generated by m elements (a theorem of Lyndon). This theorem implies that we have a crude estimate

$$|H^2(G, A)| \leq |A|^m$$

for any finite G-module A (P. Neumann [2]).

If we set $I = I_F$ and $\Gamma I_N = P$, then we get

(*) $\cdots \subset P^3 \subset IP^2 \subset P^2 \subset IP \subset P \subset I \subset \Gamma.$

We have shown that I and P are ideals of Γ and that they are free right Γ-modules as well as free left Γ-modules. In fact, all terms P^m and IP^m of the above series are free. We have the following lemma. If J is a right ideal of Γ which is a free Γ-module, then JP is also a free Γ-module. If J and P are freely generated by the sets of elements $\{u_i\}$ and $\{v_j\}$, respectively, then JP is generated by elements of the form $u_i \gamma v_j$ where $\gamma \in \Gamma$. But, P is a two-sided ideal, so γv_j is a sum of elements of the form $v_k \rho_k$ where $\rho_k \in \Gamma$. Thus, as a right Γ-module, JP is generated by the elements $\{u_i v_j\}$. It is easy to see that JP is freely generated by these elements.

The argument in the part (b) can be used to prove that if J is a right ideal of Γ freely generated by a set $\{u_i\}$, then J/JP is a free Δ-module and is freely generated by $\{u_i + JP\}$. Thus, the series

$$\to P^2/P^3 \to IP/IP^2 \to P/P^2 \to I/IP \to \Delta \to \mathbb{Z} \to 0$$

is a free resolution of the trivial $\mathbb{Z}G$-module $\mathbb{Z}$, where the arrows (up to the fourth one from the right) are induced by the inclusions of the series (∗). This is the "resolution by relations" due to Gruenberg [1, 2].)

9. Let P be a Γ-module. We call P *projective* if the following condition is satisfied. Whenever we have a diagram

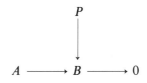

where A and B are Γ-modules, $\alpha: A \to B$ and $\beta: P \to B$ are Γ-homomorphisms, and the row is exact, then there is a Γ-homomorphism $f: P \to A$ such that $f\alpha = \beta$. Prove the following propositions.

(a) Theorem 7.18 holds for projective modules (if F is assumed to be projective).

(b) A Γ-module P is projective if and only if P is a direct summand of a free Γ-module.

(c) Let P be a projective Γ-module. Then, there is a free Γ-module T such that $P + T \cong T$.

(*Hint.* Use (b) in proving the part (c). We can choose the free Γ-module T to satisfy an additional condition $T + T \cong T$.)

10. An exact sequence $\mathscr{P}$

$$\to P_n \to P_{n-1} \to \cdots \to P_1 \to P_0 \to \mathbb{Z} \to 0$$

where $\mathbb{Z}$ is the trivial Γ-module and all terms P_n are projective Γ-modules is said to be a *projective resolution*. When $P_k = \{0\}$ for all $k \geq n_0$, the projective resolution $\mathscr{P}$ is said to be of finite length. Show that if $\mathscr{P}$ is a projective resolution of finite length, then there is a free resolution $\mathscr{F}$ of the trivial Γ-module $\mathbb{Z}$ which has the same length as P.

(*Hint.* Suppose $P_k = \{0\}$ for $k > n_0$. From Exercise 9(c), we can prove that there is a free Γ-module T such that $P_i + T$ are free for $i = 0, 1, \ldots, n_0$. Construct a free resolution from these free modules. In this chapter, we defined the cohomology dimension in terms of free resolutions. But, Exercise 10 shows that this definition coincides with the usual one which uses projective resolutions. Thus, $cd(G) \leq n$ is equivalent to the condition $H^k(G, A) = \{0\}$, $(k \geq n + 1)$ for any $\mathbb{Z}G$-module A (Nakayama–Hattori [1], Theorem 9.3.).

11. Let G be a group.

(a) Prove that $cd(H) \leq cd(G)$ for any subgroup $H \subset G$.

(b) Show that if $cd(G)$ is finite, G is torsion-free; that is, the only element of finite order in G is the identity element.

(*Hint.* A free resolution of the trivial $\mathbb{Z}G$-module $\mathbb{Z}$ is also a free resolution of the subgroup H. This proves (a). Example 2 and the part (a) prove (b). If both $cd(G)$ and the index $|G:H|$ are finite, then the equality $cd(H) = cd(G)$ holds (a theorem of Serre [1]).)

§8. Applications of Cohomology Theory:
   The Schur–Zassenhaus Theorem

Definition 8.1. Let H and F be two groups. An extension G of H by F is said to be a **split extension** if there is a subgroup K of G such that $HK = G$ and $H \cap K = \{1\}$. In this case, we also say that the extension **splits**. The subgroup K which satisfies the above conditions is called a **complement** of H.

A split extension is nothing but a semidirect product of H and K. The Isomorphism Theorem shows that $K \cong HK/H = G/H = F$. Thus, a complement K is isomorphic to F.

(8.2) *Let G be an extension of H by F. If a subgroup L of G contains H, then L is an extension of H by L/H. If the extension G splits over H, then L does too. If K is a complement of H in G, then $K \cap L$ is a complement of H in L.*

Proof. The proof of the first part is trivial. To prove the remaining part, suppose that the extension G splits over H, and let K be a complement of H. Then, since $G = HK$, we get $L = HK \cap L$, so, by Theorem 3.14 of Chapter 1, $L = H(L \cap K)$ holds. As $H \cap (L \cap K) = \{1\}$, $L \cap K$ is a complement of H in L. $\square$

(8.3) *Let G be an extension of H by F, and let f be the factor set associated with G. The extension G splits over H if and only if f is equivalent to the trivial factor set g such that $g(\sigma, \tau) = 1$ for all σ and τ of F. If H is abelian, the extension G splits if and only if the factor set f is a coboundary.*

Proof. If the extension splits, then there is a complement K. In this case, we can choose the set K as a transversal, and the factor set g with respect to K satisfies $g(\sigma, \tau) = 1$ for all σ and τ of F. So, f is equivalent to the trivial factor set. The converse also holds. By (7.10), if f is equivalent to the trivial factor set, then G is isomorphic to a split extension. $\square$

When the normal subgroup H is abelian, the theorems on cohomology groups may be applied to find sufficient conditions for an extension to split.

Theorem 8.4. *Let A be an abelian normal subgroup of a group G. Let L be a subgroup of finite index such that $A \subset L \subset G$ and $|G:L| = j$. Suppose that the mapping $a \to a^j$ is an automorphism of A. Then, the extension G splits over A if and only if L splits over A.*

Proof. It follows immediately from (8.2) that if G splits over A, then L splits over A. Conversely, suppose that L splits over A. By assumption, the mapping $\theta : a \to a^j$ is an automorphism of A. Clearly, G acts on A by conjugation. Since A is abelian, this induces an action of G/A on A. Moreover, θ is a G/A-isomorphism of A, and it induces an automorphism θ_* of $H^2(G/A, A)$. Let c be the class containing the factor set associated with the extension G. Then, the image $\rho^* c$ under the restriction $\rho^*(G/A, L/A)$ is trivial. This means that c belongs to Ker ρ^*. By Cor. 1 of Theorem 7.26, we have $jc = 0$. Thus, we get $\theta_* c = jc = 0$. Since θ_* is an automorphism, we must have $c = 0$. By (8.3), the extension G splits over A. □

As a particular case of Theorem 8.4, we see that if $j = |G : A|$ is finite and if the mapping $a \to a^j$ is an automorphism of A, then the extension splits over A. If A is a finite abelian group of order prime to j, then A is j-regular. Theorem 8.4 is applicable to this situation, and the extension splits. This can be generalized to the case in which a normal subgroup A is not necessarily abelian.

(8.5) *Let H be a normal subgroup of a finite group G. If the order $|H|$ is prime to the index $|G : H|$, then the extension splits over H.*

Proof. We will prove the above statement by contradiction. Suppose that the proposition (8.5) is false, and let G be a counterexample of minimal order. Thus, G possesses a normal subgroup H such that the order $|H|$ is prime to the index $|G : H|$. Yet there must not be a complement to H. Furthermore, for any finite group of order smaller than $|G|$, the proposition (8.5) holds. We will prove a series of lemmas under these assumptions.

(a) *The group H is not abelian.*

By the remark preceding (8.5), if H were abelian, then the extension would split over H.

(b) *If a subgroup L satisfies $G = HL$, then we have $L = G$.*

Suppose that a proper subgroup L satisfies $G = HL$. Then, by the Second Isomorphism Theorem, $L \cap H$ is a normal subgroup of L and $L/L \cap H \cong G/H$. Hence, $|L \cap H|$ is prime to $|L : L \cap H|$. Since L is proper, $|L| < |G|$ and L possesses a complement of $L \cap H$. If K is a complement of $L \cap H$, then

$$(L \cap H)K = L \quad \text{and} \quad K \cap (L \cap H) = \{1\}.$$

We have

$$HK = H(L \cap H)K = HL = G$$

and $K \cap H = \{1\}$. This means that K is a complement of H in G, thus contradicting the assumption. Hence, we must conclude that $L = G$.

(c) *H is a minimal normal subgroup of G.*

Let M be a minimal normal subgroup of G such that $M \subset H$. Then, $M \neq \{1\}$ and (8.5) is applicable to G/M. Since the order of H/M is prime to the index, there is a complement of H/M. By the Correspondence Theorem, a complement has the form L/M and satisfies $HL = G$. By Part (b), we have $L = G$. But, as $L \cap H = M$, we get $M = L \cap H = G \cap H = H$.

(d) *There is a prime number p such that H is a p-group.*

Let p be a prime number which divides $|H|$, and let S be an S_p-subgroup of H. By Theorem 2.7, we have $G = HN_G(S)$. So, from Part (b), we get $N_G(S) = G$. This shows that $S \lhd G$, and by Part (c), $H = S$.

(e) *H is abelian.*

Theorem 1.6 and (d) show that $Z(H) \neq \{1\}$. Since $Z(H)$ char H, (6.14) of Chapter 1 gives us $Z(H) \lhd G$. Hence, by (c), we get $Z(H) = H$. So, H is abelian.

As (e) contradicts (a), we conclude that the counterexample G cannot exist and that (8.5) holds. □

Theorem 8.6. *Let G be a finite group, and let A be an abelian normal subgroup of G. The extension splits over A if and only if any S_p-subgroup S of G splits over $S \cap A$ for any prime divisor p of $|G|$.*

Proof. Assume that the extension G splits over A, and let K be a complement of A in G. Let A_p be the S_p-subgroup of A, and let P be an S_p-subgroup of K. Since A is abelian, A_p is a characteristic subgroup of A. So, $A_p \lhd G$ and $A_p P$ is a subgroup of G. We have

$$A_p \cap P \subset A \cap K = \{1\},$$

so $A_p P$ is a split extension. By (3.11) of Chapter 1, we see that $A_p P$ is an S_p-subgroup of G. If S is any S_p-subgroup of G, then S is conjugate to $A_p P$ and $S \supset A_p$. Then, S is a split extension of A_p.

Conversely, assume that any S_p-subgroup S splits over $S \cap A$ for any prime divisor p of $|G|$, and let T be a complement of $S \cap A$ in S. Then, $T(S \cap A) = S$ and $T \cap A = \{1\}$. We have $TA = T(S \cap A)A = SA$. So, T is a complement of A in AS. Let c be the cohomology class of the factor set associated with the extension. Then, $\rho^*(G/A, AS/A)c$ is 0. So, by Corollary 1 of Theorem 7.26, $jc = 0$ where $j = |G : SA|$. Since S is an S_p-subgroup, j is prime to p. This means that the order of the class c is prime to p for all prime divisors of $|G|$. Since the order of c is divisible by only the prime numbers in $\pi(G)$, we get $c = 0$. By (8.3), the extension G splits over A. □

The preceding theorem, as well as Theorem 8.9, which will be proved later, is one of the theorems of Gaschütz [1], and it fails to hold without the commutativity assumption on A.

EXAMPLE. The following counterexample due to Baer is one of the simplest. Let $L = SL(2, 3)$. As stated at the end of Chapter 1, L has a chief series $L \supset Q \supset Z \supset \{1\}$. Set

$$Q = \langle \sigma, \tau \rangle \quad \text{and} \quad R = \langle \rho \rangle \ (\sigma^4 = \rho^3 = 1).$$

Then, Z is the center of L and is generated by σ^2. Let $C = \langle \gamma \rangle$ be a cyclic group of order 4, and let G be the central product of L and C in which σ^2 is identified with γ^2. We will prove two propositions.

(a) $P = CQ$ is an S_2-subgroup of G and splits over Q.

(b) For every prime number p, any S_p-subgroup S splits over $S \cap Q$, but G does not split over Q.

Set $\langle \sigma \gamma \rangle = C_1$. Then, $(\sigma \gamma)^2 = \sigma^2 \gamma^2 = 1$ and $|C_1| = 2$. We have $P = C_1 Q$ and $C_1 \cap Q = \{1\}$. So, C_1 is a complement of Q, and so (a) holds.

Since $|G| = 48$, the prime divisors of $|G|$ are 2 and 3. If $p = 2$, we know from (a) that all S_2-subgroups split over Q. If $p = 3$, we have $S \cap Q = \{1\}$ and certainly, all S_3-subgroups split over $S \cap Q$.

We need only to show that the extension G does not split over Q. The factor group G/Q is a cyclic group of order 6. If the extension were split, then, by Sylow's theorem, there would be a complement containing ρ. Let K be such a complement. By (a), $P = CQ$ is an S_2-subgroup, and $G = PR$. So, by Theorem 3.14 of Chapter 1, we get

$$K = PR \cap K = R(P \cap K).$$

This shows that $|P \cap K| = 2$. Furthermore, we have $P \cap K \subset C_G(\rho)$. On the other hand, from the structure of G, we see that $C_P(\rho) = \langle \gamma \rangle$, and $\langle \gamma \rangle$ contains only one subgroup of order 2. This gives us $P \cap K = \langle \gamma^2 \rangle = Z$. We get

$$Z \subset Q \cap K = \{1\},$$

and this is a contradiction.   □

A relationship between group extensions and one-dimensional cohomology groups is given by the following theorem.

(8.7) *Let A be an abelian group, and let G be an extension of A by a group F. Let A_0 be the totality of automorphisms of G which leave every element of A and every coset of A invariant :*

$$A_0 = \{\varphi \in \text{Aut } G \mid \varphi(a) = a \quad (a \in A), \qquad \varphi(xA) = xA \quad (x \in G)\}$$

Then, $A_0 \cong Z^1(F, A)$, and by this isomorphism, $B^1(F, A)$ corresponds to the set of inner automorphisms by the elements of A.

Proof. It is obvious that A_0 is a subgroup of Aut G. Let T be a transversal of A in G. If $\varphi \in A_0$, then there is an element $d(\sigma)$ of A such that

$$\varphi(t(\sigma)) = t(\sigma)d(\sigma) \quad \text{for each} \quad \sigma \in F.$$

Applying φ to the formula (7.1), we get $d(\sigma)^\tau d(\tau) = d(\sigma\tau)$. Thus, d is a crossed homomorphism. The mapping $\varphi \to d$ is a homomorphism from A_0 into $Z^1(F, A)$. This is proved as follows. If $\varphi' \to d'$, then

$$(\varphi\varphi')(t(\sigma)) = \varphi'(\varphi(t(\sigma))) = \varphi'(t(\sigma)d(\sigma))$$
$$= t(\sigma)d'(\sigma)d(\sigma).$$

Conversely, if d is an element of $Z^1(F, A)$, then an easy computation shows that the mapping φ defined by $\varphi(t(\sigma)a) = t(\sigma)d(\sigma)a$ is an automorphism of G. In fact, φ is an element of A_0. Thus, $Z^1(F, A)$ is isomorphic to A_0. If an automorphism φ of A_0 is induced by the conjugation of element a of A, then we have

$$\varphi(t(\sigma)) = a^{-1}t(\sigma)a = t(\sigma)a^{-\sigma}a.$$

So, $d(\sigma) = a^{-\sigma}a \in B^1(F, A)$. Conversely, if $d(\sigma) = a^{-\sigma}a$, then the corresponding automorphism is the inner automorphism by the element a. $\square$

It may be obvious to the reader, but we will note here that in $Z^1(F, A)$ and $B^1(F, A)$, the group A is considered to be a $\mathbb{Z}F$-module with respect to the action of F and that A is not just an abelian group.

Corollary. *Let G be a split extension of an abelian group A. If K_1 and K_2 are two complements of A, then there is an automorphism φ of A_0 such that $\varphi(K_1) = K_2$.*

Proof. For each $\sigma \in G/A$, let $t(\sigma)$ be the unique element of K_1 contained in the coset σ. Similarly, there is a unique element of K_2 contained in σ. This element has the form $t(\sigma)d(\sigma)$ for some $d(\sigma) \in A$. Then, we have

$$t(\sigma)t(\tau) = t(\sigma\tau),$$

and

$$t(\sigma)d(\sigma)t(\tau)d(\tau) = t(\sigma\tau)d(\sigma\tau).$$

Thus, we get $d(\sigma)^\tau d(\tau) = d(\sigma\tau)$ and $d \in Z^1(F, A)$. By (8.7), there is an element φ of A_0 such that $\varphi(K_1) = K_2$. $\square$

For a split extension of a nonabelian group, the assertion of the preceding corollary is not necessarily true (see Exercise 1).

(8.8) *Let G be a split extension of an abelian group A, and set $F = G/A$. Any two components of A are conjugate in G if and only if $H^1(F, A) = \{0\}$.*

Proof. Let K_1 and K_2 be two complements of A. Then, by the Cor. of (8.7), $K_2 = \varphi(K_1)$ for some element φ of A_0. (For an explanation of the notation, see (8.7).) By (8.7), φ corresponds to an element d of $Z^1(F, A)$. If $H^1(F, A) = \{0\}$, then φ is an element of $B^1(F, A)$. So, φ is introduced by the conjugation of an element a of A. Then, we have $K_2 = \varphi(K_1) = a^{-1} K_1 a$.

Conversely, by (3.15) of Chapter 1, if K_2 is conjugate to K_1, then there is an element a of A such that $K_2 = a^{-1} K_1 a$. In this case, $\varphi = \delta a$, and we have $H^1(F, A) = \{0\}$. □

Theorem 8.9. *Let G be an extension of an abelian group A, and let L be a subgroup of G such that $A \subset L \subset G$ and $|G : L| = j < \infty$. Assume that the mapping $a \to a^j$ $(a \in A)$ is an automorphism of A. If L splits over A and if any two complements of A in L are conjugate in L, then the extension G splits over A and any two complements are conjugate in G. In this case, if H is a complement of A in L, then there is a complement K in G such that $H = L \cap K$.*

Proof. By Theorem 8.4, G splits over A. By assumption and by (8.8), we have $H^1(L/A, A) = \{0\}$. Thus, any cohomology class c of $H^1(G/A, A)$ is in the kernel of the restriction mapping. By Corollary 1 of Theorem 7.26, we have $jc = 0$ which implies that $c = 0$ as in the proof of Theorem 8.4. Thus, we get $H^1(G/A, A) = \{0\}$, and by (8.8), any two complements of A are conjugate in G.

Let H be a complement of A in L, and let K be a complement of A in G. Then, by (8.2), $K \cap L$ is a complement of A in L. So, by assumption, $K \cap L$ is conjugate to H in L. Hence, there is an element x of L which satisfies $H = (L \cap K)^x$. Since

$$(L \cap K)^x = L^x \cap K^x, \quad \text{we have} \quad H = L \cap K^x.$$

As K^x is also a complement of A in G, the last assertion follows. □

The key point of the preceding proof is that $H^1(L/A, A) = \{0\}$ implies that $H^1(G/A, A) = \{0\}$. Since $H^1(L/A, A)$ may be different from $\{0\}$ even if $H^1(G/A, A) = \{0\}$, a given complement of A in L need not be contained in a complement in G (see Exercise 2).

The next theorem which is due to Zassenhaus is one of the fundamental theorems of finite group theory; a particular case of the theorem in which N is contained in the center $Z(G)$ of G had been proved by Schur in 1902 (using essentially the same methods).

Theorem 8.10. *Let G be a finite group, and let N be a normal subgroup of G. Assume that the order $|N|$ is relatively prime to the index $|G : N|$. Then, the extension splits over N, and any two complements are conjugate in G.*

Proof. We have already seen in (8.5) that the extension splits over N. Let K_1 and K_2 be two complements of N. We need to show that K_1 is conjugate to K_2. Suppose that the theorem fails, and let G be a minimal counterexample to the theorem so that K_1 is not conjugate to K_2 in G. We will prove a series of lemmas under these hypotheses.

(a) *The normal subgroup N is minimal in G.*

Let M be a minimal normal subgroup contained in N. Since $N \neq \{1\}$, we have $M \neq \{1\}$. The factor group G/M contains a normal subgroup N/M whose index is relatively prime to $|N/M|$. Moreover, the two subgroups MK_1/M and MK_2/M are complements of N/M. By the minimality of $|G|$, they are conjugate. So, there is an element x of G such that $MK_1^x = MK_2$. The subgroup MK_2 is an extension of M such that $|M|$ is prime to $|K_2 M : M|$, while MK_2 contains the two complements K_1^x and K_2. But, by assumption, K_1^x and K_2 are not conjugate in G. This means that the theorem fails for MK_2. Hence, the minimality of $|G|$ implies that $MK_2 = G$ and $M = N$.

(b) *The group N is not solvable, and it is certainly not abelian.*

If N were solvable, then, by Corollary 2 of Theorem 3.14, N would contain an abelian characteristic subgroup A different from $\{1\}$. By (a), we would have $A = N$. If N were abelian, Theorem 8.9 proves that K_1 and K_2 would be conjugate, contrary to the hypothesis.

(c) *The factor group G/N is simple.*

Let L be a maximal normal subgroup of G such that $N \subset L$. By definition, we have $L \neq G$. By (8.2), $L \cap K_1$ and $L \cap K_2$ are complements of N in L. So, by the minimality of $|G|$, they are conjugate. This means that there is an element y of L such that

$$L \cap K_2 = (L \cap K_1)^y = L \cap K_1^y.$$

Set $D = L \cap K_2 = L \cap K_1^x$. Then, by the Second Isomorphism Theorem, D is a normal subgroup of K_2 and K_1^x. Thus, $N_G(D)$ contains both K_2 and K_1^x. By Theorem 3.14 of Chapter 1, we have

$$N_G(D) = N_0 K_2 = N_0 K_1^x \quad \text{where} \quad N_0 = N \cap N_G(D) \lhd N_G(D).$$

Hence, K_2/D and K_1^x/D are complements of $N_0 D/D$ in $N_G(D)/D$. The group $N_G(D)/D$ and its normal subgroup $N_0 D/D$ satisfy all the assumptions of the theorem, and yet K_2/D and K_1^x/D cannot be conjugate. So, again by the minimality of $|G|$, we have $G = N_G(D)/D$. In particular, we have $D = \{1\}$. This occurs only when $L = N$. Therefore, N is a maximal normal subgroup, and by Theorem 4.15 of Chapter 1, the factor group G/N is simple.

(d) *For some prime number p, G/N is a cyclic group of order p, and both K_1 and K_2 are S_p-subgroups of G.*

By (b), N is nonsolvable. So, the Feit–Thompson Theorem [3] shows that $|N|$ is even. Since $|G/N|$ is prime to $|N|$ by assumption, $|G/N|$ must be odd.

Again by the Feit–Thompson Theorem, G/N is solvable. On the other hand, G/N is simple by (c). So, it must be a cyclic group of prime order p (Theorem 2.24 of Chapter 1 and Cor. 2 of Theorem 3.14).

Clearly, both K_1 and K_2 are S_p-subgroups of G. Then, by Sylow's theorem, they must be conjugate in G. This is contrary to our hypothesis. Thus, Theorem 8.10 holds. ☐

The preceding proof of the conjugacy of the complements uses the Feit–Thompson Theorem on the solvability of groups of odd order. At the time of this writing, there is no known proof which does not use this theorem. The Feit–Thompson Theorem is a fundamental result in finite group theory. However, its proof is so lengthy and difficult that we cannot present it in this book. It is highly desirable to have a simplified proof of the Feit–Thompson Theorem. This is certainly one of the many challenging major problems in finite group theory today.

As we can see from the proof, Theorem 8.10 can be proved without invoking the Feit–Thompson Theorem under the additional assumption that either N or G/N is solvable. Even this weak form of Theorem 8.10 is useful. It is indispensable in the proofs of many other theorems, including the Feit–Thompson Theorem. After the step (c) we can prove, without using the Feit–Thompson Theorem, that $C_G(N) = \{1\}$ and that N is a simple group. This is a result due to P. Hall and has been presented in Huppert [3], pp. 129–30. It follows from this theorem that $G \subset \operatorname{Aut} N$ and that G/N is isomorphic to a subgroup of $\operatorname{Out} N$. Thus, Theorem 8.10 may be proved if the Schreier conjecture we mentioned earlier (Chapter 1, p. 48) is true.

The next result is a corollary of Theorem 8.10. Not only is it a useful theorem, but also its proof illustrates a typical application of Theorem 8.10.

Corollary. *Let N be a normal subgroup of a group G. Assume that $|N|$ is prime to $|G : N|$, and set $\pi = \pi(G/N)$. If U is any π-subgroup of G, then N has a complement which contains U. A subgroup K of G is a complement of N if and only if K is maximal among the π-subgroups of G.*

Proof. By Theorem 8.10, there is a complement H of N. Set $V = UN \cap H$. Then, by (8.2), V is a complement of N in UN. By assumption, U is a π-subgroup, so U is also a complement of N in UN. By applying Theorem 8.10 to UN, we see that U and V are conjugate in UN. Thus, there is an element x of UN such that $U = V^x$. Then, we have

$$U = V^x = (UN \cap H)^x = UN \cap H^x.$$

Clearly, H^x is another complement of N in G and $U \subset H^x$. ☐

As another application of the fundamental theorem 8.10, Theorem 2.9 can be generalized as follows.

Theorem 8.11. *Let Q be an operator group acting on a group G such that $|Q|$ is relatively prime to $|G|$. Then, for any prime divisor p of $|G|$, the generalized Sylow's theorem (Theorem 2.9 (i), (ii), and (iii)) holds.*

Proof. The proof of Theorem 2.9 can be adapted to the present situation if we consider the complements isomorphic to Q instead of the S_q-subgroups and if we use Theorem 8.10 in place of Sylow's theorem. ☐

We have proved here the generalized Sylow's theorem (i) for an arbitrary operator group Q using Theorem 8.10. The statement (i) is, in fact, equivalent to the conjugacy of the complements. So, if Theorem 2.9 (i) holds for an arbitrary Q (such that $(|Q|, |G|) = 1$), then the conjugacy of any two complements can easily be proved without using the Feit–Thompson Theorem.

An analogue of Theorem 2.7 holds also.

(8.12) *Let H be a normal subgroup of finite order. Suppose that H contains two subgroups L and K such that $L \lhd H$, $KL = H$, $K \cap L = \{1\}$, and $|L|$ is prime to $|K|$. Then, we have $G = HN_G(K)$.*

Proof. The key point in the proof of Theorem 2.7 was the conjugacy of S and S^σ for any automorphism σ. In the situation in (8.12), the conjugation by an element of G induces an automorphism σ of H. The assumptions imply that L char H and that K is a complement of L. So, K^σ is a complement of $L^\sigma = L$, and by Theorem 8.10, K^σ is conjugate to K in H. The remaining argument is the same as in the proof of Theorem 2.7. ☐

Let Q be an operator group acting on a group G. If H is a Q-invariant normal subgroup of G, Q induces an action on the factor group G/H and becomes an operator group of G/H (Definition 8.10 of Chapter 1). In many cases, it may become necessary to study the relationship between $C_G(Q)$ and $C_{G/H}(Q)$. Certainly, we have

$$C_G(Q)H/H \subset C_{G/H}(Q).$$

It is also clear that the equality will not hold in general. But, the following useful theorem does hold.

Theorem 8.13. *Let Q be an operator group acting on a group G. Let H be a Q-invariant normal subgroup such that H and Q are finite and of relatively prime orders. Then, we have*

$$C_G(Q)H/H = C_{G/H}(Q).$$

Proof. Consider the semidirect product of Q and G with respect to the action of Q on G. We consider Q and G as subgroups of the semidirect product GQ.

Let L be the subgroup of G corresponding to $C_{G/H}(Q)$. Then, L is a Q-invariant subgroup such that $C_{G/H}(Q) = L/H$. By (8.11) of Chapter 1, H is a normal subgroup of GQ, and GQ/H is the semidirect product of G/H and Q with respect to the induced action of Q on G/H. Thus, L/H is centralized by Q, and HQ is a normal subgroup of LQ. We can apply (8.12) to LQ and its normal subgroup HQ, and we get

$$LQ = HQ \cdot N = HN,$$

where $N = N_{LQ}(Q)$. So, by Theorem 3.14 of Chapter 1, $L = H(N \cap L)$ holds. Since LQ is a semidirect product, (8.9) of Chapter 1 gives us

$$N \cap L \subset C_L(Q) \subset C_G(Q).$$

So, we have $L \subset HC_G(Q)$. On the other hand, $C_G(Q)$ is obviously contained in L. Therefore, we have the equality $L = HC_G(Q)$. $\square$

Corollary. *Suppose that a π'-group Q acts on a π-group G. Let*

$$H = \langle g^{-1}g^{\sigma} \mid g \in G, \sigma \in Q \rangle.$$

Then we have $G = C_G(Q)H$.

Proof. We will prove that H is a Q-invariant normal subgroup of G. For any g, $x \in G$ and $\sigma, \tau \in Q$, we have

$$(g^{-1}g^{\sigma})^x = x^{-1}g^{-1}g^{\sigma}x = (gx)^{-1}(gx)^{\sigma}(x^{-1}x^{\sigma})^{-1} \in H,$$

$$(g^{-1}g^{\sigma})^{\tau} = (g^{-1}g^{\tau})^{-1}(g^{-1}g^{\sigma\tau}) \in H.$$

This proves the assertion that H is a Q-invariant normal subgroup of G. By definition, Q acts trivially on the factor group G/H, that is, Q centralizes G/H. By Theorem 8.13, we have $C_G(Q)H = G$. $\square$

The subgroup H defined here is an example of the commutator subgroups which we will later study in detail (in Chapter 4). The above corollary is a particular case of Theorem 1.6 of Chapter 4.

We note that the preceding corollary is a generalization of (5.17) to general groups.

Automorphism Groups of p-Groups

As an application of cohomology theory to group theory, we will prove the following theorem of Gaschütz [3] on the automorphisms of p-groups.

Theorem 8.14. *Let G be a finite p-group. If $|G| > p$, we have $|\mathrm{Aut}\, G : \mathrm{Inn}\, G| \equiv 0 \,(\mathrm{mod}\, p)$.*

Proof. We will prove the theorem in three steps.

(a) First consider the case when G possesses a maximal subgroup M such that $C_G(M) = Z(G)$. By the Cor. of Theorem 1.6, M is a normal subgroup of G, and G/M is cyclic of order p. Since $M \cap Z(G) \neq \{1\}$, there is a homomorphism φ from G into $M \cap Z(G)$ such that Ker $\varphi = M$. Consider the function σ defined by $\sigma(x) = x\varphi(x)$ for all $x \in G$. It is easy to verify that σ is an automorphism of order p. We will show that σ is not inner. Suppose that σ is inner. Then, there is an element g of G such that $x\varphi(x) = x^g$. If $x \in M$, we have $\varphi(x) = 1$. So, $x^g = x$ for all $x \in M$. This means that the element g lies in $C_G(M)$. But, by assumption, we have $C_G(M) = Z(G)$. Hence, g lies in $Z(G)$ and $x\varphi(x) = x$ for all x. This contradicts the definition of φ. Therefore, σ is not an inner automorphism.

(b) Let A be an abelian normal subgroup which is maximal among the abelian normal subgroups of G. We choose A to be noncyclic if it is possible to do so. The Corollary of Theorem 1.12 shows that $C_G(A) = A$. Suppose that we have $H^1(G/A, A) \neq \{0\}$. Then, there is a crossed homomorphism d which does not belong to $B^1(G/A, A)$. By (8.7), d corresponds to an automorphism φ which belongs to the group A_0 of (8.7). The order of φ is a power of p. We will show that φ is not inner. If φ were an inner automorphism of G, then we would have $\varphi(x) = x^g$ for some element g of G. Since $\varphi \in A_0$, we would get $\varphi(a) = a$ for all $a \in A$. Thus, we would have $a^g = a$ for all $a \in A$, and so the element g would belong to $C_G(A)$. Since $C_G(A) = A$, g lies in A. But, this would mean that $\varphi \in B^1(G/A, A)$ by (8.7). This contradicts the definition of φ. Thus, φ is not inner.

(c) We assume that $H^1(G/A, A) = \{0\}$. By Example 3, §7, for any subgroup U of G/A, we have

$$H^1(U, A) = H^2(U, A) = \{0\}.$$

In particular, we have $H^2(G/A, A) = \{0\}$ which implies that G splits over A. So, there is a complement K of A in G. Let M be a maximal subgroup which contains K. By Part (a), we may assume that $C_G(M) \neq Z(G)$. We will show that under these assumptions, the structure of G is uniquely determined.

First, we will show that $C_G(M)$ is not contained in A. If $C_G(M)$ were contained in A, then $C_G(M)$ would centralize both M and A. So, $C_G(M)$ would lie in the center of $G = AK = AM$, contrary to the assumption. Thus, we have $AC_G(M) \neq A$, and there is a subgroup B such that $A \subset B \subset AC_G(M)$ and $|B : A| = p$ (Theorem 1.9, §1). It follows from the definition of B that $B = AC_B(M)$. So, both A and M normalize B, and we conclude that $B \lhd AM = G$.

Clearly, we have $B \cap M \lhd G$. We will show that $B \cap M$ is abelian but not cyclic. Set $P = B \cap K$. Then, as $G = AK$, Theorem 3.14 of Chapter 1 shows that $B = AP$ and $|P| = p$. Again by Theorem 3.14 of Chapter 1, we have $B \cap M = (A \cap M)P$. Clearly, $A \cap M$ centralizes $AC_G(M)$, and $(A \cap M)P = (A \cap M) \times P$. This proves that $B \cap M$ is a noncyclic abelian normal subgroup. It follows from the definition of A that A is not cyclic.

We will prove next that $A \cap M$ is cyclic and that one of the maximal subgroups of B is cyclic. For the proofs of these propositions, we will need simple lemmas on abelian p-groups. Let X be an abelian p-group of finite order, and let $\Omega_1(X)$ be the set of elements of order at most p. Then, $\Omega_1(X)$ is an elementary abelian subgroup of X. Let $|\Omega_1(X)| = p^d$. With these notations, the following two lemmas hold.

(i)   *The group X is cyclic if and only if $d = 1$.*
(ii) *If a maximal subgroup Y of X does not contain $\Omega_1(X)$, then there are exactly p^{d-1} subgroups of order p which are not contained in Y.*

The first lemma follows immediately from the Fundamental Theorem (see also Exercise 10, §5). The second one may be proved as follows. By assumption, we have $|\Omega_1(Y)| = p^{d-1}$. So, there are exactly $p^d - p^{d-1}$ elements of order p outside Y. Each element of order p generates a subgroup of order p, while each subgroup of order p contains exactly $p - 1$ elements of order p. Thus, there are exactly $(p^d - p^{d-1})/(p - 1)$ subgroups of order p which are not contained in Y.

We return to the proof of Theorem 8.14, and continue to use the notations introduced earlier in the proof. Since $B = AP$ and $A \cap P = \{1\}$, P is a complement of A in B. We have shown that $H^1(B/A, A) = \{0\}$ under our assumptions. So, all the complements are conjugate. We have also shown that $B \cap M$ is an abelian normal subgroup of B and that $|B : B \cap M| = p$. Then, P has exactly p conjugates in B, and all the conjugate subgroups of P are contained in $B \cap M$. Thus, A also has exactly p complements, and they are all contained in $B \cap M$. By applying Lemmas (i) and (ii) to the abelian p-group $B \cap M$, we get

$$|\Omega_1(B \cap M)| = p^2 \quad \text{and} \quad |\Omega_1(A \cap M)| = p.$$

Furthermore, $A \cap M$ is cyclic. As shown before, $A \cap M$ is contained in the center of B. Since $B/A \cap B$ is not cyclic, there is a maximal subgroup C of B different from A and from $B \cap M$ (Example, §3 of Chapter 1). By (2.26) of Chapter 1, C is abelian. But, as all the complements of A are contained in $B \cap M$, we get $\Omega_1(C) \subset A \cap M$. So, by Lemma (i), C is cyclic.

We have shown that B contains a maximal subgroup C which is cyclic, as well as two abelian maximal subgroups A and $B \cap M$ which are not cyclic. We will show that $B = G$ and that G is a dihedral group of order 8.

Since A is not cyclic, there is an element $a \in A - M$ of order p. Choose a suitable generator y of P so that the element $x = ay^{-1}$ belongs to C. This is possible, and we have $\langle x \rangle = C$. Let p^n be the order of x. Then, we have $n \geq 2$. Since $C \triangleleft B$, we have an integer m such that $y^{-1}xy = x^m$. Since $y^p = (xy)^p = 1$, we get

$$(xy)^p = y^{-p}xy^p y^{-(p-1)}xy^{p-1} \cdots y^{-1}xy = 1.$$

So, the integer m satisfies $m^p \equiv 1 \pmod{p^n}$ and

$$1 + m + m^2 + \cdots + m^{p-1} \equiv 0 \pmod{p^n}.$$

Since $m^p \equiv m \pmod{p}$ by Fermat's theorem, we get

$$m = 1 + kp^\lambda \qquad (k, p) = 1.$$

The group B is nonabelian, so we have $1 \le \lambda < n$. The binomial theorem gives us

$$m^p \equiv 1 + kp^{\lambda+1} + k^2(p-1)p^{2\lambda+1}/2 \pmod{p^{\lambda+2}}.$$

If p were odd, we would have $m^p \equiv 1 + kp^{\lambda+1} \pmod{p^{\lambda+2}}$. Since $m^p \equiv 1 \pmod{p^n}$, n would be equal to $\lambda + 1$. On the other hand, we have $n \ge 2$ and $2\lambda \ge n$. So, we would have

$$0 \equiv 1 + m + m^2 + \cdots + m^{p-1} \equiv p + k(p-1)p^n/2 \pmod{p^n}.$$

This is clearly a contradiction. Thus, we have $p = 2$.

In this case, $1 + m$ is divisible by p^n. Thus, we have $x^y = x^{-1}$. On the other hand, x^2 belongs to $A \cap M$ which is the center of B. This implies that $x^2 = x^{-2}$ and that $n = 2$. Thus, we have $|A| = 4$. Since A is an abelian normal subgroup which satisfies $C_G(A) = A$, the factor group G/A is isomorphic to a subgroup of Aut A (Theorem 6.11 of Chapter 1). By (9.14) of Chapter 1, Aut A is isomorphic to $GL(2, 2)$, so we have $|G : A| = 2$ and $G = B$. Clearly, the group G is a dihedral group of order 8.

We can easily see that Theorem 8.14 holds for the dihedral group of order 8. The function

$$x \to x^{-1}, \qquad y \to xy$$

can be extended to an automorphism of G which is of order 2 and is not inner. Thus, Theorem 8.14 holds for the dihedral group of order 8. This completes the proof of Theorem 8.14. $\square$

Exercises

1. Let H be a nonabelian simple group, and let $G = H \times H_1$ where H_1 is isomorphic to H. Show that G is a split extension of H by G/H, where $G/H \cong H$ and has complements which are not normal in G, as well as one which is normal. Let K_1 and K_2 be two complements which are not normal in G. Show that there is an automorphism σ of G such that $K_2 = K_1^\sigma$.

(*Hint.* Use Exercise 4, §4.)

2. Let F be a field such that its characteristic is not two. Let G be the group of matrices of the form

$$\begin{pmatrix} A & 0 \\ u & 1 \end{pmatrix}$$

where $A \in GL(2, F)$ and u ranges over a 2-dimensional vector space over F. We define the two subgroups L and H of G as follows: L is the set of all matrices with

$$A = \begin{pmatrix} 1 & 0 \\ \lambda & 1 \end{pmatrix} \qquad (\lambda \in F)$$

and H consists of these elements of L which satisfy $\lambda = 0$.

(a) Show that $H \lhd G$ and that G is a split extension of H by $GL(2, F)$.
(b) Show that L possesses a complement of H which is not contained in any complement of H in G.
(c) Show that all complements of H in G are conjugate.

(*Hint.* (b) If a subgroup U of L satisfies $L = HU$ and $H \cap U = \{1\}$, and if U is contained in a complement of H in G, then $C_G(U)$ contains an element of order 2. We note that the assertion (b) is false if $|F| = 2$. If the characteristic of F is two, then (b) still holds provided $|F| > 2$. But, the proof requires a different method.

(c) Consider the subgroup Z of G such that Z/H is the center of G/H. Then, all complements of H in Z are conjugate (by Theorem 8.10). Use (8.2) to complete the proof.)

3. Let F be an operator group acting on a group N. A function f defined on F which takes on values in N is said to be a *crossed homomorphism* if f satisfies

$$f(\sigma\tau) = f(\sigma)^\tau f(\tau) \quad \text{for all} \quad \sigma \text{ and } \tau \text{ of } F.$$

Let $Z^1(F, N)$ be the totality of crossed homomorphisms from F to N with respect to a fixed action of F. Two crossed homomorphisms f and g are said to be *equivalent* if there is an element x of N such that

$$f(\sigma) = x^{-\sigma} g(\sigma) x \quad \text{for all} \quad \sigma \in F;$$

in this case, we write $f \sim g$.

(a) Show that the relation $\sim$ is an equivalence relation; that is, (i) $f \sim f$, (ii) $f \sim g$ implies $g \sim f$, and (iii) $f \sim g$ and $g \sim h$ imply $f \sim h$. (We denote the set of equivalent classes by $H^1(F, N)$.)

(b) Let G be the semidirect product of N and F with respect to the given action of F. Show that there is a one-to-one correspondence between the set of complements of N in G and the set $Z^1(F, N)$.

(c) Using the same notations as in (b), show that there is a one-to-one correspondence between $H^1(F, N)$ and the conjugacy classes of the complements of N in G. Also prove that the complements of N are conjugate in G if and only if $H^1(F, N)$ consists of a single element.

(d) Show that if F and N are finite and if $(|F|, |N|) = 1$, then $H^1(F, N)$ consists of a single element.

(*Hint.* See the proof of the Cor. of Theorem 8.7. For the proof of (c), (3.15) of Chapter 1 is necessary. Part (d) is a restatement of the conjugacy part of Theorem 8.10. If N is nonabelian, both $Z^1(F, N)$ and $H^1(F, N)$, as defined here, are just sets and are not equipped with any operations.)

4. Let Q be an operator group acting a group G, and let H be a Q-invariant normal subgroup. Suppose that both Q and H are finite and that $(|Q|, |H|) = 1$. Show that a Q-invariant coset with respect to H contains a Q-invariant element.

(*Hint.* If a coset Hx is Q-invariant, we have $x^\sigma = d(\sigma)x$ for any $\sigma \in Q$ and some element $d(\sigma)$ of H. Show that $d \in Z^1(Q, H)$, and use Exercise 3(d). This is equivalent to Theorem 8.13, and with the proof indicated above, it is a theorem of Glauberman [1].)

5. Let G be an extension of an abelian group A, and let $F = G/A$.

(a) Let K be a normal subgroup of G such that $K \subset A$. Show that the canonical homomorphism θ from A to A/K is a $\mathbb{Z}F$-homomorphism and induces a homomorphism θ_*

$$H^n(F, A) \to H^n(F, A/K).$$

In particular, show that if $c \in H^2(F, A)$ is the cohomology class associated with the extension G, then the class $\theta_* c$ is the class associated with the extension G/K over A/K.

(b) Suppose that A is a torsion abelian group in which the orders of the elements are bounded (from above). Let A_p be the p-primary component for each prime number p, and let B_p be the subgroup such that $A = A_p \times B_p$. Show that G splits over A if and only if G/B_p is a split extension of A/B_p for every prime number p.

(c) Assume that A satisfies the same conditions as in (b), and also assume that the extension G splits over A. Show that the complements of A are conjugate in G if and only if complements of A/B_p are conjugate in G/B_p for every prime number p.

(*Hint.* (b) It follows from Exercise 2, §7 that

$$H^2(F, A) = \sum_p H^2(F, A/B_p).$$

A similar formula holds for 1-dimensional cohomology groups. Both (b) and (c) are results of Gaschütz [1] and D. G. Higman [2]. If G is finite, we can prove (b) using Theorem 8.6.)

6. Let A be an abelian normal subgroup of a finite group G. For each prime $p \in \pi(A)$, we assume that the following condition is satisfied. $(G)_p$: there is a p'-subgroup Q of G such that $QA \lhd G$ but QA_0 is not normal in G for any normal subgroup A_0 satisfying

$$A_0 \subset A, \qquad A_0 \lhd G, \quad \text{and} \quad |A : A_0| = p^n > 1.$$

(a) Show that the extension G splits over A and that the complements of A are conjugate in G.

(b) Furthermore, suppose that there is a finite group H which contains G and A as normal subgroups. Show that there are complements of A in H and that they are conjugate.

(c) Let P be the property of finite groups defined by the following: a group X is a P-group $\Leftrightarrow X$ is the direct product of S_p-subgroups. (It will be proved later that a finite group X is a P-group if and only if X is nilpotent (see Chapter 4).) Show that the property P satisfies the two conditions of the Corollary of (4.20). Define $O^P(X)$ as in Definition 4.21. Suppose that $O^P(N)$ is abelian for a normal subgroup N of G. Then, show that there are complements of $O^P(N)$ in G and that the complements are conjugate.

(*Hint.* (a) By Exercise 5, we may assume that A is a p-group. Then, by (8.12), we have $G = AN_G(Q)$. Set $K = N_G(Q)$, and show that K is a complement of A. By (5.17), we get $A = C_A(Q) \times A_0$ where A_0 is the subgroup of A generated by the commutator $[u, a]$ where $u \in Q$ and $a \in A$ (see Definition 3.11). For any element x of G, we have $Q^x \subset QA$. So, by Theorem 8.10, there is an element a of A such that $Q^x = Q^a$. We may even choose $a \in A_0$. Then, we have $QA_0 \lhd G$. Also, we can show that $A_0 \lhd G$ (see the proof of the Cor. of Theorem 8.13). The condition $(G)_p$ gives us $A = A_0$. This implies that K is a complement of A (see (8.9) of Chapter 1). The conjugacy of the complements can be proved by a method similar to the one in Exercise 2(c).

(b) Show that the condition $(H)_p$ holds for some p'-subgroup R such that $Q \subset R \subset G$. We can choose such a p'-subgroup R so that RA/A is the maximal normal p'-subgroup of G/A (see the Cor. of Theorem 8.10).

(c) First, consider the special case when $G = N$. For each prime $p \in \pi(A)$, show that there is a p'-subgroup satisfying the condition $(G)_p$. Any maximal p'-subgroup Q of G will do. The general case follows from Part (b).

The special case of (c) is a theorem of Schenkman [1]. This exercise and the proof indicated above are due to G. Higman [2].)

§9. Central Extensions, Schur's Multiplier

In this section, we will study central extensions of finite groups (see Definition 9.1). This is a concept which originated during the research on the problem concerning projective representations of finite groups. Early in this century, in 1904 to be exact, Schur discovered the main theorems about both the central extensions and the projective representations. Since then, these results have been very effectively applied to various other problems on finite groups, and in light of the tremendous recent advances, the theory of central extensions has been recognized as one of the fundamental tools for the investigation of finite groups. Its significance and relevance to the study of simple groups will be discussed in Chapter 6. In this section, we will present Schur's results following the work of Yamazaki [1], but the relationship between central extensions and projective representations will not be explored in any great depth. (See Exercise 2 at the end of this section.)

Definition 9.1. A **central extension** of a group G is an exact sequence

$$1 \to Z \to H \to G \to 1$$

such that Z (or more precisely, the image of Z in H) belongs to the center of H. Sometimes, we will say that the group H is a central extension of G.

A central extension is determined by a pair (H, Z) consisting of a group H and a subgroup Z of the center of H such that $H/Z \cong G$. Such a central extension is often denoted by (H, Z).

For the most part, we are interested in the central extensions of a finite group G, but it is often convenient to allow these central extensions to be infinite. The main problem in this branch of group theory is that of finding all the possible central extensions of a given finite group G.

(9.2) *Let G be a given group. Then, G has a presentation $G = F/K$ where F is a free group. For a given presentation $G = F/K$, set $N = \langle [x, k] \mid x \in F, k \in K \rangle$. Then, the following propositions hold.*

(1)   *The group N is a normal subgroup of F.*
(2)   *We have $K/N \subset Z(F/N)$.*
(3)   *The short exact sequence $1 \to K/N \to F/N \to G \to 1$ is a central extension of G.*
(4)   *Suppose that the exact sequence $1 \to Z' \to H' \to G' \to 1$ is a central extension of a group G'. If a homomorphism $\gamma: G \to G'$ is given, then there are homomorphisms κ and φ such that the diagram*

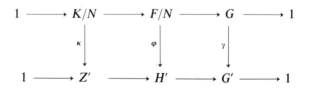

*is commutative. (The homomorphism φ is called a **lifting** of γ.)*

Proof. By Theorem 6.5, there is a presentation of a group G. So, we have $G \cong F/K$ for some free group F and a normal subgroup K of F. The group N is generated by the commutators

$$[x, k] = x^{-1}k^{-1}xk$$

of $x \in F$ and $k \in K$ (Definition 3.11). If g is an element of F, then we have $[x, k]^g = [x^g, k^g]$. So, the assertion (1) follows immediately.

The elements kN of K/N commute with any element xN of F/N because $[xN, kN] = N$. Thus, the proposition (2) holds.

By the Cor. of Theorem 5.3, Chapter 1, we have an isomorphism

$$G \cong F/K \cong (F/N)/(K/N).$$

This shows that (3) holds.

Suppose that we are given a central extension (H', Z') of G' and a homomorphism $\gamma: G \to G'$. Since F is free, we have a proposition analogous to (7.18). This proposition gives us a mapping $f: F \to H'$ such that the diagram

is commutative. Then, f maps the group K into Ker $(H' \to G')$. We will show that $N \subset$ Ker f. Consider a commutator $[x, k]$ where $x \in F$ and $k \in K$. Then, we have

$$f([x, k]) = [f(x), f(k)].$$

Since $f(k) \in Z' =$ Ker$(H' \to G')$, and since Z' is a subgroup of the center of H', we get $[f(x), f(k)] = 1$. Thus, f maps any generator of N into 1. Therefore, we have $N \subset$ Ker f. By the Cor. of Theorem 5.3, Chapter 1, f induces a homomorphism φ from F/N into H' such that the diagram

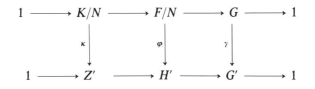

is commutative. Here, the mapping κ is the restriction of φ on K/N. $\square$

Definition 9.3. The central extension

$$1 \to K/N \to F/N \to G \to 1$$

defined in (9.2) is said to be a **free central extension** of G. If F is a free group of rank n, we call F/N a free central extension of **rank** n.

A free central extension is determined by a presentation of G such that $G = F/K$. So, a free central extension of G is not uniquely determined.

In the theory of central extensions, the exact sequence (7.29) is important. We will consider a central extension (H, Z) of a group G and study the modules over the group rings. But, we will consider only those modules on which the actions of the groups are trivial; that is, those modules for which each element of these groups acts as the identity mapping. Thus, the one-dimensional cohomology group $H^1(H, A)$ coincides with $\mathrm{Hom}(H, A)$. Since the action is trivial, the $\mathbb{Z}H$-homomorphisms are just the usual homomorphisms, and we write simply Hom, omitting the subscript $\mathbb{Z}H$.

The exact sequence (7.29) associated with the central extension (H, Z) of G is written as follows:

$$0 \to \mathrm{Hom}(G, A) \to \mathrm{Hom}(H, A) \to \mathrm{Hom}(Z, A) \to H^2(G, A).$$

The last arrow represents the mapping γ defined on page 216. The concrete form of the mapping γ is given by the next theorem.

(9.4) *Let (H, Z) be a central extension of a group G, and let the corresponding fundamental exact sequence be given by the above sequence. Choose a transversal $T = \{t_\rho\}$ of Z in H such that $Z \cap T = \{1\}$, and let $\{f(\sigma, \tau)\}$ be the associated factor set. Then, for any $\varphi \in \mathrm{Hom}(Z, A)$, the element $\gamma\varphi$ is the cohomology class of $H^2(G, A)$ which contains the factor set $\{\varphi f(\sigma, \tau)\}$.*

Proof. The proof of (7.29) shows that the homomorphism γ is uniquely determined once a set $\{a_\sigma\}$ of elements of A is chosen. From the definition of these elements $\{a_\sigma\}$, we see that we may choose $a_\tau = 0$ for any $\tau \in T$ (because the action is trivial). Thus, for any element

$$\sigma = \lambda t_\rho \quad (\lambda \in Z, \, t_\rho \in T),$$

we can define $a_\sigma = \varphi(\lambda)$. According to the definition of γ, the class $\gamma\varphi$ is the cohomology class of

$$\alpha(u, v) = a_{uv} - a_u v - a_v.$$

We will compute

$$\alpha(u, v) \quad \text{for} \quad u = \lambda t_\sigma \quad \text{and} \quad v = \mu t_\tau \, (\lambda, \mu \in Z, \, t_\sigma, t_\tau \in T).$$

By (7.1), we get $t_\sigma t_\tau = t_{\sigma\tau} f(\sigma, \tau)$. So, we have $uv = \lambda\mu f(\sigma, \tau)t_{\sigma\tau}$ and

$$\alpha(u, v) = \varphi(\lambda\mu f(\sigma, \tau)) - \varphi(\lambda) - \varphi(\mu) = \varphi f(\sigma, \tau). \quad \square$$

(9.5) *If H is a free central extension of G, then the mapping γ associated with the extension H is surjective.*

Proof. There is a normal subgroup Z of $Z(H)$ such that $H/Z \cong G$. Choose a transversal $\{t_\sigma\}$ of Z, and define the corresponding factor set $\{f(\sigma, \tau)\}$.

Let $\{a(\sigma, \tau)\}$ be a factor set in a cohomology class of $H^2(G, A)$. By Theorem 7.6, there is an extension H' of A by G such that the factor set associated with the extension H' is equivalent to $\{a(\sigma, \tau)\}$. Since the action of G on A is trivial, H' is a central extension of G. Then, by (9.2), the identity mapping $G \to G$ can be lifted to a homomorphism from H into H'. Let us denote this homomorphism by θ. Then, the commutative diagram

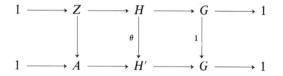

shows that $\theta(t_\sigma)$ belongs to the coset of H' which corresponds to the element σ of G. Thus, $\{\theta(t_\sigma)\}$ is a transversal of A in H'. We have

$$\theta(t_\sigma)\theta(t_\tau) = \theta(t_{\sigma\tau})\theta f(\sigma, \tau).$$

This implies that the factor set $\{\theta f(\sigma, \tau)\}$ is equivalent to the given factor set $\{a(\sigma, \tau)\}$. Let φ be the restriction of θ on Z. Then, $\varphi \in \mathrm{Hom}(Z, A)$, and by (9.4), $\gamma(\varphi)$ is the class of $\{a(\sigma, \tau)\}$. The mapping γ is thus surjective. $\quad \square$

Definition 9.6. Let $\mathbb{C}^*$ be the multiplicative group of nonzero complex numbers. Consider $\mathbb{C}^*$ as a trivial G-module (although $\mathbb{C}^*$ is a multiplicative group). The cohomology group $H^2(G, \mathbb{C}^*)$ is called the **Schur multiplier** of G, and we denote it by $M(G)$.

The Schur multiplier $M(G)$ will usually be considered as a multiplicative group, and the identity is usually denoted by 1, although it may sometimes be more convenient to conform with the additive notation commonly used in cohomology theory.

When $A = \mathbb{C}^*$, we use the notation $Z^* = \mathrm{Hom}(Z, \mathbb{C}^*)$ introduced in §5. In this case, Theorem 5.15 holds, and this facilitates the study of the connections between the exact sequence (7.29) and the structure of a central extension.

(9.7) *Let (H, Z) be a central extension of a group G, and let*

$$0 \to \mathrm{Hom}(G, \mathbb{C}^*) \to \mathrm{Hom}(H, \mathbb{C}^*) \overset{\rho}{\to} Z^* \overset{\gamma}{\to} M(G)$$

be the fundamental exact sequence associated with the extension H. Let H' be the commutator subgroup of H, and set $D = H' \cap Z$. Then, we have

$$\text{Im } \rho = \text{Ker } \gamma = D^{\perp} \quad \text{and} \quad \text{Im } \gamma \cong D^*,$$

where $D^{\perp} = \{\varphi \mid \varphi \in Z^*, \varphi(D) = 1\}$. If G is a finite group of order g, then D is a torsion group, and the order of any element of D divides g.

Proof. By the exactness of the above sequence, we get $\text{Im } \rho = \text{Ker } \gamma$.

Let θ be an element of $\text{Hom}(H, \mathbb{C}^*)$. Then, since $\mathbb{C}^*$ is abelian, we get $\theta(x) = 1$ for any element x of H' (Theorem 3.13). Hence, we have $H' \subset \text{Ker } \theta$. The mapping ρ is the restriction $\rho^*(H, Z)$. So, every element of $\text{Im } \rho$ is trivial on D. Thus, we have $\text{Im } \rho \subset D^{\perp}$.

Conversely, if $\varphi \in D^{\perp}$, then we have $D \subset \text{Ker } \varphi$. Combined with the isomorphism $H'Z/H' \cong Z/D$, φ induces a homomorphism φ' from $H'Z/H'$ into $\mathbb{C}^*$. By the injectivity of $\mathbb{C}^*$ (Lemma, §5), φ' can be extended to a homomorphism ψ' of H/H' into $\mathbb{C}^*$. Clearly, we can regard ψ' as an element of $\text{Hom}(H, \mathbb{C}^*)$ which we will write as ψ. Then, by definition, we have $\rho\psi = \varphi$. This proves that $D^{\perp} \subset \text{Im } \rho$. Thus, we have $D^{\perp} = \text{Im } \rho$. By the exactness of the sequence and Theorem 5.15, we have $\text{Im } \gamma \cong Z^*/D^{\perp} \cong D^*$.

We have shown that $D^* \cong \text{Im } \gamma$. So, D^* is isomorphic to a subgroup of $M(G)$. If G is a finite group of order g, then Cor. 2 of Theorem 7.26 proves that $M(G)$ is a torsion group and that the order of any element of $M(G)$ divides g. So, D^* has the same property. Theorem 5.15 (iv) shows that D is also a torsion group and the orders of its elements are divisors of g. $\square$

If G is finite, then the group D defined in (9.7) is indeed a finite group which is isomorphic to a subgroup of $M(G)$. This assertion is a particular case of the following theorem due to Schur [1] and Baer [1].

Theorem 9.8. *Suppose that the center of a group H has finite index. Then, the commutator subgroup H' of H is finite.*

Proof. Let Z be the center of H, and let $G = H/Z$. Then, H is a central extension of the finite group G. Suppose that H is finitely generated. Then, Z is a subgroup of finite index in a finitely generated group H. By Cor. 1 of Theorem 6.9, Z is finitely generated. So, by Cor. 2 of Theorem 5.2, the torsion subgroup $T(Z)$ of Z is finite. As we have shown in (9.7), we get $Z \cap H' \subset T(Z)$. So, $Z \cap H'$ is finite. On the other hand, the Isomorphism Theorem gives us $H'/H' \cap Z \cong H'Z/Z$. Since $G = H/Z$ is finite, $H'/H' \cap Z$ is finite, too. Thus, H' itself is finite.

Consider the general case. Let us choose a transversal T of Z, and let H_1 be the subgroup generated by T. Let x and y be arbitrary elements of H, and choose two representatives s and t of the cosets xZ and yZ, respectively.

Then, we have elements u and v of Z such that $x = su$ and $y = tv$. Since Z is the center of H, we have

$$[x, y] = [su, tv] = [s, t].$$

Thus, H' coincides with H'_1 which is finite, as we saw in the first step.   □

If the index of the center is given, then we can give an upper bound of the order of H' (see the last remark in Example 3 at the end of this section).

The following theorem of Schur is the fundamental theorem about the multipliers of finite groups.

Theorem 9.9. *Let (H, Z) be a free central extension of rank n of a finite group G. Then, the torsion subgroup $T(Z)$ is isomorphic to the Schur multiplier $M(G)$ of G. Let H' be the commutator subgroup of H, and set $D = Z \cap H'$. Then, Z/D is a free abelian group of rank n, and $D = T(Z) \cong M(G)$. The multiplier $M(G)$ is a finite group. Furthermore, the order of any element of $M(G)$ divides $|G|$.*

Proof. Consider the fundamental exact sequence associated with the extension H and use the same notations as in (9.7). Since H is free, the mapping γ is surjective by (9.5), and we have $D^* \cong \gamma(Z^*) = M(G)$. By Theorem 9.8, D is finite. So, by Theorem 5.15, we have $D \cong D^* \cong M(G)$.

We need to show that D coincides with the torsion subgroup $T(Z)$. By assumption and from (9.2), we can see that there is a free group F of rank n, as well as a normal subgroup K of F, such that

$$F/K = G, \qquad H = F/N, \quad \text{and} \quad Z = K/N$$

where $N = [K, F]$. Let F' be the commutator subgroup of F. Then, the factor group F/F' is a free abelian group of rank n (p. 166). Clearly, F' contains N. So, by (3.12), F'/N is the commutator subgroup H' of H. By the Isomorphism Theorem, we have $H/H' \cong F/F'$. Thus, H/H' is a free abelian group of rank n. The definition of D and the Isomorphism Theorem give us $Z/D \cong ZH'/H'$. The right side is a subgroup of the free abelian group H/H', so it is free by Theorem 5.11. Thus, the group Z/D is a free abelian group. As D is finite, it must be the torsion subgroup of Z. Hence, we have $D = T(Z)$. Since the group ZH'/H' is of finite index in H/H', the rank of ZH'/H' must coincide with the rank of H/H' (Theorem 3.20). So, the rank of Z/D is equal to n.

The last assertion of Theorem 9.9 follows from (9.7) or from Corollary 2 of Theorem 7.26.   □

The fact that $M(G)$ is a finite group if G is finite can be proved directly, without using the preceding theorems. To do this, it suffices to show that each class of $M(G)$ contains a cocycle f such that $f(\sigma, \tau)$ is a g-th root of unity for all σ and τ (where $g = |G|$). Let $\{a(\sigma, \tau)\}$ be an element of $Z^2(G, \mathbb{C}^*)$. Since a is a cocycle, we have

$$a(\sigma, \tau\rho)a(\tau, \rho) = a(\sigma\tau, \rho)a(\sigma, \tau).$$

If we take the product of the above formulas as ρ ranges over G, then we get

$$a(\sigma, \tau)^g = d_{\sigma\tau}^{-1} d_\sigma d_\tau,$$

where $d_\sigma = \prod_\rho a(\sigma, \rho)$. This proves Cor. 2 of Theorem 7.26 which asserts that a^g is a coboundary. In the above formula, each d_σ is an element of $\mathbb{C}^*$, so there is a g-th root e_σ in $\mathbb{C}^*$. The cocycle b defined by

$$b(\sigma, \tau) = e_{\sigma\tau} a(\sigma, \tau) e_\sigma^{-1} e_\tau^{-1}$$

is in the same cohomology class as a, and since $e_\sigma^g = d_\sigma$ for all σ, we have $b(\sigma, \tau)^g = 1$. Thus, for the cocycle b, every element $b(\sigma, \tau)$ is a g-th root of unity. The number of distinct g-th roots of unity is g. So, $M(G)$ must be finite. This is the original argument of Schur [1].

It can also be proved by the above method that the order d of an element of $M(G)$ is a divisor of g. In fact, d^2 divides g. But, we need the representation theory of finite groups to prove this theorem. For this proof, we refer the reader to the original paper of Schur [1] or Huppert [3], p. 635.

Theorem 9.9 can be used to compute the Schur multipliers of various groups (see Examples 1 and 2 of this section and Chapter 3, §2).

Suppose that a central extension (H, Z) of a group G contains a proper subgroup L such that $ZL = H$. Then, H is a central product of Z and L, and we can study the structure of H using the properties of L. It can be proved from the Isomorphism Theorem that $(L, L \cap Z)$ is a central extension of G. So, if H is finite, we can reduce the problems concerning the structure of H to the similar problems about the group L which has a smaller order than H. We will define a few terms here to facilitate all further discussions.

Definition 9.10. A central extension (H, Z) of a finite group G is said to be **irreducible** if there is no proper subgroup L having the property $H = ZL$. A central extension is said to be **primitive** if it is irreducible and also satisfies the equality $|H' \cap Z| = |M(G)|$. Following Schur, a central extension H is called a **representation group** of G if H is primitive and $|H| = |G| \cdot |M(G)|$.

(9.11) Let $(\bar{F}, \bar{K})$ be a free central extension of a finite group G. Any irreducible central extension (H, Z) of G is a homomorphic image of $\bar{F}$; that is, there is a subgroup $\bar{J}$ of $\bar{K}$ such that $H \cong \bar{F}/\bar{J}$ and $Z \cong \bar{K}/\bar{J}$.

Proof. By (9.2), the identity mapping $G \to G$ can be lifted to a homomorphism φ from $\bar{F}$ to H such that the diagram

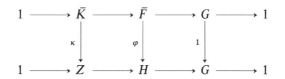

is commutative. We will consider $\bar{K}$ as a subgroup of $\bar{F}$ and let κ be the restriction of φ on $\bar{K}$. Similarly, we regard Z as a subgroup of H. Since the above diagram is commutative, the image $\varphi(\bar{F})$ intersects with every coset of H. Thus, we have $Z\varphi(\bar{F}) = H$. By assumption, H is irreducible, so we get $\varphi(\bar{F}) = H$. The commutativity of the above diagram implies that we have Ker $\varphi \subset \bar{K}$. Let $\bar{J} = \text{Ker } \varphi$. Then, we have $H = \bar{F}/\bar{J}$, and we conclude from the Isomorphism Theorem that $\bar{K}/\bar{J}$ corresponds to Z.   □

It follows from the preceding theorem that all the irreducible central extensions of a fixed finite group G can be studied simultaneously by looking at a free central extension and the normal subgroups contained in its center. In the rest of this section, we will consider the presentation $F/K = G$ to be fixed, where G is any finite group and F is a *finitely generated* free group. Define the subgroup N as in (9.2). Set $\bar{F} = F/N$ and $\bar{K} = K/N$. Then, by (9.2), $(\bar{F}, \bar{K})$ is a free central extension of G. Let (H, Z) be any irreducible central extension of G. By (9.11), there is a normal subgroup $\bar{J}$ of $\bar{F}$ such that $\bar{J} \subset \bar{K}$, $\bar{F}/\bar{J} = H$, and $\bar{K}/\bar{J} = Z$. This subgroup $\bar{J}$ is called the subgroup corresponding to the central extension H and is written as $\bar{J} = \bar{J}(H)$. Let $J(H)$ be the subgroup of F which corresponds to $\bar{J}$. Then, we have $J(H)/N = \bar{J}$. We will continue to use these notations till the end of this section.

If J is a subgroup of F such that $N \subset J \subset K$, then the factor group F/J is a central extension of G, but F/J need not be irreducible. A necessary and sufficient condition for F/J to be irreducible is given by the following theorem.

(9.12) *Using the notations introduced above, the central extension F/J is irreducible if and only if any maximal subgroup of F which contains JF' also contains K.*

The notation F' denotes the commutator subgroup of F. By the Correspondence Theorem, the above condition is equivalent to the following one in the factor group $\bar{F} = F/N$.

A maximal subgroup $\bar{M}$ of $\bar{F}$ such that $\bar{J}\bar{F}' \subset \bar{M}$ contains $\bar{K}$.

By (3.12), $\bar{J}\bar{F}'/\bar{J}$ is the commutator subgroup of $H = \bar{F}/\bar{J}$. So, again by the Correspondence Theorem the above condition can be stated in the group H as follows.

($*$)   A maximal subgroup M of H satisfying $M \supset H'$ contains Z.

Proof. Suppose that the group $H = F/J$ is not irreducible. Then, there is a proper subgroup L of H which satisfies $LZ = H$. Since $Z \subset Z(H)$, not only is $L \lhd H$, but also the group H/L is abelian. Thus, by Theorem 3.13, L contains the commutator subgroup H' of H. The free group F is finitely generated, and so is H. Then, H/L is also finitely generated. By Theorem 5.2, H/L has a maximal subgroup. The Correspondence Theorem shows that H has a maximal subgroup M which contains L (also see Exercise 5 of Chapter 1, §2). Clearly, $M \supset H'$, but M does not contain Z. Hence, the condition ($*$) is violated.

Conversely, suppose that the condition ($*$) does not hold. Then, there is a maximal subgroup M such that M does not contain Z. We get $H = MZ$, so H is not irreducible. □

(9.13) *Let (H, Z) be an irreducible central extension of a finite group G. Then, we have $|H' \cap Z| \leq |M(G)|$. The extension H is primitive if and only if $J(H) \cap F' = N$ where $J(H)$ is the subgroup of F corresponding to the extension H.*

Proof. The definition of $J(H)$ shows that $F/J(H) \cong H$. So, by the Correspondence Theorem, the subgroup H' corresponds to $J(H)F'$. Hence, the subgroup $H' \cap Z$ corresponds to $J(H)F' \cap K$ which, by Theorem 3.14 of Chapter 1, is equal to $J(H)(F' \cap K)$. Let $D = F' \cap K$. Since $(\bar{F}, \bar{K})$ is free, Theorem 9.9 asserts that $\bar{D} = D/N$ is isomorphic to $M(G)$. On the other hand, we have

$$H' \cap Z = J(H)D/J(H) \cong D/D \cap J(H).$$

So, in general, we get $|H' \cap Z| \leq |M(G)|$. The extension H is primitive if and only if $|H' \cap Z| = |M(G)|$. This is the case if and only if $D \cap J(H) = N$. Clearly, we have $D \cap J(H) = J(H) \cap F'$. □

Theorem 9.14. (1) *An irreducible central extension of a finite group is finite.*

(2) *Any irreducible central extension of a finite group G is a homomorphic image of a primitive central extension of G. If H is an irreducible central extension of G, then there is a primitive central extension H_0 and a subgroup N_0 of $Z(H_0)$ such that*

$$H \cong H_0/N_0 \quad and \quad |H : H'| = |H_0 : H_0'|.$$

(3) *If the central extension (H, Z) of G is irreducible, then, for any subgroup $Z_1 \subset Z$, the factor group H/Z_1 is also an irreducible central extension of G.*

Proof. (1) Let (H, Z) be an irreducible central extension of G. By Theorem 9.8, the commutator subgroup H' of H is finite. We need to show that H/H' is finite. By (9.11), H is a homomorphic image of a finitely generated group, so H/H' is finitely generated. By Theorem 5.2, H/H' is a direct product of a finite number of cyclic groups. So, we need to show that the number of maximal subgroups of H/H' is finite. Any maximal subgroup of H/H' corresponds to a maximal subgroup of H which contains H'. Let M be such a maximal subgroup. By (9.12), H satisfies the condition ($*$). Thus, the maximal subgroup M contains ZH'. On the other hand, by Cor. 3 of (3.12), H/ZH' is isomorphic to G/G' which is finite. This implies that there are only a finite number of such maximal subgroups M. Thus, H/H' is finite, and so is H.

(2) Let $\bar{J}(H)$ be the subgroup of $\bar{F}$ corresponding to the central extension (H, Z). Since the free group F is finitely generated, and H is finite by (1), $J(H)$ is also finitely generated (see Cor. 1 of Theorem 6.9). Set $\bar{D} = \bar{K} \cap \bar{F}'$. By Theorem 9.9, $\bar{D}$ is the torsion subgroup $T(\bar{K})$ of $\bar{K}$. So, we have $\bar{J}(H) \cap \bar{D} = T(\bar{J}(H))$, and by Cor. 2 of Theorem 5.2, $T(\bar{J}(H))$ is a direct factor of $\bar{J}(H)$. There is a subgroup $\bar{J}_0$ of $\bar{J}(H)$ such that $\bar{J}(H) = T(\bar{J}(H)) \times \bar{J}_0$. The subgroup $\bar{J}_0$ satisfies $\bar{J}_0 \bar{F}' = \bar{J}(H)\bar{F}'$ and $\bar{J}_0 \cap \bar{F}' = \{1\}$. Set $H_0 = \bar{F}/\bar{J}_0$. Then, the commutator subgroup H_0' is $\bar{J}_0 \bar{F}'/\bar{J}_0$. By (9.12), H_0 is an irreducible central extension of G. Since $\bar{J}_0 \cap \bar{F}' = \{1\}$, (9.13) shows that H_0 is primitive. Clearly, $H_0/N_0 \cong H$ where $N_0 = \bar{J}(H)/\bar{J}_0$, and (2) is proved.

(3) Let J_1 be the subgroup of F corresponding to the subgroup Z_1. Then, we have $J(H) \subset J_1 \subset K$. Let M be a maximal subgroup of F which contains $J_1 F'$. Then, M contains $J(H)F'$. Since H is irreducible, (9.12) shows that $M \supset K$. Again from (9.12), we see that F/J_1 is irreducible. Since $F/J_1 \cong H/Z_1$, (3) is proved. $\square$

(9.15) *Let (H, Z) be a primitive central extension of a finite group G.*

(1) *We have $H' \cap Z \cong M(G)$. In particular, we have*

$$|H| \geq |G| \times |M(G)|.$$

(2) *The following three conditions are equivalent:*
   (a) *The extension H is a representation group of G.*
   (b) $Z \subset H'$.
   (c) $|G : G'| = |H : H'|$.

Proof. Let $D = H' \cap Z$. If (H, Z) is primitive, we have $|D| = |M(G)|$. On the other hand, (9.7) shows that D^* is isomorphic to some subgroup of $M(G)$. Since D is finite, D^* is isomorphic to D. Thus, D is isomorphic to $M(G)$.

By definition, H is a representation group of G if $|Z| = |M(G)|$. So, by the first part (1), (a) is equivalent to (b). By Correspondence Theorem and Cor. 3 of (3.12), the condition (b) is equivalent to (c). $\square$

(9.16) *Any finite group has a representation group.*

Proof. The pair $(G, 1)$ is an irreducible central extension of G. So, by Theorem 9.14 (2), there is a primitive central extension H_0 and a subgroup N_0 of $Z(H_0)$ such that

$$G \cong H_0/N_0 \quad \text{and} \quad |H_0 : H_0'| = |G : G'|.$$

By (9.15) (2), H_0 is a representation group of G. □

The primitive central extensions of a finite group G are not necessarily unique. In fact, if $G \neq G'$, then the orders of the primitive central extensions are not bounded (see Exercise 1). But, as (9.15) shows, the group $H' \cap Z$ of a primitive central extension (H, Z) is always isomorphic to $M(G)$. There are several other properties which are common to all primitive central extensions of a group G.

Theorem 9.17. *Let* (H_1, Z_1) *and* (H_2, Z_2) *be two primitive central extensions of the same finite group* G. *Then, the following hold :*

(1) $H' \cap Z_1 \cong H_2' \cap Z_2 \cong M(G)$.
(2) $H_1' \cong H_2'$.

Proof. The first assertion has been proved in (9.15). Let $J_i = J(H_i)$ be the subgroup of the free group F corresponding to H_i. Then, by (9.13), we have $J_i \cap F' = N$ for $i = 1, 2$. By Cor. 3 of (3.12), we have $H_i' \cong F'J_i/J_i$. Hence, by the Isomorphism Theorem, we get

$$H_i' \cong F'J_i/J_i \cong F'/J_i \cap F' = F'/N.$$

This proves (2).   □

Theorem 9.17 is a theorem of Schur. In fact, the groups H_1 and H_2 are isoclinic (as P. Hall implied in his paper [6]). Following P. Hall, any two groups H_1 and H_2 are said to be **isoclinic** if there is an isomorphism φ from $H_1/Z(H_1)$ to $H_2/Z(H_2)$ such that the mapping

$$[x, y] \rightarrow [\varphi(xZ), \varphi(yZ)] \qquad (Z = Z(H_1))$$

induces an isomorphism from the commutator subgroup H_1' to H_2'. We will return to the property of isoclinism later (Chapter 4, §4). It follows from (4.31) of Chapter 4, and (9.13) of this section that a primitive central extension H_i is isoclinic to the free central extension $\bar{F}$. In particular, H_1 and H_2 are isoclinic, and the factor groups by their centers are isomorphic. These results were rediscovered by Gaschütz–Neubüser–Yen [1].

As we mentioned earlier, the theory of central extensions has been developed in connection with the theory of the projective representations of finite groups. Any irreducible projective representation of a finite group G is obtained from the ordinary irreducible representation of a primitive central extension of G. This is the main theorem of Schur. See Exercise 2 at the end of this section for a more precise statement of the theorem and a brief outline of its proof.

When the finite group G has the property that $G' = G$, the theory of central extensions is easy to handle.

Theorem 9.18. *Let G be a finite group which satisfies the property $G' = G$, and let (H, Z) be a central extension of G.*

(1)   *Set $H_1 = H'$ and $Z_1 = Z \cap H'$. Then, (H_1, Z_1) is an irreducible central extension of G.*

(2)   *The group H is a central product of an irreducible central extension H_1 and an abelian group Z.*

(3)   *The extension (H, Z) is irreducible if and only if $H = H'$.*

(4)   *The extension (H, Z) is primitive if and only if H is the representation group of G.*

(5)   *The primitive central extension of G is uniquely determined. Any irreducible central extension of G is a homomorphic image of the representation group.*

(6)   *If a central extension (H, Z) of G is irreducible, then the representation group of H is (isomorphic to) the representation group of G. Hence, we have $|M(H)| = |M(G)|/|Z|$. If H is the representation group of G, then we have $M(H) = \{1\}$.*

Proof. (1) Since $G = G'$, we have $ZH' = H$ by Cor. 3 of (3.12). The Isomorphism Theorem shows $H_1/Z_1 \cong H/Z = G$. Thus, H_1 is a central extension of G. If there is a subgroup M of H_1 which satisfies $MZ_1 = H_1$, then we have $H = ZH_1 = ZM$. So, $M \lhd H$, and the factor group H/M is abelian. Thus, $M \supset H'$, and so $M = H_1$ and H_1 is irreducible.

(2) As $H = ZH_1$, the assertion follows immediately from the definition of a central product.

(3) If H is irreducible, we have $H = H'$ because $H = ZH'$. Conversely, if $H = H'$, the extension (H, Z) is irreducible by (1).

(4) If H is primitive, H is irreducible by definition, and $H = H'$ by (3). It follows from Theorem 9.17 (1) that $H' \cap Z \cong M(G)$. Thus, we get $|H| = |G| \cdot |M(G)|$. By Definition 9.10, H is a representation group of G. The converse also holds since, by definition, any representation group is primitive.

(5) By Theorem 9.17 (2), the isomorphism class of the commutator subgroup is uniquely determined by the group G. The second half of the proposition (5) follows from Theorem 9.14 (2).

(6) Let K be the representation group of H. Since $H = H'$ by (3), we can apply Parts (3) and (4) to H, and we get $K = K'$. Let $\alpha: K \to H$ and $\beta: H \to G$ be the canonical mappings, and let φ be the composite of α and β. Let L be the kernel of φ. We will prove that (K, L) is a central extension of G. Consider the commutator $[x, z]$ where $x \in K$ and $z \in L$. Then, we have $\varphi(z) = 1$. Thus,

$\alpha(z)$ belongs to Ker $\beta = Z$. Since $Z \subset Z(H)$, we have $[x, z]^\alpha = 1$. This shows that

$$[x, z] \in \text{Ker } \alpha \subset Z(K).$$

It is easy to check that $[xy, z] = [x, z][y, z]$. Thus, the mapping $x \to [x, z]$ is a homomorphism from K into $Z(K)$. Since $Z(K)$ is abelian and $K = K'$, we get $[x, z] = 1$ for all pairs (x, z) with $x \in K$, $z \in L$. This proves that L lies in $Z(K)$. Thus, (K, L) is a central extension of G. Since $K = K'$, Part (3) shows that (K, L) is irreducible, and Part (5) gives us $|L| \leq |M(G)|$. On the other hand, by (5), the representation group of G is an irreducible central extension of H. So, the representation group of G is a homomorphic image of K. Thus, we have $|L| \geq |M(G)|$, and therefore, K is the representation group of G. We conclude that

$$|M(H)| \times |Z| = |M(G)|. \qquad \square$$

EXAMPLE 1. Suppose that a finite group G is generated by n elements $x_1, \ldots, x_n$ with r defining relations $\rho_1 = 1, \ldots, \rho_r = 1$. The integers n and r are considered to be fixed throughout Example 1. In this case, $G = F/K$ for a free group F of rank n. By Cor. 2 of Theorem 6.6, K is the smallest normal subgroup of F which contains the words $\rho_1, \ldots, \rho_r$ of F. Thus, K is generated by $\rho_1, \ldots, \rho_r$ and all the conjugates of $\rho_1, \ldots, \rho_r$. Let N be the normal subgroup defined by (9.2). Then, K/N is contained in the center of F/N. So, the group K/N is generated by the images of $\rho_1, \ldots, \rho_r$ in K/N.

Set $D = K \cap F'$ as before. The fundamental theorem (Theorem 9.9) shows that the factor group K/D is a free abelian group of rank n. Also, the fundamental theorem proves that D/N is isomorphic to the Schur multiplier $M(G)$. Let s be the minimal number of generators of $M(G)$. Then, the abelian group K/N cannot be generated by fewer than $n + s$ elements, so we have $r \geq n + s$. In particular, this implies the inequality $r \geq n$ for any finite group.

If a finite group G is generated by n elements with n defining relations, then the above inequality gives us $s = 0$. Thus, we have $M(G) = \{1\}$. In particular, the multiplier of any finite cyclic group is $\{1\}$. We will give an example when $n = 2$:

$$G = \langle x, y \,|\, x^m = y^2, \, y^{-1}xy = x^{-1} \rangle.$$

In order to prove that the group G with the above presentation is finite, we will compute $y^{-1}x^k y$. The second relation gives us $y^{-1}x^k y = x^{-k}$. So, for $k = m$, we get

$$x^{-m} = y^{-1}x^m y = y^{-1}y^2 y = y^2 = x^m.$$

This shows that $x^{2m} = 1$ and $|G| \leq 4m$. It is almost obvious that $|G| = 4m$. If $m = 2$, the group G is the quaternion group defined in §4, p. 140. In general, when m is even, G is called a *generalized quaternion group*. The multiplier of any generalized quaternion group is trivial. If $n \geq 4$, there is no known example of a finite group which can be generated by n elements with n defining relations. It is not known whether the special linear group $SL(2, 5)$ (more generally, whether any finite group with a trivial multiplier) can be generated by n elements with exactly n defining relations.

If we have $r = n + 1$ for a finite group G, then $M(G)$ is cyclic. The dihedral group of order $2m$ is an example of such a group. It is defined by

$$G = \langle x, y \, | \, x^m = y^2 = 1, \, y^{-1}xy = x^{-1} \rangle.$$

This presentation of G determines a free central extension (H, Z). The subgroup Z of $Z(H)$ is generated by the images of the words x^m, y^2, and $y^{-1}xyx$. We will denote their images by u, v, and w, respectively. The preceding result shows that there is a relation among u, v, and w. We have

$$x^m = y^{-1}x^m y = (wx^{-1})^m = w^m x^{-m}$$

which gives the relation $u^2 = w^m$. If m is an odd integer, we may write $m = 2k + 1$. Set $t = uw^{-k}$. Then, $Z = \langle t, v \rangle$ where $t^2 = u^2 w^{-2k} = w$. So, Z is a free abelian group, and we get $M(G) = \{1\}$ by Theorem 9.9. On the other hand, if m is even, we set $m = 2k$ and $t = uw^{-k}$. Then, we have $Z = \langle t, v, w \rangle$ where $t^2 = 1$. Thus,

$$Z = \langle t \rangle \times \langle v, w \rangle \quad \text{and} \quad M(G) \cong \langle t \rangle.$$

In this case, it is easy to verify that the dihedral group of order $4m$, as well as the generalized quaternion group of order $4m$, is a representation group of G. The Schur multiplier $M(G)$ is a cyclic group of order 2. The number of isomorphism classes of the representation groups of G is more than 1.

EXAMPLE 2. As we have seen in Example 4 of §6, the group $PSL(2, 5)$ is defined by

$$G_1 = \langle x, y \, | \, x^5 = y^3 = (xy)^2 = 1 \rangle.$$

The center of the free central extension of G_1 associated with the above presentation is generated by the images of x^5, y^3, and $(xy)^2$. Let the images of these elements be u, v, and w, respectively. Again, by Theorem 9.9, there is a relation among these generators. The relation is proved to be $w^{30} = u^{12}v^{20}$. In order to find this relation, we start from $w = xyxy$ (in F/N). Since u, v, and w are commutative with x and y, we get $w^2 = xwyxy = x^2yxy^2xy$. For w^3, we insert w between x^2 and y. We continue this process to get

$$w^5 = x^5y(xy^2)^4xy = u(xy^2)^5.$$

The last equality follows because y commutes with w^5 and u. For w^{10}, we insert w in between x and y^2 to get

$$w^{10} = u(xwy^2)^5 = u(x^2 yxy^3)^5 = uv^5(x^3 y)^5.$$

Similarly, $w^{15} = uv^5(x^4 yxy^2)^5$. Finally, we get $w^{30} = u^{12}v^{20}$. Set

$$t = u^6 v^{10} w^{-15}, \qquad r = u^2 v^3 w^{-5}, \qquad s = uv^2 w^{-3}.$$

Then, we have $t^2 = 1$ and $\langle r, s, t \rangle = \langle u, v, w \rangle$. By Theorem 9.9, the elements r and s generate a free abelian group of rank 2. Thus, the multiplier $M(G_1)$ of $PSL(2, 5)$ is isomorphic to $\langle t \rangle$. By the Cor. of (9.10), Chapter 1, the special linear group $SL(2, 5)$ coincides with its commutator subgroup. Hence, by Theorem 9.18, $SL(2, 5)$ is an irreducible central extension of $PSL(2, 5)$. So, $SL(2, 5)$ is a representation group of $PSL(2, 5)$. Thus, we have $|M(G_1)| = 2$. (In particular, we have $t \neq 1$.)

The group $G_0 = \langle x, y \,|\, x^5 = y^3 = 1, (xy)^2 \in Z(G_0) \rangle$ which was discussed at the end of Example 4, §6, is also a central extension of $G_1 = PSL(2, 5)$. By a computation similar to the one above, we see that the center $Z(G_0)$ is generated by the images of x^5, y^3, and $(xy)^2$. In this case, we have $u = v = 1$. Hence, we get the relation $w^{30} = 1$. This implies that $Z(G_0)$ is a cyclic group and that its order is at most 30. Clearly, $Z(G_0)$ contains a subgroup J_0 such that $G_0/J_0 \cong SL(2, 5)$. Since $SL(2, 5)$ coincides with its own commutator subgroup, we have $G_0 = G_0' J_0$. By Theorem 9.18, G_0' is an irreducible central extension of G_1, so we have $G_0' = SL(2, 5)$. Comparing the orders, we get $J_0 \cap G_0' = \{1\}$ and $G_0 = J_0 \times G_0'$. Since the order of $Z(G_0')$ is 2, we see that $|J_0|$ divides 15. We will show that $|J_0|$ is indeed 15.

Let ζ be a primitive fifteenth root of unity, and let x and y be the elements of the direct product $\langle \zeta \rangle \times SL(2, 5)$ defined by

$$x = \left(\zeta^3, \begin{bmatrix} 1 & 1 \\ 0 & 1 \end{bmatrix} \right), \qquad y = \left(\zeta^5, \begin{bmatrix} -1 & -1 \\ 1 & 0 \end{bmatrix} \right).$$

Then, x and y satisfy all the relations for G_0. Thus, we have

$$G_0 \cong \langle \zeta \rangle \times SL(2, 5) \qquad (\zeta^{15} = 1).$$

In general, the preceding direct method of finding the Schur multiplier of a given finite group G is difficult to apply. An alternate approach is to use the theorems discussed in §7 (Theorem 7.26 and its corollaries, Theorem 7.27, and others) to study the multipliers of Sylow subgroups or other suitably chosen subgroups of G, and to use the results to determine the multiplier $M(G)$. We will give an example of this alternate approach. If $M(S) = \{1\}$ for any Sylow subgroup S of G, then, by Cor. 3 of Theorem 7.26, we conclude that $M(G) = \{1\}$. The group $G = SL(2, 5)$ discussed in Example 2 has the property that an S_p-subgroup of G is cyclic for any odd prime p and an S_2-subgroup is a quaternion group. In Example 1, we have shown that $M(S) = \{1\}$ for any Sylow subgroup S of G. Thus, we have $M(G) = \{1\}$. Since $G = G'$, Theorem 9.18 (6) proves that G is a representation group of $PSL(2, 5)$. Also by this method, we can conclude that $|M(G_1)| = 2$, just as we proved in Example 2. In general, if a finite group G has the property that an S_p-subgroup is cyclic for $p > 2$ and an S_2-subgroup is a generalized quaternion 2-group, then we have $M(G) = \{1\}$. For any prime number p, the group $SL(2, p)$ has this property (see §6, Chapter 3).

EXAMPLE 3. (Green [1]) *Let P be a p-group of order p^n. Then, the multiplier $M(P)$ is a p-group, and its order is at most p^m where $m = n(n - 1)/2$.*

We will prove this theorem following Wiegold [1]. We need the following lemma.

Lemma. *Let p be a prime number, and let G be a group. If $|G : Z(G)| = p^n$, then the commutator subgroup G' is a p-group, and its order is at most p^m (where m is defined as before).*

Proof. We have shown that G' is finite (in Theorem 9.8), but this special case can be proved without using Theorem 9.8. We will use induction on n. If $n \leq 1$, then G is abelian by (2.26) of Chapter 1. So, we have $G' = \{1\}$, and the Lemma holds. Assume that $n > 1$. By Theorem 1.4, the center of $G/Z(G)$ is not $\{1\}$. So, there is an element g of G such that $g \notin Z(G)$, but $[g, x] \in Z(G)$ for all x in G. As in the proof of Theorem 9.18 (6), we have

$$[g, xy] = [g, x][g, y] \qquad (x, y \in G).$$

Thus, the mapping $\varphi: x \to [g, x]$ is a homomorphism form G into the center $Z(G)$. Let N be the image under $\varphi: N = \text{Im } \varphi$. Since $\langle Z(G), g \rangle \subset \text{Ker } \varphi$, N is a p-group whose order is at most p^{n-1}. Consider the factor group $\bar{G} = G/N$. Then, the center of $\bar{G}$ contains $Z(G)/N$ and gN. So, we have $|\bar{G} : Z(\bar{G})| = p^l$ where $l < n$. By the inductive hypothesis, the commutator subgroup $\bar{G}'$ of $\bar{G}$ is a p-group, and its order is at most p^k where $k = l(l - 1)/2$. Let H be the subgroup of G which corresponds to $\bar{G}'$ according to the Correspondence Theorem. Then, H is a p-group such that its order is at most p^{k+n-1}. By Corollary 3 of (3.12), G' is contained in H. Thus, G' is a p-group, and

$$|G'| \leq |H| \leq p^{k+n-1} \leq p^m \qquad (m = n(n - 1)/2). \quad \square$$

Let P be a p-group of order p^n, and let (G, Z) be a representation group of P. Then, Z is isomorphic to the multiplier $M(P)$, and $Z \subset G'$ (see (9.15)). The preceding Lemma applied to G shows that $M(P)$ is a p-group and that its order is at most p^m.

The upper bound given here is the best possible. That is, there is an abelian p-group of order p^n whose multiplier has order p^m (see Exercise 4(d)).

We have shown that a cyclic group or a generalized quaternion 2-group has a trivial Schur multiplier. But, as a rule, a finite p-group has a nontrivial multiplier. Let P be a finite p-group such that $|P : \Phi(P)| = p^d$. Then, by Theorem 1.16, P can be generated by d elements, but not by fewer than d elements. So, d is the minimal number of generators of P. There is a free central extension (H, Z) of rank d. Let r_0 be the minimal number of generators of Z. Then, it can be shown that the inequality

$$r_0 > d^2/4$$

holds. From the general theory, we see that Z is a direct product of $M(P)$ and a free abelian group of rank d (see Theorems 5.2 and 9.9). So, if $d \geq 4$, we have $r_0 - d > 0$. This implies that $M(P) \neq \{1\}$. Thus, a p-group P with $d \geq 4$ has a nontrivial multiplier.

Incidentally, if a finite p-group P has a presentation with d generators and r relations, then, as we have seen in Example 1, we get $r \geq r_0$. Thus, we have

$$r > d^2/4.$$

This is the inequality used by Golod–Safarevic [1] in constructing a counterexample to the problem of the class field tower. For the proof of the above inequality, the reader is referred to Huppert [3] III, §18 and p. 643.

As the final remark of this section, we note that if

$$|G : Z(G)| = n = \prod p^e,$$

then the upper bound of $|G'|$ is given by

$$|G'| \leq \prod p^f \qquad (f = e(e - 1)/2).$$

This follows easily from the above Lemma and Corollary 3 of Theorem 7.26 which asserts that the p-primary component of $M(G/Z(G))$ is a subgroup of the multiplier of the S_p-subgroup of $G/Z(G)$.

Exercises

1. Let G be a finite group. Show that the orders of the primitive central extensions of G are bounded if and only if $G = G'$.

(*Hint.* By Theorem 9.14 (2), it suffices to show that the orders of the irreducible central extensions are not bounded for a group with $G \neq G'$. Let (H, Z) be a free central extension of G. Then, H/H' is free abelian. Since ZH'/Z is the subgroup corresponding to G' and since $G \neq G'$, we have $H \neq ZH'$. Let $H/ZH' = A \times B$ where B is cyclic and $B \neq \{1\}$, and let C/ZH' be the subgroup corresponding to the factor A. Then, $H/C \cong B$ is cyclic. By

Exercise 2, §5, there is a set of free generators $\{u_1, \ldots, u_n\}$ of H/H' such that C/H' is generated by $\{au_1, u_2, \ldots, u_n\}$ for some positive integer $a > 1$. For any natural number k and any prime divisor p of a, let C_k/H' be the subgroup generated by $\{ap^k u_1, u_2, \ldots, u_n\}$. Set $J_k = C_k \cap Z$. Then, we get $ZH' \cap C_k = J_k H'$. This implies that the central extension H/J_k is irreducible. The order of $|H/J_k|$ is $|G| p^k$ and is not bounded as $k \to \infty$. If $G \neq G'$, there are primitive central extensions such that none of their homomorphic images are representation groups of G. Thus, the representation groups do not control all the properties of central extensions.)

2. Let G be a finite group, and let F be a field. A *representation* ρ of G over F is a homomorphism from G into the group $GL(n, F)$. That is, for any element σ of G, $\rho(\sigma)$ is an $n \times n$ nonsingular matrix with coefficients in F, and for any σ and τ of G, we have

$$\rho(\sigma\tau) = \rho(\sigma)\rho(\tau).$$

A mapping φ from G into $GL(n, F)$ is said to be a *projective representation* with respect to $\{C_{\sigma, \tau}\}$ $(C_{\sigma, \tau} \in F^*)$ if $\varphi(\sigma)$ is nonsingular and if, for any σ and τ of G, we have

$$\varphi(\sigma)\varphi(\tau) = C_{\sigma, \tau} \varphi(\sigma\tau).$$

If $C_{\sigma, \tau} = 1$ for all $\sigma, \tau \in G$, then the projective representation becomes the representation defined earlier. For each $\sigma \in G$, $\varphi(\sigma)$ may be considered to be an invertible linear transformation of the (fixed) n-dimensional vector space V over F. The projective representation φ is said to be *irreducible* if the only subspaces of V which are left invariant by all $\varphi(\sigma)$ $(\sigma \in G)$ are V and $\{0\}$.

(a) Show that the set of elements $\{C_{\sigma, \tau}\}$ associated with a projective representation φ is an element of $Z^2(G, F^*)$, where G acts trivially on F^*. Let φ' be defined by $\varphi'(\sigma) = d_\sigma \varphi(\sigma)$ for any $\sigma \in G$, where $d_\sigma \in F^*$. Show that φ' is a projective representation of G with respect to $\{C'_{\sigma, \tau}\}$, where $\{C'_{\sigma, \tau}\}$ belongs to the same cohomology class as $\{C_{\sigma, \tau}\}$. We say that φ and φ' are *associated*.

(b) Assume that the field F is algebraically closed. Let ρ be an irreducible representation of G. Show that the elements of the center $Z(G)$ are represented by scalar matrices; that is, if $\sigma \in Z(G)$, then $\rho(\sigma) = \lambda I$ where $\lambda \in F^*$ and I is the identity matrix of $GL(n, F)$. Let Z be a subgroup of $Z(G)$, and let $\{t_\sigma\}$ be a transversal of Z in G where σ ranges over the elements of G/Z. Show that the mapping φ defined on G/Z by $\varphi(\sigma) = \rho(t_\sigma)$ is an irreducible projective representation of G/Z.

(c) Let (H, Z) be a primitive central extension of a finite group G. Let φ be an irreducible projective representation of G over the field $\mathbb{C}$ of complex numbers. Show that there is an irreducible representation ρ of H over $\mathbb{C}$ such that, for a transversal $\{t_\sigma\}$ of Z in H, the projective representation φ' of G defined by $\varphi'(\sigma) = \rho(t_\sigma)$ is associated with φ.

(*Hint.* (a) This is proved in the same way as (7.5). According to the Part (a), the classes of mutually associated projective representations correspond to the elements of $H^2(G, F^*)$. The class of representations corresponds to the identity of $H^2(G, F^*)$.

(b) Apply Schur's Lemma (p. 159) to the irreducible FG-module V. Since any element $\sigma \in Z(G)$ gives rise to an FG-endomorphism $\rho(\sigma)$, the endomorphisms $\{P(\sigma) | \sigma \in Z(G)\}$ generate an algebraic extension field of F. But, as an algebraically closed field, F does not have any proper algebraic extension.

(c) Let $\{f(\sigma, \tau)\}$ be the factor set associated with the extension H of Z. Since (H, Z) is primitive, the mapping γ of the exact sequence (9.7) is surjective (see (9.15)). Thus, by (9.4), there is an element λ of Z^* such that λf is in the cohomology class of $\{C_{\sigma, \tau}\}$. So, we have $\{d_\sigma\}$ such that

$$\lambda f(\sigma, \tau) = d_{\sigma\tau}^{-1} C_{\sigma, \tau} d_\sigma d_\tau.$$

Let $\{t_\sigma\}$ be the transversal of Z which satisfies (7.1). Then, the mapping ρ defined by

$$\rho(x t_\sigma) = \lambda(x) d_\sigma \varphi(t_\sigma) \qquad (x \in Z)$$

is a representation of H which satisfies the required property.

Part (c) is one of the main theorems of Schur. It is important in that it reduces problems concerning projective representations to corresponding problems on the representations of a representation group.)

3. Let (H, Z) be a central extension of a finite group G.

(a) Show that if H is finite, $|M(G)|$ is a divisor of $|Z| \times |M(H)|$.

(b) Show that if $M(H) = \{1\}$, $M(G)$ is a homomorphic image of some subgroup of Z. Thus, if $M(H) = \{1\}$ and Z is finite, $M(G)$ is isomorphic to a subgroup of Z.

(*Hint.* A free central extension F of H contains a normal subgroup N such that F/N is isomorphic to a free central extension of G.)

4. Let G and H be two groups. In this exercise, we will describe the multiplier $M(G \times H)$ of the direct product $G \times H$. First, we will define the *tensor product* of G and H as the group generated by all $g \otimes h$ ($g \in G$ and $h \in H$) which satisfy the following relations

$$gg' \otimes h = (g \otimes h)(g' \otimes h), \qquad g \otimes hh' = (g \otimes h)(g \otimes h')$$

for any $g, g' \in G$ and $h, h' \in H$. The tensor product is denoted by $G \otimes H$.

(a) Show that the tensor product is abelian and that we have

$$G \otimes H \cong (G/G') \otimes (H/H').$$

(b) Let A be a finite abelian group, and let $A = A_1 \times \cdots \times A_n$ be a decomposition into cyclic factors. Set $a_i = |A_i|$. Similarly, let

$$B = B_1 \times \cdots \times B_m \quad \text{with} \quad B_j = |B_j|.$$

Let $d_{ij} = (a_i, b_j)$ be the greatest common divisor of a_i and b_j. Show that the tensor product $A \times B$ is the direct product of mn cyclic groups D_{ij} of order d_{ij}.

(c) Let G and H be two finite groups. Show that

$$M(G \times H) \cong M(G) \times M(H) \times (G \otimes H).$$

(d) Let A be a finite abelian group such that $A = A_1 \times \cdots \times A_n$ where A_i is a cyclic group of order a_i. Let A_{ij} be a cyclic group of order (a_i, a_j). Show that the Schur multiplier $M(A)$ is the direct product of the cyclic groups A_{ij}.

(Hint. (a) We can expand $gg' \otimes hh'$ in two different ways to prove the commutativity. The relations imposed on the tensor product show that the mapping $g \to g \otimes h$ is a homomorphism from G into the abelian group $G \otimes H$. Thus, $g \otimes h = 1$ for any element g of G'. For abelian groups A and B, the tensor product $A \otimes B$ defined here coincides with the tensor product of two $\mathbb{Z}$-modules in linear algebra ([IK], pp. 306–315, [AS], pp. 296–313). In particular, we have

$$(A + B) \otimes C = A \otimes C + B \otimes C.$$

(c) We can construct a free central extension $(\bar{F}, \bar{L})$ of $G \times H$ as follows. For each element g of G, choose a symbol u_g; similarly, for each h of H, choose v_h. Let F be the free group on the set

$$\{u_g, v_h\} \ (g \in G - \{1\} \text{ and } h \in H - \{1\}).$$

For $g, g' \in G$ and $h, h' \in H$, set

$$u(g, g') = u_{gg'}^{-1} u_g u_{g'}, \qquad v(h, h') = v_{hh'}^{-1} v_h v_{h'}$$
$$w(g, h) = u_g^{-1} v_h^{-1} u_g v_h = [u_g, v_h].$$

Let L be the subgroup of F generated by the elements $u(g, g')$, $v(h, h')$, $w(g, h)$, and their conjugates. Then, L is a normal subgroup of F and $F/L \cong G \times H$. Let $N = \langle [x, y] | x \in L, y \in F \rangle$, $\bar{F} = F/N$, and $\bar{L} = L/N$. Then, $(\bar{F}, \bar{L})$ is a free central extension of $G \times H$.

We will determine the structure of $\bar{L}$. We will associate the symbol $U(g, g')$ with each pair (g, g') of the elements of G. Let U be the abelian group generated by the elements $U(g, g')$ and the following relations

$$U(1, 1) = 1, \qquad U(gg', g'')U(g, g') = U(g, g'g'')U(g', g'')$$

for any three elements g, g', g'' of G. Note that the elements $u(g, g')$ satisfy similar relations in $\bar{L}$. Similarly, we associate the symbol $V(h, h')$ with each pair of elements h and h' of H, and let V be the abelian group generated by all the $V(h, h')$ with the relations

$$V(1, 1) = 1, \qquad V(hh', h'')V(h, h') = V(h, h'h'')V(h', h'').$$

Let $J = U \times V \times (G \otimes H)$. Furthermore, let $W(g, h) = g \otimes h$. As remarked earlier, the mapping $U(g, g') \to u(g, g')$ can be extended to a homomorphism from U into $\bar{L}$. Similarly, $V(h, h') \to v(h, h')$ and $W(g, h) \to w(g, h)$ can be extended to homomorphisms. Thus, there is a homomorphism φ from J into $\bar{L}$. On the other hand, the function f defined by

$$f(gh, g'h') = U(g, g')V(h, h')W(g', h)$$

is proved to be a factor set with respect to the trivial action of $G \times H$ on J. Let M be the extension of J associated with the factor set f. By definition, the elements of the group M are pairs (gh, j) where $g \in G$, $h \in H$, and $j \in J$. The definition of the product in M shows that M is generated by the set $\{(g, 1), (h, 1)\}$ where g ranges over G and h ranges over H. The mappings $u_g \to (g, 1)$ and $v_h \to (h, 1)$ can be extended to a homomorphism ψ from $\bar{F}$ into M, and ψ is indeed surjective. It is easy to verify that the composite $\varphi\psi$ is the identity mapping on J and that $\psi\varphi$ is also the identity on $\bar{L}$. Thus, $\bar{L} \cong J$. The torsion subgroup of $\bar{L}$ is $M(G \times H)$, while the torsion subgroup of J is $M(G) \times M(H) \times (G \otimes H)$.)

5. Schur [2] has shown that the number of isomorphism classes of the representation groups of a finite group G does not exceed $|G \otimes M(G)|$. In order to prove this theorem we need the following concepts and notations. Set $Z = M(G)$, $Z^* = \mathrm{Hom}(Z, \mathbb{C}^*)$ and

$$C = \{c \in Z^2(G, Z) | \text{ for all } \varphi \in Z^*, \varphi(c) \in B^2(G, \mathbb{C}^*)\}.$$

The cohomology classes represented by the elements of C form a subgroup of $H^2(G, Z)$ which we will denote by D. Let $\mathscr{A}$ be the totality of the extensions A of Z by G/G' such that A is abelian. We define an equivalence relation in $\mathscr{A}$ by Definition 7.15, and we let $\mathrm{Ext}(G/G', Z)$ be the set of equivalence classes. If $A \in \mathscr{A}$, $[A]$ denotes the equivalence class in $\mathrm{Ext}(G/G', Z)$ which contains A. For any $f \in Z^2(G, Z)$, let $E(f)$ be the extension of Z by G constructed from the factor set f. Let (H, Z) be a fixed representation group of G, and let f_0 be the factor set associated with the extension H. Using these notations, show that the following propositions hold.

(a) Let (H_1, Z) be any representation group of G. Then, there is an element c of C such that $H_1 \cong E(f_0 + c)$.
(b) If $c \in C$, then $A(c) = E(c)/E'(c)$ belongs to $\mathscr{A}$.

(c)   If $A \in \mathscr{A}$, then $A \cong A(c)$ for some $c \in C$.
(d)   Let c and c' be elements of C. Then, we have

$$c \equiv c' (\text{mod } B^2(G, Z)) \Leftrightarrow [A(c)] = [A(c')].$$

Thus, we get $|D| = |\text{Ext}(G/G', Z)|$.

(e)   If $A \in \mathscr{A}$, the factor set associated with the extension of Z by G/G' satisfies $g(\sigma, \tau) = g(\tau, \sigma)$. The proposition (7.16) holds for all symmetric factor sets as well.
(f)   We have $|\text{Ext}(G/G', Z)| = |G \otimes M(G)|$.
(g)   Propositions (a), (d), and (f) give us Schur's Theorem which we mentioned at the beginning.

(*Hint*. (a) Let f_1 be the factor set associated with the extension H_1. Then, by (9.4) and (9.7), any element of $M(G)$ can be written as φf_0 for some $\varphi \in Z^*$. Similarly, an element of $M(G)$ has the form ψf_1 with $\psi \in Z^*$. Thus, for any $\varphi \in Z^*$, there is an element $\varphi' \in Z^*$ such that

$$\varphi f_0 \equiv \varphi' f_1 \qquad (\text{mod } B^2(G, \mathbb{C}^*)).$$

Clearly, $\varphi \to \varphi'$ is an automorphism of Z^*. Let θ be the dual mapping of this automorphism. Then, θ is an automorphism of Z such that $\varphi(z^\theta) = \varphi'(z)$ for all $z \in Z$. We see that f_1^θ is a factor set and that $f_1^\theta = f_0 + c$ where $c \in C$. Clearly, θ can be extended to an isomorphism from H_1 onto $E(f_1^\theta)$. This proves (a).

(b) From the definition of $c \in C$, we know that $\varphi(c) \in B^2(G, \mathbb{C}^*)$ for any $\varphi \in Z^*$. So, by (9.4) and (9.7), we have $E(c)' \cap Z = \{1\}$.

(c) The group $E(c)$ is isomorphic to a subgroup of $A(c) \times G$ (see (4.20)).

(d) If $E(c)$ is equivalent to $E(c')$, then this induces an isomorphism from $A(c)$ onto $A(c')$ which defines the equivalence between the extensions. Similarly, the converse holds (see the hint for (c)).

(e) This is proved as in the discussion at the beginning of §7.

(f) Let $G/G' = A_1 \times \cdots \times A_r$ be a decomposition into the direct product of cyclic groups A_i of order a_i. For any $A \in \mathscr{A}$, choose elements $u_1, \ldots, u_r$ of A such that u_i corresponds to a generator of A_i. Then, the class $[A]$ depends on the classes U_i of Z/Z^{a_i} ($i = 1, 2, \ldots, r$) such that $u_i^{a_i} \in U_i$. (Construct the factor set using $u_i^{a_i}$.) Then, we have

$$|\text{Ext}(G/G', Z)| = \prod |Z/Z^{a_i}| = |(G/G') \otimes Z|.$$

Incidentally, we can define addition in $\text{Ext}(G/G', Z)$ so as to make it an abelian group. In this case, we have the isomorphism

$$\text{Ext}(G/G', Z) \cong G \otimes Z.)$$

6. Even when the group G is not necessarily finite, we may define **irreducible**, as well as **free**, central extensions according to Definition 9.10, or Definition 9.3. If $G = G'$, we can generalize Theorem 9.18.

Show that the following propositions hold. (Always assume that $G = G'$, but G may be infinite.)

(a) Any central extension (H, Z) of G is a central product of Z and H'. The extension $(H', Z \cap H')$ is irreducible.
(b) A central extension (H, Z) is irreducible if and only if $H = H'$.

The commutator subgroup of a free central extension is called the **representation group**. (If G is finite and $G' = G$, then this definition coincides with the previous definition.)

(c) If a central extension (R, C) of a group G is a representation group of G, then R is irreducible, and $C^* \cong M(G)$.
(d) Any irreducible central extension of G is a homomorphic image of a representation group. Let R be a representation group of G, and let ρ be the canonical mapping from R onto G. Let H be any irreducible central extension of G with the canonical mapping $\eta: H \to G$. Then, there is a unique surjective homomorphism φ of R onto H such that $\varphi\eta = \rho$.
(e) The isomorphism class of the representation groups is uniquely determined by G.
(f) If R is a representation group of G, then we have $M(R) = \{1\}$.

(*Hint.* Parts (a), (b), and (f) are proved as in Theorem 9.18. By (a), a representation group is irreducible. As in (9.5) and (9.7), we can show that $C^* \cong M(G)$. By definition, R is the commutator subgroup of a free central extension F of G. By (9.2), there is a homomorphism ψ from F to H such that $\psi\eta$ is the canonical mapping from F onto G. This mapping ψ induces a homomorphism φ of the commutator subgroup R. Thus, a mapping φ satisfying $\varphi\eta = \rho$ exists. Suppose that $\varphi': R \to H$ satisfies $\varphi'\eta = \rho$. Then, for any $x \in R$, $\varphi(x)$ and $\varphi'(x)$ are contained in the same coset of $\mathrm{Ker}\ \eta$. So, we have $\varphi(x)\varphi'(x)^{-1} \in Z(H)$. Therefore, the mapping $x \to \varphi(x)\varphi'(x)^{-1}$ is a homomorphism of R into $Z(H)$. Then, as $R = R'$, we get $\varphi(x) = \varphi'(x)$. This proves the uniqueness. This also proves the last assertion of (d) from which Part (e) follows as in the proof of (5.7).

We can define the representation group as a central extension which has the property stated in (d). We defined the multiplier $M(G)$ as $H^2(G, \mathbb{C}^*)$ following Schur's original definition. Thus, we obtained $C^* = M(G)$ in (c). Actually, some authors prefer C as the definition of $M(G)$. If G is finite, we have $C^* \cong C$, and so no real distinction arises.)

§10. Wreath Products

When two groups G and H are given, there are several ways of obtaining a third group from them. The direct product, the free product, and the semi-direct product are a few of the methods we have studied so far. Each of these has many useful applications in various branches of group theory. In this section, we will introduce yet another concept, called the wreath product, and we will give a few of its applications. The wreath product is particularly useful in constructing examples of groups which exhibit some curious properties. For the many interesting results which have been obtained in this direction, the readers are refered to P. Hall's papers, particularly [8, 11].

We will define the twisted wreath product, a concept originally due to B. H. Neumann [2]. The ingredients for this construction are:

a group A, a group G, a subgroup H of G, and the action φ of H on A.

We will write the action of $h \in H$ exponentially as

$$\varphi(h): a \to a^h \qquad (a \in A).$$

First, we will define *the base group* B. An element b of B is a function on G which takes on values in A such that, for any $h \in H$, we have

(10.1) $$b(xh) = b(x)^h \qquad (x \in G).$$

Let B be the totality of such functions. If b_1 and b_2 are in B, let

$$b_3(x) = b_1(x)b_2(x).$$

We will call b_3 the **product** of b_1 and b_2, and we write $b_3 = b_1 b_2$. It is easy to verify that b_3 satisfies (10.1) and that $b_3 \in B$. Thus, the multiplication is defined in B, and we can quickly prove that the set B forms a group with respect to the operation just introduced. The identity is the function b_0 such that $b_0(x) = 1$ for all x of G. The inverse of b is the function b' which satisfies $b'(x) = b(x)^{-1}$ for all $x \in G$.

Secondly, we will determine the action of G on the base group B. For this purpose, suppose we have $u \in G$ and $b \in B$. Then, let b^u be defined by the following formula:

(10.2) $$b^u(x) = b(ux).$$

If $b \in B$, then b^u satisfies the formula (10.1), too. Thus, we get $b^u \in B$. It is also easy to prove the formula corresponding to (8.2), Chapter 1:

$$(b_1 b_2)^u = b_1^u b_2^u, \qquad b^{uv} = (b^u)^v, \qquad b^1 = b.$$

This shows that the group G does indeed act on B.

Definition 10.3. The semidirect product W of B and G with respect to the above action of G on B is called the **twisted wreath product**. We write $W = (A, G, H, \varphi)$ or

$$W = A \operatorname{Wr}_H G$$

(10.4) *With the preceding notations, the base group B is isomorphic to a complete direct product $\prod A_\lambda$ where λ ranges over the set Λ of left cosets of H and where, for each λ, A_λ is a group isomorphic to the group A (see Definition 4.10).*

Proof. Let $\{t_\lambda\}$ ($\lambda \in \Lambda$) be a left transversal of H in G. Then, every element of G can be written in the form $t_\lambda h$ ($h \in H$). By (10.1), an element b of B is determined by the elements $\{b(t_\lambda)\}$ $\{\lambda \in \Lambda\}$. So, we have

$$B = \prod A_\lambda \ (\lambda \in \Lambda) \quad \text{where} \quad A_\lambda \cong A. \quad \square$$

If the index $|G : H|$ is infinite, we can define a group which has properties similar to $A \operatorname{Wr}_H G$ but is smaller. Instead of the base group B, we may use the restricted direct product (see Definition 4.10). We only need to take those functions b of B which satisfy $b(x) = 1$ for all but a finite number of elements x in G. The totality of such functions forms a group B_{res} which is isomorphic to the restricted direct product of the groups A_λ (in the same notation as in (10.4)). We will denote this smaller group by $A \operatorname{wr}_H G$ to distinguish it from the group defined earlier. The symbol Wr with the capital W signifies that the base group is the complete direct product (and therefore, large). If $|G : H|$ is finite, there is no difference between Wr and wr.

Definition 10.5. A twisted wreath product with $H = \{1\}$ is called the **complete wreath product** and will be denoted simply by $A \operatorname{Wr} G$. When the base group is restricted, we write $A \operatorname{wr} G$ and call it simply the **wreath product** of A and G.

Suppose that $|G : H|$ is finite, and let A be an FH-module over a field F. Then, we note that the base group B with the action of G defined above is isomorphic to the induced FG-module A^G (Curtis–Reiner, §43).

We will prove the following theorem to illustrate an application of the wreath products.

(10.6) *There is an infinite p-group whose center is $\{1\}$ (Schmidt [1]). There is an infinite p-group which contains a proper subgroup equal to its normalizer.*

Proof. Let A be any p-group, $A \neq \{1\}$, and let K be an infinite p-group. Set $G = A \operatorname{wr} K$. By definition, G is a semidirect product of the base group B and K, where B is the restricted direct product of the groups isomorphic to A. We can consider B and K to be subgroups of G (see §8, Chapter 1). It is clear that G is a p-group and that K is a proper subgroup of G. We will show that $Z(G) = \{1\}$ and $N_G(K) = K$.

If an element x of G does not belong to B, we have $x = kb$ with $k \in K - \{1\}$ and $b \in B$. Then, x transforms a direct factor $A_u = \{b \mid b(t) = 1$

for $t \neq u\}$ into A_v, where $v = k^{-1}u$. Thus, $Z(G)$ is contained in B. Now, by (8.9) of Chapter 1, we get

$$N_G(K) = KC_B(K).$$

Since $Z(G) \subset C_B(K)$, we need only to prove that $C_B(K) = \{1\}$.

Let b be an element of $C_B(K)$. Then, for any $k \in K$, we have

$$b(1) = b^k(1) = b(k).$$

Thus, b takes on a constant value as k ranges over K. But since the base group B is a restricted direct product, we have $b(k) = 1$ for all except possibly for a finite number of elements k of K. As K is infinite, we must have $b(k) = 1$ for all $k \in K$. Hence, we get $C_B(K) = \{1\}$. $\square$

The above theorem shows that the fundamental theorems on finite p-groups, Theorems 1.4 and 1.6, do not hold for infinite p-groups. In fact, it has been shown in Novikov–Adjan [1] that, for a prime number $p \geq 4381$, there is a finitely generated infinite simple p-group in which every element satisfies $x^p = 1$.

The preceding proof utilized very effectively the property that the base group is the restricted direct product. There are, however, cases when the complete direct product is more useful. The following example illustrates this point.

(10.7) *Let W be the twisted wreath product $A \mathrm{Wr}_H G$. We consider G and the base group B as subgroups of W. If a complement K of B contains the subgroup H, then K is conjugate to G in W.*

Proof. Any element of K can be written as ub_u where $u \in G$ and $b_u \in B$. Thus, we have the function $u \to b_u$ from G to B which satisfies $b_{uv} = b_u^v b_v$. By assumption, K contains H, so we get $b_h = 1$ for all $h \in H$. Let us define a function b on G by

$$b(x) = b_x(1).$$

Then, $b(xh) = b_{xh}(1) = b_x(1)^h = b(x)^h$. Thus, we have $b \in B$. On the other hand, when we evaluate both sides of the relation $b_{uv} = b_u^v b_v$ at $x = 1$, we get

$$b_u(v) = b(uv)b(v)^{-1} = b^u(v)b(v)^{-1}.$$

This gives us $b_u = b^u b^{-1}$. Then, we have

$$ub_u = u(b^u)b^{-1} = bub^{-1}.$$

This proves that $G = b^{-1}Kb$. $\square$

If $H = \{1\}$, W is the complete wreath product. In this case, any complement of the base group is conjugate to G (Theorem of P. Neumann [1]). If the base group B is restricted, a complement of B need not be conjugate to G.

Originally, the wreath product was defined when G was a permutation group.

Definition 10.8. Let A and G be groups, and let X be a G-set. Let B be the totality of functions defined on X which take on values in A. We can define the product of two elements b and b' of B and the action of G on B as before:

$$bb'(x) = b(x)b'(x), \qquad b^g(x^g) = b(x).$$

The semidirect product of B and G with respect to the action just defined is called the **general wreath product**.

It is easy to prove that the base group B is isomorphic to the complete direct product of the groups A_x, where A_x is a group isomorphic to A for each $x \in X$. Any group G may be considered to be a G-set via the left translation $x \to g^{-1}x$. In this case, the general wreath product defined from the group A and the G-set G according to Definition 10.8 coincides with the complete wreath product defined in Definition 10.5.

The general wreath product W is defined using a G-set X. So, W is determined by the permutation group G, not by the abstract group G, although the notation $W = A \text{ Wr } G$ is sometimes used. The complete wreath product defined in Definition 10.5 is also called the **standard wreath product**. In this book, the term wreath product refers to the standard one, and it will be explicitly stated so when we mean the general wreath product rather than the standard one.

A general wreath product occurs naturally in the centralizer of an element of a symmetric group (see (2.13) of Chapter 3).

If a group G contains a normal subgroup A, we can embed G in the complete wreath product $A \text{ Wr } \bar{G}$ where $\bar{G} = G/A$. This is a theorem of Kaloujnine–Krasner [1]. Here, we will prove the following generalization due to B. H. Neumann [2]. First, we will introduce some new notations. Let A, G, and $\bar{G}$ be defined as above. Suppose that a subgroup U of G is given and satisfies $A \cap U = \{1\}$. Set $H = AU/A$. Then, there is an isomorphism ρ from H onto U such that the mapping ρ is the inverse of the isomorphism induced by the canonical mapping from G into $\bar{G}$. For each $h \in H$, the conjugation by the element $\rho(h)$ defines an action of H on A. Let this action be φ. Then, for each $h \in H$, we have

$$\varphi(h)a = \rho(h)^{-1}a\rho(h).$$

The collection $(A, \bar{G}, H, \varphi)$ defines the twisted wreath product W according to Definition 10.3. Let B denote the base group of W. With these notations, we have the following theorem.

Theorem 10.9. *There is an isomorphism θ from G into W such that*

$$\theta(G) \cap B = \theta(A), \qquad \theta(G)B = W.$$

If G contains a complement V of A such that $U \subset V$, then there is an isomorphism θ which satisfies $\theta(V) = \bar{G}$ as well as the above two conditions.

Proof. Let T be a left transversal of the subgroup AU of G. Then, by the analogue of (3.5), Chapter 1, for the left transversal, TU is a left transversal of A. We will choose T to satisfy the condition $T \cap AU = \{1\}$.

Let σ be the canonical homomorphism from G onto $\bar{G}$, and let τ denote the representative function from the group $\bar{G}$ onto TU defined by the rule that, for any element x of $\bar{G}$,

$$\tau(x) = X \cap TU$$

where X is the coset of A corresponding to x. By the definitions, we have

$$\rho(\sigma(u)) = u, \qquad \sigma(\tau(x)) = x, \quad \text{and} \quad \tau(x\sigma(u)) = \tau(x)u$$

for any $u \in U$ and $x \in \bar{G}$. Furthermore, for $g \in G$, the element $f(g, x)$ defined by

$$g\tau(x) = \tau(\sigma(g)x)f(g, x)$$

belongs to A, and by expanding the expression $(gh)\tau(x)$ in two different ways, we get

$$f(gh, x) = f(g, \sigma(h)x)f(h, x).$$

The conditions imposed on U give us $\tau(xh) = \tau(x)\rho(h)$ for all h of H. Thus, we get

$$f(g, xh) = f(g, x)^{\rho(h)}.$$

This shows that the function $f(g)$ defined on $\bar{G}$ by $f(g)(x) = f(g, x)$ is an element of the base group B of the twisted wreath product W. We now define the function θ on G by

$$\theta(g) = (\sigma(g), f(g)).$$

We will show that θ is an isomorphism from G into the group W. First, the definition of the product in W gives us

$$(\sigma(g), f(g))(\sigma(g'), f(g')) = (\sigma(g)\sigma(g'), f(g)^{\sigma(g')}f(g')).$$

On the other hand, by the definitions of f and the product in B, we get

$$(f(g)^{\sigma(g')}f(g'))(x) = f(g, \sigma(g')x)f(g', x).$$

The right side is equal to $f(gg', x)$. So, we have

$$\theta(g)\theta(g') = \theta(gg').$$

This proves that the function θ is a homomorphism. If $\theta(g)$ is the identity of W, we have $\sigma(g) = 1$ and $f(g) = 1$. Then, g is an element of A and satisfies

$$g\tau(x) = \tau(x)$$

for all $x \in \bar{G}$. This shows that $g = 1$ and that θ is an isomorphism.

Clearly, we have $\theta(G)B = W$. By definition, $\theta(G) \cap B$ is the set of images of those elements g of G which satisfy $\sigma(g) = 1$. Then, we have $\theta(G) \cap B = \theta(A)$.

The embedding θ defined above depends on the choice of the transversal T. If G has a subgroup V which satisfies the conditions stated in Theorem 10.9, then we can choose T so that $TU = V$. In this case, the function f satisfies the property that $f(v, x) = 1$ for any $v \in V$. Hence, we get $f(v) = 1$ for all $v \in V$, and

$$\theta(v) = (\sigma(v), 1).$$

Thus, the image $\theta(V)$ coincides with the subgroup of W which is canonically identified with $\bar{G}$. □

Corollary. *Let $G = G_0 \supset G_1 \supset \cdots \supset G_r = \{1\}$ be a subnormal series with factor groups $H_i = G_{i-1}/G_i$ ($i = 1, 2, \ldots, r$). Then, G is isomorphic to a subgroup of the repeated complete wreath product*

$$W = ((H_r \text{ Wr } H_{r-1}) \text{ Wr } H_{r-2} \cdots) \text{ Wr } H_1.$$

Proof. Since $G_1 \triangleleft G$, G is isomorphic to a subgroup of G_1 Wr H_1 by Theorem 10.9. In general, the following lemma holds.

Let A_1 and A_2 be two groups. If A_1 is isomorphic to a subgroup of A_2, then A_1 Wr G is isomorphic to a subgroup of A_2 Wr G.

The corollary follows easily by induction on r. □

As a particular case of this Corollary, we note that any finite solvable group is isomorphic to a subgroup of the repeated wreath product of cyclic groups of prime order.

In §8, we proved the main theorem (Theorem 8.10) by applying the cohomology theory. We will give here a cohomology-free proof of the theorems of Gaschütz (Theorems 8.4 and 8.9) which are the bases of the proof of the main theorem. This method is closely related to the remark of Wielandt (Wielandt–Huppert [1]).

We will prove a lemma about the twisted complete wreath product

$$W = (A, G, H, \varphi).$$

We will assume that the following additional conditions hold.

(a) The group A is abelian.
(b) The action φ of H can be extended to an action ψ of G on A.
(c) The index $|G : H|$ is finite.

(10.10) *Under these assumptions, there is a G-homomorphism d from the base group B of W onto A. If $B_0 = \mathrm{Ker}\ d$, B_0 is a normal subgroup of W, and the factor group W/B_0 is isomorphic to the semidirect product of A and G with respect to the action ψ of G on A.*

Proof. For any element $b \in B$, the function on G defined by $\psi(x^{-1})b(x)$ $(x \in G)$ takes on a constant value over each left coset xH of H (because $\psi(h^{-1})b(xh) = b(x)$). Let T be a left transversal of H in G. Define d by the formula

$$d(b) = \prod_{t \in T} \psi(t^{-1})b(t).$$

Since $|T| = |G : H|$ is finite by (c), $d(b)$ is a well-defined element of A which is independent of the choice of T, as well as of the order of the product (because A is abelian by (a)). The definition of the product in B shows that d is a homomorphism from B into A. Furthermore, for any element g of G, we have

$$d(b^g) = \prod \psi(t^{-1})b^g(t) = \prod \psi(t^{-1})b(gt)$$
$$= \psi(g) \prod \psi(t^{-1}g^{-1})b(gt) = d(b)^g$$

since gT is another transversal of H in G. This proves that d is a G-homomorphism. Then, B_0 is a normal subgroup of W.

We will show that d is surjective. If $a \in A$, we define a function b_a on G by the formulas

$$b_a(h) = a^h \quad (h \in H), \qquad b_a(x) = 1 \quad (x \notin H).$$

It follows immediately that $b_a \in B$ and $d(b_a) = a$. Thus, d is surjective, and $B/B_0 \cong A$.

The mapping $(g, b) \rightarrow (g, d(b))$ is easily shown to be a homomorphism from W onto the semidirect product of A and G with respect to the action ψ of G on A. Clearly, the kernel of this mapping is the subgroup B_0. □

A proof of Theorem 8.4. We need only to show that the extension over A splits when L has a complement U of A. Set $H = L/A$. Then, $L = AU$, and the hypotheses of Theorem 10.9 and (10.10) are satisfied. Thus, there is an isomorphism θ from G into $W = A \,\mathrm{Wr}_H (G/A)$ and a G-homomorphism d from the base group B of W onto A. If $a \in A$, we have $\theta(a) = (1, f(a))$ where $f(a)$ is the function on G/A defined by

$$f(a)(x) = f(a, x) = \tau(x)^{-1} a \tau(x)$$

for any $x \in G/A$. Then, from the definition of d, we get

$$d(f(a)) = a^j \qquad (j = |G : H|).$$

By assumption, A is j-regular. Hence, we have $B_0 \cap \theta(A) = \{1\}$ and $B_0\, \theta(A) = B$. This gives us $B_0 \cap \theta(G) = \{1\}$ and $B_0\, \theta(G) = W$. Then, by the Second Isomorphism Theorem, we get $\theta(G) \cong W/B_0$. By Theorem 10.9, θ is an isomorphism and by (10.10), W/B_0 is isomorphic to the semidirect product. Therefore, G splits over A. □

A proof of Theorem 8.9. Suppose that the complements of A in L are conjugate. By Theorem 8.4, G has a complement V. Then, by (8.2), $V \cap L$ is a complement of A in L. So, by assumption, $V \cap L$ is conjugate to U in L, and there is an element u of L such that $(V \cap L)^u = U$. If we choose V^u in place of V, we may assume that the complement V satisfies $V \cap L = U$. By the last assertion of Theorem 10.9, we have an embedding θ such that $\theta(V) = \bar{G}$. Let V_1 be an arbitrary complement of A in G. Then, $B\theta(V_1) = W$ and

$$B \cap \theta(V_1) = B \cap \theta(G) \cap \theta(V_1) = \{1\}.$$

As before, there is an element x of L such that $V_1^x \cap L = U$. Thus, $\theta(V_1^x)$ is a complement of the base group B, and it contains $\theta(U) = H$. By (10.7), $\theta(V_1^x)$ is conjugate to $\bar{G} = \theta(V)$ in W. Therefore, in the factor group W/B_0, the images of $\theta(V)$ and $\theta(V_1^x)$ are conjugate. Since $\theta(G) \cong W/B_0$, V_1 and V are conjugate in G. □

Exercises

1. (a) Let P be a group theoretical property which satisfies the following two conditions.
  (1) If a group X is a P-group, the direct power (the complete direct power) of X is also a P-group. (A *direct power* of X is a direct product in which all of the factors are isomorphic to X.)
  (2) If X and Y are P-groups, then an extension of Y by X is a P-group.

Prove that, if G and H are P-groups, G wr H (G Wr H) is a P-group.

(b) Let π be a set of prime numbers. Show that if G and H are π-groups, so is G wr H.

(c) Prove that G wr H is finitely generated if both G and H are finitely generated..

(d) Show that $\mathbb{Z}$ wr $\mathbb{Z}$ is generated by two elements, but that its base group B is not finitely generated.

(e) Prove that $\mathbb{Z}$ Wr $\mathbb{Z}$ is not finitely generated.

(*Hint*. Any finitely generated group is at most countably infinite.)

2. Let $W = A$ Wr G, and consider G and the base group B as subgroups of W. Set

$$\Delta = \Delta(A) = \{b_a \mid b_a(x) = a \text{ (constant)}\}.$$

(a) Show that Δ is a subgroup of B isomorphic to A and that $C_W(G) = Z(G)\Delta$. Furthermore, show that

$$Z(W) = Z(\Delta) = \Delta(Z(A)) \quad \text{if} \quad A \neq \{1\}.$$

(b) Assume that G is a finite cyclic group of order n. Show that every element of Δ is an n-th power of some element of W.

(c) Either assume that G is a finite cyclic group of order n and that the order of any element of A divides n, or assume that G is an infinite cyclic group. Then, prove that every element of Δ is a commutator of two elements of W.

(*Hint*. (b) Let a be an arbitrary element of A. Define the element b of B by

$$b(1) = a \quad \text{and} \quad b(x) = 1 \quad \text{for} \quad x \neq 1.$$

Let $G = \langle z \rangle$, and compute $(bz^{-1})^n$.

(c) For each b_a of Δ, we can solve the equation $b_a = [c, z]$ for $c \in B$. The proposition (c) is due to B. H. Neumann and H. Neumann [1].)

3. Let C_n be a cyclic group of order n. For any group A, we define a sequence of groups $A_1 = A, A_2, A_3, \ldots$ by

$$A_{n+1} = A_n \text{ wr } C_n,$$

and we consider the isomorphism

$$A_n \to \Delta(A_n) \subset A_{n+1}.$$

We regard A_n as a subgroup of A_{n+1} by identifying A_n with its image $\Delta(A_n)$, and set $G = \bigcup A_n$.

Show that, for any element g of G and for any integer m, there is an element x of G such that $x^m = g$.

(*Hint.* By Exercise 2(b), every element of A_n is an n-th power of some element of A_{n+1}. A group which satisfies the conclusion of this exercise is said to be *divisible* or *algebraically closed*. Exercise 3 asserts that every group is embedded in a divisible group. This is a theorem of B. H. Neumann [1], and the above proof is due to Baumslag [1].)

4. Show that there is an infinite p-group G which satisfies $G' = G$.

(*Hint.* Use Exercise 2(c), and construct the p-group G by a method similar to that of Exercise 3.)

5. Let W be the wreath product of a cyclic group of order n and a cyclic group of order m. Show that there are elements x and y of order m whose product xy has order n.

(*Hint.* Let $A = \langle a \rangle$ and $G = \langle y \rangle$, where the element a has order n. In the base group of $W = A$ wr G, we define an element b by

$$b(y) = a, \qquad b(y^2) = a^{-1}, \qquad b(y^i) = 1 \qquad (3 \le i \le m).$$

Then, the element yb is of order m. Let $x^{-1} = yb$.)

6. Any group G which is a countable set is isomorphic to a subgroup of a group generated by two elements. Prove this theorem (due to G. Higman, B. H. Neumann, and H. Neumann [1]) in the following steps.

(a) Let the group G be a countable set: $G = \{g_1, \ldots\}$. Let $W = G$ Wr $\mathbb{Z}$. Every element of $\Delta(G)$ is a commutator of two elements of W, that is, $g_i = [u_i, v_i]$ $(u_i, v_i \in W)$. Let H be the subgroup of W generated by the elements $u_1, v_1, u_2, v_2, \ldots$. Then, H is a countable set, and its commutator subgroup H' contains a subgroup which is isomorphic to G.

(b) Let $1, h_1, h_2, \ldots$ be all the elements of H. Consider $W = H$ Wr $\mathbb{Z}$. Let an element u of the base group be defined as $u(2^r) = h_r$ and $u(x) = 1$ if x is not a power of 2. If z is a generator of the infinite cyclic group $\mathbb{Z}$, then the subgroup $\langle z, u \rangle$ contains a subgroup isomorphic to the commutator subgroup H' of H.

(c) Any group which is a countable set is embedded in a group that is generated by two elements.

(*Hint.* By the definitions of the elements u and z, the conjugation by the element z^k sends u into the element v of B such that $v(m) = u(m + k)$. Let $I(v)$ be the set of all integers m such that $v(m) \ne 1$. Then, $I(v) = \{2^n - k\}$ $(n = 1, 2, \ldots)$. Set $v_r = z^{-k}uz^k$ where $k = 2^r - 2$. Let r and s be two distinct positive integers, and consider the commutator $w = [v_r, v_s]$. Clearly, we have $w(m) \ne 1$ only when $m \in I(v_r) \cap I(v_s)$. The crucial points in the proof are

$$I(v_r) \cap I(v_s) = \{2\} \quad \text{and} \quad w(2) = [h_r, h_s].$$

It follows from these properties that the set of the elements of $B \cap \langle z, u \rangle$ which take on the value 1 for $m \neq 2$ contains a subgroup which is isomorphic to the commutator subgroup of H.

This proof is due to P. Hall [11]. Any finite group G is isomorphic to a subgroup of the symmetric group $\Sigma(G)$ on the set G. Since $\Sigma(G)$ is generated by two elements (see Chapter 3, §2), the most interesting case of Exercise 6 occurs when G is countably infinite.)

Chapter 3

Some Special Classes of Groups

§1. Torsion-free Abelian Groups

In this section, we will study the class of infinite abelian groups which are torsion-free. The main purpose is to show that the structure of such groups is not so simple as we might envision. In particular, we will show that there are examples of groups in which the Remark–Schmidt Theorem does not hold. For further investigation of abelian groups, torsion-free and otherwise, the readers should consult the standard reference, the book by Fuchs [1].

Convention 1.1. *Throughout this section, except for the last example, we will consider only torsion-free abelian groups which will be written additively. Thus, an additive group A always means an abelian group with $T(A) = \{0\}$.*

By this convention, the only element of finite order in A is the identity. It follows that, for any element a of A and for any natural number n, there is at most one element x of A such that $nx = a$. This is proved as follows. If $nx = ny$, then we have $n(x - y) = 0$. Hence, as $x - y$ is of finite order, we get $x - y = 0$, or $x = y$.

Definition 1.2. A finite subset $\{a_1, \ldots, a_m\}$ of an additive group A is said to be **free** if a relation

$$n_1 a_1 + \cdots + n_m a_m = 0$$

with integral coefficients $n_i \in \mathbb{Z}$ implies that $n_1 = 0, n_2 = 0, \ldots, n_m = 0$ (see Chapter 2 §5, Exercise 1). If an additive group A contains a free subset of m elements, but no free subset of $m + 1$ or more elements, then we say that the group A is of rank m. If an additive group A is of rank m for some natural number m, then A is said to be of **finite rank**.

It is clear that the rank is uniquely determined.

(1.3) *An additive group A of rank m is isomorphic to some subgroup of the direct sum of m copies of $\mathbb{Q}$, the additive group of the rational numbers. Thus, there is* *an isomorphism φ from A into the m-dimensional vector space V over the field $\mathbb{Q}$ of rational numbers such that the set $\{a_1, \ldots, a_k\}$ of elements of A is free if and only if the elements $\{\varphi(a_1), \ldots, \varphi(a_k)\}$ of V are linearly independent over $\mathbb{Q}$.*

Proof. Let V be the totality of row vectors $(r_1, \ldots, r_m)$ where $r_i \in \mathbb{Q}$. We will define the mapping $\varphi \colon A \to V$. Suppose that a subset $\{b_1, \ldots, b_m\}$ of A is free. If a is any element of A, the set $\{a, b_1, \ldots, b_m\}$ is not free. There is a relation

$$na + n_1 b_1 + \cdots + n_m b_m = 0$$

for some integers $n, n_1, \ldots, n_m$, and at least one of the coefficients is not zero. Since $\{b_1, \ldots, b_m\}$ is free, we must have $n \neq 0$. Let

$$\varphi(a) = (-n_1/n, \ldots, -n_m/n).$$

If there is another relation $n'a + n'_1 b_1 + \cdots + n'_m b_m = 0$, then

$$0 = n'(na) - n(n'a) = \sum (n'n_i - nn'_i)b_i.$$

Since $\{b_1, \ldots, b_m\}$ is free, we get $n'n_i = nn'_i$ for all i. Thus, if $n' \neq 0$, then $n_i/n = n'_i/n'$ for all i. This means that the mapping φ is defined uniquely and does not depend on the particular relation.

Suppose $a' \in A$ and

$$\varphi(a') = (-l_1/l, \ldots, -l_m/l) \qquad (l_i, l \in \mathbb{Z}).$$

If there is a relation $l'a' + l'_1 b_1 + \cdots + l'_m b_m = 0$, then the definition of φ gives us $l'_i/l' = l_i/l$ for all i. Hence, we have

$$l(l'a' + l'_1 b_1 + \cdots + l'_m b_m) = l'(la' + l_1 b_1 + \cdots + l_m b_m) = 0.$$

This implies that $la' + l_1 b_1 + \cdots + l_m b_m = 0$ (see the remark after (1.1)). Hence, we get

$$ln(a + a') + (ln_1 + nl_1)b_1 + \cdots = 0.$$

This shows that

$$\varphi(a + a') = (-(ln_1 + nl_1)/ln, \ldots).$$

Therefore, we have $\varphi(a + a') = \varphi(a) + \varphi(a')$, and so the mapping φ is a homomorphism. It is obvious that φ is an injection.

If the subset $\{a_1, \ldots, a_k\}$ of A is not free, there is a relation

$$n_1 a_1 + \cdots + n_k a_k = 0$$

for some integers n_i, and at least one of them is not zero. Then, we get

$$n_1 \varphi(a_1) + \cdots + n_k \varphi(a_k) = 0.$$

Thus, the set $\{\varphi(a_1), \ldots, \varphi(a_k)\}$ is not linearly independent. Conversely, if $\{\varphi(a_1), \ldots, \varphi(a_k)\}$ is not linearly independent over $\mathbb{Q}$, then there are rational numbers $r_1, \ldots, r_k$ such that $\sum r_i \varphi(a_i) = 0$ and such that at least one of $r_1, \ldots, r_k$ is not zero. We can write these rational numbers r_i with a common denominator as $r_i = n_i/n$ where $n_1, \ldots, n_k$, and n are integers. Then, not all n_i are zero, and we have

$$\varphi(\sum n_i a_i) = \sum n_i \varphi(a_i) = 0.$$

This implies that $\sum n_i a_i = 0$, and the set $\{a_1, \ldots, a_k\}$ is not free. $\square$

Corollary. *Let A be an additive group of rank m. Any free subset of A can be enlarged to a free subset consisting of m elements.*

Proof. Let $\{a_1, \ldots, a_k\}$ be a free subset which is not contained in any free subset of $k + 1$ elements. Then, for any $a \in A$, the set $\{a, a_1, \ldots, a_k\}$ is not free. Hence, if we let φ be the isomorphism given in (1.3), $\varphi(a)$ is linearly dependent on $\varphi(a_1), \ldots, \varphi(a_k)$. Thus, we have dim $V \le k$. But, the dimension of V over $\mathbb{Q}$ is equal to the rank m of A. So, any maximal free subset contains exactly m elements. $\square$

Definition 1.4. Let p be a prime number. The **p-height** of an element a of an additive group A is defined as follows. If A contains an element x such that $p^k x = a$ for every natural number k, then the p-height of a is infinite. On the other hand, if $p^k x = a$ has a solution for $k = n$, but not for $k \ge n + 1$, then the p-height of a is n.

The p-height of an element a is denoted by $H_p(a)$.

Definition 1.5. Let π be the set of all prime numbers. Let f and g be functions defined on π taking on integral values ≥ 0 and the symbol ∞. We call f and g **equivalent** if $f(p) \ne g(p)$ implies that both $f(p)$ and $g(p)$ are not equal to ∞ and if the set of such prime numbers p is finite. An equivalence class with respect to this relation is called a **type**.

Let A be an additive group, and let $a \in A$. As a function of $p \in \pi$, the p-height $H_p(a)$ is a function of the sort considered in Definition 1.5. We call it the **height** of a and denote it by $H(a)$.

Here, we will present a collection of the properties of the height function which will be necessary in later discussions.

(1.6) *Let a be an element of an additive group A, and let $p \in \pi$.*

(1) *For any natural number m, we have $H_p(a) \leq H_p(ma)$.*
(2) *If p does not divide m, then $H_p(a) = H_p(ma)$.*
(3) *For any integer m, we define $e_p(m) = e$ as the highest exponent of p which divides m :*

$$m = p^e m', \qquad (p, m') = 1.$$

If $H_p(a) < \infty$, then we have $H_p(ma) = H_p(a) + e_p(m)$. If $H_p(a) = \infty$, then $H_p(ma) = H_p(a)$.

(4) *There is an element x of A which satisfies $mx = a$ if and only if $e_p(m) \leq H_p(a)$ for all $p \in \pi$.*

Proof. The first two propositions are special cases of the third. We will prove (3). If $p^k x = a$, then

$$p^k(mx) = ma = p^{k+e}(m'x).$$

Hence, $H_p(a) = \infty$ implies that $H_p(ma) = \infty$, while $H_p(a) = k < \infty$ implies that $H_p(ma) \geq k + e$. Assume that $H_p(ma) \geq l + e_p(m)$ for some natural number $m = p^e m'$. Since m' is relatively prime to p, we can find integers u and v such that

$$up^l + vm' = 1.$$

By assumption, there is an element y such that $p^{l+e}y = ma$. Hence, we get $p^l y = m'a$ by (1.1). So, we have

$$a = (up^l + vm')a = p^l(ua + vy).$$

This means that $l \leq H_p(a)$. So, $H_p(ma) = H_p(a) + e_p(m)$.
   (4) If $mx = a$, then $p^e(m'x) = a$. So, we have $e_p(m) \leq H_p(a)$ for any p. In order to prove the converse, assume that $e_p(m) \leq H_p(a)$ for all $p \in \pi$. Then, if we set $e = e_p(m)$, there is an element y_p such that $p^e y_p = a$. Let $m = p^e m_p$. Then, the integers $\{m_p\}$ are relatively prime. So, there is a set of integers $\{l_p\}$ such that $\sum l_p m_p = 1$, where p ranges over the set of prime divisors of m. Therefore, we get

$$m(\sum l_p y_p) = \sum l_p m_p p^e y_p = (\sum l_p m_p)a = a,$$

and so $mx = a$ has a solution in A.   □

Theorem 1.7. *Let A be an additive group of rank 1.*

(1) *If $0 \neq a \in A$, then the height $H(a)$ belongs to a uniquely determined type. We denote it by* Typ A.

(2) *Let B be another additive group of rank 1. Then,*

$$A \cong B \Leftrightarrow \text{Typ } A = \text{Typ } B.$$

(3) *Let T be any type. Then, there is an additive group A of rank 1 such that* Typ $A = T$.

Proof. (1) If a and b are two nonidentity elements of A, we have integers m and n such that

$$ma = nb, \qquad (m \neq 0, n \neq 0).$$

By (1.6) (3) and Definition 1.5, the heights of a and ma belong to the same type. Similarly, we have $H(nb) = H(b)$. Hence, the type of $H(a)$ is equal to the type of $H(b)$.

(2) Let $\{u\}$ be a free set in A. We will consider the element u to be fixed and will construct an isomorphism φ from A into $\mathbb{Q}$ as in (1.3.) Let a be an arbitrary nonidentity element of A. If $\varphi(a) = m/n$ where $(m, n) = 1$, then we have $na = mu$, as shown in the proof of (1.3). We will show that $e_p(n) \leq H_p(u)$ for all primes p. If $e_p(n) = 0$ or $H_p(u) = \infty$, then it is trivial to prove that the inequality is satisfied. On the other hand, if $e_p(n) > 0$ and $H_p(u) < \infty$, then we have $e_p(m) = 0$ and

$$e_p(n) \leq H_p(mu) = H_p(u)$$

by (1.6) (3).

Conversely, if $e_p(n) \leq H_p(u)$ for all primes p, then there is an element x of A satisfying $nx = u$ $((1.6)(4))$. By the definition of φ, we have $\varphi(x) = 1/n$ and $\varphi(mx) = m/n$. Thus, for a reduced fraction $r = m/n$, we know that r belongs to $\varphi(A)$ if and only if its denominator n satisfies $e_p(n) \leq H_p(u)$ for all primes p.

Suppose that $A \cong B$. Then, we clearly have Typ $A =$ Typ B. Conversely, suppose that Typ $A =$ Typ B. Then, if $0 \neq a \in A$ and $0 \neq b \in B$, the height $H(a)$ of a is equivalent to the height $H(b)$ of b. Let $d(a, b)$ be the set of prime numbers p such that $H_p(a) \neq H_p(b)$. Then, by definition, $d(a, b)$ is a finite set and, for $p \in d(a, b)$, both $H_p(a)$ and $H_p(b)$ are finite. Suppose that $H_p(a) > H_p(b)$, and set $H_p(a) = H_p(b) + e$. Then, we know that $e > 0$, and from (1.6) (3), we get

$$H_p(p^e b) = H_p(b) + e = H_p(a).$$

If q is any prime number not equal to p, then $H_q(p^e b) = H_q(b)$. Thus, $d(a, p^e b)$ has one fewer prime than does $d(a, b)$. By repeating this process, we can conclude that there are natural numbers m and n such that $H_p(ma) = H_p(nb)$ for all p. We will use these elements ma and nb to define the isomorphism into

$\mathbb{Q}$ as in the first part of the proof. Then, the image of A coincides with the image of B. This proves that $A \cong B$.

(3) Let $f(p)$ be a function representing the type T. Let A be the totality of rational numbers r which can be written as $r = m/n$ where m and n are integers, $n > 0$, and $e_p(n) \leq f(p)$ for all prime numbers p. It is easily verified that A forms an additive group. Furthermore,

$$p^k x = 1 \quad \text{has a solution in } A \Leftrightarrow 1/p^k \in A \Leftrightarrow k \leq f(p).$$

Thus, we get $H_p(1) = f(p)$. By the definition of Typ A, we get Typ $A = T$.

$\square$

Corollary. *There are uncountably many isomorphism classes of additive groups of rank 1.*

Torsion-free additive groups of rank 1 are completely classified by Theorem 1.7. However, the classification of additive groups of higher rank has not yet been completed.

EXAMPLE 1. The direct sum of infinite cyclic groups is always a free abelian group (see Chapter 2, Theorem 5.9, Corollary 1). But, the complete direct product of infinite cyclic groups is not necessarily free. In fact, *if A is the complete direct product of a countable number of infinite cyclic groups, then A is not free, but any countable subgroup of A is free.* We will prove the above proposition which is one of the many interesting theorems due to Baer.

An element of A can be expressed as an infinite sequence $(a_1, a_2, \ldots)$ of integers $a_1, a_2, \ldots$. Let p be a fixed prime number, and let J be the totality of the sequences $v = (a_i)$ with the following property: for any natural number k, the number of components a_i which are not divisible by p^k is finite. The set J is not countable, but it is certainly a subgroup of A. Let A_0 be the set of $v = (a_i)$ such that all but a finite number of the components a_i are zero. Clearly, A_0 is a subgroup of J, and A_0 is countable. Furthermore, it is easy to see that

$$J = \langle A_0, pJ \rangle,$$

where pJ is the set of pv with $v \in J$. Thus, the factor group J/pJ is a countable set. This implies that J is not free. If J were free, then J would be a direct sum of uncountably many infinite cyclic groups, and the factor group J/pJ would also be uncountable. Thus, the subgroup J is not free. By Theorem 5.11 of Chapter 2, A is also not free.

In order to prove that any countable subgroup of A is free, we need three lemmas.

Lemma 1. *Let X be an arbitrary additive group. Let $\mathcal{R}_n$ be the set of subgroups of rank n which contain a fixed subset Y. If $\mathcal{R}_n$ is not empty, it is an inductively ordered set. Thus, $\mathcal{R}_n$ has a maximal element.*

Proof. This is almost obvious. Let $\{U_\lambda\}$ be a linearly ordered family of subgroups of rank n. Then, the union $\cup$ of the sets U_λ is a subgroup of X. Any subset of U consisting of $n + 1$ elements must be contained in a member U_λ of the family. So, the set is not free because U_λ is of rank n. Thus, the subgroup U is also of rank n. This shows that the set $\mathcal{R}_n$ is inductively ordered. By Zorn's lemma, $\mathcal{R}_n$ has a maximal element. $\square$

Lemma 2. *Let X be an additive group which is a countable set. Then, X is free if and only if every subgroup of finite rank is free.*

Proof. It is clear that every subgroup of finite rank of a free additive group is free (Theorem 5.11, Chapter 2). To prove the converse, assume that every subgroup of finite rank is free. By assumption, X is a countable set, so let $X = \{a_1, a_2, \ldots\}$. The subgroup $\langle a_1 \rangle$ has rank at most 1. Let B_1 be a subgroup which is maximal among subgroups of rank 1 containing a_1. Then, the subgroup $\langle B_1, a_2 \rangle$ has rank at most 2. Let B_2 be a subgroup which is maximal among subgroups of rank 2 containing $\langle B_1, a_2 \rangle$. Repeat this process. Suppose that a subgroup B_{k-1} has been defined as a subgroup of rank $k - 1$ which is maximal among those subgroups of rank $k - 1$ containing $\langle B_{k-2}, a_{k-1} \rangle$. Set $C = \langle B_{k-1}, a_k \rangle$. Suppose that there is no subgroup of rank k which contains C. Then, the subgroup C is of rank $k - 1$. Since B_{k-1} is maximal, we get $a_k \in B_{k-1}$. Then, for any n, the subgroup $\langle B_{k-1}, a_n \rangle$ contains C, and we have $a_n \in B_{k-1}$ as before. This implies that $X = \{a_1, a_2, \ldots\}$ is equal to B_{k-1}, which is free by assumption.

For every k, we may assume that there is a subgroup of rank k which contains $\langle B_{k-1}, a_k \rangle$. By Lemma 1 there is a maximal element B_k among those subgroups of rank k. By definition, we have

$$B_1 \subset B_2 \subset \cdots \subset B_n \subset \cdots, \qquad \bigcup B_k = X.$$

Each B_k is free by our assumption. Furthermore, the factor group B_k/B_{k-1} is torsion free by maximality of B_{k-1}. Hence, B_k/B_{k-1} is an infinite cyclic group, and for an element b_k of B_k, we have

$$B_k = B_{k-1} + \langle b_k \rangle.$$

The set $\{b_1, b_2, \ldots\}$ is clearly a free generating set of X. Therefore, the group X is free. $\square$

Lemma 3. *Let X be an additive group which is a countable set. A necessary and sufficient condition for X to be free is the following : for each element $x \neq 0$ of X, there is an element φ of $\mathrm{Hom}(X, \mathbb{Z})$ such that $\varphi(x) \neq 0$.*

Proof. A free additive group is a direct sum of cyclic groups which are isomorphic to $\mathbb{Z}$. If $\{\varphi_\lambda\}$ is the family of projections into the summands, then $\varphi_\lambda \in \text{Hom}(X, \mathbb{Z})$, and for any $x \neq 0$, there is a member φ_λ such that $\varphi_\lambda(x) \neq 0$.

Conversely, let X be an additive group satisfying the condition of Lemma 3. It suffices to show that any subgroup Y of finite rank n is free. Use induction on n. If $n = 0$, then Y is certainly free. So, assume that $n > 0$, and let y be a nonzero element of Y. There is a homomorphism $\varphi \in \text{Hom}(X, \mathbb{Z})$ such that $\varphi(y) \neq 0$. Set

$$Z = Y \cap \text{Ker } \varphi.$$

Then, we get $Z \cap \langle y \rangle = \{0\}$. Therefore, $\langle Z, y \rangle = Z + \langle y \rangle$ is a direct sum. This implies that Z is of finite rank. In fact, it is of rank at most $n - 1$. So, by the inductive hypothesis, Z is free. The factor group Y/Z is isomorphic to a subgroup of $\text{Im } \varphi$. So, it is infinite cyclic. Hence, by choosing a suitable element u of Y, we get

$$Y = Z + \langle u \rangle.$$

Thus, Y is a direct sum of infinite cyclic groups, and therefore, Y is free. $\square$

The lemma 3 implies the freeness of a countable subgroup of A. Let φ_i be the mapping which sends $v = (a_1, a_2, \ldots)$ into the i-th coordinate a_i. Clearly, φ_i belongs to $\text{Hom}(A, \mathbb{Z})$, and for any $v \neq 0$, there is at least one φ_i such that $\varphi_i(v) \neq 0$. If X is any countable subgroup of A, the restriction of φ_i on X belongs to $\text{Hom}(X, \mathbb{Z})$. So, the lemma 3 proves that X is free. $\square$

The group A illustrates that the lemmas 2 and 3 are not necessarily true for general additive groups without the countability assumption.

EXAMPLE 2. In Chapter 2, §4, we proved the fundamental theorem on direct product decomposition (Theorem 4.8). One of the assumptions of the theorem was the existence of a principal series. We will construct an example to show that Theorem 4.8 is no longer true without a chain condition. This example is due to Jónsson [1].

We will construct an additive group G of rank 3 such that

$$G = A + B = C + D$$

where A, B, C, and D are directly indecomposable, $A \cong C$, but $B \not\cong D$.

Let l be a prime number ≥ 5. Let π_1 and π_2 be infinite sets of prime numbers such that

$$l \notin \pi_1 \cup \pi_2, \qquad \pi_1 \cap \pi_2 = \varnothing \text{ (the empty set)}.$$

Furthermore, let x, y, and z be three real numbers which are linearly independent over the field of rational numbers. Define A and B as follows:

$$A = \langle x/n \rangle, \qquad B = \langle y/n, z/m, (y + z)/l \rangle,$$

where n ranges over those square free natural numbers which are divisible only by prime numbers in π_1, and similarly, where m ranges over all square free positive integers whose prime factors are in π_2. Set

$$G = A + B,$$

and prove that G does not satisfy the conclusion of the Krull–Remak–Schmidt Theorem.

(1)   *The groups A and B are directly indecomposable.*

Proof. The group A is of rank 1, so it is clearly indecomposable. An element of B has the form $ry + sz$ where r and s are rational numbers. By definition, the denominator of r is square free and divisible only by prime numbers in $\pi_1 \cup \{l\}$. Similarly, the denominator of s is square free and is divisible only by prime numbers in $\pi_2 \cup \{l\}$. Hence, if $H_p(ry + sz) > 0$ for some prime $p \in \pi_1$, then p must divide the numerator of s. So, if $H_p(ry + sz) > 0$ for all $p \in \pi_1$, then we must have $s = 0$ (as π_1 contains infinitely many distinct prime numbers).

Suppose that the group B is decomposable, and set $B = B_1 + B_2$. With respect to this decomposition, we write $y = y_1 + y_2$ ($y_i \in B_i$). If $p \in \pi_1$, we have $H_p(y) > 0$. This implies that $H_p(y_i) > 0$ for $i = 1, 2$. Then, as shown above, $y_1 = r_1 y$ and $y_2 = r_2 y$ for some rational numbers r_1 and r_2. Thus, we have $r_1 y_2 = r_2 y_1$. Since $B_1 \cap B_2 = \{0\}$, we get either $y_1 = 0$ or $y_2 = 0$. This means that the element y is contained in one of the summands. In the same way, we can prove that the element z is also contained in one of the summands. If $B_1 \neq 0$ and $B_2 \neq 0$, then y and z are contained in different summands. So, suppose that $y \in B_1$ and $z \in B_2$. The group B contains the element $(y + z)/l$. If $(y + z)/l = b_1 + b_2$ ($b_i \in B_i$), then $z = lb_2$ and $z/l = b_2 \in B_2$. But, this is a contradiction since $l \notin \pi_2$. Thus, the group B is directly indecomposable. $\square$

We will now define the subgroups C and D of G. We choose integers a, b, c, and d which satisfy

$$ad - bc = 1, \qquad c \equiv 0, \qquad a \not\equiv \pm 1 \pmod{l}.$$

Since $l \geq 5$, such integers always exist. Let

$$x' = ax + by,$$
$$y' = cx + dy.$$

This can be solved for x and y:

$$x = dx' - by', \qquad y = -cx' + ay'.$$

Let

$$C = \langle x'/n \rangle, \qquad D = \langle y'/n, z/m, (ay' + z)/l \rangle,$$

where n and m are restricted in the same way as before. We have

$$(ay' + z)/l = acx/l + bcy/l + (y + z)/l.$$

Since $c \equiv 0 \pmod{l}$, we have $(ay' + z)/l \in G$. This shows that both C and D are subgroups of G. Clearly, we get $G = C + D$. As in (1), both C and D are directly indecomposable. It is obvious that $A \cong C$. However, we will prove the following proposition.

(2)   *The two groups B and D are not isomorphic.*

Proof. Suppose that they are isomorphic, and let φ be an isomorphism from B onto D. Then, the p-height of $\varphi(y)$ is the same as the p-height of y. Since the p-height of y is 1 for $p \in \pi_1$ but is 0 for all other prime numbers, we must have $\varphi(y) = \pm y'$. Similarly, we get $\varphi(z) = \pm z$. Thus,

$$\varphi((y + z)/l) = (\pm y' \pm z)/l.$$

Since $(ay' + z)/l \in D$, we get $(a + 1)y'/l \in D$ or $(a - 1)y'/l \in D$. But, since $a \not\equiv \pm 1 \pmod{l}$, this is a contradiction. Hence, B and D are not isomorphic. $\square$

EXAMPLE 3. In this example, we will consider abelian groups which are not necessarily torsion-free. Let $GL(n, \mathbb{Z})$ be the totality of $n \times n$ matrices with integral entries and determinant ± 1. This set forms a group with respect to matrix multiplication. It is considered to be an operator group acting on the free abelian group L of the totality of row vectors $(a_1, \ldots, a_n)$ $(a_i \in \mathbb{Z})$. (In fact, the group $GL(n, \mathbb{Z})$ is isomorphic to Aut L (see Exercise 3, Chapter 2, §5).) If an n-dimensional vector space V over the field $\mathbb{Q}$ of rational numbers is identified as the set of row vectors $(r_1, \ldots, r_n)$, then L is a subgroup of V, and $GL(n, \mathbb{Z})$ can be regarded as a group of linear transformations on V. The following theorem holds.

Any abelian subgroup of GL(n, $\mathbb{Z}$) is finitely generated.

A particular case of this theorem is a part of the unit theorem of Dirichlet in algebraic number theory. For the theorem of Dirichlet, the reader should consult Iyanaga [1], page 249. We will derive the finite generation theorem from the theorem of Dirichlet.

Let A be an abelian subgroup of $GL(n, \mathbb{Z})$. We consider A to be an operator group acting on the vector space V over $\mathbb{Q}$. Then, V is a $\mathbb{Q}A$-module where $\mathbb{Q}A$ is the group ring of A over $\mathbb{Q}$.

(a) Suppose that V is irreducible as a $\mathbb{Q}A$-module. Let Δ be the ring of all the $\mathbb{Q}A$-endomorphisms of V. By Schur's lemma, Δ is a finite dimensional division algebra over $\mathbb{Q}$. Since the operator group A is abelian, we have $A \subset \Delta$. Hence, A is contained in the center of Δ. Let F be the center of Δ. Then, F is an extension of finite degree over $\mathbb{Q}$, i.e. F is an algebraic number field. If $a \in A$, the action of a has the characteristic polynomial $P(t)$ which is monic and has integral coefficients. Since $P(a) = 0$, a is an algebraic integer of F. Using the multiplicative notation, we note that a^{-1} is also in A. So, $A \subset F$ is a multiplicative group consisting of units in the ring of algebraic integers. By the theorem of Dirichlet to which we referred earlier, the totality of such units forms a finitely generated abelian group. Since A is one of its subgroups, A itself is finitely generated by the Cor. to (3.18) of Chapter 2.

(b) Suppose that V is not irreducible as a $\mathbb{Q}A$-module. Let U be a $\mathbb{Q}A$-invariant subspace of V such that $0 < \dim U < \dim V$. As before, consider A to be an operator group acting on the free abelian group L which is contained in V. With this notation, we have the following lemma.

The subgroup $L \cap U$ is a direct summand of L, and its rank r is equal to $\dim U$.

Proof. Clearly, we have $r \leq \dim U$. Any element u of U is a linear combination of the elements of L with rational coefficients. Thus, if m is a common denominator of these coefficients, we get $mu \in L$. So, if $u_1, \ldots, u_d$ is a basis of U over $\mathbb{Q}$ ($d = \dim U$), then there are integers $n_1, \ldots, n_d$ such that $n_1 u_1, \ldots, n_d u_d \in L \cap U$. Clearly, these elements are free in $L \cap U$, so we have $d \leq r$. Thus, we get $r = \dim U$.

Next, consider the factor group $L/L \cap U$. Since U is a subspace, $L/L \cap U$ is torsion-free. Hence, it is free, and $L \cap U$ is a direct summand of L.   □

(c) As a corollary of the above lemma, we have the following proposition. A set of free generators of $L \cap U$ can be enlarged to a set of free generators of L. We can write linear transformations using this basis. As $L \cap U$ is A-invariant, each element of A has the form

$$\begin{pmatrix} \alpha(a) & 0 \\ \gamma(a) & \beta(a) \end{pmatrix}$$

when $\alpha(a) \in GL(r, \mathbb{Z})$, $\beta(a) \in GL(n - r, \mathbb{Z})$, and $\gamma(a)$ belongs to the set M of $(n - r) \times r$ matrices with integral entries. When the element a ranges over A, the set of matrices $\{\alpha(a)\}$ and $\{\beta(a)\}$ form abelian groups A_1 and A_2, respectively.

We will prove the assertion on finite generation by using induction on n.

By the inductive hypothesis, both A_1 and A_2 are finitely generated. The definition of the multiplication of matrices shows that the mapping

$$\theta: a \rightarrow (\alpha(a), \beta(a))$$

is a homomorphism from A into the direct sum $A_1 + A_2$. Set $B = \mathrm{Ker}\ \theta$. Then, B is a subgroup consisting only of matrices of the form

$$\begin{pmatrix} I & 0 \\ Z & I \end{pmatrix} \quad (Z \in M),$$

and B is isomorphic to a subgroup of M. The group M is finitely generated, as is its subgroup B by the Cor. to (3.18) of Chapter 2. On the other hand, the group A/B is isomorphic to Im θ which is a subgroup of $A_1 + A_2$. Since both A_1 and A_2 are finitely generated, any subgroup of $A_1 + A_2$ is also finitely generated. Thus, both B and A/B are finitely generated, and so is A.   □

§2. Symmetric Groups and Alternating Groups

In this section, we will consider the symmetric group Σ_n on a finite set X, where the letter n always stands for $|X|$. By (7.2) of Chapter 1, the structure of the symmetric group is determined by n alone. It is customary to choose X as the set of the first n natural numbers $\{1, 2, \ldots, n\}$. In this case, a permutation σ is expressed by either

$$\sigma = \begin{pmatrix} 1 & 2 & \cdots & n \\ \sigma(1) & \sigma(2) & \cdots & \sigma(n) \end{pmatrix}$$

or

(2.1) $\sigma = (a_{11} a_{12} \cdots a_{1r_1})(a_{21} \cdots a_{2r_2}) \cdots (a_{s1} \cdots a_{sr_s}).$

In (2.1), $a_{ij} \in X$, $a_{ij} = a_{kl}$ implies $i = k$ and $j = l$, and

$$\sigma(a_{ij}) = a_{ij+1} \quad (j < r_i), \qquad \sigma(a_{ir_i}) = a_{i1}.$$

The subgroup $G = \langle \sigma \rangle$ generated by the permutation σ is a permutation group on X (Definition 7.3 of Chapter 1). In the notation of (2.1), the set X_i of elements $\{a_{i1}, \ldots, a_{ir_i}\}$ within a pair of parentheses is an orbit under the action of G. The decomposition of X into $X_1, \ldots, X_s$ corresponds to the one in (7.8) of Chapter 1. Thus, the expression (2.1) is uniquely determined by the arrangement of the various orbits $X_1, \ldots, X_s$ and by the choice of the first elements a_{i1} from X_i. In particular, the set of the lengths of the orbits $\{r_1, \ldots, r_s\}$ is unique.

Definition 2.2. The arrangement of the orbit lengths in increasing order of magnitude $r_1 \le r_2 \le \cdots \le r_s$ is said to be the **type** of the permutation σ. The decomposition (2.1) is called the **cycle decomposition** of σ.

(2.3) *Let (2.1) be the cycle decomposition of a permutation σ. Let ρ be the permutation such that $\rho(a_{ij}) = b_{ij}$ for all i and j. Then*

$$\rho^{-1}\sigma\rho = (b_{11} \cdots b_{1r_1}) \cdots (b_{s1} \cdots b_{sr_s})$$

is the cycle decomposition of the permutation $\rho^{-1}\sigma\rho$.

Proof. This is easily verified (see (7.2) of Chapter 1).   □

Theorem 2.4. *Two permutations σ and ρ are conjugate in Σ_n if and only if they have the same type.*

Proof. This theorem follows from (2.3). First of all, we know that conjugate permutations have the same type. Conversely, if σ and τ have the same type, we can easily construct a permutation ρ which satisfies $\tau = \rho^{-1}\sigma\rho$.   □

In the cycle decomposition (2.1), it is customary to omit terms with $r_i = 1$. For example, we write

$$\begin{pmatrix} 1 & 2 & 3 & 4 & 5 & 6 \\ 2 & 3 & 1 & 6 & 5 & 4 \end{pmatrix} = (1 \quad 2 \quad 3)(4 \quad 6)(5) = (1 \quad 2 \quad 3)(4 \quad 6).$$

In this section, the symmetric group is assumed to act on $X = \{1, 2, \ldots, n\}$ unless the contrary is explicitly stated. So, all the numbers not appearing in a cycle decomposition are supposed to be fixed by the permutation. Following this convention, we may omit terms with $r_i = 1$ from the type of the permutation. Thus, we may say σ is of type r, instead of using the more cumbersome notation, type $(1, 1, \ldots, 1, r)$.

Definition 2.5. For a natural number $r > 1$, a permutation of type r is said to be an **r-cycle**. A permutation is called a **cycle** if it is an r-cycle for some r. A 2-cycle is also called a **transposition**.

(2.6) *Any permutation can be decomposed into a product of cycles in which any two distinct factors do not move a common letter. This decomposition is unique once the order of the cycles in the product is determined. Any permutation can be decomposed into a product of transpositions. The decomposition into a product of transpositions is not unique, but the numbers of transpositions appearing in these decompositions are always all even or all odd. Thus, the parity of the number of transpositions is independent of the way a permutation is written as a product of transpositions.*

Proof. For any subset $\{a_1, \ldots, a_r\}$ of r distinct elements of X, $(a_1 \cdots a_r)$ defines an r-cycle which moves a_r to a_1 and a_i into a_{i+1} for $i < r$. In fact, the decomposition (2.1) of σ gives a decomposition of σ into a product of an r_1-cycle $(a_{11} a_{12} \cdots a_{1r_1})$, an r_2-cycle $(a_{21} \cdots a_{2r_2}) \ldots$, etc., and all these cycles act on disjoint subsets $X_1, X_2, \ldots, X_s$. Conversely, if $\sigma = \sigma_1 \cdots \sigma_s$ is the product of the cycles $\sigma_1, \sigma_2, \ldots, \sigma_s$ which operate on the disjoint subsets $X_1, X_2, \ldots, X_s$, then those subsets are the orbits of the group $\langle \sigma \rangle$ (which are of length > 1). Hence, $\{X_i\}$, and therefore $\{\sigma_i\}$, are uniquely determined by σ. The formula

$$(a_1 \cdots a_r) = (a_1 a_2)(a_1 a_3) \cdots (a_1 a_r)$$

shows that any permutation can be written as a product of transpositions. It remains to verify the invariance of the parity.

The permutation σ can be considered to be a change of the coordinate axes in a vector space of dimension n over $\mathbb{R}$. Thus, σ can be considered as the linear transformation $\theta(\sigma)$ which sends the i-th coordinate to the $\sigma(i)$-th coordinate. Then, the mapping θ is an isomorphism from Σ_n into $GL(V)$. The transposition is mapped to an element of determinant -1. Hence, σ is a product of an even number of transpositions if and only if $\det(\theta(\sigma)) = 1$. This proves the assertion. $\square$

We remark that in (2.1), every cycle in the right side does commute with σ. We will add two more obvious corollaries.

Corollary 1. *The symmetric group is generated by the set of transpositions.*

Corollary 2. *The totality of permutations which can be written as a product of an even number of transpositions forms a normal subgroup of index 2.*

Definition 2.7. A permutation σ is said to be **even** if σ is the product of an even number of transpositions. Otherwise, σ is called an **odd permutation**. The normal subgroup consisting of all even permutations is called the **alternating group** on n letters, and we denote it by A_n.

The order of A_n is $n!/2$.

(2.8) (i) *The alternating group A_n is generated by the totality of 3-cycles.*

(ii) *If $n \geq 5$, A_n is generated by the totality of permutations of type $(2, 2)$.*

(iii) *If $n = 4$, the set of permutations of type $(2, 2)$ generates a normal subgroup of order 4 in Σ_4.*

(iv) *The group A_5 is isomorphic to $PSL(2, 5)$.*

Proof. By the definition of A_n, the alternating group A_n is generated by elements which are products of two transpositions. These elements are

$$(ab)(cd), \quad (ab)(ac) = (abc).$$

If different letters represent distinct numerals, then we have

$$(abc)(cad) = (ab)(cd).$$

This proves (i).

If $n \geq 5$, then for any given (abc), we can find d and e such that

$$(abc) = (ab)(de) \cdot (de)(ac).$$

This proves (ii). If $n = 4$, then there are exactly 3 permutations of type $(2, 2)$. It is easy to verify that, together with the identity 1, they form a subgroup of order 4. By (2.3), this subgroup is normal in Σ_4.

The isomorphism of (iv) is proved as follows. If $x = (1\ 2\ 3\ 4\ 5)$ and $y = (1\ 4\ 2)$, then we have $(xy)^2 = 1$. So, by Chapter 2, §6, page 178, the subgroup $\langle x, y \rangle$ is a homomorphic image of the group $PSL(2, 5)$. Since $PSL(2, 5)$ is simple (see the Cor. of Theorem 9.10 of Chapter 1), we get $PSL(2, 5) \cong \langle x, y \rangle$. By comparing the orders, we get $\langle x, y \rangle = A_5 \cong PSL(2, 5)$. □

(2.9) *The isomorphism* $\Sigma_4 \cong PGL(2, 3)$ *holds. The normal subgroups of* Σ_4 *are* $\Sigma_4, A_4, \{1\}$, *and the subgroup of order 4 generated by the permutations of type* $(2, 2)$. *We have* A_4 char Σ_4, *and* $A_4 \cong PSL(2, 3)$.

Proof. We have shown at the end of Chapter 1 that all the proper normal subgroups of $GL(2, 3)$ are $SL(2, 3)$, Q, and its center Z. We will show that $GL(2, 3)$ contains a subgroup of index 4. An S_3-subgroup S is not normal. The normalizer $N_G(S)$ contains the center Z, so the index divides 8. By Sylow's theorem, the index of $N_G(S)$ is congruent to 1 modulo 3. Hence, it must be 4. Therefore, there is a homomorphism φ from $GL(2, 3)$ into Σ_4 ((7.16) of Chapter 1). Then, Ker φ is a normal subgroup such that

$$Z \subset \text{Ker } \varphi \subset N_G(S),$$

so we have $Z = \text{Ker } \varphi$. By comparing the orders, we get

$$\Sigma_4 \cong GL(2, 3)/Z = PGL(2, 3).$$

The assertion on normal subgroups follows from the Correspondence Theorem. □

(2.10) *If* $n \neq 4$, *any normal subgroup* $N \neq \{1\}$ *of* Σ_n *contains* A_n. *If* $n \neq 4$, A_n *is the unique minimal normal subgroup of* Σ_n. *For all* n, A_n *is a characteristic subgroup of* Σ_n.

Proof. By assumption, N contains a permutation $\sigma \neq 1$. We will show that there is a transposition τ which does not commute with σ. Choose i such that

$\sigma(i) \neq i$. Then $\tau = (ij)\,(j \neq i, \sigma(i))$ does not commute with σ. The commutator $\rho = [\sigma, \tau]$ is an element of N, and it is a product of the two transpositions τ and τ^σ. Hence, ρ is either a 3-cycle or of type (2, 2). By Theorem 2.4, permutations of the same type are conjugate in Σ_n. So, N either contains all the 3-cycles or all the permutations of type (2, 2). Hence, by (2.8), $A_n \subset N$ if $n \neq 4$. The remaining assertions are easy to verify. $\square$

Theorem 2.11. *If $n \neq 4$, the alternating group A_n is simple.*

Proof. Let N be a proper normal subgroup of A_n. Since $\Sigma_n = \langle A_n, \tau \rangle$ for any transposition τ, we get $N^\tau \neq N$ by (2.10). The groups $\langle N, N^\tau \rangle$ and $N \cap N^\tau$ are τ-invariant normal subgroups of A_n. Hence, by (2.10),

$$A_n = NN^\tau, \qquad N \cap N^\tau = \{1\}, \quad \text{and} \quad A_n = N \times N^\tau.$$

In particular, $|A_n| = |N|^2$. Then, $n \geq 5$ and $|N|$ is even. Hence, by Sylow's theorem, N contains an element σ of order 2. The cycle decomposition of σ shows that σ commutes with a transposition ρ. Then, $N \cap N^\rho$ contains $\sigma \neq 1$, and this is a contradiction. $\square$

(The order of A_n is $n!/2$ and hence is never a perfect square. This follows easily from a famous theorem of number theory: there is a prime number between m and $2m$ for $m \geq 2$. Thus, the above proof had actually been finished after the first half.)

The Structure of the Centralizer of a Permutation

We will determine the structure of the centralizer of a permutation σ of type $(r_1, \ldots, r_s)$. In this section, we will not omit those r_i which are 1, so we have $r_1 + r_2 + \cdots + r_s = n$.

Let (2.1) be the cycle decomposition of σ. Then, by (2.3), a permutation ρ commutes with σ if and only if

$$\sigma = (b_{11} \cdots b_{1r_1}) \cdots (b_{s1} \cdots b_{sr_s})$$

where $\rho(a_{ij}) = b_{ij}$ for all i and j.

Suppose that the number of l-cycles in the decomposition (2.1) is exactly a_l for $l = 1, 2, \ldots$. Let Y_l be the set of letters a_{ij} with $r_i = l$. Clearly, Y_l is σ-invariant. Moreover, Y_l is $C_G(\sigma)$-invariant. The following lemma is obvious.

Lemma. *A permutation ρ commutes with σ if and only if each Y_l is ρ-invariant and if the restriction ρ_l of ρ on Y_l commutes with the restriction σ_l of σ on Y_l.*

Since $Y_i \cap Y_j = \emptyset$ for $i \neq j$, the permutation ρ is uniquely determined by giving its restrictions on Y_i. Hence, we have

(2.12) $C_G(\sigma) = C_1 \times \cdots \times C_n$

where C_i is the centralizer of σ_i in $\Sigma(Y_i)$. Thus, the study of the structure of centralizers is reduced to the case in which σ is a product of m distinct l-cycles for a fixed l.

We will use the following notation:

$$\sigma = (a_{10} a_{11} \cdots a_{1, l-1})(a_{20} \cdots a_{2, l-1}) \cdots (a_{m0} \cdots a_{m, l-1})$$
$$X_i = \{a_{i0}, a_{i1}, \ldots, a_{i, l-1}\} \quad (i \in Y)$$
$$Y = \{1, 2, \ldots, m\}.$$

The second index ranges over $\{0, 1, \ldots, l-1\}$. It is convenient to identify this set with $\mathbb{Z}/(l) = Z_l$, the cyclic group of order l, and to use notation such as $a_{i, l+j} = a_{ij}$.

(2.13) $C(\sigma) = Z_l$ wr $\Sigma(Y)$, *where the wreath product is a general wreath product when $\Sigma(Y)$ is considered to be permutation group on Y (Definition 10.8 of Chapter 2).*

Proof. Let W be the wreath product of the right side. We will define an isomorphism θ from $C(\sigma)$ onto W. For any element ρ of $C(\sigma)$, we will define a permutation $\varphi(\rho)$ and a function f_ρ from Y into Z_l by

$$\rho(a_{i0}) = a_{jk}, \quad j = \varphi(\rho)i, \quad k = f_\rho(j).$$

Since $\rho \in C(\sigma)$, we get

$$\rho(a_{ir}) = \rho(\sigma^r(a_{i0})) = a_{j, k+r}.$$

This shows that $\varphi(\rho)$ is indeed a permutation on Y and that ρ is determined by $\varphi(\rho)$ and f_ρ. Let

$$\theta(\rho) = (\varphi(\rho), f_\rho).$$

Then, θ maps $C(\sigma)$ into W. If ρ and τ are elements of $C(\sigma)$, we have

$$\rho\tau(a_{i0}) = \tau(a_{jk}) = a_{s, t+k}$$

where $s = \varphi(\tau)j$ and $t = f_\tau(s)$. Thus,

$$\varphi(\rho\tau) = \varphi(\rho)\varphi(\tau), \quad f_{\rho\tau}(s) = f_\tau(s) + f_\rho(j).$$

On the other hand, the group $\Sigma(Y)$ acts on the base group of W by

$$f_\rho^{\varphi(\tau)}(s) = f_\rho(j)$$

(Definition 10.8 of Chapter 2). Thus, we have $f_{\rho\tau} = f_\tau + f_\rho^{\varphi(\tau)}$. This proves that the mapping θ is a homomorphism. If $\rho \in \text{Ker } \theta$, then $\varphi(\rho) = 1$ and $f_\rho(j) = 0$. This means that $\rho(a_{i0}) = a_{i0}$ for all i. Hence, $\rho(a_{ij}) = a_{ij}$ for all i and j, and $\rho = 1$.

For any $\varphi \in \Sigma(Y)$ and for any function $f: Y \rightarrow Z_l$, let ρ be the permutation defined by

$$\rho(a_{ir}) = a_{j,\, k+r}, \qquad j = \varphi(i), \qquad k = f(j).$$

Then, as $\sigma^t(a_{ir}) = a_{i,\, r+t}$, ρ commutes with σ. Furthermore, we have $\varphi(\rho) = \varphi$ and $f_\rho = f$. Thus, the mapping θ is an isomorphism from $C(\sigma)$ onto W. $\square$

Corollary. *If the cycle decomposition of σ contains exactly a_t t-cycles, then the order of $C(\sigma)$ is*

$$\prod (a_t\,!) t^{a_t}$$

where $n = a_1 + 2a_2 + \cdots + na_n$.

This follows from (2.12) and (2.13).

Defining Relations

The symmetric group Σ_n is generated by the set of transpositions (2.6), but the formula

$$(i\ i+k) = (i\ i+1)(i+1,\ i+k)(i\ i+1) \qquad (k > 1)$$

shows that Σ_n is generated by the $n-1$ transpositions

$$(1\ 2),\ (2\ 3),\ \ldots,\ (i\ i+1),\ \ldots,\ (n-1\ n).$$

Set $t_i = (i\ i+1)$. Then, these elements satisfy the relations

$$t_i^2 = (t_i t_{i+1})^3 = (t_i t_j)^2 = 1$$

for all i and j such that $|i - j| > 1$.

(2.14) *The above relations among the set $\{t_i\}$ of generators are the defining relations of the symmetric group Σ_n.*

Proof. Let G be the group defined by $\{t_i\}$ and the above set of relations. By (6.7) of Chapter 2, Σ_n is a homomorphic image of G. In order to prove that G is isomorphic to Σ_n, we need only to show that $|G| \le |\Sigma_n|$. Use induction on n. If $n = 1$, $\Sigma_1 = \{1\}$ and the proposition is trivial to prove. We may assume that $H = \langle t_2, t_3, \ldots, t_{n-1} \rangle$ is isomorphic to Σ_{n-1}. So, it is sufficient to show that $|G:H| \leq n$. We will employ the coset enumeration discussed in Chapter 2, §6. Let $H_1 = H$, $H_1 t_1 = H_2$, $H_2 t_2 = H_3, \ldots$, and $H_{n-1} t_{n-1} = H_n$. We will show that

$$H_{j+1} t_j = H_j, \qquad H_i t_j = H_i \quad \text{if} \quad i \neq j, j+1.$$

Since $t_j^2 = 1$, the first is obvious. From the given relations, we get

$$t_i t_{i+1} t_i = t_{i+1} t_i t_{i+1}, \qquad t_i t_j = t_j t_i \quad \text{if} \quad |j - i| > 1.$$

If we set $u_i = t_1 t_2 \cdots t_{i-1}$, then $H_i = Hu_i$ for all $i \leq n - 1$. Thus, if $j \geq i + 1$, then t_j commutes with u_i. So,

$$H_i t_j = Hu_i t_j = Ht_j u_i = Hu_i = H_i$$

as $t_j \in H$. If $j < i - 1$, then $u_i = u_j t_j t_{j+1} v$ and v commutes with t_j. Hence, we have

$$H_i t_j = Hu_j t_j t_{j+1} t_j v = Hu_j(t_{j+1} t_j t_{j+1})v$$
$$= Ht_{j+1} u_j t_j t_{j+1} v = H_i.$$

Since $G = \langle t_1, \ldots, t_{n-1} \rangle$, we have $|G : H| \leq n$. $\square$

The alternating group A_n is generated by the totality of 3-cycles. It is easy to see that the 3-cycles $(1\ 2\ a)$ for $a = 3, \ldots, n$ generate A_n. But, in order to find the defining relations for A_n, a more convenient set of generators is $\{s_i\}$ where

$$s_1 = (1\ 2\ 3), \qquad s_2 = (1\ 2)(3\ 4), \ldots, s_{n-2} = (1\ 2)(n-1\ n).$$

The following relations hold among these elements:

$$s_1^3 = s_i^2 = (s_{i-1} s_i)^3 = (s_j s_k)^2 = 1 \qquad (i > 1, |j - k| > 1).$$

(2.15) *The above relations are the defining relations for A_n.*

Proof. Use induction on n. If n is small, the proposition is trivial. We may assume that $H = \langle s_1, \ldots, s_{n-3} \rangle \cong A_{n-1}$. Define the subsets H_i starting from $H_n = H$ by

$$H_i = H_{i+1} s_{i-1} \quad (i > 1) \quad \text{and} \quad H_1 = H_2 s_1.$$

Then, by a method similar to the proof of (2.14), we can prove that the index of H is at most n. We need to verify that $H_2 s_i = H_1$, $H_i s_j = H_i$ if $i > 2$, and $j \neq i - 1, i - 2$. This proves that $\langle H, s_{n-2} \rangle = A_n$. $\square$

According to a general theorem (see Exercise 4, Chapter 2, §6), if a presentation of a group G is given, then we can find a presentation of any of its subgroups. The presentation of A_n given in (2.15) is obtained from the presentation (2.14) of Σ_n following the general procedure.

The Automorphism Group

A product of two 3-cycles has one of the following four types:

$$(a\ b\ c)(a\ b\ d) = (a\ d)(b\ c), \qquad (a\ b\ c)(a\ d\ b) = (b\ c\ d),$$

$$(a\ b\ c)(a\ d\ e) = (a\ b\ c\ d\ e), \qquad (a\ b\ c)(d\ e\ f).$$

Here, different letters represent distinct numerals. Thus, if a product of 3-cycles is of order 2, then we must have the first case. This remark is crucial in the following proof.

(2.16) *If an automorphism θ of the alternating group A_n sends 3-cycles into 3-cycles, then there is an element ρ of Σ_n such that $\theta(\sigma) = \rho^{-1}\sigma\rho$ for all $\sigma \in A_n$.*

Proof. Set $u_{i-2} = (1\ 2\ i)$. By assumption, the elements $v_i = \theta(u_i)$ are 3-cycles. If $i \neq j$, the product $u_i u_j$ is of order 2. So, the order of $v_i v_j$ is 2. By the remark preceding (2.16), we have $v_1 = (a\ b\ c)$ and $v_2 = (a\ b\ d)$. Consider v_3. If v_3 fixes a, then by comparing it with v_1 and v_2, we get $v_3 = (b\ c\ \cdot) = (b\ d\ \cdot)$. This is impossible. So, the cycle of v_3 contains a, and we have $v_3 = (a\ b\ e)$. Thus, in general, there are $a_1, a_2, \ldots, a_n$ such that

$$\theta(1\ 2\ i) = (a_1\ a_2\ a_i).$$

Since $a_i \neq a_j$ for $i \neq j$, we can define a permutation $\rho \in \Sigma_n$ by $\rho(i) = a_i$. By (2.3), we get

$$\rho^{-1}(1\ 2\ i)\rho = (a_1\ a_2\ a_i).$$

Since A_n is generated by the 3-cycles u_i, we conclude that $\theta(\sigma) = \rho^{-1}\sigma\rho$ for any $\sigma \in A_n$. $\square$

(2.17) *If $n \geq 3$ and $n \neq 6$, then any automorphism of A_n is obtained by a conjugation of an element of Σ_n. We have*

$$\mathrm{Aut}\ A_n = \Sigma_n.$$

Proof. Let θ be an automorphism of A_n. Consider the image of a 3-cycle σ. If $n \leq 5$, then any element of order 3 is a 3-cycle. So, $\theta(\sigma)$ is a 3-cycle. The proposition (2.17) follows from (2.16). Assume that $n \geq 6$. The centralizer of σ

has order $3(n-3)!/2$. The image $\theta(\sigma)$ has order 3. So, the cycle decomposition of $\theta(\sigma)$ is a product of r disjoint 3-cycles for some r. By the Corollary of (2.13), the centralizer of $\theta(\sigma)$ has order

$$3^r r! \, (n-3r)!/2.$$

Since σ and $\theta(\sigma)$ have isomorphic centralizers, we get

$$3^r r! \, (n-3r)! = 3(n-3)!$$

which implies that a product of $3(r-1)$ consecutive integers is equal to $3^{r-1}r!$. If $r > 1$, this happens only when $r = 2$ and $n = 6$. Thus, if $n > 6$, $\theta(\sigma)$ is a 3-cycle, and (2.17) follows from (2.16).   □

(2.18) *If $n \geq 2$ and $n \neq 6$, any automorphism of the symmetric group Σ_n is inner:* $\text{Aut } \Sigma_n = \Sigma_n$.

Proof. Set $H = \text{Aut } \Sigma_n$. We can consider Σ_n as a subgroup of H by identifying $\text{Inn } \Sigma_n$ with Σ_n. By (6.7) of Chapter 1 and the subsequent remark, Σ_n is a normal subgroup of H and the conjugation by an element θ of H induces the automorphism θ in Σ_n.

By (2.9), A_n is a characteristic subgroup of Σ_n. Hence, by (6.14) of Chapter 1, A_n is a normal subgroup of H. Set $C = C_H(A_n)$. Then, $C \triangleleft H$ and $\Sigma_n \cap C = \{1\}$. (The last equality follows immediately, either from computation or from (2.10).) By (3.16) of Chapter 1, C centralizes Σ_n. Then, $C = \{1\}$ by the remark in the first paragraph of the proof.

If $\sigma \in H$, the conjugation by σ induces an automorphism of A_n. By (2.16), there is an element ρ of Σ_n such that $\sigma^{-1}\tau\sigma = \rho^{-1}\tau\rho$ for any element τ of A_n. Hence, $\sigma\rho^{-1} \in C$. But, $C = \{1\}$, so $\sigma = \rho \in \Sigma_n$. Thus, we get $\text{Aut } \Sigma_n = \Sigma_n$.

□

The proposition (2.18) can be proved directly by showing that the image of a transposition is also a transposition. The above proof can be generalized to the automorphism group of an arbitrary simple group. See Exercises 2 and 3 of Chapter 1, §6.

The case in which $n = 6$ (the case which was excluded in (2.17) and (2.18)) is indeed special since we have $\text{Aut } A_6 \neq \Sigma_6$. The influence of this exceptional behavior is very deep, and this is one of the reasons that the theory of finite simple groups is so complicated.

We will construct an exceptional automorphism of A_6. We will begin with the following general theorem.

(2.19) *Let G_1 be the stabilizer of 1 in the symmetric group Σ_n on $X = \{1, 2, \ldots, n\}$, and H be a subgroup of Σ_n with $|\Sigma_n : H| = n$. The following propositions hold.*

(i)   Some automorphism θ of Σ_n maps H onto $G_1 : G_1 = \theta(H)$.
(ii) If $n \neq 6$, H is conjugate to G_1 in Σ_n, and H is the stabilizer of some element of X.
(iii) If $n = 6$, there are subgroups of index n which are not conjugate to G_1.
      We have

$$\text{Aut } \Sigma_6 = \text{Aut } A_6 \quad \text{and} \quad |\text{Aut } \Sigma_6 : \text{Inn } \Sigma_6| = 2.$$

Proof. Let $G = \Sigma_n$, and consider G as a permutation group on the cosets of H. By (7.16) of Chapter 1, there is a homomorphism θ from G into Σ_n such that Ker $\theta \subset H$. By Theorem 2.10, we have Ker $\theta = \{1\}$. Thus, θ is an automorphism of Σ_n. Clearly, θ maps H to the stabilizer of k. Even if $k \neq 1$, we can find a conjugation by a suitable element of Σ_n such that there is an automorphism θ which satisfies $\theta(H) = G_1$. This proves (i), and (ii) follows from (2.18).

In order to prove (iii), we will consider the symmetric group Σ_5 of 5 letters. Its order is 120. By Sylow's Theorem, or by just counting the number of 5-cycles, we can see that Σ_5 contains exactly six S_5-subgroups. Hence, the group Σ_5 acts on the set $\mathfrak{S}$ of S_5-subgroups (as in Example 3 of Chapter 1, §7). This gives a homomorphism φ from Σ_5 into Σ_6. As before, φ is an injection and the image $H = \varphi(\Sigma_5)$ is a subgroup of index 6. However, Σ_5 acts transitively on $\mathfrak{S}$. Thus, H is not a stabilizer of any letter. This shows that H is not conjugate to G_1 in Σ_6. This proves that Aut $\Sigma_6 \neq$ Inn Σ_6.

If an automorphism θ of A_6 maps a 3-cycle σ into a 3-cycle σ', then θ maps any conjugate element of σ into a conjugate of σ'. Hence, θ maps 3-cycles into 3-cycles. In this case, by (2.16), θ is induced by an element of Inn Σ_6. Since A_6 contains two conjugacy classes C_1 and C_2 of elements of order 3, any automorphism of A_6 either fixes both C_1 and C_2 or exchanges them. Hence, we have $|\text{Aut } A_6 : \text{Inn } \Sigma_6| \leq 2$. As we saw in the proof of (2.18), Aut Σ_6 is contained in Aut A_6. Therefore, we have

$$\text{Aut } \Sigma_6 = \text{Aut } A_6, \quad |\text{Aut } \Sigma_6 : \text{Inn } \Sigma_6| = 2. \quad \square$$

(2.20) *Any exceptional automorphism of A_6 exchanges the two conjugacy classes of elements of order 3. Any exceptional automorphism of Σ_6 maps transpositions to permutations of type $(2, 2, 2)$.*

Proof. Any exceptional automorphism of A_6 sends a 3-cycle into a permutation of type $(3, 3)$. So, any exceptional automorphism of Σ_6 maps a transposition to a permutation of type $(2, 2, 2)$. $\square$

The Schur Multiplier

(2.21) *The Schur multiplier $M(\Sigma_n)$ of the symmetric group is a cyclic group of order 2 if $n \geq 4$ and of order 1 for $n \leq 3$. If $n \geq 4$ but $n \neq 6$, there are two* *isomorphism classes of the representation groups. They are distinguished by the property that the elements corresponding to transpositions have order 2 or 4. If $n = 6$, there is a unique isomorphism class of the representation groups.*

Proof. From the general theory discussed in Chapter 2, §9, we can compute the Schur multiplier once a presentation of Σ_n is given. We will use the presentation (2.14). Then, the center of a free central extension of Σ_n is generated by

$$u_i = t_i^2, \qquad v_i = (t_i t_{i+1})^3, \qquad w_{ij} = [t_i, t_j]$$

for $|i - j| > 1$ and $i, j = 1, 2, \ldots, n - 1$. If $n \geq 4$, Σ_n contains an element which transforms (t_1, t_3) into (t_i, t_j). Then, in the free central extension, the corresponding element transforms $t_1 \rightarrow t_i a$ and $t_3 \rightarrow t_j b$ where a and b are in the center. Hence, the same element transforms w_{13} into w_{ij}. Since these elements w_{ij} are central, we get $w_{13} = w_{ij}$. We may set $w = w_{ij}$. If $n \leq 3$, there is no element corresponding to w. We will study possible relations among those generators u_i, v_i, and w.

Since w is central, we get $w^2 = [t_i, t_j]^2 = [t_i^2, t_j] = 1$. Secondly,

$$t_i t_{i+1} t_i = v_i t_{i+1}^{-1} t_i^{-1} t_{i+1}^{-1}$$

implies that $u_i u_{i+1} u_i = v_i^2 u_{i+1}^{-1} u_i^{-1} u_{i+1}^{-1}$ or $u_i^3 u_{i+1}^3 = v_i^2$. Set $z_1 = u_1$ and $z_{i+1} = (u_i u_{i+1})^{-1} v_i$. Then, we get

$$z_{i+1}^2 = u_i u_{i+1} \quad \text{and} \quad z_{i+1}^3 = v_i.$$

Thus, $\langle z_1, z_2, \ldots, z_{n-1}, w \rangle = \langle u_i, v_j, w \rangle$. By the general theory, the center is an abelian group of rank $n - 1$. Hence, $z_1, z_2, \ldots, z_{n-1}$ must be free, and the torsion subgroup of the center is generated by w. This shows that $|M(\Sigma_n)| \leq 2$, and that $|M(\Sigma_n)| = 1$ if $n \leq 3$.

In order to show that $M(\Sigma_n) = 2$ when $n \geq 4$, we need to construct a representation group of Σ_n. Following Schur [3], we will construct the 'Hauptdarstellung'. Consider the 2×2 matrices

$$A = \begin{pmatrix} 0 & 1 \\ 1 & 0 \end{pmatrix} \quad B = \begin{pmatrix} 0 & 1 \\ -1 & 0 \end{pmatrix} \quad C = \begin{pmatrix} 1 & 0 \\ 0 & -1 \end{pmatrix} \quad I = \begin{pmatrix} 1 & 0 \\ 0 & 1 \end{pmatrix}$$

and construct tensor products (of m factors)

$$M_1 = A \otimes \cdots \otimes A, \qquad M_2 = A \otimes \cdots \otimes A \otimes B, \qquad M_3 = A \otimes \cdots \otimes A \otimes C$$

$$\cdots,$$

$$M_{2r} = A \otimes \cdots \otimes A \otimes B \otimes I \otimes \cdots \otimes I$$

where A appears $m - r$ times. Similarly, we define M_{2r+1} by replacing the middle B by C. These matrices satisfy

$$-M_{2r}^2 = M_{2r+1}^2 = E, \qquad M_k M_l = -M_l M_k \quad (l \neq k)$$

where E is the identity matrix. (Note that $(X, Y) \to X \otimes Y$ is a homomorphism from the direct product to the tensor product.)

Set $n = 2m + 1$ or $n = 2m + 2$. Define $n - 1$ matrices

$$T_\lambda = a_{\lambda-1} M_{\lambda-1} + b_\lambda M_\lambda,$$

for $\lambda = 1, 2, \ldots, n - 1$, which satisfy

$$T_\lambda^2 = -E, \qquad T_\lambda T_{\lambda+1} + T_{\lambda+1} T_\lambda = E, \qquad T_\lambda T_\mu + T_\mu T_\lambda = 0$$

for $|\lambda - \mu| > 1$. The coefficients a_λ, b_λ are complex numbers which will be determined next. Suppose, for example, that $a_0 = 0$ (and $M_0 = E$). The remaining a_λ and b_λ are successively determined by the equations

$$a_{\lambda-1}^2 - b_\lambda^2 = (-1)^{\lambda-1} = 2a_\lambda b_\lambda.$$

It is easy to check that

$$a_i^2 = (-1)^i i/2(i+1), \qquad b_i^2 = (-1)^i (i+1)/2i$$

and that, with a suitable choice of square roots, a_i and b_i satisfy the required equations. Then, the matrices $T_1, \ldots, T_{n-1}$ satisfy

$$T_\lambda^2 = -E, \qquad (T_\lambda T_{\lambda+1})^3 = -E, \qquad (T_\lambda T_\mu)^2 = -E$$

for $|\lambda - \mu| > 1$. Let T be the group generated by $T_1, \ldots, T_{n-1}$. Then, $-E$ is an element of the center of T, and the group $T/\langle -E \rangle$ is a homomorphic image of the symmetric group Σ_n by (2.14). Since $-E = [T_\lambda, T_\mu]$, $-E$ is an element of the commutator subgroup of T. So, by (2.10), $T/\langle -E \rangle$ is isomorphic to Σ_n. From the discussions of Chapter 2, §9, particularly from (9.15), we can see that T is a representation group of Σ_n. Thus, $M(\Sigma_n)$ $(n \geq 4)$ is a cyclic group of order 2.

By the general theory, any representation group is a homomorphic image of the free central extension we constructed in the first half of this proof. The group T is indeed isomorphic to the factor group by the subgroup

$$\langle z_1 w, z_2 w, \ldots, z_{n-1} w \rangle.$$

In this group T, all the transpositions correspond to elements of order 4. Any representation group corresponds to a subgroup of index 2 in the center of the free extension. Thus, the isomorphism class of a representation group is determined by a homomorphism

$$\langle u_i, v_j, w \rangle \rightarrow \langle w \rangle$$

such that $w \rightarrow w$. Consider the element t_1 to be fixed. Then, changing t_i to $t_i w$ if necessary, we may assume that $v_j \rightarrow w$ for all j. As $z_{i+1}^3 = v_i$, z_i must go to w for $i > 1$. Thus, any representation group is isomorphic to either the group T constructed earlier or to the factor group T^* by the subgroup

$$\langle z_1, z_2 w, z_3 w, \ldots, z_{n-1} w \rangle.$$

In the group T^*, we have $t_1^2 = u_1 = z_1 \rightarrow 1$. Thus, transpositions are represented by elements of order 2. If $n \neq 6$, then any automorphism of Σ_n carries the class of transpositions into itself. Hence, T and T^* are not isomorphic. But, if $n = 6$, an exceptional automorphism of Σ_6 exchanges the class of transpositions and the class of permutations of type $(2, 2, 2)$. In the representation group T, we have $(T_1 T_3 T_5)^2 = E$. So, an element of type $(2, 2, 2)$ corresponds to an element of order 2. This shows that T is isomorphic to T^*. $\square$

In any case, the commutator subgroup of a representation group gives an irreducible central extension of the alternating group because the permutations of type $(2, 2)$ correspond to elements of order 4. So, $M(A_n)$ $(n \geq 4)$ is of order at least 2. In fact, we have the following theorem of Schur.

(2.22) *The Schur multiplier of A_n $(n \geq 4)$ is of order 2 if $n \neq 6$ or 7.*

Proof. Use the presentation (2.15) of A_n to construct a free central extension. Then, its center is generated by the following elements:

$$u_1 = s_1^3, \ v_1 = (s_1 s_2)^3, \ w_k = (s_1 s_k)^2 \quad (k > 2)$$

$$u_i = s_i^2, \ v_i = (s_i s_{i+1})^3, \ w_{ij} = [s_i, s_j]$$

where $i > 1$ and $|i - j| > 1$. As in the case of Σ_n, w_{ij} is independent of i and j. Set $w = w_{ij}$. We have

$$w^2 = 1 \quad \text{and} \quad v_i^2 = u_i^3 u_{i+1}^3 \quad (i > 1).$$

The definitions of w_k and u_k give us $u_k^{-1} s_1^{-1} w_k = s_k^{-1} s_1 s_k$. So, we get

$$u_1^2 u_k^3 = w_k^3 \quad (k > 2).$$

The definitions of v_1 and u_2 give us $s_1 s_2 s_1 = v_1 u_2^{-1} s_2 s_1^{-1} s_2^{-1}$. Hence, we get

$$(s_1 s_2 s_1)^3 = v_1^3 u_2^{-3} u_1^{-1}.$$

On the other hand, we have $s_2 s_1^2 = u_2 u_1 (s_1 s_2)^{-1}$. So,

$$(s_1 s_2 s_1)^3 = s_1 (s_2 s_1^2)^3 s_1^{-1} = u_2^3 u_1^3 v_1^{-1}.$$

Thus, we get $u_1^4 u_2^6 = v_1^4$. But, if $n \geq 6$, $w = v_1^2 u_1^{-2} u_2^{-3}$ holds. To prove this, we compute

$$s_4 s_1 s_2 s_4^{-1} = w_4 s_1^{-1} s_4^{-1} s_2 s_4 u_4^{-1}$$

$$= w_4 s_1^{-1} s_2 w u_4^{-1}$$

$$= s_2 (s_1 s_2)^{-1} s_2^{-1} u_2 w_4 w u_4^{-1}.$$

Then, the cube of both sides gives us

$$v_1^2 = u_2^3 w_4^3 w u_4^{-3} = u_2^3 u_1^2 w.$$

If $n \leq 5$, the element w has not been defined. So we define $w = v_1^2 u_1^{-2} u_2^{-3}$. Then, in all cases, we have $w^2 = 1$.

If $n = 4$, the center of the free central extension is generated by u_1, v_1, and u_2. The only relation among the generators is $w^2 = 1$. If $z_1 = u_1$ and $z_2 = (u_1^{-1} v_1) w u_2^{-1}$, then we have $z_2^2 = w u_2$ and $z_2^3 = u_1^{-1} v_1$. Hence, the center is generated by z_1, z_2, and w. As before, we conclude that $M(A_4) = \langle w \rangle$.

If $n = 5$, the center is generated by u_1, v_1, w_3, u_2, v_2 and u_3. Among these generators, we have the relations $w^2 = 1$, $v_2^3 = u_2^3 u_3^3$ and $u_1^2 u_3^3 = w_3^3$. Set

$$z_1 = u_1 (u_3^{-1} w_3)^{-1}, \qquad z_2 = v_2 (u_2 u_3)^{-1}, \qquad z_3 = (v_1 u_1^{-1})(w u_2^{-1}).$$

Then, the center is generated by z_1, z_2, z_3, and w. This proves that $M(A_5) = \langle w \rangle$. (See Example 2 of Chapter 2, §9 and (2.8). We have used a lemma: if x and y commute and satisfy $x^3 = y^2$, then $\langle x, y \rangle$ is a cyclic group generated by $u = x^{-1} y$. This is clear since $u^2 = x$ and $u^3 = y$.)

If $n \geq 7$, the generators satisfy even more relations. If $k \geq 5$, we have

$$s_3 s_1 s_k s_3^{-1} = w_3 s_1^{-1} s_3^{-1} s_k s_3 u_3^{-1} = w_3 s_1^{-1} s_k w u_3^{-1}.$$

So, using $s_1^{-1} s_k = u_k s_k^{-1} (s_k^{-1} s_1^{-1}) s_k$, we get

$$w_k = w_3^2 w_k^{-1} u_k^2 u_3^{-2}, \quad \text{or} \quad (w_k u_k^{-1})^2 = (w_3 u_3^{-1})^2.$$

On the other hand, $(w_k u_k^{-1})^3 = u_1^2 = (w_3 u_3^{-1})^3$. So, we have

$$w_3 u_3^{-1} = w_k u_k^{-1} \quad \text{if} \quad k \geq 5.$$

If $n \geq 8$, then a similar argument shows that $w_4 u_4^{-1} = w_6 u_6^{-1}$. Hence, for all $k \geq 3$, $w_k u_k^{-1} = w_3 u_3^{-1}$ holds. Set

$$z_1 = u_1 (w_3 u_3^{-1})^{-1}, \qquad z_2 = (v_1 u_1^{-1})(w u_2)^{-1}, \qquad z_{i+1} = v_i (u_i u_{i+1})^{-1}$$

($i = 2, 3, \ldots, n - 3$). Then, the elements $z_1, \ldots, z_{n-2}$, and w generate the center of the free central extension. This proves that $M(A_n) = \langle w \rangle$ as before.

$\square$

Exceptional Cases. When $n = 6$, the relations are

$$w^2 = 1, \qquad v_2^2 = (u_2 u_3)^3, \qquad v_3^2 = (u_3 u_4)^3$$

$$w = v_1^2 u_1^{-2} u_2^{-3}, \qquad u_1^2 = (w_3 u_3^{-1})^3 = (w_4 u_4^{-1})^3.$$

Set $k = w u_3 w_4 u_4^{-1} w_3^{-1}$. Then, we get $k^3 = w$, and the center of the free central extension is generated by

$$z_1 = u_1(u_3 w_3^{-1}), \qquad z_2 = (v_1 u_1^{-1})(w u_2)^{-1},$$

$$z_3 = v_2(u_2 u_3)^{-1}, \qquad z_4 = v_3(u_3 u_4)^{-1},$$

and k. Therefore, $M(A_6)$ is a cyclic group generated by k.

When $n = 7$, we have one more relation: $w_3 u_3^{-1} = w_5 u_5^{-1}$. Then, if k is defined as before, then $M(A_7) = \langle k \rangle$ where $k^6 = 1$.

In both cases, the Schur multiplier is indeed of order 6. But, we will omit the actual construction of a representation group and refer the readers to the original paper of Schur [3].

Exercises

1. Let Σ_n be the symmetric group on the set $X = \{1, 2, \ldots, n\}$.

(a) Suppose that a product σ of m transpositions is given, and let $a \in X$. Show that there is an integer $k \geq 0$ such that the permutation σ can be written as a product of $m - 2k$ transpositions where the particular number a appears at most once in the new expression.

(b) Let (2.1) be the cycle decomposition of a permutation σ, and set $N(\sigma) = n - s$, where s is the number of cycles in (2.1). Show that, for any transposition τ, we have $N(\sigma\tau) = N(\sigma) \pm 1$.

(*Hint.* (a) If ρ is a permutation which fixes a, then

$$(a\ b)\rho(a\ c) = \rho(a\ b^\rho)(a\ c) = \rho \quad \text{or} \quad \rho(a\ c)(b^\rho\ c).$$

Thus, we can reduce the number of times a appears. (This proposition is taken from Ito [3].)

(b) We have

$$(a \cdots b\ c \cdots d)(a\ c) = (a \cdots b)(c \cdots d).$$

We remark that either proposition can be used to show that a product of an odd number of transpositions is never equal to the identity permutation. The invariance of parity in (2.6) will follow from these propositions.)

2. (a) If two even permutations σ and ρ are of the same type, and if there is an odd permutation which commutes with σ, then σ and ρ are conjugate in the alternating group A_n: $\sigma = \tau^{-1}\rho\tau$ for some even permutation τ.

(b) Let $\sigma \in A_n$, and let $(r_1, \ldots, r_s)$ be its type (all the terms $r_i = 1$ are to be included). Show that the centralizer $C(\sigma)$ in Σ_n is contained in A_n if and only if $\{r_1, \ldots, r_s\}$ consists of distinct, odd integers. Also prove that in this case, the permutations of type $(r_1, \ldots, r_s)$ are divided into two conjugacy classes of A_n.

3. (a) Let N be a normal subgroup of a group G, let $\sigma \in N$, and let C be the conjugacy class of G which contains the element σ. Show that the set C is divided into a disjoint union of conjugacy classes $C_1, \ldots, C_n$ of N, where $n = |G : C_G(\sigma)N|$. If $\{t_i\}$ is a right transversal of the subgroup $C_G(\sigma)N$, then we can make $C_i = t_i^{-1}Ct_i$.

(b) Let O be an orbit of a G-set X. For a normal subgroup N of G, we regard O as an N-set. Show that O can be decomposed into a union of N-orbits $O_1, \ldots, O_n$ where $n = |G : S_G(x)N|$ with $x \in O$. Show that $O_i = t_i(O_1)$ for some transversal $\{t_i\}$ of $S_G(x)N$.

4. The simplicity of the alternating group A_n ($n \neq 4$) can be proved in various ways. Complete the following sketch of two proofs.

(a) By (2.8) (iv), A_5 is simple. Proceed by induction on n. Suppose that $n \geq 6$, and let $N \neq \{1\}$ be a normal subgroup of A_n. Let $\sigma = (ab \cdots c) \cdots$ be an element $\neq 1$ of N. Here, $(ab \cdots c)$ is the first cycle in the cycle decomposition, and the letter c is the last one in the cycle. If the length is 2, we have $b = c$. In any case, there is a letter d which is different from the three letters a, b, and c. For the 3-cycle $\rho = (a\ b\ d)$, the commutator $\tau = [\sigma, \rho]$ is a nonidentity element of N, and it is a product of two 3-cycles with at least one common letter. Hence, τ moves at most 5 letters, and so there is a letter e which is fixed by τ. Thus, $S(e) \cap N \neq \{1\}$ for the stabilizer $S(e)$ of e. But, $S(e) \cong A_{n-1}$. So, by the inductive hypothesis, we get $S(e) \cap N = S(e)$. Thus, N contains the stabilizer of any letter. In particular, N contains all the 3-cycles, and $N = A_n$.

(b) We can prove by computation that N contains a 3-cycle. The commutator τ which appeared in (a) is a 3-cycle, a 5-cycle or of type $(2, 2)$. We have the following formulas:

$$[(a\ b\ c\ d\ e), (a\ b\ c)] = (a\ b\ d),$$

$$[(a\ b)(c\ d), (a\ e\ b)] = (a\ b\ e).$$

This shows that $N \neq \{1\}$ contains a 3-cycle.

(The method (b) does not use induction, but we need to know that all 3-cycles are conjugate in A_n. In (a), this conjugacy is not needed.)

5. (a) Let X be a subset of Σ_n such that the elements of X are transpositions. Show that if X generates Σ_n, then we have $|X| \geq n - 1$.

(b) Prove that the groups Σ_n and A_n can be generated by two elements.

(c) Verify that the n-cycle $(1\ 2\ \cdots\ n)$ and a transposition $(1\ k+1)$ generate Σ_n if and only if k is relatively prime to n.

6. Prove the following propositions.

   (a) The mapping of A_6 which maps $(1\ 2\ 3)$, $(1\ 2)(3\ 4)$, $(1\ 2)(4\ 5)$, $(1\ 2)(5\ 6)$
        onto

$$(a\ b\ c)(d\ e\ f),\ (a\ d)(b\ e),\ (a\ c)(e\ f),\ (a\ b)(d\ e),$$

   respectively, can be extended to an exceptional automorphism of A_6.
   (b) Aut A_6 is a non-split extension of A_6.
   (c) Aut A_6 contains an element of order 10.
   (d) Aut A_6/A_6 is an elementary abelian group of type $(2, 2)$.

   (*Hint.* (a) It suffices to show that the images satisfy the defining relations
   (2.15). (b) Any exceptional automorphism σ of A_6 induces an exceptional
   automorphism of Σ_6. Thus, σ does not leave any odd permutation of order 2
   invariant. (c) If we consider the normalizer of an S_5-subgroup, we see that the
   centralizer is strictly larger than the S_5-subgroup. (d) An element of order 10
   and A_6 generate a subgroup of order $|\Sigma_6|$, but the subgroup is certainly not
   isomorphic to Σ_6.
   In (a), let $\{a, b, c, d, e, f\} = \{1, 2, 4, 6, 5, 3\}$ in this order, and let ρ be the
   exceptional automorphism obtained in (a). Then, ρ^2 is the conjugation by the
   element $(1\ 6\ 2\ 5)(3\ 4)$. If σ is the inner automorphism by $(3\ 4)(5\ 6)$, then we
   can show that

$$(\rho\sigma)^2 = \rho^4, \quad \text{or} \quad \sigma^{-1}\rho\sigma = \rho^3.$$

   Hence, the subgroup $H = \langle A_6, \rho \rangle$ contains no element of order 2 outside of
   A_6. If we denote by τ the inner automorphism induced by $(3\ 4)$, then the
   relation $\tau^{-1}\rho\tau = \rho^{-3}$ holds. So, $\langle \rho, \sigma, \tau \rangle$ is an S_2-subgroup of Aut A_6. Note
   that the square of any exceptional automorphism is in Inn A_6. The reader
   should verify that the description of the exceptional automorphism given in
   (a) follows naturally from the defining relations (2.15).)

7. Let H be a proper subgroup of the symmetric group Σ_n. If $H \neq A_n$ and
$n \neq 4$, show that $|\Sigma_n : H| \geq n$. Also show that if $n \neq 4$, the index of any
proper subgroup of A_n is at least n.
   (*Hint.* Use (7.16) of Chapter 1 and (2.10) of this chapter. If $n = 4$, there is a
subgroup of index three.)

8. Let G_1 be the stabilizer of 1 in the alternating group A_n, and let H be a
subgroup of index n. Show that there is an automorphism θ of A_n such that
$\theta(H) = G_1$. Furthermore, prove that if $n \neq 6$, H is the stabilizer of some letter.

9. (i) Show that a simple group of order 60 is isomorphic to A_5.
   (ii) Show that there is no simple group of order 120.

(*Hint.* Let G be a simple group of order 60 or 120. Then, by Sylow's theorem, the index of the normalizer of an S_5-subgroup must be 6. Thus, there is an isomorphism from G into Σ_6. The conclusion follows from Exercises 7 and 8.)

10. Prove the following statements.

    (i)   Let n be an odd integer > 1. Then, there is no simple group of order $2n$.

   (ii)   If n is a power of an odd prime, there is no simple group of order $4n$.

 (iii)   Show that A_5 is the only noncyclic simple group of order at most 120.

 (iv)   Let G be a nonsolvable group of order ≤ 120. Then, $|G| = 60$ or 120. Furthermore, there are exactly three isomorphism classes of the nonsolvable groups of order 120.

(*Hint.* (i) Let G be a simple group of order $2n$. The regular representation ρ of (7.4), Chapter 1, has the property that some element of $\rho(G)$ is a product of n transpositions. Thus, $\rho(G)$ contains some odd permutations. Hence, G is not simple since the set of even permutations in $\rho(G)$ forms a normal subgroup of index 2.

(ii) A Sylow subgroup of order n has index 4.

(iii) We can eliminate many cases by applying Theorem 1.9 of Chapter 2, §1, Example 2 of Chapter 2, §2 and parts (i) and (ii) of this exercise. The remaining possibilities are $105 = 3 \cdot 5 \cdot 7$ (if odd), 60 and $84 = 4 \cdot 3 \cdot 7$ (if even but not divisible by 8), and $72 = 8 \cdot 9$ and 120 (if divisible by 8). Consider each case separately.

$g = 72$:   By Sylow's theorem, the normalizer of an S_3-subgroup has index 1 or 4.

$g = 84$:   By Sylow's theorem, an S_7-subgroup is normal.

$g = 105$:   Again by Sylow's theorem, a simple group of order 105 contains exactly 15 S_7-subgroups and 21 S_5-subgroups. Therefore, there would be exactly $15(7 - 1)$ elements of order 7 and $21(5 - 1)$ elements of order 5. But, this would give us too many elements since $15(7 - 1) + 21(5 - 1) > 105$.

$g = 60$ or 120: See Exercise 9.

(iv) Consider all possible composition series. Apply Schur's theory of central extension for nonsolvable groups order 120.

The order of the smallest nonablian simple group which is greater than 120 is $168 = |PSL(2, 7)|$. To prove this, we have to eliminate $g = 3 \cdot 5 \cdot 11$, $4 \cdot 5 \cdot 7$, $4 \cdot 3 \cdot 13$, $4 \cdot 3 \cdot 11$, or $16 \cdot 9$. The first three cases can be eliminated by applying Sylow's theorem to the largest prime factor. The last two cases can be handled by the method used when $g = 105$. In general, if an S_p-subgroup S of a group G is abelian and $N_G(S) = S \neq G$, then G is not simple. This is a special case of Burnside's theorem, which we will discuss later, and it can be applied to the last two cases as well as to the case when $g = 105$. See Chapter 5, §2, where we will prove in particular that the group of square-free order is solvable. If we are willing to apply the Feit–Thompson theorem which asserts the solvability of finite groups of odd order, then we need only to consider even numbers as possible candidates for the orders of nonabelian simple groups. There is another famous theorem of Burnside asserting the solvability of groups of orders of the form $p^a q^b$ (p, q are prime, Chapter 5, Theorem 4.25). The application of these results will eliminate many possibilities. The orders of all nonabelian simple groups up to 1,000 are 60, 168, 360, 504, and 660, and in each case, a simple group of this order is isomorphic to the corresponding $PSL(2, q)$. The proof of this result is tedious and should be attempted only after studying the standard methods in finite group theory. At the time of this writing, all simple groups of order less than a million are known (Beisiegel–Stingl [1]), but we have to call upon all the available resources in group theory to prove this assertion.)

§3. Geometry of Linear Groups

In this section, we will continue the study of linear groups which we began in §9 of Chapter 1. As before, let F be an arbitrary but fixed field, and consider a group consisting of linear transformations of a vector space V over F. It is convenient to define the projective space associated with V and to consider the group as a transformation group on the projective space.

Definition 3.1. Let V be an n-dimensional vector space over the field F. Let $\mathscr{P}$ be the set of all the 1-dimensional subspaces of V. A subset α of $\mathscr{P}$ is said to be a **subspace** of $\mathscr{P}$ if there is a subspace U of V such that α coincides with the set of 1-dimensional subspaces of U. In this case, we define dim $\alpha = d$ if the subspace U has dimension $d + 1$. The set $\mathscr{P}$ on which the concept of subspaces is defined as above is called the $(n - 1)$-dimensional **projective space** on V or a projective space defined over F.

It follows from the definition that an element of $\mathscr{P}$ is a 0-dimensional subspace of $\mathscr{P}$. We will use geometric terms: an element of $\mathscr{P}$ is called a *point*, a 1-dimensional subspace a *line*, a 2-dimensional one a *plane*, etc. The empty subset is the (-1)-dimensional subspace corresponding to the subspace $\{0\}$ of V.

Let α and β be two subspaces of the projective space on V. Then, α and β correspond to subspaces U and W of V. Clearly, the point set $\alpha \cap \beta$ is the subspace of $\mathscr{P}$ corresponding to $U \cap W$. If $\{\alpha_i\}$ ($1 \le i \le k$) are subspaces of $\mathscr{P}$, the smallest subspace of $\mathscr{P}$ which contains all α_i is called *the subspace spanned by* $\{\alpha_i\}$, and we will denote it by $\alpha_1 \cup \alpha_2 \cup \cdots \cup \alpha_k$. If each α_i corresponds to a subspace U_i of V, then $\alpha_1 \cup \cdots \cup \alpha_k$ corresponds to the subspace $U_1 + U_2 + \cdots + U_k$ generated by $U_1, \ldots, U_k$. Thus, for any two subspaces α and β, we have

$$(3.2) \qquad \dim \alpha + \dim \beta = \dim(\alpha \cup \beta) + \dim(\alpha \cap \beta).$$

Any 1-dimensional subspace of the vector space V is a set of vectors of the form λu for $u \neq 0$ and $\lambda \in F$. Thus, points of $\mathscr{P}$ are equivalence classes with respect to the equivalence relation

$$u \sim v \Leftrightarrow v = \lambda u \quad \text{for} \quad \lambda \in F - \{0\}$$

defined on the set of nonzero vectors of V. Therefore, the various points P_i of the projective space $\mathscr{P}$ on V may be represented by vectors v_i of the equivalence classes corresponding to the points P_i, and many problems concerning the set of the points P_i may be reformulated as problems about the set of the vectors v_i of V. We can easily study the subset $\{v_i\}$ of V using the additive structure of V. A basic fact about this translation is the following lemma.

(3.3) *Let* $\{P_0, P_1, \ldots, P_d\}$ *be a set of* $d + 1$ *points of* $\mathscr{P}$, *and let* $\alpha = P_0 \cup P_1 \cup \cdots \cup P_d$. *Then, we have* dim $\alpha \leq d$. *Let* u_i *be a nonzero vector of* V *which represents the point* P_i. *Then,* dim $\alpha = d$ *if and only if the set of vectors* u_0, $u_1, \ldots, u_d$ *is linearly independent.*

Proof. By definition, α corresponds to the subspace of V which is generated by u_0, $u_1, \ldots, u_d$. So, the assertions follow immediately from elementary results of linear algebra. $\square$

If dim $\alpha = d$, we say that $\{P_0, P_1, \ldots, P_d\}$ is **independent**. If a set of points is independent, each of its subsets is also independent.

Definition 3.4. Let $\mathscr{P}$ be a d-dimensional projective space. A set Σ of $d + 1$ points $P_0, P_1, \ldots, P_d$ is said to be a **frame** of $\mathscr{P}$ if $\{P_0, P_1, \ldots, P_d\}$ is independent.

Let $\Sigma = \{P_0, P_1, \ldots, P_d\}$ be a frame of $\mathscr{P}$, and let u_i be a nonzero vector of the associated vector space V which represents the point P_i. Then, the set of vectors $u_0, u_1, \ldots, u_d$ is linearly independent. Since dim $V = d + 1$, the set $\{u_i\}$ is a basis of V over F. Let $\Sigma' = \{P'_0, P'_1, \ldots, P'_d\}$ be another frame of $\mathscr{P}$, and let u'_i be a nonzero vector of V representing the point P'_i. Then, $\{u'_i\}$ is another basis of V. So, there is an element σ of $GL(V)$ such that $\sigma(u_i) = u'_i$ for $i = 0, 1, \ldots, d$. Clearly, σ maps a subspace of V to another subspace, and for subspaces U and W such that $U \subset W$, we have $\sigma(U) \subset \sigma(W)$. Thus, σ induces a mapping of $\mathscr{P}$ into itself which sends subspaces into subspaces such that $\alpha \subset \beta$ implies $\alpha^\sigma \subset \beta^\sigma$. Hence, an element of $GL(V)$ induces an automorphism of the projective space $\mathscr{P}$ on V. The earlier discussion proves that $GL(V)$ induces a group of automorphisms of $\mathscr{P}$ which is transitive on the set of frames of $\mathscr{P}$.

In general, a permutation σ of the points of $\mathscr{P}$ is said to be a **collineation** of $\mathscr{P}$ if σ maps subspaces into subspaces and satisfies

$$\alpha \subset \beta \Leftrightarrow \alpha^\sigma \subset \beta^\sigma$$

for any two subspaces α and β. Each element of $GL(V)$ induces a collineation of $\mathscr{P}$. The totality of collineations induced by the elements of $GL(V)$ forms a group which is called the **projective general linear group** on $\mathscr{P}$ and is denoted by $PGL(\mathscr{P})$ or $PGL(d, F)$. Similarly, $PSL(\mathscr{P})$ is defined as the set of collineations induced by the elements of $SL(V)$.

These groups act transitively on $\mathscr{P}$. In fact, they satisfy much stronger conditions of transitivity. To prove this point, we will start with the following definition.

Definition 3.5. A permutation group G which acts on a set X is said to be k-**transitive** if the following condition is satisfied. Let $\{x_1, \ldots, x_k\}$ and $\{x'_1, \ldots, x'_k\}$ be two subsets of k distinct elements of X. Then, there is an element σ of G such that

$$x_i^\sigma = x'_i \qquad (i = 1, 2, \ldots, k).$$

A permutation group G is said to be **multiply transitive** if G is k-transitive for some $k \geq 2$.

It follows from the definition that a k-transitive permutation group is k'-transitive for any $k' \leq k$. In particular, a k-transitive permutation group is always transitive (as $k \geq 1$).

(3.6) *The projective general linear group $PGL(\mathscr{P})$ is transitive on the set of $d + 1$ independent points of $\mathscr{P}$ ($d = \dim \mathscr{P}$). The group $PGL(\mathscr{P})$ is isomorphic to the factor group of $GL(V)$ by its center. As a permutation group on the points of $\mathscr{P}$, $PSL(\mathscr{P})$ is 2-transitive.*

Proof. The first statement has already been proved. Let $\pi(\sigma)$ be the collineation of $\mathscr{P}$ induced by an element $\sigma \in GL(V)$. Then, π is a homomorphism from $GL(V)$ onto $PGL(\mathscr{P})$. Its kernel $\mathrm{Ker}\,\pi$ consists of those elements of $GL(V)$ which leave all the 1-dimensional subspaces of V fixed. By (9.8) of Chapter 1, $\mathrm{Ker}\,\pi$ is the center Z of $GL(V)$.

Let P_1 and P_2 be two different points of $\mathscr{P}$. If u_i is a nonzero vector of V representing the point P_i, then u_1 and u_2 are linearly independent vectors in V. Similarly, let any two different points Q_1 and Q_2 of $\mathscr{P}$ be represented by the linearly independent vectors w_1 and w_2, respectively. Then, there is an element σ of $SL(V)$ such that $u_1^\sigma = w_1$ and $u_2^\sigma = \lambda w_2$ for some $\lambda \in F$. It is clear that $\pi(\sigma)$ sends P_i into Q_i ($i = 1, 2$). Thus, $PSL(\mathscr{P})$ is 2-transitive on the set of points of $\mathscr{P}$. $\square$

For further geometric investigation of $PGL(\mathscr{P})$, we will introduce the concept of the flag space.

Definition 3.7. A nested sequence of nontrivial subspaces α_i of $\mathscr{P}$

$$A: \alpha_1 \supset \alpha_2 \supset \cdots \supset \alpha_k$$

is said to be a **flag** of rank k, if $\alpha_i \neq \alpha_{i+1}$ for every $i = 1, 2, \ldots, k - 1$. Let B be another flag. If A is a refinement of B (that is, if the subspaces appearing in the flag B are all members of the sequence A), we say that the flag B is contained in the flag A and write $B \subset A$. The partially ordered set consisting of all the flags in $\mathscr{P}$ is called the **flag space** and is denoted by $\Delta(\mathscr{P})$.

We include the empty sequence (the case when $k = 0$) in $\Delta(\mathscr{P})$. By definition, each subspace of a flag is nontrivial. So, $\alpha_1 \neq \mathscr{P}$ and $\alpha_k \neq \varnothing$. If dim $\mathscr{P}$ is d, then the rank of any flag does not exceed d.

In general, a partially ordered set X is called a **simplex** of rank d if X is isomorphic to the partially ordered set formed of all the subsets of a set of d elements with respect to the containment relation. With this definition, the following property holds.

(I) *Let $\Delta(\mathscr{P})$ be the flag space of a projective space $\mathscr{P}$ of dimension d. Any flag is contained in a maximal flag. Let A be a maximal flag. Then, the rank of A is equal to $d = $ dim $\mathscr{P}$, and the partially ordered set formed of all the flags contained in A is a simplex of rank d.*

Proof. Since a maximal subspace of a vector space has codimension one, the first property is obvious. If A is the sequence

$$\alpha_1 \supset \alpha_2 \supset \cdots \supset \alpha_d,$$

any flag $B \subset A$ corresponds to a unique subset of $\{\alpha_i\}$. Thus, the second property holds. $\square$

Note that a flag of rank $d - 1$ corresponds to a face of the simplex.

Definition 3.8. Let $\Delta(\mathscr{P})$ be the flag space of a projective space $\mathscr{P}$ of dimension d. Two maximal flags A and B are said to be **adjacent** if $A \neq B$ and if there is a flag of rank $d - 1$ such that $C \subset A$ and $C \subset B$. Let A' and B' be arbitrary flags of $\Delta(\mathscr{P})$. We say that A' and B' are **connected** if there is a finite sequence $\{C_i\}$ $(i = 1, 2, \ldots, m)$ of maximal flags C_i such that $A' \subset C_1$, $B' \subset C_m$, and C_i and C_{i+1} are adjacent for each i.

We will prove later that any two flags of $\Delta(\mathscr{P})$ are connected (Cor. of (IV)). Before doing that, we need to study the relations between frames and flags.

Definition 3.9. Let Σ be a frame of the projective space $\mathscr{P}$ (cf. Definition 3.4), and let $A: \alpha_1 \supset \alpha_2 \supset \cdots \supset \alpha_k$ be a flag of $\mathscr{P}$. If, for each subspace α_i, there is a subset Σ_i of Σ such that α_i is the subspace spanned by the points of Σ_i, then we say that the frame Σ is **adapted** to the flag A, or that the frame Σ **supports** the flag A. For a fixed frame Σ, we denote by Σ_A, or simply by Σ, the set of flags supported by Σ.

(3.10) *Let $A: \alpha_1 \supset \alpha_2 \supset \cdots \supset \alpha_k$ be a flag of $\mathscr{P}$, and let Σ be a frame which supports A. Suppose that the subspace α_i is spanned by the points of the subset Σ_i of Σ. Then, the following propositions hold.*

    (i) $\Sigma_1 \supset \Sigma_2 \supset \cdots \supset \Sigma_k$,
    (ii) $\dim \alpha_i = |\Sigma_i| - 1$,
    (iii) $\alpha_i \cap \Sigma = \Sigma_i$.

Proof. By assumption, α_i is spanned by the points of Σ_i. So, we have $\dim \alpha_i \leq |\Sigma_i| - 1$. By definition, the points of Σ are independent. Hence, we have

$$|\Sigma_i| \leq |\alpha_i \cap \Sigma| \leq \dim \alpha_i + 1.$$

Thus, we have equalities everywhere. This proves (ii) and (iii). The first proposition follows from (iii). ☐

By Lemma (3.10), we can describe the set of flags supported by a given frame Σ. In Σ, we choose a nested sequence of subsets

$$\Sigma_1 \supset \Sigma_2 \supset \cdots \supset \Sigma_k$$

in such a way that $\Sigma_1 \neq \Sigma$, $\Sigma_k \neq \varnothing$, and $\Sigma_i \neq \Sigma_{i+1}$, for $i = 1, 2, \ldots, k - 1$. If α_i is the subspace spanned by the points of Σ_i, then $A: \alpha_1 \supset \alpha_2 \supset \cdots \supset \alpha_k$ is a flag supported by Σ. On the other hand, by (3.10), any flag of Σ_Δ is obtained in this way. In particular, if $|\Sigma_i| = d - i + 1$, the corresponding flag is maximal. Therefore, Σ_Δ satisfies the following property.

(II) *Any flag of Σ_Δ is contained in a maximal flag which is a member of Σ_Δ.*

The following remark is useful in subsequent discussions when we attempt to prove some propositions by induction on the dimension of a projective space. If α is a subspace of a projective space $\mathscr{P}$, then α corresponds to a subspace U of the vector space V associated with $\mathscr{P}$. Any points or subspaces of $\mathscr{P}$ which are contained in α correspond to subspaces of U. Thus, all the points and subspaces which are contained in α form a projective space associated with the vector space U, and any subspace of a projective space is also a projective space. In this way, any flag of α may be considered as a flag of $\mathscr{P}$. If a flag A of $\mathscr{P}$ contains α, then the members of A which are contained in α form a flag of the projective space α.

We will prove next that the set Σ_Δ of flags supported by the frame Σ is connected.

(III) *Let A and B be two flags of Σ_Δ. Then, there are maximal flags C_0,*

$C_1, \ldots, C_m$ such that $C_i \in \Sigma_\Delta$, $A \subset C_0$, $B \subset C_m$, and C_{i-1} is adjacent to C_i for all i. In other words, the set Σ_Δ is connected.

Proof. We prove this proposition by induction on the dimension. Set $d = \dim \mathscr{P}$. Clearly, we may assume that both A and B are maximal. We write

$$A: \alpha_1 \supset \alpha_2 \supset \cdots \supset \alpha_d \quad \text{and} \quad B: \beta_1 \supset \cdots \supset \beta_d.$$

Assume first that $\alpha_1 = \beta_1$. Then, the set of subspaces contained in α_1 forms a projective space $\mathscr{P}_1$ of dimension $d - 1$. Clearly, the set $\Sigma_1 = \Sigma \cap \alpha_1$ is a frame of $\mathscr{P}_1$ which supports two flags

$$A_1: \alpha_2 \supset \cdots \supset \alpha_d, \qquad B_1: \beta_2 \supset \cdots \supset \beta_d.$$

Then, by the inductive hypothesis, A_1 and B_1 are connected in $(\Sigma_1)_\Delta$. From any maximal flag of $\mathscr{P}_1$, we get a maximal flag of $\mathscr{P}$ by adding α_1. Thus, we see easily that A and B are connected in $\mathscr{P}$.

Assume that $\alpha_1 \neq \beta_1$. In this case, Σ contains exactly one point not in α_1, and this point is contained in β_1. Similarly, there is exactly one point of Σ not in β_1, and this point is in α_1. Thus,

$$A_2: \alpha_1 \supset \alpha_1 \cap \beta_1 \quad \text{and} \quad B_2: \beta_1 \supset \alpha_1 \cap \beta_1$$

are elements of Σ_Δ. By (II), there is a maximal flag C such that $C \in \Sigma_\Delta$ and $A_2 \subset C$. If

$$C: \alpha_1 \supset \alpha_1 \cap \beta_1 = \gamma_1 \supset \gamma_2 \supset \cdots \supset \gamma_{d-1}$$

then the flag D defined by $\beta_1 \supset \gamma_1 \supset \gamma_2 \supset \cdots \supset \gamma_{d-1}$ is a maximal flag which is supported by Σ. By definition, C and D are adjacent and are certainly connected. The first part of this proof shows that A and C are connected because both begin with α_1. Similarly, D and B are connected. Since connectedness is a transitive relation, A and B are connected. □

(IV) *Let A and B be two flags of $\mathscr{P}$. Then, there is a frame Σ which supports both A and B.*

Proof. Again, we will prove this by induction on the dimension. We may enlarge A and B, and assume that both are maximal flags. Set

$$A: \alpha_1 \supset \cdots \supset \alpha_d \quad \text{and} \quad B: \beta_1 \supset \cdots \supset \beta_d.$$

Let k be the largest integer such that $\beta_k \nsubseteq \alpha_1$. If $\beta_1 = \alpha_1$, then we set $k = 0$ and $\beta_0 = \mathscr{P}$. Define

$$\gamma_{i+1} = \alpha_1 \cap \beta_i \quad (i < k) \quad \text{and} \quad \gamma_j = \beta_j \quad (j > k).$$

Then, $C: \gamma_2 \supset \gamma_3 \supset \cdots \supset \gamma_d$ is a maximal flag of the projective space α_1. The subspaces $\alpha_2, \ldots, \alpha_d$ form a maximal flag A_1 of α_1. So, by the inductive hypothesis, there is a frame Σ_1 of α_1 which supports both C and A_1. By definition, we get $\beta_i = \gamma_{i+1} \cup \beta_k$ if $i \leq k$. Set $\Sigma = \{\Sigma_1, P\}$ for any $P \in \beta_k$, $P \notin \beta_{k+1}$, and Σ is a frame of $\mathscr{P}$ which supports both A and B. $\quad\square$

Corollary. *The flag space $\Delta(\mathscr{P})$ is connected.*

Proof. By (IV), if A and B are two flags, then there is a frame Σ such that both A and B belong to Σ_Δ. By (III), the set Σ_Δ is connected. So, A and B are connected. $\quad\square$

The frame is a set of $d + 1$ points of $\mathscr{P}$. So, if Σ and Σ' are frames, there is a $1 - 1$ correspondence σ between the points of Σ and those of Σ'. For any flag

$$A: \alpha_1 \supset \cdots \supset \alpha_k$$

supported by Σ, α_i is the subspace spanned by $\alpha_i \cap \Sigma$ (3.10). Let α_i' be the subspace spanned by $\sigma(\alpha_i \cap \Sigma)$. Then, we have dim $\alpha_i =$ dim α_i' and

$$A': \alpha_1' \supset \alpha_2' \supset \cdots \supset \alpha_k'$$

is a flag supported by Σ'. We will write $A' = A^\sigma$.

(V) *Let A and B be two flags of $\mathscr{P}$. Suppose that Σ and Σ' are two frames which support both A and B. Then, there is a $1 - 1$ mapping from Σ onto Σ' such that $A^\sigma = A$ and $B^\sigma = B$.*

Proof. First, we will prove that if a frame Σ is adapted to the subspaces α and β, then Σ is adapted to both $\alpha \cap \beta$ and $\alpha \cup \beta$. By assumption, α is spanned by $\Sigma \cap \alpha$. Thus, we have

$$|\Sigma \cap \alpha| = \dim \alpha + 1.$$

Similarly, $|\Sigma \cap \beta| = \dim \beta + 1$. Suppose that $\Sigma \cap \alpha$ and $\Sigma \cap \beta$ contain exactly k common elements. Clearly, $\alpha \cup \beta$ contains $\Sigma \cap \alpha$ and $\Sigma \cap \beta$. Since Σ is independent, we get

$$\dim \alpha + 1 + \dim \beta + 1 - k \leq \dim(\alpha \cup \beta) + 1.$$

We conclude from (3.2) that $\dim(\alpha \cap \beta) + 1 \leq k$. But, $\alpha \cap \beta$ contains k independent points. So, we must have the equality. Thus, $\alpha \cap \beta$ is spanned by $\Sigma \cap (\alpha \cap \beta)$ and $\alpha \cup \beta$ is spanned by $\Sigma \cap (\alpha \cup \beta)$. Therefore, Σ is adapted to both $\alpha \cap \beta$ and $\alpha \cup \beta$.

We will prove (V) by induction on the dimension. Suppose that A consists of the subspaces $\{\alpha_i\}$ and B consists of $\{\beta_j\}$. Set $\alpha_0 = \alpha_1 \cup \beta_1$. Then, both Σ and Σ' are adapted to α_0. We may assume that $\alpha_0 = \mathscr{P}$. Set $\gamma = \alpha_1 \cap \beta_1$. From the set of subspaces $\{\gamma \cap \beta_i\}$, we extract the distinct, nonempty subspaces to form the flag $C: \gamma \supset \gamma_1 \supset \gamma_2 \supset \cdots$. By the first part, both Σ and Σ' support C. By the inductive hypothesis, there is a $1 - 1$ mapping τ from $\Sigma \cap \alpha_1$ to $\Sigma' \cap \alpha_1$ such that $A^\tau = A$ and $C^\tau = C$. Similarly, Σ and Σ' support both B and C. So, again by the inductive hypothesis, there is a $1 - 1$ mapping ρ from $\Sigma \cap \beta_1$ to $\Sigma' \cap \beta_1$ such that $B^\rho = B$ and $C^\rho = C$. We define a mapping σ from Σ to Σ' as follows: σ has the same effect as τ on $\Sigma \cap \alpha_1$, but $\sigma = \rho$ on $\Sigma \cap (\beta_1 - \gamma)$. Then, σ is $1 - 1$ and leaves both A and B invariant. $\qquad\square$

Let Σ be a frame, and let A be a maximal flag supported by Σ. If $\alpha_1 \supset \alpha_2 \supset \cdots \supset \alpha_d$ are the subspaces of A, then it follows from (3.10) that each α_i is spanned by the set $\Sigma \cap \alpha_i$. Also, $\Sigma \cap \alpha_i$ is obtained from $\Sigma \cap \alpha_{i-1}$ by removing a single point. Thus, the flag A corresponds to an arrangement of the points of Σ in linear order $P_0, P_1, \ldots, P_d$ such that

$$\Sigma \cap \alpha_i = \{P_i, P_{i+1}, \ldots, P_d\}.$$

(There are exactly $(d + 1)!$ maximal flags supported by Σ.) As shown before, $PGL(\mathscr{P})$ acts transitively on the set of $d + 1$ independent points of $\mathscr{P}$. Hence, we have the following theorem.

Theorem 3.11. *The projective general linear group $PGL(\mathscr{P})$ of a projective space $\mathscr{P}$ acts on the flag space $\Delta(\mathscr{P})$ of $\mathscr{P}$. Furthermore, $PGL(\mathscr{P})$ acts transitively on the set of pairs (Σ, A) where Σ is a frame and A is a maximal flag supported by Σ.*

The flag space $\Delta(\mathscr{P})$ is the most natural object on which $PGL(\mathscr{P})$ acts. We can study many properties of $PGL(\mathscr{P})$ by investigating the way $PGL(\mathscr{P})$ acts on $\Delta(\mathscr{P})$. Objects similar to $\Delta(\mathscr{P})$ have been defined for other classical and exceptional groups of Lie type, and they have been recognized as an essential tool in the study of the properties of these groups. In the 1960's, Tits defined a concept which generalized that of the flag space of a projective space, developed an elegant theory of buildings, or the Tits system, and gave a complete classification of the buildings of rank ≥ 3 under some finiteness assumption. This theory has been applied in many branches of Mathematics. We have to refer to the paper of Tits [2] for the details, but as a brief introduction to his beautiful theory, we will touch upon two points closely connected with group theory. We begin with these basic definitions.

Let Δ be a set endowed with a partial order which will be denoted by $A \supset B$. The set Δ is said to be a **complex** if the following two conditions are satisfied:

(a) For any two elements A and B of Δ, there is a greatest lower bound.
(b) For any $A \in \Delta$, the subset of Δ which consists of the elements contained in A forms a simplex.

A greatest lower bound of A and B is denoted by $A \cap B$ and is defined to be the element C such that $C \subset A$, $C \subset B$, and C is the largest element which satisfies these conditions. A complex Δ has a smallest element. A subset Γ of Δ is said to be a *subcomplex* of Δ if

$$A \in \Gamma, \qquad B \subset A \Rightarrow B \in \Gamma.$$

If A is an element, the set of elements of Δ which are contained in A is a simplex by definition. If the simplex under A is of rank r, we call A an element of *rank r*. An element of rank one is called a **vertex**.

In this book, we will consider only those complexes which satisfy the following property: Any element is contained in a maximal element which is of rank d. A complex satisfying this condition is called a *complex of rank d*. In a complex of rank d, we can define the *adjacency* of two maximal elements, and the *connectedness* of two elements in exactly the same way as in Definition 3.8. A connected complex of rank d is what Tits called a *chamber complex*. Between complexes, we consider mappings f satisfying the properties that the image of a maximal element is maximal and that, for any element A, the restriction of f on the simplex under A is an isomorphism.

Let $\mathscr{P}$ be a projective space of dim d. Then, the property (I) of the flag space $\Delta(\mathscr{P})$ says that $\Delta(\mathscr{P})$ is a complex of rank d. The property (II) shows that the set Σ_Δ of the flags supported by a frame Σ is also a complex of rank d and that Σ_Δ is a subcomplex of $\Delta(\mathscr{P})$. By the property (III), Σ_Δ is connected.

In order to discuss the difference between $\Delta(\mathscr{P})$ and Σ_Δ, we introduce the following concept.

Definition 3.12. Let Γ be a complex of rank d. If any element A of rank $d - 1$ is contained in exactly two maximal elements, then Γ is said to be **thin**. If there are at least 3 maximal elements containing any element of rank $d - 1$, then Γ is called **thick**.

Let Σ be a frame of $\mathscr{P}$, and let

$$A: \alpha_1 \supset \alpha_2 \supset \cdots \supset \alpha_{d-1}$$

be a flag of rank $d - 1$ which is supported by the frame Σ. Set $\alpha_0 = \mathscr{P}$ and $\alpha_d = \varnothing$. Then, by (3.10), $\alpha_{i-1} - \alpha_i$ contains exactly one point of Σ for all except one value of i, and the exceptional place, $\alpha_{k-1} - \alpha_k$, contains exactly two points P and Q of Σ. Let α'_k be the subspace spanned by α_k and P, and let α''_k be the one spanned by α_k and Q. Then,

$$A': \alpha_1 \supset \cdots \supset \alpha_{k-1} \supset \alpha'_k \supset \alpha_k \supset \cdots,$$

$$A'': \alpha_1 \supset \cdots \supset \alpha_{k-1} \supset \alpha''_k \supset \alpha_k \supset \cdots$$

are the only two maximal flags which are supported by Σ and contain A. Thus, the complex Σ_Δ is thin. On the other hand, the subspace U_k of the underlying vector space corresponding to α_k has codimension 2 in the subspace U_{k-1} corresponding to α_{k-1}. So, there are at least 3 subspaces of codimension 1 between U_{k-1} and U_k. Thus, there are at least 3 maximal flags which contain A, and so the complex $\Delta(\mathscr{P})$ is a thick complex of rank d.

Definition 3.13. A complex Δ of rank d is said to be a **building** of rank d if there is a collection $\mathscr{A}$ of subcomplexes such that $(\Delta, \mathscr{A})$ satisfies the following four conditions:

(B1) Δ is thick.
(B2) Every element of $\mathscr{A}$ is a connected, thin complex of rank d.
(B3) For any two elements A and B of Δ, there is an element of $\mathscr{A}$ which contains both A and B.
(B4) Let A and A' be any two elements of Δ. If Σ and Σ' are elements of $\mathscr{A}$ which contain both A and A', there is an isomorphism φ from Σ onto Σ' which satisfies $\varphi(B) = B$ for all $B \subset A$ as well as $\varphi(B') = B'$ for all $B' \subset A'$.

A subcomplex which is a member of $\mathscr{A}$ is called an **apartment**, a maximal element of Δ is called a **chamber**, and any element of rank $d - 1$ is called a **wall**.

The flag space $\Delta(\mathscr{P})$ of a projective space $\mathscr{P}$ is a building in which each apartment is the subset Σ_Δ of a frame Σ. In fact, $\Delta(\mathscr{P})$ is a thick complex of rank d as remarked earlier. For each frame Σ, the set Σ_Δ satisfies (B2) by (II) and (III). The remaining Axioms (B3) and (B4) are satisfied by properties (IV) and (V), respectively.

(3.14) *Let $(\Delta, \mathscr{A})$ be a building. Then, any two apartments of $\mathscr{A}$ are isomorphic.*

Proof. Let Σ and Σ' be two apartments, and let $A \in \Sigma$ and $A' \in \Sigma'$. By Axiom (3), there is an apartment Σ'' which contains both A and A'. By Axiom (B4) applied here under the assumption that the two elements of Δ coincide, Σ is isomorphic to Σ''. Similarly, Σ'' is isomorphic to Σ' because $\Sigma' \cap \Sigma'' \ni A'$. So, Σ is isomorphic to Σ' by the transitivity. $\square$

In studying buildings, the following concept is quite useful. Let A and B be two elements of a building Δ. A sequence $\{C_i\}$ $(i = 0, 1, \ldots, m)$ of chambers is said to be a **gallery** of length m joining A and B if $A \subset C_0$, $B \subset C_m$, and the chambers C_{i-1} and C_i have a common wall for all $i = 1, \ldots, m$.

As the Corollary of (IV), any building is connected; that is, any two elements are connected by a gallery. The minimal length of all the galleries connecting A and B is called the *distance* between A and B, and is denoted by

$$\text{dist}(A, B).$$

A gallery $\{C_i\}$ may 'stammer', i.e. we may have $C_i = C_{i+1}$ for some i. The shortest gallery between A and B never stammers.

The following proposition is basic to our study of buildings.

(3.15) *Let C be a chamber of a building Δ, and let Σ and Σ' be two apartments which contain C. Then, there is a unique isomorphism φ from Σ onto Σ' such that $\varphi(B) = B$ for all $B \subset C$.*

Proof. Axiom (B4) proves that such an isomorphism exists. Let φ' be another isomorphism from Σ onto Σ' such that $\varphi'(B) = B$ for all $B \subset C$. We will prove that $\varphi(A) = \varphi'(A)$ for $A \in \Sigma$ by induction on the distance $\mathrm{dist}(A, C)$ defined in Σ (see Exercise 1 (a)). If $\mathrm{dist}(A, C) = 0$, then A is contained in C. So, we have $\varphi(A) = A = \varphi'(A)$. Suppose that $\mathrm{dist}(A, C) = m > 0$. Then, there is a gallery $\{C_i\}$ of length m such that $C_0 = C$, $A \subset C_m$, and $C_i \in \Sigma$ for all i. By the inductive hypothesis, for any element $D \subset C_{m-1}$, we have $\varphi(D) = \varphi'(D)$. In particular, if W is the common wall of C_{m-1} and C_m, we get $\varphi(W) = \varphi'(W)$. Set $W' = \varphi(W)$. Then, since φ is an isomorphism, W' is a wall of Σ'. By (B2), Σ' is thin. So, there are exactly two chambers of Σ' which contain W'. One of them is $\varphi(C_{m-1}) = \varphi'(C_{m-1})$, so the other must be $\varphi(C_m) = \varphi'(C_m)$. The chamber C_m is a simplex and contains a unique vertex P which is not contained in the wall W. Since $\varphi(P)$ is the only vertex of $\varphi(C_m)$ not contained in W', we must have $\varphi(P) = \varphi'(P)$. The element contained in a chamber is determined by the set of vertices in it. So, $\varphi(A) = \varphi'(A)$. Thus, we have $\varphi = \varphi'$. $\square$

Let Σ be a fixed apartment of a building Δ, and let C be a fixed chamber of Σ. We can define a mapping ρ from Δ into Σ which is called the *retraction* and is denoted by

$$\rho = \mathrm{retr}(\Sigma, C).$$

Let A be any element of Δ. By (B3), there is an apartment Σ' which contains C and A, and by (B4), there is an isomorphism φ' from Σ' onto Σ which leaves fixed every B such that $B \subset C$. If we choose another apartment Σ'' which contains C and A, then we obtain an isomorphism φ'' form Σ'' into Σ which satisfies $\varphi''(B) = B$ for $B \subset C$. We will show that $\varphi'(A) = \varphi''(A)$. By (B4), there is an isomorphism θ from Σ' onto Σ'' which leaves invariant every vertex of C, as well as every vertex of A. Then, φ' and $\varphi'' \circ \theta$ are two isomorphisms from Σ' onto Σ which leave every vertex of C fixed. By (3.15), we must have $\varphi' = \varphi'' \circ \theta$. Hence, we have

$$\varphi'(A) = \varphi''(\theta(A)) = \varphi''(A).$$

We define $\rho(A)$ to be the common value $\varphi'(A)$. The retraction has the following properties.

(3.16) *For a fixed chamber C and a fixed apartment Σ containing C, set*
$\rho = \mathrm{retr}(\Sigma, C)$.

   (i)   *If $A \in \Sigma$, then we have $\rho(A) = A$.*
   (ii) *If $A \subset B$, then $\rho(A) \subset \rho(B)$.*
   (iii) *If B is a chamber of Δ, then $\rho(B)$ is a chamber of Σ. In general, B and $\rho(B)$ have the same rank.*
   (iv) *If two chambers B and B' are adjacent, then $\rho(B)$ and $\rho(B')$ are either identical or adjacent.*
   (v)   *If $\rho(A) \subset C$, then we have $\rho(A) = A \subset C$.*

Proof. If $A \in \Sigma$, then the identity mapping on Σ may be used to define ρ. Hence, $\rho(A) = A$, and (i) holds. Since ρ is defined by using isomorphisms among apartments, the propositions (ii) and (iii) are obvious. If B and B' are adjacent, they have a common wall W. So, by (ii) and (iii), $\rho(W)$ is a common wall of $\rho(B)$ and $\rho(B')$. Since ρ is defined by isomorphisms which leave every vertex of C fixed, no vertex outside C is mapped into C. Therefore, the proposition (v) holds. □

(3.17) *Let Σ be an apartment, C be a chamber, and A be an element of a building Δ such that*

$$A \in C \subset \Sigma.$$

Let $G = \{C_i\}$ be a shortest gallery joining A and a chamber D of Δ.

   (i) *If $D \in \Sigma$, then all chambers C_i of G are chambers of Σ.*
   (ii) *If $\rho = \mathrm{retr}(\Sigma, C)$, then $\rho(G)$ is a shortest gallery joining A and $\rho(D)$.*

Proof. (i) Let G be a gallery of length m. Then, by definition, the last chamber C_m of G is $D \in \Sigma$. Suppose that (i) is not true. Then, there is an integer i such that $C_i \notin \Sigma$ but $C_{i+1} \in \Sigma$. By the definition of a gallery, C_i and C_{i+1} have a common wall B. Since Σ is thin, B is contained in exactly two chambers of Σ. Clearly, C_{i+1} is one of them. Let E be the other chamber of $\sum$ which contains B. Consider the image $\sigma(G)$ where $\sigma = \mathrm{retr}(\Sigma, E)$. By (3.16), we have $\sigma(A) = A$ and $\sigma(D) = D$. Hence, $\sigma(G)$ is a gallery joining A and D. But, $\sigma(C_i)$ is a chamber of Σ which contains B. Since $\sigma(C_i) \neq E$ by (3.16) (v), we get

$$\sigma(C_i) = C_{i+1} = \sigma(C_{i+1}).$$

Thus, $\sigma(G)$ stammers. This contradicts the assumption that G is a shortest gallery joining A and D.

   (ii) By the axiom (B3) of a building, there is an apartment Σ' which contains C and D. By the first part of this proposition, every chamber of the gallery G belongs to Σ'. By the definition of ρ, the image $\rho(B)$ of any element $B \in \Sigma'$ is given by an isomorphism from Σ' onto Σ which leaves every vertex of C invariant. So, $\rho(G)$ is a gallery without repetitions and is a shortest one joining A and $\rho(D)$. $\square$

An apartment contained in a building has the prominent property that it has many automorphisms. We will need the following concept. Let Σ be a thin complex of rank d. A mapping φ from Σ into itself is said to be a **folding** if $\varphi^2 = \varphi$ and if any chamber of $\varphi(\Sigma)$ is the image of exactly two chambers of Σ by φ.

(3.18) *Let Δ be a building, and let Σ be an apartment of Δ. Assume that two chambers C and C' of Σ are adjacent. Then, there are foldings φ and φ' of Σ such that*

$$\varphi(C') = C \quad and \quad \varphi'(C) = C'.$$

We have $\Sigma = \varphi(\Sigma) \cup \varphi'(\Sigma)$. Furthermore, $\varphi(\Sigma)$ and $\varphi'(\Sigma)$ have no chamber in common. The mapping σ defined by

$$\sigma(A) = \begin{cases} \varphi'(A) & if \quad A \in \varphi(\Sigma) \\ \varphi(A) & if \quad A \in \varphi'(\Sigma) \end{cases}$$

is an automorphism of order 2 of Σ.

Proof. Let B be the common wall of C and C'. By Axiom (B1) of a building, there is a third chamber C'' which contains B. By Axiom (B3) there is an apartment Σ' which contains C and C''. Let φ be the restriction of the product

$$\mathrm{retr}(\Sigma', C)\mathrm{retr}(\Sigma, C')$$

on the apartment Σ. Clearly, $\varphi(\Sigma) \subset \Sigma$ and $\varphi(C) = C$. But, we have $\varphi(C') \neq C'$. This may be proved as follows. By definition, $C'' \neq C'$ and C' is not contained in Σ'. So, the image of C' by $\mathrm{retr}(\Sigma', C)$ is different from C', and $\varphi(C') \neq C'$. Since $\varphi(B) = B$, $\varphi(C')$ is a chamber of Σ which contains B. Thus, we have $\varphi(C') = C$. Similarly, for apartment Σ'' which contains C' and C'', the restriction φ' of

$$\mathrm{retr}(\Sigma'', C')\mathrm{retr}(\Sigma, C)$$

on Σ satisfies $\varphi'(C') = C' = \varphi'(C)$.

We will prove that both φ and φ' are foldings of Σ. Let D be a chamber of Σ, and let $G = \{C_i\}$ be a gallery of the shortest length m joining the wall B and the chamber D. Then, we have $C_m = D$, and all members of G are chambers of Σ by (3.17) (i). The first member C_0 of G is a chamber of Σ which contains B. Since Σ is thin, C_0 is either C or C'. We will prove that if $C_0 = C$, then for any $E \subset D$

(i) $\varphi(E) = \varphi(\varphi'(E)) = E$ and (ii) $\varphi'(D) \neq D$,

and similarly, if $C_0 = C'$, then

(i)' $\varphi'(E) = \varphi'(\varphi(E)) = E$ and (ii)' $\varphi(D) \neq D$.

We will use induction on m. If $m = 0$, both assertions (i) and (ii) follow immediately from the definition with $D = C$. Assume that $m > 0$. Let F be the common wall of C_{m-1} and $C_m = D$. By the inductive hypothesis, we have

$$\varphi(F) = \varphi(\varphi'(F)) = F \quad \text{and} \quad \varphi'(C_{m-1}) \neq C_{m-1} = \varphi(C_{m-1}).$$

Then, $\varphi(D)$ is a chamber of Σ which contains F. So, $\varphi(D)$ is either C_{m-1} or C_m. But, by the definition of φ and (3.17) (ii), $\varphi(G)$ is a gallery of the shortest length joining B and $\varphi(D)$. This means that the gallery $\varphi(G)$ does not stammer, so we have $\varphi(C_m) \neq \varphi(C_{m-1}) = C_{m-1}$. Hence, $\varphi(D)$ must be equal to D. Since D has exactly one vertex outside F, φ leaves every vertex of D invariant. Hence, for any $E \subset D$, we have $\varphi(E) = E$. Similarly, we get $\varphi(\varphi'(E)) = E$. This proves (i). If $C_0 = C'$ we can prove (i)' by exchanging C and C'.

Assume again that $C_0 = C$ and $m > 0$. Suppose that (ii) is false. Then, we have $\varphi'(D) = D$. Apply (i)' to the gallery $\varphi'(G)$ which begins with $\varphi'(C) = C'$. By (i)', φ' leaves every vertex of D invariant. In particular, $\varphi'(F) = F$ for the common wall of C_{m-1} and D. Then, $\varphi'(C_{m-1})$ is a chamber of Σ which contains F. This means that $\varphi'(C_{m-1})$ is either C_{m-1} or D. But $\varphi'(C_{m-1}) \neq C_{m-1}$ by the inductive hypothesis. So, we have $\varphi'(C_{m-1}) = D = \varphi'(D)$, and the gallery $\varphi'(G)$ stammers. We have arrived at a contradiction as before. This proves (ii). Similarly, (ii)' holds when $C_0 = C'$.

Let A be an arbitrary element of Σ, and let G be a shortest gallery joining B and A. Then, as remarked earlier, $\varphi(G)$ is a shortest gallery joining B and $\varphi(A)$. Since the first chamber of $\varphi(G)$ is C, we have $\varphi(\varphi(A)) = \varphi(A)$ by (i). Hence, we have $\varphi^2 = \varphi$.

Let H be an element of Σ, and let K be a chamber of Σ which contains H. Let G be a shortest gallery joining B and K. If G begins with C, then (i) and (ii) give us

$$\varphi(H) = \varphi(\varphi'(H)) = H \quad \text{and} \quad \varphi'(K) \neq K,$$

while, if G begins with C', we have

$$\varphi'(H) = \varphi'(\varphi(H)) = H \quad \text{and} \quad \varphi(K) \neq K.$$

Thus, we have $\Sigma = \varphi(\Sigma) \cup \varphi'(\Sigma)$, and since $\varphi^2 = \varphi$ and $\varphi'^2 = \varphi'$ hold, $\varphi(\Sigma) \cap \varphi'(\Sigma)$ contains no chamber. Suppose that $K \in \varphi(\Sigma)$. Then there is a chamber L such that $\varphi(L) = K$. Certainly, either $L = \varphi(L)$ or $L \neq \varphi(L)$. In the former case, we clearly have $L = K$, while in the latter case, we have

$$L = \varphi'(\varphi(L)) = \varphi'(K).$$

Since $\varphi(K) = K$, we get $\varphi'(K) \neq K$. Thus, any chamber of $\varphi(\Sigma)$ is the image by φ of exactly two chambers of Σ. Hence, φ is a folding. Similarly, we can prove that φ' is also a folding of Σ.

Since $\varphi(A) = A = \varphi'(A)$ for any $A \in \varphi(\Sigma) \cap \varphi'(\Sigma)$, the mapping σ is defined without ambiguity. If $A \in \varphi(\Sigma)$, then, by (i), we have $A = \varphi(\varphi'(A))$. Similarly, $A \in \varphi'(\Sigma)$ implies that $A = \varphi'(\varphi(A))$. This shows that σ^2 is the identity, so σ is an automorphism of order 2. $\square$

Definition 3.19. The automorphism σ defined in (3.18) is called the **reflection** with respect to the wall $B = C \cap C'$. The group of automorphisms generated by reflections is called the **Weyl group** of the apartment Σ.

The flag space $\Delta(\mathscr{P})$ of a projective space $\mathscr{P}$ of dimension d is a building in which apartments are defined by frames. In this case, a frame consists of $d + 1$ independent points, and a reflection corresponds to an exchange of two points while keeping the remaining $d - 1$ fixed. Thus, the Weyl group of the apartment in $\Delta(\mathscr{P})$ is the symmetric group of degree $d + 1$.

(3.20) (i) *Let Σ be an apartment of a building Δ. Let C be a fixed chamber of Σ. If $G = \{C_i\}$ is a gallery of chambers of Σ starting with $C_0 = C$, then there is a mapping α defined on Σ which depends on G, leaves every vertex of C invariant, and satisfies $\alpha(C_i) = C$ for all chambers of G.*

(ii) *Let B be any element of Σ. If φ and ψ are two mappings of Σ into Σ such that $\varphi(B) \subset C$, $\psi(B) \subset C$, and both φ and ψ leave every vertex of C invariant, then $\varphi = \psi$ on every vertex of B.*

Proof. (i) Use induction on the length m of G. If $C_1 = C$, we may remove C_1 to obtain a gallery of shorter length. If $C_1 \neq C$, then there is a folding φ which maps C_1 onto C. Then, after removing $\varphi(C_1)$, $\varphi(G)$ is a gallery of shorter length. So, by the inductive hypothesis, there is a mapping β defined on Σ such that β leaves every vertex of C invariant and $\beta(\varphi(C_i)) = C$. The composite $\alpha = \beta \circ \varphi$ satisfies the required property.

(ii) Let $G = \{C_i\}$ be a gallery joining C and B. Then, there is a gallery G' through all the chambers of $\varphi(G)$ and $\psi(G)$. By (i), there is a mapping α which maps every chamber of G' to C and leaves every vertex of C invariant. Consider $\varphi\alpha$ and $\psi\alpha$ in place of φ and ψ. We may assume that

$$\varphi(C_i) = \psi(C_i) = C$$

for all i. We can prove that $\varphi = \psi$ on every vertex of C_i in the same way as in the proof of (3.15). $\square$

Definition 3.21. The common value in (3.20) (ii) will be denoted by $\rho_C(B)$. Two elements B and B' of Σ are said to have the same **type** if $\rho_C(B) = \rho_C(B')$.

The property of having the same type defines an equivalence relation on the elements of Σ. By (3.20), this equivalence relation does not depend on the particular chamber C which we considered fixed in the above discussion. In the flag space of a projective space, two elements of rank one have the same type if and only if they are subspaces of the same dimension. If C and C' are two adjacent chambers of Σ, then the folding of Σ which sends C' onto C maps any element of C' onto an element of the same type. In general, a folding does not change the type of an element. Thus, a reflection also keeps the type of the element invariant, so any element of the Weyl group is a type-preserving automorphism of Σ.

(3.22) *Let Σ be an apartment of a building, and let W be the Weyl group of Σ. Let C be a chamber of Σ, and let S be the set of reflections with respect to the walls of C.*

(1) *We have $W = \langle S \rangle$.*
(2) *The Weyl group is the totality of type-preserving automorphisms of Σ.*
(3) *The Weyl group W is simply transitive on the set of chambers of Σ.*

*(A permutation group G on X is **simply transitive** if for any x and y of X, there is a unique element σ of G such that $x^\sigma = y$.)*

Proof. Set $W_1 = \langle S \rangle$, and let W_2 be the totality of type-preserving automorphisms of Σ. It is clear that W_2 is a group and that we have

$$W_1 \subset W \subset W_2.$$

We will show that W_1 is transitive on the set of chambers of Σ. Let D be a chamber, and let $m = \mathrm{dist}(C, D)$. Then, there is a gallery of length m

$$C_0 = C, C_1, \ldots, C_m = D.$$

By induction on m, we will prove that there is an element w of W_1 such that $w(C) = D$. If $m = 0$, this is trivial. Suppose $m > 0$. Now, C and C_1 have a common wall B. Hence, the reflection σ with respect to B is an element of S, and

$$\sigma(C_1) = C, \sigma(C_2), \ldots, \sigma(C_m) = \sigma(D)$$

is a gallery of length $m - 1$. By the inductive hypothesis, there is an element w_1 of W_1 such that $w_1(C) = \sigma(D)$. Thus, $w_1 \sigma$ is an element of W_1 which satisfies $w_1 \sigma(C) = D$. So, the group W_1 is transitive on the chambers.

We will prove next that the group W_2 is simply transitive. Suppose that two elements w and w' of W_2 satisfy $w(C) = w'(C)$. Then, for $w'' = w'w^{-1} \in W_2$, we get $w''(C) = C$. By assumption, w'' is type-preserving. Since the chamber C contains exactly one element of each type, w'' fixes every element contained in C. By (3.15), w'' must be the identity. Thus, $w(C) = w'(C)$ implies that $w = w'$. This means that the group W_2 is simply transitive on the set of chambers.

We can now prove (3.22) quickly. Let $w_2 \in W_2$. Then $w_2(C)$ is a chamber. Since W_1 acts transitively, there is an element w_1 of W_1 such that $w_1(C) = w_2(C)$. But, W_2 is simply transitive and $W_1 \subset W_2$. Hence, we get $w_1 = w_2 \in W_1$. Thus, $W_2 \subset W_1$ and we have

$$W_1 = W = W_2. \quad \square$$

We will prove later that the Weyl groups of buildings form a very special class of groups, and we will determine all possible finite Weyl groups in §4.

The concept of a building greatly facilitates the study of its group of automorphisms. Let Δ be a building, let Σ be an apartment of Δ, and let C be a chamber of Σ. Let G be a group consisting of automorphisms of Δ. We assume that the following two conditions are satisfied.

(A) Each element of G is a type-preserving automorphism of Δ.

(B) If Σ' is an apartment of Δ and $C' \in \Sigma'$, then there is an element σ of G such that $\sigma(\Sigma) = \Sigma'$ and $\sigma(C) = C'$.

By Theorem 3.11, the action of $PGL(\mathscr{P})$ on $\Delta(\mathscr{P})$ satisfies these properties. Thus, the following argument applies to the group $PGL(\mathscr{P})$.

By (B), the group G acts transitively on the set of apartments as well as on the set of chambers. Let N be the stabilizer of Σ, and let B be the stabilizer of C:

$$N = S_G(\Sigma), \qquad B = S_G(C).$$

Clearly, each element σ of N induces an automorphism of Σ which will be denoted by $\omega(\sigma)$. By (A), $\omega(\sigma)$ is type-preserving. So, it is an element of the Weyl group W of Σ. Thus, ω is a homomorphism of N into the Weyl group W of Σ. Furthermore, by (B), $\omega(N)$ is transitive on the chambers of Σ. By (3.22) (3), we have $\omega(N) = W$. The kernel of ω is equal to the stabilizer of the chamber C:

$$\text{Ker } \omega = B \cap N.$$

Thus, we get $N/B \cap N \cong W$.

Let g be an element of G. Then, $g(C)$ is a chamber, and there is an apartment Σ' which contains both $g(C)$ and C (by Axiom (B3)). The condition (B) proves the existence of an element b of G which sends the pair (Σ, C) to the pair (Σ', C). Thus, b leaves C invariant, so we have $b \in B$. Since $g(C) \in \Sigma'$, the chamber $b^{-1}(g(C))$ is contained in Σ. By (3.22) (3), the Weyl group W is transitive. As $\omega(N) = W$, we can find an element n of N such that

$$n^{-1}(b^{-1}(g(C)) = C.$$

This shows that the element $b_1 = gb^{-1}n^{-1}$ belongs to B. Thus, we have $G = \langle B, N \rangle$. More precisely, we get $g = b_1 nb$ and

(3.23) $G = BNB.$

This equality shows that G is a union of the double cosets with respect to (B, B) and that we can take an element of N as the representative from each (B, B) double coset.
    The double coset BnB is determined by $\omega(n)$:

$$\omega(n) = \omega(n') \Rightarrow BnB = Bn'B.$$

We will show shortly that the converse is also true. Meanwhile, for $w = \omega(n)$, we define

$$C(w) = BwB = BnB.$$

Sometimes, it is more convenient to consider the double cosets as being indexed by the elements of the Weyl group. We also use notations such as $Bw = Bn$ when $w = \omega(n)$.
    The most important property of a group G of automorphisms will be stated in Theorem 3.25. First, we need the following definition.

Definition 3.24. Let G be a group. Two subgroups B and N of G are said to be a (B, N)-**pair** of G if the following conditions are satisfied.

(T1)   We have $G = \langle B, N \rangle$ and $B \cap N \triangleleft N$.
(T2)   The group $W = N/B \cap N$ has a distinguished set of generators S consisting only of elements of order 2: $W = \langle S \rangle$.
(T3)   For any $s \in S$ and $w \in W$, we have

$$sBw \subset BwB \cup BswB.$$

(T4)   We have $sBs \not\subseteq B$ for any $s \in S$.

When G, B, N, and S satisfy the above conditions, we say that the quadruple (G, B, N, S) is a **Tits system**. The group W is called the Weyl group of the Tits system.

Theorem 3.25. *Let Δ be a building, and let G be a group of automorphisms of Δ. Let Σ be an apartment of Δ, and let C be a chamber of Σ. Suppose that the two* *conditions (A) and (B) of page 326 are satisfied. Then, (G, B, N, S) is a Tits system, where $B = S_G(C)$, $N = S_G(\Sigma)$, and S is the set of the reflections with respect to the walls of C.*

Proof. The first two properties have been proved. We will prove the condition (T3).

We denote by Cg the image of the chamber C by an element g of G. If $n \in N$, then Cn depends only on $\omega(n)$. So, we write $C\omega(n) = Cn$.

Let $s \in S$ and $w \in W$. Then, C and Cs are adjacent, so let D be the common wall of C and Cs. If $b \in B$, b leaves every vertex of C invariant. Hence, Csb contains D, and $Csbw$ contains Dw. By Axiom (B3), there is an apartment Σ' which contains both C and $Csbw$. Let $G = \{C_i\}$ be a shortest gallery joining Dw and C ($C_m = C$). Then, both Dw and C are contained in Σ and Σ'. By (3.17) (i), all members C_i of G are chambers of $\Sigma \cap \Sigma'$. The condition (B) on G gives us an element b' of G such that b' sends the pair (Σ', C) to the pair (Σ, C). Then, we have $Cb' = C$ and $b' \in B$. Since b' leaves every vertex of C invariant, b' fixes the common wall E of C_{m-1} and C_m ($C_m = C$). Therefore, $C_{m-1}b'$ is a chamber which contains E. Since only two chambers of Σ contain E, we have $C_{m-1}b' = C_{m-1}$, and b' leaves every vertex of C_{m-1} invariant. By repeating this process, we see that b' fixes all elements contained in C_0, and in particular, b' fixes Dw.

As we have shown, $Dw \subset Csbw$ and $Csbw$ is a chamber of Σ'. Since $Dwb' = Dw$, we get

$$Dw = Dwb' \subset Csbwb' \in \Sigma'b' = \Sigma.$$

Since Σ is an apartment, there are exactly two chambers of Σ which contain Dw. They are Cw and Csw. Hence, we get

$$Csbwb' = Cw \quad \text{or} \quad Csbwb' = Csw.$$

This implies that $sbw \in BwB \cup BswB$, and (T3) holds.

As before, let D be the common wall of C and Cs. Since Δ is a thick complex, there is a chamber C' different from C and Cs such that $D \subset C'$. Let Σ' be an apartment which contains both C and C'. Then, by (B), there is an element b of G which maps the pair (Σ, C) to the pair (Σ', C). Then, the element b belongs to B. The image of Cs is a chamber of Σ' which contains D. So, we must have $Csb = C'$. As $C' \neq Cs$, we get $sbs \notin B$. This proves (T4). $\square$

We will prove the converse of Theorem 3.25 in the rest of this section. Our aim is to show that a group G with a (B, N)-pair can be considered to be a group of automorphisms of a building such that G satisfies the conditions (A) and (B). We will need to study some of the basic properties of groups with (B, N)-pairs.

As before, we write $C(w) = BwB$. The condition (T3) can be written as

$$C(s)C(w) \subset C(w) \cup C(sw) \qquad (s \in S, w \in W).$$

We note that $C(w^{-1}) = C(w)^{-1}$. By taking the inverses of both sides, we conclude that

$$C(w)C(s) \subset C(w) \cup C(ws).$$

In general, if $s_1, s_2, \ldots, s_n \in S$, then we get

(3.26) $C(s_1 \cdots s_n)C(w) \subset \bigcup C(s_i \cdots s_k w)$

where the right side is the union of the double cosets with indices $i, \ldots,$ and k such that $1 \leq i < \cdots < k \leq n$.

Take a subset J of S, and define

$$W_J = \langle J \rangle \quad \text{and} \quad P_J = BW_J B.$$

By (3.26), the set P_J is a subgroup of G such that $P_J \supset B$. Also, we have

$$W_S = W \quad \text{and} \quad G = BWB.$$

It is important to realize that the converse is also true; that is, any subgroup of G which contains B determines a subset J of S and coincides with the subgroup P_J. We will need a few auxiliary lemmas in order to prove the above statement.

Let W be the Weyl group of a Tits system (G, B, N, S). Any element w of W can be written as a product

$$w = s_1 s_2 \cdots s_n$$

of elements $s_1, \ldots, s_n$ of S. This expression is not unique. We define the *length* of w to be the minimal number of factors and we denote it by $l(w)$. In what follows, the letters w and w' always stand for elements of the Weyl group W, and $s, s_i, \ldots$ represent elements of the distinguished generating set S.

(3.27) *The following propositions hold.*

   (i) $C(w) = C(w')$ *implies that* $w = w'$.
   (ii) *If* $l(sw) \geq l(w)$, *then* $sBw \subset C(sw)$.
   (iii) *If* $l(sw) \leq l(w)$, *then* $sBw \cap BwB \neq \emptyset$.
   (iv) *Let* $l = l(w)$, *and let* $w = s_1 s_2 \cdots s_l$ *be a shortest expression of* w. *Then, for each* i, *we have*

$$s_i B \subset \langle B, wBw^{-1} \rangle.$$

   Furthermore, we have $C(w) \subset \langle B, wBw^{-1} \rangle$.

Proof. (i) Suppose $l(w) \leq l(w')$, and use induction on $l = l(w)$. If $l = 0$, then $w = 1$. In this case, we have $C(w) = B = C(w')$ and $w' = 1$. Assume that $l > 0$.

Then, we can write $w = sw''$ with $l(w'') = l - 1$. By assumption, we have $sw''B \subset C(w')$. So, by (T3),

$$w''B \subset sBw'B \subset C(w') \cup C(sw').$$

This implies that $C(w'') = C(w')$ or $C(w'') = C(sw')$. By the inductive hypothesis (applied to w''), we get $w'' = w'$ or $w'' = sw'$. But, as $l(w') \geq l$, we get $w' \neq w''$. Hence, we have $w'' = sw'$, or $w' = sw'' = w$.

(ii) Again, use induction on $l = l(w)$. We can write $w = w'r$ where $l(w') = l - 1$ and $r \in S$. If $l(sw') < l - 1$, then we would get $l(sw) = l(sw'r) < l$. This contradicts the assumption that $l(sw) \geq l$. Thus, we have $l(sw') \geq l - 1$. Hence, by the inductive hypothesis, we get

$$sBw' \subset C(sw').$$

Then, $sBw = sBw'r \subset C(sw')C(r) \subset C(sw') \cup C(sw'r)$. On the other hand, (T3) implies that $sBw \subset C(w) \cup C(sw)$. If $C(w)$ and $C(sw')$ were to have a common element, we would have $w = sw'$ by (i). Then, $l(sw) = l(w') = l - 1$, and this contradicts the assumption. So, $C(w)$ and $C(sw')$ have no common element. This implies that $sBw \subset C(sw)$.

(iii) By (T3) and (T4), we get $sBs \cap BsB \neq \varnothing$. Hence, $sBsw \cap BsBw \neq \varnothing$ for any w. By assumption, if we set $sw = w'$, then we get $l(sw') \geq l(w')$, and so it follows from (ii) that sBw' is contained in $C(sw')$. Since $sw' = w$, we have

$$sBw \cap BwB \neq \varnothing.$$

(iv) By assumption, we have $l(s_1 w) < l(w)$, so from (iii) we get

$$s_1 Bw \cap BwB \neq \varnothing.$$

This implies that $s_1 B \subset BwBw^{-1}B \subset \langle B, wBw^{-1} \rangle$. We may apply the argument to $w_1 = s_1 w$ and s_2 to get

$$s_2 B \subset \langle B, w_1 Bw_1^{-1} \rangle \subset \langle B, wBw^{-1} \rangle.$$

By repeating this process, we can prove that $s_i B \subset \langle B, wBw^{-1} \rangle$. The last assertion follows immediately. $\square$

Theorem 3.28. Let (G, B, N, S) be a Tits system.

(i) A subgroup P which contains B determines a subset J of S such that $P = P_J$. We have

$$J = \{s \mid s \in S, C(s) \subset P\}.$$

The correspondence between P_J and J is one-to-one.

(ii) *Let J and K be two subsets of S, and set $I = J \cap K$. Then, we have*

$$P_I = P_J \cap P_K.$$

(iii) *If the subgroup P_J contains a conjugate subgroup B^g of B, then g is an element of P_J.*

(iv) *We have $N_G(P_J) = P_J$.*

(v) *If P_J is conjugate to P_K in G, then we have $P_J = P_K$.*

Proof. (i) Any subgroup which contains B is a union of the double cosets with respect to B. Suppose that $C(w) \subset P$, and let $w = s_1 \cdots s_n$ be a shortest expression of w as a product of elements $s_1, \ldots, s_n$ of S. Then, from (3.27) (iv), we get

$$C(s_i) \subset \langle B, C(w) \rangle.$$

This implies that $C(s_i) \subset P$. Let the subset J of S be defined as stated in (i). Then, for each $C(w) \subset P$, the factors appearing in a shortest expression of w belong to J. Thus, $C(w) \subset P$ if and only if $w \in W_J$. This implies that $P = P_J$.

It remains to show that $P_J = P_K$ if and only if $J = K$. By (3.27) (i), the correspondence between $C(w)$ and w is one-to-one. Hence, we merely need to show that $W_J = W_K$ implies $J = K$. First, we will show that $J = S \cap W_J$. Each element of S is a reflection with respect to a wall of the chamber C. Let us consider the action of $s \in S$ on vertices of C. Since s is a reflection, s moves exactly one vertex of C. If $t \in S$ such that $s \neq t$, then they are reflections with respect to different walls. So, if $s \in J$ and $t \notin J$, then s fixes the vertex moved by t. Hence, the vertex moved by t is left invariant by all elements of W_J. So, t is not contained in W_J. This proves that $J = S \cap W_J$. (Also see Theorem 4.5 (ii).) It is obvious now that $W_J = W_K$ implies $J = K$. Thus, the correspondence between P_J and J is one-to-one.

(ii) By (i), the subgroup $P_J \cap P_K$ has the form P_I. Again by (i), we get $I = J \cap K$.

(iii) Since $G = BWB$, the element g^{-1} belongs to $C(w)$ for some $w \in W$. By assumption, $B^g \subset P_J$, so we have $wBw^{-1} \subset P_J$. It follows from (3.27) (iv) that $C(w) \subset \langle B, wBw^{-1} \rangle$, whence we conclude that g is an element of $\langle B, wBw^{-1} \rangle \subset P_J$.

The last two properties are corollaries of (iii). □

Theorem 3.29. *Let (G, B, N, S) be a Tits system. Let Δ be the totality of the right cosets of all subgroups P of G such that $P \supset B$. We consider Δ to be a partially ordered set such that for any two elements Pg and $P'g'$ of Δ, $Pg < P'g'$ if and only if $P'g'$ is a subset of Pg. Let Σ be the subset defined by*

$$\Sigma = \{Pn\}$$

where $n \in N$ *and* P *ranges over all the subgroups which contain* B. *Set* $\mathcal{A} = \{\Sigma g\}$. *Then,* Δ *is a building, and* $\mathcal{A}$ *is the set of its apartments. A chamber of* Δ *is a coset of the subgroup* B. *Furthermore, two elements* Pg *and* $P'g'$ *have the same type if and only if* $P = P'$.

The group G *acts on* Δ *via right multiplication. Each element of* G *induces a type-preserving automorphism of* Δ, *and* G *is transitive on the set of the pairs consisting of an apartment and one of its chambers. Thus,* G *is a group of automorphisms of* Δ, *and* G *satisfies* (A) *and* (B).

Proof. (a) To begin with, we will show that Δ is a complex. Any element of Δ can be written as Pg where P is a subgroup containing B. Since we have $Pg < Bg$, it is almost obvious that Bg is a maximal element. To prove this assertion, we need only to show that the coset B is maximal since right multiplication is certainly an automorphism of Δ. Suppose that $B < Pg$. Then, Pg is a subset of B. Hence, $g \in B$ and $P \subset B$. This implies that $Pg = B$. Thus, B is maximal in Δ, and hence so is Bg.

Consider the set $[B] = \{Pg \mid Pg < B\}$. If $Pg \in [B]$, then B is a subset of Pg. Hence, $Pg = P$ is a subgroup of G which contains B. By Theorem 3.28, $P = P_J$ for a subset J of S. Thus, an element of $[B]$ corresponds uniquely to a subset J of S, and $P_J < P_K$ if and only if K is a subset of J. Thus, $[B]$ is a simplex of rank d such that $d = |S|$.

It is clear that the two elements Pg and P' of Δ have a greatest lower bound. It is given by the subgroup $P'' = \langle P', Pg \rangle$. Thus, any pair of the elements of Δ has a greatest lower bound. This completes the proof that Δ is a complex of rank d, where $d = |S|$.

(b) Next, we will prove that the subset Σ is a subcomplex of rank d. By definition, Σ consists of all cosets of the form Pn $(n \in N)$. If $P'g' < Pn$, then Pn is a subset of $P'g'$. Hence, $P'g'$ contains n, and $P'g' = P'n \in \Sigma$. This proves that Σ is a subcomplex. Since $Pn < Bn$, Σ is a subcomplex of rank d.

(c) An element of rank $d - 1$ which is contained in B is a subgroup of the form $P = \langle B, s \rangle$ $(s \in S)$. Hence, we have

$$P < Bg \Leftrightarrow g \in P = B \cup BsB.$$

Therefore, $P < Bg \in \Sigma$ if and only if $g = 1$ or $g = s$. This proves that Σ is a thin complex. On the other hand, by (T3), we get $sB \neq Bs$. Thus, $|P : B| \geq 3$ and Δ is thick. Let W be the Weyl group of the Tits system. If $w = s_1 \cdots s_n \in W$ $(s_i \in S)$, then the sequence

$$B, Bs_n, Bs_{n-1}s_n, \ldots, Bs_1 s_2 \cdots s_n$$

is a gallery joining the two elements B and Bw of Σ. This proves that Σ is connected. Thus, Axioms (B1) and (B2) of a building are satisfied.

(d) Let Pg and $P'g'$ be two elements of Δ. Since $G = BWB$, we can find elements $b, b' \in B$ and $n \in N$ such that $g'g^{-1} = b'nb$. Then, we have

$$Pg = Pbg, \quad P'g' = P'nbg.$$

This proves that Pg and $P'g'$ are contained in a member Σbg of $\mathscr{A}$. This proves Axiom (B3) of a building.

(e) In order to prove Axiom (B4), we may suppose that $\Sigma = \{Pn\}$ $(n \in N)$. Assume that $\Sigma' = \Sigma g$ contains Pn and $P'n'$ $(n, n' \in N)$. By definition, there are elements m and m' of N such that

$$Pn = Pmg \quad \text{and} \quad P'n' = P'm'g.$$

It suffices to show that we can adjust g so as to make $n = m$ and $n' = m'$. Define g' by $m'g = g'n'$. Then, we get $P' = P'g'$ and $g' \in P'$. If $m = m''m'$ and $n = n''n'$, then $Pm''g' = Pn''$. Hence, we get $m'' \in Pn''P'$. From (3.26) and (3.27) (i), we conclude that

$$m'' = wn''w' \quad (w \in P \cap N, \ w' \in P' \cap N).$$

If we define g'' by the formula $n'g'' = w'g'n'$, then we get

$$Png'' = Pn''n'g'' = Pn''w'g'n'$$

$$= Pwn''w'g'n' = Pm''g'n' = Pn,$$

$$P'n'g'' = P'w'g'n' = P'g'n' = P'n',$$

and $\Sigma g'' = \Sigma n'g'' = \Sigma w'g'n' = \Sigma g'n' = \Sigma m'g = \Sigma g$. Thus, $\Sigma \to \Sigma g''$ satisfies the required property. This completes the proof that Δ is a building of rank d such that $\mathscr{A}$ is the set of the apartments of Δ. (Actually, the set of the apartments is always uniquely determined by the building Δ if the Weyl group is finite (Tits [2], 3.26).)

(f) The remaining assertions are now obvious. There is a one-to-one correspondence between the elements of $[B]$ and the types of the elements of Δ. So, Pg and $P'g'$ have the same type if and only if $P = P'$. By definition, G acts transitively on the set of the apartments. Clearly, N is contained in the stabilizer of the apartment $\{Pn\}$. All the chambers of this apartment Σ are of the form Bn $(n \in N)$. So, N is transitive on the set of the chambers in Σ. Thus, G is a group of automorphisms of Δ, and G satisfies the conditions (A) and (B). $\square$

The stabilizer of the chamber B in G is precisely B, while the stabilizer of the apartment $\Sigma = \{Pn\}$ is $H_1 N$, where

$$H_1 = \bigcap_{n \in N} n^{-1}Bn.$$

Clearly, N normalizes H_1, so $H_1 N$ is a subgroup. It is easy to see that the pair $(B, H_1 N)$ is also a (B, N)-pair which defines the same building Δ.

Suppose that a group G is a group of automorphisms of a building Δ, and suppose that G satisfies the two conditions (A) and (B). Let Σ be an apartment of Δ, and let C be a chamber of Σ. Set $B = S_G(C)$ and $N = S_G(\Sigma)$. Then, by Theorem 3.25, (G, B, N, S) is a Tits system. We can prove that if we construct a building starting from this Tits system (G, B, N, S) by the method of Theorem 3.29, then we get a building isomorphic to the original building Δ. The chamber C contains exactly one element of each type. So, the stabilizer of any element $A \subset C$ is a subgroup P of G which contains B. Let D be an element of Δ. If D has the same type as A, then the set of elements of G which moves A to D is a coset Pg. It is easy to see that the mapping $D \to Pg$ is an isomorphism from Δ to the building constructed in Theorem 3.29.

The apartment Σ is defined in Theorem 3.29 as the set of cosets which contain elements of N. Let ω be the canonical mapping from N onto the Weyl group. Then, $(3.27)(i)$ is equivalent to

$$\omega(BnB \cap N) = \omega(n),$$

and we have

$$\omega(P_J n \cap N) = W_J \omega(n).$$

Thus, elements of Σ correspond to the cosets of W_J. In particular, the apartment is uniquely determined by the Weyl group since an apartment is isomorphic to the totality of the cosets of all subgroups W_J $(J \subset S)$.

Tits [2] has given a complete classification of buildings of rank ≥ 3 which have finite Weyl groups. Thus, all finite buildings of rank ≥ 3 are known. This theorem has been used in the identification problem of finite simple groups of Lie type (see, for example, Suzuki [7]).

We have shown that the group $PGL(\mathscr{P})$ acts on the flag space $\Delta(\mathscr{P})$ and that its action satisfies the conditions (A) and (B). Thus, $PGL(\mathscr{P})$ has a decomposition (3.23). The usefulness of this decomposition was first noticed by Bruhat in his work [1] on representation theory, and the decomposition $G = BWB$ is called a **Bruhat decomposition** of the group G with a (B, N)-pair.

As we shall remark later, most of the finite simple groups are analogues of simple Lie groups over finite fields. Any simple group of this type has a Tits system.

Exercises

1. (a) Let A and B be two elements of a building Δ, and let Σ be an apartment containing A and B. Let $\text{dist}_\Sigma(A, B)$ be the distance between A and B defined by using only those galleries consisting of chambers of Σ. Show that

$$\text{dist}_\Sigma(A, B) = \text{dist}(A, B).$$

(b) Let W be the Weyl group, and let S be the set of the reflections with respect to the walls of a fixed chamber C of Σ. Let the length of $w \in W$ be $l(w)$. Show that

$$\text{dist}(C, w(C)) = l(w).$$

(*Hint*. (a) Use (3.17) (i).)

2. Let V be an m-dimensional vector space over a field F. A bilinear form $f(u, v)$ on V is said to be *alternating* if

$$f(u, u) = 0 \qquad \text{for all} \quad u \in V.$$

A bilinear form $f(u, v)$ is a function on $V \times V$ taking on values in F such that

$$f(\lambda u, \mu v) = \lambda \mu f(u, v) \qquad (\lambda, \mu \in F)$$
$$f(u_1 + u_2, v) = f(u_1, v) + f(u_2, v)$$
$$f(u, v_1 + v_2) = f(u, v_1) + f(u, v_2)$$

for all u, u_i, v, and v_i of V. If f is alternating, we have

$$f(u, v) = -f(v, u)$$

for all u and v. We will assume that the alternating form f is *nondegenerate*; that is, if $f(x, v) = 0$ for all $x \in V$, then $v = 0$.
For any subspace U of V, let

$$U^{\perp} = \{v \mid f(u, v) = 0 \qquad \text{for all} \quad u \in U\}.$$

Show that the following propositions hold.

(a) For any subspace U of V, $U^{\perp}$ is a subspace and

$$\dim U^{\perp} + \dim U = \dim V.$$

(b) $\dim V = 2n$ is even.
(c) There is a basis $\{u_1, \ldots, u_{2n}\}$ of V such that

$$f(u_i, u_{i+n}) = 1,$$
$$f(u_i, u_j) = 0 \qquad \text{if} \quad |i - j| \neq n.$$

(A basis $\{u_i\}$ with the above property is said to be a *symplectic basis* of V.)
(d) A subspace U of V is said to be *totally isotropic* if it satisfies $U \subset U^{\perp}$. The dimension of a totally isotropic subspace is at most n. Also, there are totally isotropic subspaces of dimension n.

3. Let V be a vector space of even dimension $2n$, and let f be a nondegenerate alternating form defined on V. Consider the set of all F-linear transformations θ on V such that, for all u, v of V

$$f(\theta u, \theta v) = \lambda f(u, v)$$

where $\lambda = \lambda(\theta)$ is a nonzero element of F which depends only on θ and not on u or v. We will denote this set by $GS_p(f)$. Let $S_p(f)$ be the subset of $GS_p(f)$ consisting of all elements θ such that $\lambda(\theta) = 1$.

(a) Show that θ is an F-isomorphism of V and that $GS_p(f)$ forms a subgroup of $GL(V)$.
(b) Let $\{u_i\}$ be a symplectic basis of V. Show that $\theta \in S_p(f)$ if and only if $\{\theta u_i\}$ is a symplectic basis of V.
(c) Let τ be a transvection on V, and set

$$H = \mathrm{Ker}(\tau - 1) \quad \text{and} \quad U = \mathrm{Im}(\tau - 1).$$

Show that $\tau \in S_p(f)$ if and only if $H = U^{\perp}$.
(d) Suppose that u, v, w are three elements of V satisfying

$$f(u, v) \neq 0 \quad \text{and} \quad f(u, w) = f(v, w).$$

Show that there exists a transvection τ such that $\tau \in S_p(f)$, $\tau(u) = v$, and $\tau(w) = w$.
(e) Show that $S_p(f)$ is generated by transvections which are contained in $S_p(f)$. Thus, we have $S_p(f) \subset SL(V)$.

(*Hint.* (a) Since f is nondegenerate, $\theta v = 0$ implies that $v = 0$.
(e) Let $\{u_i\}$ be a symplectic basis of V. Then, for $\theta \in S_p(f)$, $\{\theta u_i\}$ is also a symplectic basis. Let G be the subgroup of $S_p(f)$ which is generated by the transvections in $S_p(f)$. Try to show that there is an element $\varphi \in G$ such that $\theta u_i = \varphi u_i$ for all i. If $f(u_1, \theta u_1) \neq 0$, then there is a transvection τ of $S_p(f)$ such that $\tau u_1 = \theta u_1$. If $f(u_1, \theta u_1) = 0$, then find an element v of V such that $f(u_1, v) \neq 0$ and $f(v, \theta u_1) \neq 0$. Use Part (d) to find an element φ_1 of G such that

$$\varphi_1 u_1 = \theta u_1 \quad \text{and} \quad \varphi_1 u_{n+1} = \theta u_{n+1}.$$

Then, $\{\varphi_1 u_2, \ldots\}$ and $\{\theta u_2, \ldots\}$ are two symplectic bases of the space $U^{\perp}$ where U is generated by θu_1 and θu_{n+1}. Use induction on n to complete the proof.

The group $S_p(f)$ is called the *symplectic group*.)

4. We will continue to study the vector space V of dimension $2n$ and the nondegenerate alternating form f on V. Let $\mathscr{P}$ be the projective space on V. For two points P and Q of $\mathscr{P}$, we define the relation $\perp$ by

$$p \perp Q \quad \text{if} \quad f(u, v) = 0$$

where u and v are elements of V which represent the points P and Q, respectively. Then, the relation $\perp$ is well defined and does not depend on the choice of the elements which represent P and Q. A subspace α of $\mathscr{P}$ is said to be *totally isotropic* if $P \perp Q$ for any two points P and Q of α. Let $\Delta(f)$ be the totality of the flags consisting only of totally isotropic subspaces. A frame $(P_1, \ldots, P_{2n})$ is said to be a *symplectic frame* if $P_i \perp P_j$ unless $|i - j| = n$. For a symplectic frame F, let $\Sigma(F)$ be the set of the flags of $\Delta(f)$ which are supported by the frame F. Prove the following statements.

(a) Let α be a subspace of $\mathscr{P}$. Define $\alpha^{\perp}$ to be the set of points P of $\mathscr{P}$ such that $P \perp Q$ for all $Q \in \alpha$. Then, $\alpha^{\perp}$ is a subspace of $\mathscr{P}$, and $\alpha^{\perp}$ corresponds to the subspace $U^{\perp}$ of V where U is the subspace of V corresponding to α.

(b) Suppose that α and β are totally isotropic subspaces of $\mathscr{P}$ and that γ is a subspace contained in α. In this case, the subspace spanned by γ and $\beta \cap \gamma^{\perp}$ is totally isotropic.

(c) Let F be a symplectic frame, and let $\alpha_1 \supset \cdots \supset \alpha_k$ be a flag of $\Sigma(F)$. Then, F supports the flag

$$\alpha_k^{\perp} \supset \alpha_{k-1}^{\perp} \supset \cdots \supset \alpha_1^{\perp} \supset \alpha_1 \supset \alpha_2 \supset \cdots \supset \alpha_k.$$

(If $\alpha_1^{\perp} = \alpha_1$, omit the term $\alpha_1^{\perp}$.)

(d) Let

$$A: \alpha_1 \supset \alpha_2 \supset \cdots \supset \alpha_n \quad \text{and} \quad B: \beta_1 \supset \beta_2 \supset \cdots \supset \beta_n$$

be two maximal elements of $\Delta(f)$. Set $\gamma = \alpha_1 \cap \beta_1$ and $\dim \gamma = n - l$. Show that there is a set $F = \{P_1, P_2, \ldots, P_{n+l}\}$ of $n + l$ independent points such that (i) F supports both A and B, (ii) $P_{l+1} \cup P_{l+2} \cup \cdots \cup P_n = \gamma$, and (iii) $P_i \perp P_j$ if $|i - j| \neq n$.

(e) Let $\mathscr{A}$ be the totality of subsets of the form $\Sigma(F)$. Then, $\Delta(f)$ is a building of rank n in which $\mathscr{A}$ is the set of the apartments.

(f) $S_p(f)$ acts on $\Delta(f)$ as a group of automorphisms which satisfies the two properties (A) and (B).

(g) Show that the Weyl group of the building $\Delta(f)$ is isomorphic to the general wreath product $Z_2 \text{ wr } \Sigma_n$, where Z_2 is the cyclic group of order 2 and the symmetric group Σ_n is considered to be a permutation group on n letters.

(*Hint.* (d) From $\alpha_1 \cap \beta_i^{\perp}$ and $\gamma \cap \beta_j$, we can construct a maximal flag of α_1. Apply (IV) to α_1.

(e) The first two axioms of buildings are proved by Exercise 2(d) and the definitions. Axiom 3 follows from Part (d). For the Axiom 4, let $\alpha_1 \supset \cdots$ and $\beta_1 \supset \cdots$ be two elements of $\Delta(f)$ supported by two symplectic frames F and F'. Use Part (b) to reduce it to the case where

$$\alpha_1 \cap \beta_1 = \alpha_1^{\perp} \cap \beta_1.$$

Then, apply (V) to $\alpha_1 \supset \cdots$ and

$$\alpha_1 \cap \beta_l^{\perp} \supset \cdots \supset \alpha_1 \cap \beta_1^{\perp} \supset \alpha_1 \cap \beta_1$$

to get a one-to-one correspondence between $F \cap \alpha_1$ and $F' \cap \alpha_1$. Extend this to an isomorphism from F to F'.

(g) By (3.22) (3), the elements of the Weyl group correspond to the chambers in $\Sigma(F)$. Let $\{P_i\}$ be a symplectic basis of V. Then, a maximal flag of totally isotropic subspaces corresponds to an arrangement in linear order of n points which contain exactly one point from each of the n sets $\{P_i, P_{n+i}\}$ ($i = 1, \ldots, n$).

The group induced on the projective space $\mathscr{P}$ by the transformations of $S_p(f)$ is called the *projective symplectic group*. From Exercise 2 (c), we can see that the structure of $S_p(f)$ does not depend on the form f but is determined uniquely by the dimension of V and by the field F. We use such notations as $S_p(f)$, $S_p(2n, F)$, and $S_p(2n, q)$ when F is a finite field of q elements, and the projective symplectic group is denoted by $PS_p(f)$, $PS_p(2n, F)$, etc. We have

$$PS_p(f) \cong S_p(f)/Z$$

where Z is the cyclic group of order 2 if the characteristic of the field F is not two, and where $Z = \{1\}$ if the characteristic is two. The above isomorphism can be proved easily by using (9.8) of Chapter 1 and Exercise 3 (c) and (e). The group $S_p(2, F)$ coincides with $SL(2, F)$. If $n > 1$, then $PS_p(2n, F)$ is always simple except when $n = 2 = F$ (see Nagao [3], p. 180).)

5. Let (G, B, N, S) be a Tits system, and let H be a normal subgroup of G. Show that if $BH = P_J$ for a subset J of S, then J and $S - J$ commute elementwise.

(*Hint.* If $s \in J$ and $r \notin J$, then $Bs \cap H \neq \emptyset$ and $r^{-1}Bsr \cap P_J \neq \emptyset$. Hence, there is an element $w \in W_J$ such that $Bsr \cap rBwB \neq \emptyset$. It follows from (T3) that $sr = rw$. By (3.27) (iv), $l(w)$ is not 3 but 1. So, w is an element of S. Each element of S is a reflection with respect to a wall of a simplex. Consider the action on the vertices. Then, it will be obvious that $w = s$.)

6. Let (G, B, N, S) be a Tits system, let U be a normal subgroup of B, let G_1 be the normal subgroup of G generated by all the conjugates of U, and let $Z = $ Core B (Definition 7.17 of Chapter 1). Assume that the following conditions are satisfied.

(i)   We have $B = UT$ where $T = B \cap N$.
(ii) The group U is solvable.
(iii) The group G_1 coincides with the commutator subgroup $D(G_1)$:
      $G_1 = D(G_1)$.
(iv) The Weyl group is *irreducible* (that is, if $J \subset S$ commutes elementwise with $S - J$, then either $J = S$ or J is empty).

Then, if a subgroup H of G is normalized by G_1, we have either $H \subset Z$ or $H \supset G_1$. In particular, the group $G_1/G_1 \cap Z$ is a simple group. Prove this theorem following the subsequent steps.

(a) We have $G_1 T = G$.
(b) Set $G_0 = G_1 H$. Define $B_0 = G_0 \cap B$, $N_0 = G_0 \cap N$, and $| T_0 = G_0 \cap T$. Then, $B = B_0 T$, $N = N_0 T$, and $T_0 \triangleleft N_0$. By setting $W_0 = N_0/T_0$, we get an isomorphism α from W onto W_0 such that

$$BwB = B\alpha(w)B_0, \qquad BwB \cap G_0 = B_0\,\alpha(w)B_0.$$

The quadruple (G_0, B_0, N_0, S_0) is a Tits system where $S_0 = \alpha(S)$.
(c) We have either $B_0 H = B_0$ or $B_0 H = G_0$.
(d) If $B_0 H = B_0$, then we get $H \subset Z$.
(e) Suppose that $B_0 H = G_0$. Then, any conjugate of U is obtained by the conjugation of an element of H. This implies that $G_0 = G_1 H = UH$. So, we get

$$U/U \cap H \cong UH/H \cong G_1/G_1 \cap H.$$

(f) The left side of the isomorphism is solvable by (ii), while the right side coincides with its commutator subgroup by (iii). Hence, $G_1 = G_1 \cap H$ or $G_1 \subset H$. This proves the assertion.

(*Hint.* (a) Both B and N normalize $G_1 T$. Apply Exercise 5 here and again for Part (c).

This problem gives an alternate proof of (9.10) of Chapter 1. The last step is an idea of Iwasawa [1]. In order to apply this proof to the symplectic group $S_p(2n, F)$, we need to verify that the conditions (i), (ii), and (iii) hold for $S_p(2n, F)$. A chamber of the building $\Delta(f)$ is a maximal flag $\alpha_1 \supset \cdots \supset \alpha_n$ of totally isotropic subspaces. So, by Exercise 4 (c), the elements of B leave the maximal flag

$$\alpha_n^{\perp} \supset \cdots \supset \alpha_2^{\perp} \supset \alpha_1 \supset \cdots \supset \alpha_n$$

invariant. The dimensions of these subspaces decrease by one at each step, so the elements of B are represented by triangular matrices. Thus, B is solvable. The set U of the elements which are represented by unipotent matrices (those triangular matrices with 1's on the main diagonal) satisfies the conditions (i) and (ii). Since B contains transvections, it is sufficient to show that any transvection is contained in the commutator subgroup of $S_p(f)$.

Let $\{u_i\}$ be a symplectic basis of V. Then,

$$\tau_\lambda: u_1 \to u_1 + \lambda u_{n+1}, \qquad u_i \to u_i \quad (i > 1)$$

is a transvection of $S_p(f)$. For any element $\mu \neq 0$ of F, let ρ be the transformation which satisfies

$$u_1 \to \mu^{-1} u_1, \qquad u_{n+1} \to \mu u_{n+1}, \qquad u_i \to u_i \quad (i \neq 1, n+1).$$

Then, $\rho \in S_p(f)$, and $[\tau_\lambda, \rho] = \tau_v$ with $v = (\mu^2 - 1)\lambda$. Any transvection of $S_p(f)$ is conjugate to τ_λ in $S_p(f)$. (This is proved in the same way as in (9.6) of Chapter 1.) So, if there is an element μ such that $\mu^2 \neq 0$, 1, any transvection is a commutator. In this way, the simplicity of $PS_p(2n, F)$ can be proved when $|F| \geq 4$. If $|F| \leq 3$, the required proposition holds only when $n > 1$ for $|F| = 3$ and when $n > 2$ for $|F| = 2$.)

§4. Coxeter Groups

In the preceding section, we defined the Weyl group of an apartment (Definition 3.19). It is the group generated by the set S of reflections with respect to the walls of a fixed chamber (3.22) (i). In this section, we will consider the defining relations of the Weyl group.

Definition 4.1. Let W be any group generated by a subset S of W. The group W is said to be a **Coxeter group** if the following two conditions are satisfied.
(1) Every element of S has order 2.
(2) For every pair of elements σ and τ of S, let $m = m(\sigma, \tau)$ be the order of the product $\sigma\tau$. Then, the relations

$$R = \{(\sigma\tau)^m = 1\}$$

are the defining relations of W with respect to the set S of generators.
We also say that (W, S) is a **Coxeter system**.

In this section, we will prove that a Weyl group of a building is a Coxeter group, and we will give a complete classification of finite Coxeter groups.
Let W be any group generated by a subset S consisting of elements of order 2. Let T be the totality of elements of W which are conjugate to some element of S. Set

$$X = \{\pm 1\} \times T.$$

For a sequence $\mathsf{s} = (s_1, \ldots, s_q)$ of elements of S, we define the sequence $\mathsf{t} = (t_1, \ldots, t_q)$ of elements of T by

$$t_i = (s_1 \cdots s_{i-1})s_i(s_1 \cdots s_{i-1})^{-1},$$

and denote it by $\mathsf{t} = \Phi(\mathsf{s})$. For any $t \in T$, let $n(\mathsf{s}, t)$ be the number of indices j such that $t_j = t$. Define

$$\eta(\mathsf{s}, t) = (-1)^{n(\mathsf{s}, t)}.$$

With this notation, we have the following lemma.

(4.2). *Let (W, S) be a Coxeter system.*

(i) *If* $\mathsf{s} = (s_1, \ldots, s_q)$ *and* $\mathsf{s}' = (s'_1, \ldots, s'_q)$ *are two sequences such that* $s_1 s_2 \cdots s_q = s'_1 s'_2 \cdots s'_q$, *then we have*

$$\eta(\mathsf{s}, t) = \eta(\mathsf{s}', t).$$

Thus, for any $w = s_1 s_2 \cdots s_q$ *of W, we can define $\eta(w, t)$.*

(ii) *Let U_w be the permutation on X defined by*

$$U_w(\varepsilon, t) = (\varepsilon \eta(w, t), w^{-1} t w).$$

Then, the mapping $w \to U_w$ is a homomorphism from W onto a permutation group on the set X.

Proof. For any $s \in S$, we define

$$U_s(\varepsilon, t) = (\pm \varepsilon, s^{-1} t s)$$

where the signature of the right side of the equation is negative only when $s = t$. Since the element s of S has order 2 we have $U_s^2 = 1$. So, U_s is a permutation on the set X. We will show that the mapping $s \to U_s$ can be extended to a homomorphism from the group W. By the definition of W, we need only to prove that the mapping preserves the defining relations R of W (Definition 6.1 of Chapter 2). Let s and s' be two elements of S. Suppose that the order of the product ss' is exactly m. Then, for any $k < m$, we get

$$(U_s U_{s'})^k (\varepsilon, t) = (\pm \varepsilon, (ss')^{-k} t (ss')^k).$$

We will investigate exactly when the sign of ε changes. By definition, the sign changes at $k + 1$ if and only if

$$(ss')^{-k} t (ss')^k = s \quad \text{or} \quad ss's.$$

In these cases, we have either $t = (ss')^{2k} s$ or $t = (ss')^{2k+1} s$. The integer k ranges from 0 to $m - 1$. If $t = (ss')^l s$ for some $l < m$, then $t = (ss')^{l+m} s$. So, a change of sign occurs twice. Thus, $(U_s U_{s'})^m = 1$ and the mapping U preserves the defining relations of W. Hence, U can be extended to a homomorphism from W into a permutation group on X. Suppose that $w = s_1 \cdots s_q$ ($s_i \in S$). Then, we have

$$U_w = U_{s_1} \cdots U_{s_q} \quad \text{and} \quad U_w(\varepsilon, t) = (\pm \varepsilon, w^{-1} t w).$$

The sign of ε will change whenever $s_j = (s_1 \cdots s_{j-1})^{-1} t (s_1 \cdots s_{j-1})$. Thus, the sign of ε agrees with $\eta(\mathsf{s}, t)$. This proves Lemma (4.2). $\square$

Let W be any group generated by a set S of elements of order 2. For any element w of W, the *length* of w is defined to be the number of factors in a shortest expression $w = s_1 s_2 \cdots s_q$ of w as a product of elements s_i of S, and we denote the length of w by $l(w)$.

(4.3) Let (W, S) be a Coxeter system. For a sequence $\mathfrak{s} = (s_1, s_2, \ldots, s_q)$ of elements s_i of S, let $w = s_1 s_2 \cdots s_q$ and $\mathfrak{t} = \Phi(\mathfrak{s}) = (t_1, \ldots, t_q)$. Then, the length $l(w)$ is q if and only if the set $\{t_1, t_2, \ldots, t_q\}$ consists of q distinct elements.

Proof. Let T_w be the set of elements t of T such that $\eta(w, t) = -1$. By definition, we clearly have $T_w \subset \{t_1, \ldots, t_q\}$. So, by considering the shortest expression of w, we get $|T_w| \leq l(w)$.

Suppose that $\{t_1, \ldots, t_q\}$ consists of q distinct elements. Then, for each i, we get $n(\mathfrak{s}, t_i) = 1$. Hence, $t_i \in T_w$ and we have $|T_w| = q$. This implies that $l(w) = q$.

Suppose that $t_i = t_j$ for some $i < j$. Then, by definition, we have

$$s_i s_{i+1} \cdots s_{j-1} = s_{i+1} \cdots s_{j-1} s_j.$$

Thus, in the expression of w, s_i and s_j are cancelled:

$$w = s_1 \cdots s_{i-1} s_{i+1} \cdots s_{j-1} s_{j+1} \cdots s_q.$$

We have $l(w) \leq q - 2$. $\quad \square$

The above two lemmas give us the following theorem (Matsumoto [1]).

Theorem 4.4. Let W be a group generated by a set S of elements of order 2. Then, (W, S) is a Coxeter system if and only if the following **exchange condition** is satisfied.

Let $s, s_1, s_2, \ldots, s_q$ be elements of S such that the product $w = s_1 s_2 \cdots s_q$ has length exactly q. If $l(sw) \leq l(w) = q$, then there is an integer j, $1 \leq j \leq q$, such that we have $s s_1 \cdots s_{j-1} = s_1 \cdots s_j$.

Proof. (a) Suppose that (W, S) is a Coxeter system. Set $s_0 = s$. If we have

$$l(s_0 s_1 \cdots s_q) \leq q < q + 1,$$

then, by (4.3), there are integers i and j such that

$$s_i s_{i+1} \cdots s_{j-1} = s_{i+1} \cdots s_j, \qquad 0 \leq i < j \leq q.$$

However, if $i > 0$, we would get $l(w) < q$. Hence, we have $i = 0$. Thus, the exchange condition holds.

(b) Conversely, suppose that the exchange condition holds for (W, S). Let

R be the set of relations in Definition 4.1 (2). We will show that a mapping from S into a group G which preserves the relations R can be extended to a homomorphism from W into G.

So, let f be such a mapping. Let w be an element of W with $l(w) = q$. By definition, we can write $w = s_1 s_2 \cdots s_q$ $(s_i \in S)$. There is only one possible extension. The first step is to show that the extension is uniquely defined regardless of how w is written. Let D_w be the set of sequences $\mathfrak{s} = (s_1, \ldots, s_q)$ of elements of S such that $w = s_1 s_2 \cdots s_q$. Define $f(\mathfrak{s})$ to be equal to

$$f(s_1) f(s_2) \cdots f(s_q).$$

We want to show that $f(\mathfrak{s}) = f(\mathfrak{s}')$ for any two sequences $\mathfrak{s}$ and $\mathfrak{s}'$ of D_w. We will use induction on q.

If we suppose that $f(\mathfrak{s}) \neq f(\mathfrak{s}')$ for some $\mathfrak{s}$ and $\mathfrak{s}'$ of D_w, then we should arrive at a contradiction. Let

$$\mathfrak{s} = (s_1, \ldots, s_q) \quad \text{and} \quad \mathfrak{s}' = (s_1', \ldots, s_q').$$

Then, we have $w = s_1 \cdots s_q = s_1' \cdots s_q'$. For $s = s_1'$, $l(sw)$ is at most $q - 1$. Hence, by the exchange condition, we have

$$ss_1 \cdots s_{j-1} = s_1 \cdots s_j \quad \text{for some} \quad j \leq q.$$

Suppose that $j < q$. Then, we have $w = ss_1 \cdots s_{j-1} s_{j+1} \cdots s_q$, and the sequence

$$\mathfrak{s}'' = (s, s_1, \ldots, s_{j-1}, s_{j+1}, \ldots, s_q)$$

belongs to D_w. By comparing the first $q - 1$ terms of $\mathfrak{s}''$ with $\mathfrak{s}$, and the last $q - 1$ terms with $\mathfrak{s}'$, we get

$$f(\mathfrak{s}) = f(\mathfrak{s}'') = f(\mathfrak{s}')$$

from the inductive hypothesis. This contradicts the assumption. Hence we have $j = q$, and the sequence $\mathfrak{s}'' = (s, s_1, \ldots, s_{q-1})$ belongs to D_w. Moreover, by the inductive hypothesis, we get $f(\mathfrak{s}'') = f(\mathfrak{s}') \neq f(\mathfrak{s})$. Next, consider $\mathfrak{s}''$ and $\mathfrak{s}$. Then, the above argument shows that

$$\mathfrak{s}^* = (s_1, s, s_1, \ldots, s_{q-2}) \in D_w \quad \text{and} \quad f(\mathfrak{s}^*) = f(\mathfrak{s}) \neq f(\mathfrak{s}'').$$

By repeating this process, we get

$$\mathfrak{u} = (u, v, u, \ldots), \qquad \mathfrak{v} = (v, u, \ldots) \in D_w \quad \text{and} \quad f(\mathfrak{u}) \neq f(\mathfrak{v}),$$

where $u = s_1$, $v = s'_1$, and u and v appear in alternating order in the sequences. But, we have

$$w = uvuv \cdots = vuvu \cdots .$$

Hence $(uv)^q = 1$. Since f is a mapping which preserves the relations R, we get $(f(u)f(v))^q = 1$. This would mean that $f(u) = f(v)$ which contradicts the assumption.

(c) Let $w \in W$ and $q = l(w)$. If $w = s_1 \cdots s_q$, then we define

$$g(w) = f(s_1)f(s_2) \cdots f(s_q).$$

By (b), the right side does not depend on how the element w is written. We will show that we have

(*) $\qquad\qquad g(sw) = f(s)g(w) \qquad (s \in S, w \in W).$

If $l(sw) = q + 1$, then we have

$$g(sw) = f(s)f(s_1) \cdots f(s_q) = f(s)f(w).$$

Assume that $l(sw) \leq q$. Then, by the exchange condition, we have

$$ss_1 \cdots s_{j-1} = s_1 \cdots s_j$$

for some integer j. Hence, we have $sw = s_1 \cdots s_{j-1}s_{j+1} \cdots s_q$. Since $l(sw) = q - 1$, we have

$$g(sw) = f(s_1) \cdots f(s_{j-1})f(s_{j+1}) \cdots f(s_q).$$

On the other hand, by (b), we have

$$f(s)f(s_1) \cdots f(s_{j-1}) = f(s_1) \cdots f(s_j)$$

and

$$f(s)g(w) = f(s)f(s_1) \cdots = f(s_1) \cdots f(s_j)f(s_j) \cdots f(s_q).$$

Since $f(s_j)^2 = 1$, we get the equality (*).

(d) Finally, we will prove that g is a homomorphism from W. We need to show that $g(uv) = g(u)g(v)$ by using induction on $l(u)$. If $l(u) = 1$, then the formula (*) proves that the assertion is valid. Assume that $l(u) > 1$, and set

$$u = su' \ (s \in S, u' \in W) \quad \text{with} \quad l(u') = l(u) - 1.$$

Then, by (∗) and the inductive hypothesis, we get

$$g(uv) = g(su'v) = f(s)g(u'v)$$
$$= f(s)g(u')g(v) = g(u)g(v).$$

Thus, (W, S) is proved to be a Coxeter system. □

Theorem 4.5. (i) *Let (W, S) be a Coxeter system. Let*

$$w = s_1 s_2 \cdots s_q \qquad (s_i \in S, \ q = l(w))$$

be a shortest expression of w. Then, the set of the distinct elements of $\{s_1, s_2, \ldots, s_q\}$ is uniquely determined by w.

(ii) *Let J be a subset of S, and let W_J be the subgroup generated by J. If $w = s_1 s_2 \cdots s_q$ is a shortest expression of an element w of W_J, then all the factors s_i belong to J.*

(iii) *In the notation of (ii), (W_J, J) is a Coxeter system.*

Proof. (i) Let D_w be the set of sequences defined in Part (b) of the proof of Theorem 4.4. For each $\mathsf{s} = (s_1, \ldots, s_q)$ of D_w, let $f(\mathsf{s})$ be the subset of S consisting of $s_1, \ldots, s_q$. As in Part (b) of the preceding proof, we can prove that $f(\mathsf{s})$ is constant.

(ii) The constant value in (i) will be denoted by S_w. Then, S_w is a subset of S and satisfies

$$S_{w^{-1}} = S_w \quad \text{and} \quad S_{sw} \subset \{s\} \cup S_w.$$

Hence, we get $S_{vw} \subset S_v \cup S_w$ for any elements v and w of W. Thus, the set $W_1 = \{w \mid S_w \subset J\}$ is a subgroup of W. Clearly, we have $W_1 = W_J$. This proves (ii).

(iii) Let w be an element of W_J. By (ii), the length of w with respect to (W_J, J) is the same as the length of w with respect to (W, S). So, the exchange condition for (W, S) implies the exchange condition for (W_J, J). Hence, by Theorem 4.4, (W_J, J) is a Coxeter system. □

We will now consider the Weyl group of a building.

(4.6). *Let Σ be an apartment of a building, and let W be the Weyl group of Σ. Let C be a fixed chamber of Σ, and let S be the set of reflections with respect to the walls of C.*

(i) *Let w_1 and w_2 be two elements of W. Then, $w_1 C$ and $w_2 C$ are adjacent chambers if and only if there is an element s of S such that $w_2 = sw_1$.*

(ii) *If $\{w_0 C, w_1 C, \ldots, w_k C\}$ is a gallery without repetitions which joins*

$w_0 C$ and $w_k C$, then there is a sequence $s_1, \ldots, s_k$ of elements of S such that $w_k = s_k \cdots s_1 w_0$. In particular,

$$l(w_k w_0^{-1}) \leq k.$$

(iii) For any $w \in W$, we have $\operatorname{dist}(C, wC) = l(w)$.

Proof. (i) If $w_1 C$ and $w_2 C$ are adjacent, then C and $w_2 w_1^{-1}(C) = w_1^{-1}(w_2 C)$ have a common wall B. Let s be the reflection with respect to B. Then, we have $sC = w_2 w_1^{-1}(C)$. By (3.22), the Weyl group is simply transitive on the set of chambers. Hence, we get $s = w_2 w_1^{-1}$. Conversely, if $w_2 = sw_1$, then $w_1(C)$ and $w_1(sC) = w_2 C$ have a common wall. Therefore, they are adjacent.

(ii) By the definition of a gallery, $w_i C$ and $w_{i+1} C$ are adjacent. So, by (i), there is an element s_{i+1} such that $w_{i+1} = s_{i+1} w_i$. Hence, we get $w_k = s_k s_{k-1} \cdots s_1 w_0$. By the definition of the length, $l(w_k w_0^{-1}) \leq k$ holds.

(iii) By (ii), we get $l(w) \leq \operatorname{dist}(C, wC)$. But, if $w = s_1 \cdots s_q$ with $q = l(w)$, then there is a gallery of length q joining C and wC. Hence, $\operatorname{dist}(C, wC) \leq q$. This implies that $l(w) = \operatorname{dist}(C, wC)$. $\square$

Theorem 4.7. *Let W be the Weyl group of an apartment Σ in a building. Let S be the set of reflections with respect to the walls of a fixed chamber of Σ. Then, (W, S) is a Coxeter system.*

Proof. By (3.22), W is generated by S. By definition, each element of S is of order 2. So, in order to prove Theorem 4.7, it is sufficient to prove that the exchange condition of Theorem 4.4 is satisfied.

Assume that $w = s_1 s_2 \cdots s_q$ ($s_i \in S$) is a shortest expression of w and that $l(sw) \leq q$ for some element s of S. By (4.6),

$$C, s_1 C, s_2 s_1 C, \ldots, s_q s_{q-1} \cdots s_1 C$$

is a shortest gallery joining C and $w^{-1}C$. By assumption, we have $l(sw) \leq q$. Hence, the distance between C and $(sw)^{-1}C$ is at most q, and so we have

$$\operatorname{dist}(C, s^{-1}(w^{-1}C)) = \operatorname{dist}(sC, w^{-1}C) \leq q.$$

Since $s \in S$, C and sC are adjacent. Let B be their common wall. Let φ be the folding which sends sC to C. Then, the actions of φ and s are identical for all chambers not in $\varphi(\Sigma)$.

Suppose that $w^{-1}C \in \varphi(\Sigma)$. Since $\operatorname{dist}(sC, w^{-1}C) \leq q$, there is a gallery G of length at most q which joins sC and $w^{-1}C$. Then, $\varphi(G)$ is a gallery joining C and $w^{-1}C$. Since $sC \notin \varphi(\Sigma)$, G contains adjacent chambers D and D' such that $D \notin \varphi(\Sigma)$ but $D' \in \varphi(\Sigma)$. Then, $\varphi(D)$ must be equal to D', and $\varphi(G)$ stammers. This contradicts the equality $\operatorname{dist}(C, w^{-1}C) = q$. Therefore, we get

$w^{-1}C \notin \varphi(\Sigma)$. In the gallery $C, s_1 C, \ldots, w^{-1}C$, there must be an integer j such that

$$s_{j-1} \cdots s_1 C \in \varphi(\Sigma), \quad \text{but} \quad s_j s_{j-1} \cdots s_1 C \notin \varphi(\Sigma).$$

Then, we have

$$s_{j-1} \cdots s_1 C = \varphi(s_j s_{j-1} \cdots s_1 C) = s(s_j \cdots s_1 C).$$

Since the Weyl group is simply transitive, we get

$$s s_1 \cdots s_{j-1} = s_1 s_2 \cdots s_j.$$

Thus, the exchange condition is satisfied. By Theorem 4.4, (W, S) is a Coxeter group. $\square$

The Classification of Finite Coxeter Groups

The aim of the remaining portion of this section is to classify all the finite Coxeter groups. So, we will assume that the set S is *finite*. Let $s_1, \ldots, s_n$ be the elements of S, and let m_{ij} be the order of the product $s_i s_j$. Then, the structure of the Coxeter group W is uniquely determined by the $n \times n$ matrix

$$M = (m_{ij}).$$

The matrix M is a symmetric matrix, and its entries are positive integers (or infinity ∞) such that

$$m_{ii} = 1, \qquad m_{ij} \geq 2 \quad (i \neq j).$$

It will follow from later discussions (Cor. 3 of (4.13)) that the converse also holds: any symmetric matrix satisfying the above conditions is a matrix associated with a Coxeter group.

The matrix associated with a Coxeter group is most conveniently represented by a graph.

Definition 4.8. A **graph** is a pair (E, S) consisting of a set S of **vertices** and a subset E of $S \times S$. An element of E is called an **edge**. If $e = (s_1, s_2) \in E$, then e is the edge which joins the two vertices s_1 and s_2.

Strictly speaking, the edge $e = (s_1, s_2)$ is an arrow from s_1 to s_2. Thus, edges are directed. Furthermore, an edge (s_1, s_1) is possible since a graph may have a *loop*. But, in this book, we will only consider undirected graphs without loops. So, a graph (E, S) satisfies the following two conditions: $(s, s) \notin E$ for all $s \in S$ and

$$(s, t) \in E \Leftrightarrow (t, s) \in E.$$

In this section, we will consider a slightly generalized graph in which each edge has some multiplicity. Thus, two vertices may be joined by a double bond, a triple bond, etc. We will represent a graph on a plane by drawing a set of points, each representing a vertex of the graph, and by joining two points s_1 and s_2 by a line segment when $e = (s_1, s_2)$ is an edge of the graph. A double bond is expressed by either of the following notations:

$$\circ \Longrightarrow \circ \quad \text{or} \quad \circ \overset{2}{\text{———}} \circ$$

Let $M = (m_{ij})$ be the matrix associated with a Coxeter system (W, S). The most important numbers are $m_{ij} - 2$ for $i \neq j$. So, we will define the graph associated with M as follows:

$$\Gamma = \Gamma_M = (E, I) \qquad I = \{1, 2, \ldots, n\}$$
$$(i, j) \in E \Leftrightarrow m_{ij} \geq 3.$$

We will define the multiplicity of the edge (i, j) to be $m_{ij} - 2$. The graph defined above is said to be *the graph associated with the Coxeter system* (W, S).

The vertex i of Γ corresponds to the generator s_i. We may identify the set of vertices of Γ with the set S. Two different vertices i and j are not joined in Γ if and only if the product $s_i s_j$ of generators has order 2. By Theorem 6.8 of Chapter 2, this is equivalent to saying that s_i commutes with s_j.

EXAMPLE. By Definition 4.1 and Theorem 2.14, the symmetric group Σ_{n+1} on $n + 1$ letters and a generating set

$$S = \{(1\ 2), (2\ 3), \ldots, (n\ n + 1)\}$$

form a Coxeter system (Σ_{n+1}, S). The corresponding graph of this system is the following:

$$\begin{array}{ccccccc} 1 & 2 & 3 & 4 & & & n \\ \circ\!\!-\!\!\circ\!\!-\!\!\circ\!\!-\!\!\circ\!\!-\!\!\circ\!-\!-\!\circ\!\!-\!\!\circ \end{array}$$

We will give another example of a Coxeter group. Consider the group generated by three elements r, s, and t of order 2 with the relations

$$(rs)^5 = (st)^3 = (rt)^2 = 1.$$

This is certainly a Coxeter group. If $rs = x$ and $st = y$, we get

$$x^5 = y^3 = (xy)^2 = 1.$$

So, by Example 4 of Chapter 2, §6, we get $\langle x, y \rangle \cong A_5$. Clearly, we have $x^s = x^{-1}$, $y^s = y^{-1}$ and $\langle x, y, s \rangle = \langle r, s, t \rangle$. Hence, $\langle x, y \rangle$ is a normal subgroup of index 2. It is easy to see that

$$x \rightarrow (1\ 2\ 3\ 4\ 5), \qquad y \rightarrow (1\ 5\ 3), \qquad xy \rightarrow (1\ 2)(3\ 4)$$

can be extended to an isomorphism from $\langle x, y \rangle$ onto A_5. The inner automorphism by the element $(1\ 5)(2\ 4)$ corresponds to the automorphism which extends $x \rightarrow x^{-1}$ and $y \rightarrow y^{-1}$. If z is the element of $\langle x, y \rangle$ which corresponds to $(1\ 5)(2\ 4)$, then the element sz^{-1} centralizes $\langle x, y \rangle$. Thus, we get

$$\langle r, s, t \rangle = \langle sz^{-1} \rangle \times \langle x, y \rangle \cong Z_2 \times A_5.$$

The graph corresponding to this Coxeter group is

As may be seen from these examples, any Coxeter group contains a normal subgroup of index 2 consisting of elements of even length. □

Let Γ be a general graph. Two vertices (p, q) of Γ are said to be *connected* in Γ if there are vertices $p_0, p_1, \ldots, p_n$ of Γ such that

$$p_0 = p, \qquad p_n = q, \quad \text{and} \quad (p_{i-1}, p_i) \in E$$

for all $i = 1, 2, \ldots, n$. A graph Γ is said to be *connected* if every pair of vertices is connected in Γ. If p is a vertex of Γ, then the set Δ of vertices of Γ which are connected to p is a subset of Γ. In this case, the set

$$E(\Delta) = \{(p, q) \mid p, q \in \Delta, (p, q) \in E\}$$

is a subset of E, and $(E(\Delta), \Delta)$ is a graph. Moreover, it is connected. We call it *the connected component* containing p.

(4.9) *Let (W, S) be a Coxeter system, and let Γ be the associated graph. If Γ is not connected, then the set S is a union of two disjoint, nonempty subsets S_1 and S_2. Let W_i be the subgroup of W generated by S_i $(i = 1, 2)$. Then, (W_i, S_i) is a Coxeter system $(i = 1, 2)$, and we have the direct product decomposition $W \cong W_1 \times W_2$. Conversely, if W is the direct product of W_1 and W_2 as above, then the corresponding graph is not connected.*

Proof. The set of vertices of Γ may be identified with the set S. By assumption, Γ is not connected. Let S_1 be the vertex set of a connected component, and set $S_2 = S - S_1$. Then, S_1 and S_2 are nonempty disjoint subsets of S. If $s \in S_1$ and $t \in S_2$, then the vertices s and t are not joined in Γ. So, s and t are two commuting elements of S. Thus, W_1 and W_2 commute elementwise. Clearly, we have $W = W_1 W_2$. Let w be an element of $W_1 \cap W_2$. Then, by Theorem 4.5 (ii), a shortest expression of w is written as a product of elements of $S_1 \cap S_2$. Since $S_1 \cap S_2$ is the empty set, we have $W_1 \cap W_2 = \{1\}$. So, we have $W = W_1 \times W_2$. Theorem 4.5 (iii) proves that (W_i, S_i) is a Coxeter system. The converse is easy to verify. $\square$

Definition 4.10. A Coxeter system (W, S) is said to be **irreducible** if the graph associated with (W, S) is connected.

By (4.9), any finite Coxeter group is a direct product of irreducible Coxeter groups. Thus, we need only to classify irreducible finite Coxeter systems.

One of the most powerful methods of investigating the structure of Coxeter groups is to represent each element of S by a geometric reflection of a vector space V with respect to a hyperplane of V, and to consider W as a group of linear transformations of V.

Definition 4.11. Let V be a vector space of dimension n over the field $\mathbb{R}$ of real numbers. Let $\{u_1, u_2, \ldots, u_n\}$ be a basis of V over $\mathbb{R}$. Let (W, S) be a Coxeter system where $S = \{s_1, \ldots, s_n\}$, and let R be the defining relations of the form

$$(s_i s_j)^m = 1 \qquad (m = m_{ij}).$$

Let B be the bilinear form defined by

$$B(u_i, u_j) = -\cos(\pi/m_{ij}) \qquad (i, j = 1, \ldots, n).$$

The form B is called the **bilinear form associated with the Coxeter system** (W, S). Define the linear transformation σ_i of V by

$$\sigma_i(x) = x - 2B(u_i, x)u_i \qquad (i = 1, 2, \ldots, n).$$

Using these notations, we have the following proposition.

(4.12) (i) *The form B is a symmetric bilinear form on V, and it satisfies*

$$B(u_i, u_i) = 1, \qquad B(u_i, u_j) \le 0 \quad (i \ne j).$$

(ii) *Let H_i be the hyperplane which is defined by*

$$H_i = \{v \mid B(u_i, v) = 0\}.$$

Then, σ_i is the reflection with respect to H_i. Furthermore, σ_i leaves the form B invariant :

$$B(\sigma_i x, \sigma_i y) = B(x, y).$$

(iii) *The order of the product $\sigma_i \sigma_j$ is precisely equal to m_{ij}.*

Proof. (i) This is obvious from the definitions.

(ii) The set H_i is certainly a hyperplane of V. We have $\sigma_i(u_i) = -u_i$. On the other hand, if $v \in H_i$, then $B(u_i, v) = 0$ and $\sigma_i(v) = v$. This proves that σ_i is the reflection with respect to H_i. Any vector w of V can be written as a sum $u + v$ where $u \in H_i$ and v is a scalar multiple of u_i. Then, we have $\sigma_i(u + v) = u - v$. If $u' + v'$ is a similar decomposition of another element w' of V, we get

$$B(w, w') = B(u, u') + B(v, v')$$

because $B(u, v') = B(u', v) = 0$. Similarly, we get

$$B(\sigma_i w, \sigma_i w') = B(u, u') + B(-v, -v').$$

This shows that $B(\sigma_i w, \sigma_i w') = B(w, w')$. Thus, σ_i leaves the form B invariant.

(iii) Suppose that m_{ij} is infinite. (This case does not occur if W is finite.) If $m_{ij} = \infty$, then we interpret $B(u_i, u_j) = -1$. In this case, $u = u_i + u_j$ is orthogonal to both u_i and u_j, and it is left invariant by both σ_i and σ_j. Since $\sigma_j(u_i) = 2u - u_i$, we have

$$\sigma_i \sigma_j(u_i) = u_i - 2u, \qquad (\sigma_i \sigma_j)^k(u_i) = u_i - 2ku.$$

This shows that the order of $\sigma_i \sigma_j$ is infinite.

Suppose that $m = m_{ij}$ is finite. Let P be the plane spanned by the two vectors u_i and u_j. Then, the restriction of B on the plane gives the quadratic form Q: for $z = xu_i + yu_j$, we have

$$Q(z) = B(z, z) = x^2 - 2xy \cos(\pi/m) + y^2$$
$$= (x - y \cos(\pi/m))^2 + y^2 \sin^2(\pi/m).$$

Thus, Q is positive definite. In particular, B is nonsingular on P. Hence, we get $V = P + H$, where H is the subspace of vectors which are orthogonal to both u_i and u_j. Since both σ_i and σ_j act trivially on H, the order of $\sigma_i \sigma_j$ is equal to the order of its restriction on P. Since Q is positive definite, σ_i and σ_j correspond to reflections with respect to lines in the usual Euclidean plane. Since

$$B(u_i, u_j) = -\cos(\pi/m) = \cos(\pi - (\pi/m)),$$

the angle between the two unit vectors u_i and u_j is $\pi - (\pi/m)$. Thus, $\sigma_i \sigma_j$ is a clockwise rotation through an angle of $2\pi/m$. Hence, the order of $\sigma_i \sigma_j$ is exactly m. $\square$

By (4.12) (i), the form B is symmetric. So, B determines the quadratic form $Q(x) = B(x, x)$. This is called *the quadratic form associated with the Coxeter system* (W, S). This quadratic form will be the focal point of the following discussion. The form B is determined from the quadratic form Q by the well-known formula:

$$Q(u + v) = Q(u) + Q(v) + 2B(u, v).$$

If $\{u_1, \ldots, u_n\}$ is the basis which we considered fixed in our definition of B, then we have

$$Q(u) = \Sigma a_{ik} x_i x_k, \qquad x_{ik} = B(u_i, u_k),$$

for any element $u = x_1 u_1 + \cdots + x_n u_n$ of V. We may refer to a_{ik} as the *coefficients* of the quadratic form Q.

By (4.12) (iii), the mapping $s_i \to \sigma_i$ preserves the defining relations R of the Coxeter group W. Thus, the mapping is extended to a homomorphism from W to a group of linear transformations of V. We call this the *canonical representation* and denote it by σ. An important property of the canonical representation is that σ is faithful. We will prove this result.

The group W acts on V via the representation σ. This will induce the contragredient representation σ^* of W on the dual space V^* of V([IK], Chapter 6, §3 or [AS], Chapter 5, §2). The dual space V^* is defined to be $\mathrm{Hom}_R(V, \mathbb{R})$. So, if $v^* \in V^*$, v^* is a linear function defined on V. We denote the value $v^*(u)$ on an element u of V by $\langle v^*, u \rangle$. Then, $\langle \, , \, \rangle$ is a function from $V^* \times V$ into $\mathbb{R}$ which is linear with respect to both variables. The contragredient representation σ^* is defined by the formula:

$$\langle \sigma^*(w)v^*, u \rangle = \langle v^*, \sigma(w^{-1})u \rangle \qquad (w \in W).$$

We write simply $w(v^*)$ for $\sigma^*(w)v^*$.

For each i, let $A_i = \{x^* \in V^* \mid x^*(u_i) > 0\}$. Then, A_i is a half-space of V^*. Let C be the intersection of all A_i for $i = 1, 2, \ldots, n$. This is the portion of space bounded by n hyperplanes passing through the origin. Hence, it is an open simplicial cone.

We will prove the next theorem of Tits following Bourbaki [1].

(4.13) *Let C be the cone of V^* defined above. If $C \cap w(C)$ is not empty for $w \in W$, then $w = 1$.*

We need the following lemma.

Lemma. *Let s_i and s_j be two elements of S, and let W_{ij} be the subgroup generated by s_i and s_j. Let A_i be the half-space defined above. If $w \in W_{ij}$, then $w(A_i \cap A_j)$ is contained in A_i or $s_i(A_i)$. If $w(A_i \cap A_j) \subset s_i(A_i)$, then we have $l(ws_i) = l(w) - 1$.*

Proof. Let P be the plane spanned by u_i and u_j. The set $P^\perp$ defined by $\{v^* \in V^* \mid \langle v^*, P \rangle = 0\}$ is a subspace of V^*, and $V^*/P^\perp$ is isomorphic to P^*. Since P is left invariant by both σ_i and σ_j, $P^\perp$ is also left invariant by the contragredient representation of W_{ij}. Thus, W_{ij} acts on $P^* = V^*/P^\perp$. The image of A_i in P^* is identified with the half-space defined in P^* with respect to u_i. Furthermore, by Theorem 4.5 (ii), the length of an element w of W_{ij} does not change when w is considered to be an element of W. Thus, we may assume that $S = \langle s_i, s_j \rangle$ and $W = W_{ij}$.

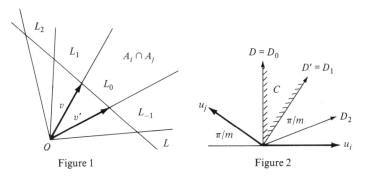

Figure 1   Figure 2

Suppose that $m = m_{ij}$ is infinite. If $\{v, v'\}$ is the dual basis of $\{u_i, u_j\}$, then we get

$$s_i v = -v + 2v', \qquad s_j v = v$$

$$s_i v' = v', \qquad s_j v' = 2v - v'.$$

We will represent the elements v and v' as vectors OR and OR' from the origin. Let L be the line through R and R'. Then, since $s_i(v - v') = -(v - v')$, s_i induces the reflection on L with respect to R'. Similarly, s_j induces the reflection with respect to R. Since $\{v, v'\}$ is the dual basis of $\{u_i, u_j\}$, A_i is the half-plane which is bounded by the line through O and R' and includes R. So, $A_i \cap A_j$ is the region between the two rays OR and OR'. Then, the image of $A_i \cap A_j$ under $w \in W$ corresponds faithfully to the image of the segment RR' on L. Let us introduce a coordinate system on L, taking R' as the origin and RR' as the unit. Then, the images of R and R' by elements of W are points with integral coordinates. Let L_n be the interval $(n, n + 1)$ on L. Then, $L_0 = L \cap (A_i \cap A_j)$ and

$$(s_i s_j)^k L_0 = L_{2k}, \qquad (s_i s_j)^k s_i L_0 = L_{-2k-1}.$$

Thus, for any $w \in W_{ij}$, $w(A_i \cap A_j)$ is contained in either A_i or $s_i A_i$. (The set $s_i A_i$ is the other half-plane bounded by the line OR'.) If $w(A_i \cap A_j) \subset s_i A_i$, then $wL_0 = L_s$ with a negative integer s. Hence, we have either $w = (s_i s_j)^k$

with $k < 0$ or $w = (s_i s_j)^k s_i$ with $k \geq 0$. In either case, we have

$$l(ws_i) = l(w) - 1.$$

Suppose that m is finite. As in the proof of (4.12) (iii), the associated quadratic form Q is positive definite. So, we may identify P with P^* and consider P to be the usual Euclidean plane. In this case, the angle between the two vectors u_i and u_j is $\pi - (\pi/m)$. Let R_0 be the ray through the origin which is orthogonal to u_i and lies in the upper half-plane. Similarly, let R_1 be the ray orthogonal to u_j lying in the right half-plane. Then, $A_i \cap A_j$ is the region bounded by R_0 and R_1. The product $s_i s_j$ is the clockwise rotation through an angle of $2\pi/m$. Let us define the rays R_i so that the angle from R_0 to R_i is $\pi i/m$. Then, $R_{2m} = R_0$ and the rays R_i divide the plane into $2m$ equal parts. Let C_i be the region between R_i and R_{i+1}. For $0 \leq k < m$, we have

$$(s_i s_j)^k C_0 = C_{2k} \quad \text{and} \quad (s_i s_j)^k s_i C_0 = C_{2(m-k)-1}.$$

The set A_i is the right half-plane and $s_i A_i$ is the left half. So, $w(A_i \cap A_j)$ is certainly contained in either A_i or $s_i A_i$. If $w(A_i \cap A_j) \subset s_i A_i$, then we must have

$$w = (s_i s_j)^k \quad (m \leq 2k), \quad \text{or} \quad w = (s_i s_j)^k s_i \quad (2k < m).$$

In the former case, a shortest expression for w is $(s_j s_i)^{m-k}$. So, in all cases, we have $l(w_i s) = l(w_i) - 1$. $\quad\square$

We now return to the proof of (4.13). Let C and A_i be defined as before. Let w be an element of W with $l(w) = n$. Let (P_n) and (Q_n) be the following propositions.

(P_n) For any i, $w(C) \subset A_i$ or $s_i A_i$. If $w(C) \subset s_i A_i$, then $l(ws_i) = l(w) - 1$.

(Q_n) Set $W_{ij} = \langle s_i, s_j \rangle$ for $i \neq j$. Then, there is an element x of W_{ij} such that $w(C) \subset x(A_i \cap A_j)$ and $l(w) = l(wx^{-1}) + l(x)$.

If $n = 0$, then $w = 1$ and the propositions (P_0) and (Q_0) are trivial to prove. So, we will proceed by induction on n.

(a) (P_n) and (Q_n) imply (P_{n+1}).

Proof. An element w with $l(w) = n + 1$ can be written as $w = w's'$ with $l(w') = n$ and $s \in S$. If $s = s_i$, then we get $w'(C) \subset A_i$ by (P_n). Hence, we have

$$w(C) = w's_i(C) \subset s_i A_i \quad \text{and} \quad l(ws_i) = l(w') = n.$$

By (Q_n), if $s = s_j \neq s_i$, then there is an element x of W_{ij} such that

$$w'(C) \subset x(A_i \cap A_j) \quad \text{and} \quad l(w') = l(w'x^{-1}) + l(x).$$

Hence, $w(C) = w's(C) \subset xs(A_i \cap A_j)$. By the lemma, we get

$$xs(A_i \cap A_j) \subset A_i \quad \text{or} \quad xs(A_i \cap A_j) \subset s_i A_i,$$

and if $xs(A_i \cap A_j) \subset s_i A_i$, then $l(xss_i) = l(xs) - 1$. So, if $W(C) \subset s_i A_i$, then we have

$$
\begin{aligned}
l(ws_i) &= l(w'ss_i) \\
&\le l(w'x^{-1}) + l(xss_i) = l(w') - l(x) + l(xs) - 1 \\
&\le l(w) - 1.
\end{aligned}
$$

This gives us $l(ws_i) = l(w) - 1$ and (P_{n+1}) holds. $\quad\square$

(b) (P_{n+1}) and (Q_n) imply (Q_{n+1}).

Proof. Let $l(w) = n + 1$. If $w(C) \subset A_i \cap A_j$, then (Q_{n+1}) holds for $x = 1$. If $w(C) \not\subset A_i \cap A_j$, then either $w(C) \not\subset A_i$ or $w(C) \not\subset A_j$. Let us assume that $w(C)$ is not contained in A_i. By (P_{n+1}),

$$w(C) \subset s_i A_i \quad \text{and} \quad l(ws_i) = l(w) - 1.$$

Apply (Q_n) to ws_i. Then, there is an element x of W_{ij} such that $ws_i(C) \subset x(A_i \cap A_j)$ and

$$l(ws_i) = l(ws_i x^{-1}) + l(x).$$

Hence, we get $w(C) \subset xs_i(A_i \cap A_j)$ and

$$
\begin{aligned}
l(w) &= l(ws_i) + 1 = l(ws_i x^{-1}) + l(x) + 1 \\
&\ge l(ws_i x^{-1}) + l(xs_i) \ge l(w).
\end{aligned}
$$

Thus, we have the equality everywhere, and so (Q_{n+1}) holds for $xs_i \in W_{ij}$. $\quad\square$

(c) *Theorem* (4.13) *holds.*

Proof. Let w be a nonidentity element of W. Then, we get $w = w's_i$ for some s_i and $l(w') = l(w) - 1$. Apply (P_n) to w'. Then, as $l(w's_i) \neq l(w') - 1$, we must have $w'(C) \subset A_i$. Hence, we get $w(C) = s_i(w'(C)) \subset s_i A_i$. Since $A_i \cap s_i A_i$ is empty, we get $C \cap w(C) = \varnothing$. □

Corollary 1. *The canonical representation σ of W, as well as the contragredient representation σ^*, is faithful.*

Proof. Suppose that $\sigma^*(w) = 1$. Then, clearly, we get $C = w(C)$ which implies that $w = 1$ by (4.13). Thus, σ^* is faithful. Suppose that $\sigma(w) = 1$. Then, by definition, we have

$$\langle \sigma^*(w)v^*, u \rangle = \langle v^*, \sigma(w^{-1})u \rangle = \langle v^*, u \rangle$$

for all $u \in V$. Thus, $\sigma^*(w) = 1$. This implies that $w = 1$, and σ is also faithful. □

Corollary 2. *The group $\sigma^*(W)$ is a discrete subgroup of $GL(V^*)$. A similar proposition holds for $\sigma(W)$, and there is no limit point of $\sigma(W)$ in $GL(V)$.*

Proof. Let C be the open simplicial cone defined earlier. Let P be a point inside C. Consider the set U of elements g of $GL(V^*)$ such that $g(P) \subset C$. Since the action of $GL(V^*)$ on V^* is continuous, the set U is an open set which contains the identity. By (4.13), we have $\sigma^*(W) \cap U = \{1\}$. Thus, $\sigma^*(W)$ is discrete.

The subgroup $\sigma(W)$ is also discrete because the correspondence between $\varphi \in GL(V)$ and $'\varphi \in GL(V^*)$ is bicontinuous.

It is almost obvious that $\sigma(W)$ does not have a limit point in $GL(V)$. Since $\sigma(W)$ is discrete, there is a neighborhood U_1 of the identity such that $\sigma(W) \cap U_1 = \{1\}$. The group operation is continuous. So, there are neighborhoods U_2 and U_3 of the identity such that $U_2^{-1}U_3 \subset U_1$. Suppose that an element p of $GL(V)$ were a limit point of $\sigma(W)$. Then, pU_2 and pU_3 are neighborhoods of p, so there would be elements x and y of $\sigma(W)$ such that $x \in pU_2$, $y \in pU_3$, and $x \neq y$. Then, we would get

$$x^{-1}y = (p^{-1}x)^{-1}(p^{-1}y) \in U_2^{-1}U_3 \subset U_1.$$

Since $x^{-1}y$ is a nonidentity element of $\sigma(W)$, this is a contradiction. □

Corollary 3. *Let $M = (m_{ij})$ be a matrix with integral coefficients m_{ij} such that $m_{ii} = 1$ and $m_{ij} = m_{ji} \geq 2$ (including $m_{ij} = \infty$) for $i \neq j$. There us a Coxeter system (W, S) such that M is the matrix associated with the Coxeter group W.*

Proof. By (4.12) (iii), the order of $\sigma_i \sigma_j$ is exactly m_{ij}. Since the representation σ is faithful, $\sigma(W)$ is the Coxeter group to which the matrix M is associated. □

Theorem 4.14. *Let (W, S) be a Coxeter system, and let Q be the quadratic form associated with (W, S). Then, W is finite if and only if the form Q is positive definite.*

Proof. (a) Assume that the form Q is positive definite. Let σ be the canonical representation of W on the vector space V. By (4.12) (ii), each element of $\sigma(W)$ leaves Q invariant. The totality of elements of $GL(V)$ which leave the form Q invariant form a group, the orthogonal group $O(Q)$ ([IK], Chapter 8, §3). Since Q is assumed to be positive definite, Q is transformed to the standard form $x_1^2 + x_2^2 + \cdots + x_n^2$. The subgroup of elements leaving this form invariant is conjugate to $O(Q)$ and is clearly a compact set. Since $\sigma(W) \subset O(Q)$, $\sigma(W)$ is a subset of a compact set, and it does not have any limit point (Corollary 2 of (4.13)). Then, $\sigma(W)$ must be a finite set. Since σ is a faithful representation of W by Corollary 1 of (4.13), W is a finite group.

(b) In the remainder of the proof, we will prove that the form associated with a finite Coxeter system is positive definite. *In order to prove the proposition, we may assume that W is irreducible.* (See Definition 4.10 of the irreducibility of W.)

Proof. Suppose that the theorem has been proved when W is irreducible. We will use induction on the number of generators. By (4.9), if (W, S) is reducible, there is a decomposition of S into a union of disjoint subsets S_1 and S_2 such that any element s of S_1 commutes with any element t of S_2. In this case, the order of st is two. Hence, the corresponding coefficient of Q vanishes. Thus, Q is the sum of the quadratic form Q_1 on those variables corresponding to S_1 and the form Q_2 on those variables corresponding to S_2. So, Q is positive definite since both Q_1 and Q_2 are.   □

(c) *Let (W, S) be an irreducible Coxeter system, and let B be the linear form associated with (W, S). Set*

$$V^0 = \{v \in V \mid B(u, v) = 0 \quad \text{for all } u \in V\}.$$

Then the following propositions hold.
  (i)   *Each element of $\sigma(W)$ fixes every element of V^0.*
  (ii) *Any $\sigma(W)$-invariant proper subspace of V is contained in V^0.*
  (iii) *If W is finite, then $V^0 = \{0\}$, and B is nondegenerate.*
  (iv) *If W is finite, the only linear transformations of V which commute with all elements of $\sigma(W)$ are $v \to \lambda v$ ($\lambda \in \mathbb{R}$). (In the terminology of representation theory, the representation σ is absolutely irreducible.)*

Proof. (i) By the definitions of σ and V^0, each reflection σ_i fixes all elements of V^0. Hence, $\sigma(W)$ centralizes V^0.

(ii) Let V' be a $\sigma(W)$-invariant subspace of V. Suppose that $u_i \in V'$. Let Γ be the graph associated with the Coxeter group (W, S). We identify the vertex set of Γ with S. If s_i is joined to s_j in Γ, then $m_{ij} \geq 3$ and $B(u_i, u_j) =$

$- \cos(\pi/m_{ij}) < 0$. Hence, we get

$$2B(u_j, u_i)u_j = u_i - \sigma_j(u_i) \in V'.$$

This implies that $u_j \in V'$. By assumption, W is irreducible, so any pair of vertices of Γ is connected. Hence, V' contains all the vertices of Γ, and so we have $V' = V$.

Let v be an element of V'. Then, as before, we get

$$2B(u_i, v)u_i = v - \sigma_i(v) \in V'.$$

If V' is $\sigma(W)$-invariant and proper, then $u_i \notin V'$. Then, we have $B(u_i, v) = 0$ for all i. Thus, V' is contained in V^0.

(iii) Set $|W| = g$. Since V is a vector space over $\mathbb{R}$, V is g-regular according to (5.16) of Chapter 2. So, by (5.17) of Chapter 2 and Part (i), V^0 is a direct summand of V. Thus, we have $V = V^0 + V^1$ where V^1 is generated by $\sigma(w)v - v$ ($w \in W, v \in V$). Then, V^1 is also $\sigma(W)$-invariant. (See the proof of the Cor. of Theorem 8.13, Chapter 2.) If V^1 is proper, then (ii) gives rise to a contradiction: $V^1 \subset V^0$. So, we have $V^1 = V$ and $V^0 = \{0\}$.

(iv) Since $\sigma_i \neq 1$, we get $\operatorname{Im}(\sigma_i - 1) = \mathbb{R}u_i$. If $\varphi \in GL(V)$ centralizes $\sigma(W)$, then we get $\varphi(u_i) = \lambda u_i$ for some $\lambda \in \mathbb{R}$. Set

$$N = \operatorname{Ker}(\varphi - \lambda \cdot 1).$$

Then, $\varphi - \lambda \cdot 1$ centralizes $\sigma(W)$, so N is a $\sigma(W)$-invariant subspace of V. By (ii) and (iii), we have either $N = V$ or $N = \{0\}$. But, as shown above, $u_i \in N$. Hence, we get $N = V$; that is, $\varphi(u) = \lambda u$ for all $u \in V$.   □

(d) *Let (W, S) be an irreducible finite Coxeter system. If B' is a $\sigma(W)$-invariant bilinear form, there we have $B' = \lambda B$ for some $\lambda \in \mathbb{R}$.*

Proof. An element u of V determines an element of V^* by sending $v \mapsto B(u, v)$. Let us denote this element by $\theta(u)$. Thus, θ is a linear mapping from V into V^*. By (c), Part (iii), B is nondegenerate, and so θ is a surjective isomorphism. So, for any bilinear form B' and any $u \in V$, there is an element $\varphi(u)$ of V such that

$$B'(u, v) = B(\varphi(u), v) \quad \text{for any} \quad v \in V.$$

The mapping φ is a linear transformation on V. We will show that if B' is $\sigma(W)$-invariant, φ centralizes $\sigma(W)$. If w is an element of W, then for all $v \in V$,

$$\begin{aligned}
B(\varphi(\sigma(w)u), v) &= B'(\sigma(w)u, v) \\
&= B'(u, \sigma(w)^{-1}v) = B(\varphi(u), \sigma(w)^{-1}v) \\
&= B(\sigma(w)(\varphi(u)), v).
\end{aligned}$$

Since B is nondegenerate by (c) (iii), we get $\sigma(w)\varphi = \varphi\sigma(w)$. By (c) (iv), $\varphi(u) = \lambda u$ for all $u \in V$. Thus, we have

$$B'(u, v) = B(\lambda u, v) = \lambda B(u, v)$$

and $B' = \lambda B$.   □

(e) *If W is finite Coxeter group, then the associated quadratic form Q is positive definite.*

Proof. By (b), we may assume that W is irreducible. Let Q_1 be a positive definite quadratic form defined on V. For example, we may choose $Q_1 = x_1^2 + \cdots + x_n^2$. Let B_1 be the bilinear form associated with Q_1. Set

$$B_2(u, v) = \sum_{w \in W} B_1(\sigma(w)u, \sigma(w)v).$$

Clearly, B_2 is $\sigma(W)$-invariant. Hence, by (d), we have $B_2 = \lambda B$ for some $\lambda \in \mathbb{R}$. By definition, we get

$$B_2(u, u) = \sum_w Q_1(\sigma(w)u).$$

Since Q_1 is positive definite, we have $B_2(u, u) \geq 0$. If $B_2(u, u) = 0$, then $Q_1(\sigma(w)u) = 0$ for all w. This implies that $u = 0$. Thus, the quadratic form Q_2 associated with B_2 is positive definite. Since $B_2 = \lambda B$, we get $Q_2 = \lambda Q$. Since Q_2 is positive definite, the coefficient of x_1^2 is positive. On the other hand, the coefficient of x_1^2 in B is 1. So, λ is a positive real number. Thus, Q is also positive definite.   □

The preceding theorem shows that in order to classify all finite Coxeter systems, it is sufficient to classify all the positive definite quadratic forms associated with Coxeter systems. We will follow the treatment of this problem due to Witt [2]. (There is another efficient method due to Dynkin for which readers are referred to Matsushima [1], Chapter 10, or Bourbaki [1], Chapter VI.)

Let (W, S) be a Coxeter system. Let $M = (m_{ij})$ be the matrix which gives the order of the product, let Γ be the graph associated with M, and let Q be the associated quadratic form. Then, any one of M, Γ, and Q uniquely determines the other two. The aim is to prove the following theorem.

Theorem 4.15. (i) *The graph of an irreducible Coxeter system whose associated quadratic form is positive definite is one of the graphs in the following list. Conversely, the quadratic form associated with any one of the graphs in the list is positive definite.*

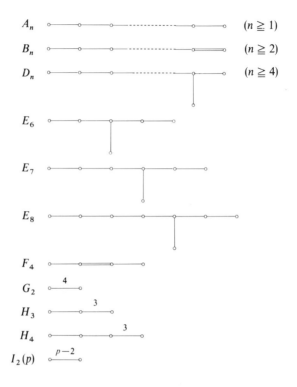

A_n	$(n \geq 1)$
B_n	$(n \geq 2)$
D_n	$(n \geq 4)$
E_6	
E_7	
E_8	
F_4	
G_2	4
H_3	3
H_4	3
$I_2(p)$	$p-2$

(ii) *The next list contains all of the graphs associated with irreducible Coxeter systems which have positive semi-definite quadratic form. Also, the quadratic form associated with any one of the graphs in the list is positive semi-definite.*

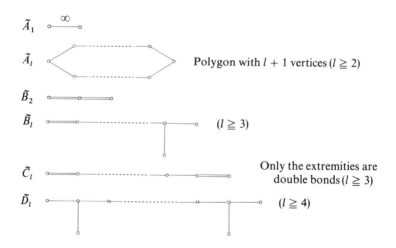

$\tilde{A}_1$	∞
$\tilde{A}_l$	Polygon with $l + 1$ vertices $(l \geq 2)$
$\tilde{B}_2$	
$\tilde{B}_l$	$(l \geq 3)$
$\tilde{C}_l$	Only the extremities are double bonds $(l \geq 3)$
$\tilde{D}_l$	$(l \geq 4)$

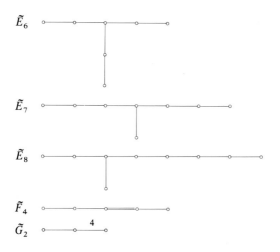

$\tilde{E}_6$

$\tilde{E}_7$

$\tilde{E}_8$

$\tilde{F}_4$

$\tilde{G}_2$

In List (i), the index on the symbol representing each graph denotes the number of vertices, while in List (ii), the index is one less than the number of vertices. (This notation is that of Bourbaki [1], except for the numbers attached to the edges which indicate the multiplicities.)

Proof. (a) First, we will show that the quadratic forms associated with the graphs in these lists are indeed positive definite or positive semi-definite. (A form Q which is not positive definite is *positive semi-definite* if $Q(u) \geq 0$ for all $u \in V$.)

We will use induction on number n of vertices.

If $n = 1$ or $n = 2$, then the form associated with $\tilde{A}_1$ is positive semi-definite (it is $(x_1 - x_2)^2$), but the forms associated with the other graphs are all positive definite, as we can see from the proof of (4.12) (iii).

Assume that $n \geq 3$. Every graph in the lists has the property that if a suitable vertex and all the edges starting from that vertex are removed, then the remaining graph is one of the graphs in List (i). So, let Γ be a graph in List (i) or (ii). Let n be the vertex of Γ such that the graph Γ' obtained by removing the vertex n and all the edges which contain n is a graph in List (i). If

$$Q = \sum a_{ik} x_i x_k \qquad (1 \leq i, k \leq n)$$

is the quadratic form associated with Γ, then the quadratic form associated with Γ' is

$$Q' = \sum{}' a_{ik} x_i x_k \qquad (1 \leq i, k \leq n - 1).$$

By the inductive hypothesis, Q' is positive definite. Hence, the principal minors of the coefficient matrix of Q' are all positive. If

$$d_j = \det |a_{ik}| \qquad (1 \leq i, k \leq j, j = 1, 2, \ldots, n - 1),$$

then all d_j are positive, and Q is equivalent to

$$d_1 x_1^2 + (d_2/d_1)x_2^2 + \cdots + (d_{n-1}/d_{n-2})x_{n-1}^2 + (d/d_{n-1})x_n^2$$

where $d = \det |a_{ik}|$ $(1 \le i, k \le n)$. ([IK], Chapter 8, §5). Thus, Q is positive definite if and only if $d > 0$, while Q is positive semi-definite if and only if $d = 0$.

The computation of d for each graph Γ is not hard. However, it is easier to compute

$$\Delta(\Gamma) = \det |2a_{ik}| \qquad (1 \le i, k \le n).$$

Suppose that we can choose the vertex n in such a way that n is on only one edge $(n - 1, n)$ and that the edge $(n - 1, n)$ is without multiplicity. Let Γ' be the graph obtained from Γ by removing the vertex n and the edge $(n - 1, n)$, and let Γ'' be the graph obtained from Γ' by removing the vertex $n - 1$ and all the edges which contain $n - 1$. Then, by assumption, the last row of the matrix $(2a_{ik})$ is $(0, \ldots, 0, -1, 2)$. Expansion along the last row gives us

(*) $$\Delta(\Gamma) = 2\Delta(\Gamma') - \Delta(\Gamma'').$$

(Note that the matrix $(2a_{ik})$ is symmetric.)

The formula (*) can be used to compute Δ by induction, starting from $\Delta(A_1) = \Delta(B_2) = 2$ and $\Delta(A_2) = 3$. The results are as follows:

$$\Delta(A_n) = n + 1, \qquad \Delta(B_n) = 2, \qquad \Delta(D_n) = 4,$$
$$\Delta(E_6) = 3, \qquad \Delta(E_7) = 2, \qquad \Delta(E_8) = 1 = \Delta(F_4),$$
$$\Delta(\tilde{B}_l) = \Delta(\tilde{D}_l) = \Delta(\tilde{E}_l) = \Delta(\tilde{F}_4) = 0.$$

Also, $\Delta(G_2) = 1$ and $\Delta(\tilde{G}_2) = 0$. In order to compute $\Delta(H_3)$, let ζ be the primitive fifth root of unity such that $\pi/2 < \arg \zeta < \pi$. Then,

$$-\cos(\pi/5) = (\zeta + \zeta^4)/2.$$

This is the negative root of $4t^2 + 2t - 1 = 0$, so

$$-\cos(\pi/5) = -(1 + \sqrt{5})/4.$$

Hence, we get

$$\Delta(I_2(5)) = (5 - \sqrt{5})/2, \qquad \Delta(H_3) = 3 - \sqrt{5},$$

and

$$\Delta(H_4) = (7 - 3\sqrt{5})/2.$$

For $\tilde{B}_2$ and $\tilde{C}_l$, we have to remove a double bond. The expansion along the last row gives us

$$\Delta(\Gamma) = 2\Delta(\Gamma') - 2\Delta(\Gamma'').$$

By computation we see that $\Delta(\tilde{B}_2) = \Delta(\tilde{C}_l) = 0$. For $\tilde{A}_1$, the sum of each row is zero. Hence, we get $\Delta(\tilde{A}_l) = 0$. This proves that the quadratic forms associated with the graphs of List (i) are positive definite, and the forms associated with the graphs of List (ii) are positive semi-definite.

(b) In order to complete the proof, we need the following properties of the quadratic form Q.

Lemma. Let $Q(x) = \sum a_{ik} x_i x_k$ be the quadratic form associated with an irreducible Coxeter system (W, S). Let B be the corresponding bilinear form. Suppose that Q is either positive definite, or positive semi-definite. Then, the following propositions hold :

(i) If $Q(x) = 0$, then we get $B(x, y) = 0$ for all $y \in V$.

(ii) If $Q(x) = 0$ for $x = \sum x_i u_i \neq 0$, then we have $x_i \neq 0$ for all i.

(iii) Suppose that, for some $m \leq n$, real numbers b_{ik} $(1 \leq i, k \leq m)$ are defined and satisfy

$$b_{ii} = 1, \qquad a_{ik} \leq b_{ik} \leq 0 \quad (i \neq k).$$

Then, the quadratic form $Q' = \sum b_{ik} \xi_i \xi_k$ is positive definite except possibly when $Q' = Q$.

Proof. (i) For any real number t and for any $y \in V$, we have

$$0 \leq Q(x + ty) = Q(x) + Q(ty) + 2B(x, ty)$$
$$= t^2 Q(y) + 2t B(x, y).$$

This implies that $B(x, y) = 0$.

(ii) Suppose $Q(x) = 0$ for $x = x_1 u_1 + \cdots + x_n u_n \neq 0$. Let J be the set of indices i such that $x_i = 0$, and let I be the complement of J. If $|x|$ denotes $\sum |x_i| u_i$ where $|x_i|$ is the absolute value of the real number x_i, then we have

$$0 \leq Q(|x|) = \sum a_{ik} |x_i| |x_k| \leq \sum a_{ik} x_i x_k = 0.$$

The third inequality follows because $a_{ii} = 1$ and $a_{ik} \leq 0$ for $i \neq k$. The above inequality forces the equality $Q(|x|) = 0$. So, by (i), $B(|x|, y) = 0$ for all $y \in V$. Take $y = u_j$ for $j \in J$. Then, we get $\sum a_{ij} |x_i| = 0$ $(j \in J)$. The left side is really a sum over $i \in I$. But, if $i \in I$, then $a_{ij} \leq 0$ and $|x_i| > 0$. So, we must have $a_{ij} = 0$ for $i \in I$ and $j \in J$. Since the Coxeter system (W, S) is irreducible, (4.9) shows that either I or J is empty. Since $x \neq 0$ by assumption, the set J must be empty. This proves the assertion.

(iii) Suppose that $Q'(y) \leq 0$ for $y = y_1 u_1 + \cdots + y_m u_m$. Then, we have

$$0 \leq Q(|y|) = \sum a_{ik}|y_i| \cdot |y_k| \leq \sum b_{ik}|y_i| \cdot |y_k|$$
$$\leq \sum b_{ik} y_i y_k = Q'(y) \leq 0.$$

This implies that $Q(|y|) = 0$. So, by (ii), either $y = 0$, or $m = n$ and all $y_i \neq 0$ $(i = 1, 2, \ldots, n)$. Furthermore, in the latter case, we have

$$\sum a_{ik}|y_i||y_k| = \sum b_{ik}|y_i||y_k|.$$

By assumption, $a_{ik} \leq b_{ik}$ for all i and k. As $y_i \neq 0$ for all i, we must have $a_{ik} = b_{ik}$ for all i and k. Thus, Q' is not positive definite only when $Q' = Q$. $\square$

(c) *Let (W, S) be an irreducible Coxeter system, let Γ be the graph associated with (W, S), and let Q be the associated quadratic form. If Q is positive definite or positive semi-definite, then Γ is contained in either List* (i) *or List* (ii).

Proof. Let Γ' be the graph obtained from Γ by removing some edges, or some vertices together with all the edges which contain the removed vertices, or both. Then, the quadratic form Q' associated with Γ' satisfies the condition of the preceding Lemma (iii). (The removal of a vertex together with all the edges containing it reduces the number of variables, while the removal of an edge increases the corresponding coefficient of the quadratic form.) Thus, by Lemma (iii), Q' is positive definite unless $Q = Q'$. So, if $\Gamma' \neq \Gamma$, then Γ' is not included in List (ii).

Suppose that the graph Γ is not contained in List (i) or List (ii). We will arrive at a contradiction. As remarked above, Γ does not contain any subgraph which is in the List (ii). Furthermore, Γ does not contain

$$Z_4 \; \circ\!\!-\!\!\overset{3}{-}\!\!-\!\!\circ\!\!-\!\!-\!\!\circ\!\!-\!\!-\!\!\circ \;, \qquad Z_5 \; \circ\!\!-\!\!\overset{3}{-}\!\!-\!\!\circ\!\!-\!\!-\!\!\circ\!\!-\!\!-\!\!\circ\!\!-\!\!-\!\!\circ$$

This can be seen by proving that $\Delta(Z_i) < 0$. In fact, using $(*)$, we have

$$\Delta(Z_4) = 2\Delta(H_3) - \Delta(A_2) = 3 - 2\sqrt{5} < 0,$$
$$\Delta(Z_5) = 2\Delta(H_4) - \Delta(H_3) = 4 - 2\sqrt{5} < 0.$$

Thus, the corresponding quadratic forms are not of definite sign.

In the following argument, we will use some terminology from graph theory. A graph is said to be a **tree** if it does not contain any closed polygon. A vertex of a tree is called a **branch point** if it is on more than two edges. If there is only one branch point in a tree, there are several branches coming out of the branch point. The number of edges in each branch is called the *length* of the branch.

The proof is done using the following scheme. The left side contains the reason why the conclusion at the right follows. Recall again that any sub-graph is not the graph in the List (ii), or Z_i. Let n denote the number of vertices of Γ.

$\Gamma \neq \tilde{A}_1, I_2(p)$ $\qquad \Rightarrow n \geq 3$

$\Gamma \not\supset \tilde{A}_l$ $\qquad\qquad \Rightarrow \Gamma$ is a tree

Case I   There is no multiple bond.

$\Gamma \neq A_n$ $\qquad\qquad \Rightarrow$ There is a branch point.

$\Gamma \not\supset \tilde{D}_l \ (l \geq 5)$ $\qquad \Rightarrow$ The branch point is unique.

$\Gamma \not\supset \tilde{D}_4$ $\qquad\qquad \Rightarrow$ There are exactly 3 branches.

Let the lengths be a, b, and $c \ (a \leq b \leq c)$.

$\Gamma \not\supset \tilde{E}_6$ $\qquad\qquad \Rightarrow$ We have $a = 1$.

$\Gamma \neq D_n$ $\qquad\qquad \Rightarrow$ We have $b \geq 2$.

$\Gamma \not\supset \tilde{E}_7$ $\qquad\qquad \Rightarrow$ We have $b = 2$.

$\Gamma \not\supset \tilde{E}_8$ $\qquad\qquad \Rightarrow$ We have $c \leq 3$.

$\Gamma \neq E_n$ $\qquad\qquad \Rightarrow$ This is a contradiction.

Case II   There is a multiple bond.

$\Gamma \not\supset \tilde{B}_2, \tilde{C}_l$ $\qquad \Rightarrow$ There is a unique edge with a multiple bond.

$\Gamma \not\supset \tilde{B}_l \ (l \geq 3)$ $\qquad \Rightarrow$ There is no branch point.

Suppose that the multiple bond is a double bond.

$\Gamma \neq B_n$ $\qquad\qquad \Rightarrow$ The multiple bond is not at either extremity.

$\Gamma \not\supset \tilde{F}_4$ $\qquad\qquad \Rightarrow$ We have $n = 4$.

$\Gamma \neq F_4$ $\qquad\qquad \Rightarrow$ This is a contradiction.

The multiple bond is more than double.

$\Gamma \not\supset \tilde{G}_2$ $\qquad\qquad \Rightarrow$ The multiplicity is three.

$\Gamma \not\supset Z_4$ $\qquad\qquad \Rightarrow$ The multiple bond occurs at the extremity.

$\Gamma \not\supset Z_5$ $\qquad\qquad \Rightarrow$ We have $n \leq 4$.

$\Gamma \neq H_n$ $\qquad\qquad \Rightarrow$ We get a final contradiction. $\square$

Theorem 4.16. *The graph associated with a finite irreducible Coxeter system is contained in the List* (i). *Conversely, each graph in the List* (i) *is associated with a finite irreducible Coxeter system.*

Proof. By Theorem 4.14, the quadratic form associated with a finite Coxeter system is positive definite. So, by Theorem 4.15 (i), the corresponding graph appears in the List (i).

Conversely, any graph Γ in List (i) or List (ii) determines the matrix $M = (m_{ik})$. By Cor. 3 of (4.13), there is a Coxeter system (W, S) having M as the matrix associated with it. Since the corresponding quadratic form is positive definite, W is finite. As Γ is connected, the system (W, S) is irreducible. $\square$

Any Coxeter system associated with the graph of type X is called a Coxeter system of type X. Then, the symmetric group on $n + 1$ letters is a Coxeter group of type A_n. There are relations between Coxeter systems and the so-called *root systems* in Euclidean space. Root systems play important roles in the study of simple Lie algebras, but we will not go into this here. For the details (and further developments) the readers are referred to Bourbaki [1].

Exercises

1. Suppose that the Coxeter system (W, S) is given by $S = \{s_i\}$ and the defining relations $R = \{(s_i s_j)^{m_{ij}} = 1\}$.

(a) Show that W contains a normal subgroup W_0 of index 2.
(b) Find a presentation of the group W_0.

(*Hint.* Compare with the Example of page 348.)

2. Let $G(l, m, n)$ be the group with the presentation

$$\langle x, y, z \mid x^l = y^m = z^n = xyz = 1 \rangle.$$

Show that $G(l, m, n)$ is finite if and only if

$$(1/l) + (1/m) + (1/n) > 1.$$

(*Hint.* Use Exercise 1 (b) and Theorem 4.16.)

3. Let (W, S) be a Coxeter group. For each $s \in S$, let

$$P_s = \{w \in W \mid l(sw) > l(w)\}.$$

Prove the following three propositions about P_s.

(a) We have $\bigcap_s P_s = \{1\}$.
(b) We have $W = P_s \cup sP_s$ and $P_s \cap sP_s = \varnothing$.
(c) If $w \in P_s$ and $ws' \notin P_s$, then we get $ws' = sw$.

(*Hint.* If a shortest expression of w starts with s, then certainly $w \notin P_s$. The assumptions of (c) say that

$$l(sw) > l(w), \qquad l(sws') \le l(ws').$$

So, we get $l(w) \le l(ws')$. If $w = s_1 \cdots s_q$ is a shortest expression of w, then $s_1 \cdots s_q s'$ is a shortest expression of ws'. Apply Theorem 4.4.)

4. Let $W = \langle S \rangle$ be a group generated by a set S consisting of elements of order 2. Suppose that for each $s \in S$, a subset P_s ($s \in S$) of W is given and that $\{P_s\}$ satisfies the following three conditions:

(1) For each $s \in S$, we have $1 \in P_s$.
(2) For each s, $P_s \cap sP_s$ is empty.
(3) If $w \in P_s$ and $ws' \notin P_s$ for some $s' \in S$, then $ws' = sw$.

Show that (W, S) is a Coxeter system and that

$$P_s = \{w \in W \mid l(sw) > l(w)\}.$$

(*Hint.* Suppose that $w = s_1 \cdots s_q$ is a shortest expression of $w \notin P_s$. Then, by (1), there is an integer j such that $s_1 \cdots s_{j-1} \in P_s$ but $s_1 \cdots s_j \notin P_s$. By (3), we get $ss_1 \cdots s_{j-1} = s_1 \cdots s_j$. So, we have $l(sw) < l(w)$. On the other hand, if $w \in P_s$, then $sw \in sP_s$ and (2) proves that $sw \notin P_s$. Hence, we get $l(w) < l(sw)$. Thus, the assumptions imply the validity of the exchange condition (Theorem 4.4), and

$$w \in P_s \Leftrightarrow l(sw) > l(w).$$

The condition (1) and (2) are weaker than (a) and (b) of Exercise 3, respectively. So, the existence of subsets P_s satisfying conditions (1), (2), and (3) is a necessary and sufficient condition for $W = \langle S \rangle$ to be a Coxeter system (W, S).)

5. Let V be a finite-dimensional Euclidean space. Let W be a finite group generated by the usual reflections with respect to hyperplanes through the origin. The problem here is to prove that W is a Coxeter group. Let $\mathscr{H}$ be the set of hyperplanes which determine the reflections contained in W. We denote by s_H the reflection with respect to the hyperplane H. The hyperplanes of $\mathscr{H}$ partition the space V. A minimal nonempty open subset bounded by hyperplanes of $\mathscr{H}$ is called a *chamber* with respect to or of the group W. Let C be a fixed chamber of W, and let S be the set of reflections with respect to the boundary hyperplanes of C.

(a) Let p be a fixed point of C. For an arbitrary $v \in V$, consider all the points $w(v)$ ($w \in \langle S \rangle$), and let $w_0(p)$ be the point closest to the point p. Show that $w_0(p)$ is contained in the closure of C.

(b) Let C' be an arbitrary chamber of W. Prove that there is an element $w \in \langle S \rangle$ such that $w(C') = C$.

(c) Let H be any hyperplane in $\mathscr{H}$. Prove that, $w(H)$ is one of the boundary hyperplanes of C for some $w \in \langle S \rangle$.

(d) Show that $W = \langle S \rangle$.

(e) Let H and H' be two hyperplanes which bound C. Set $s = s_H$ and $s' = s_{H'}$. Suppose that an element w of W satisfies the following property: the two chambers C and $w^{-1}C$ are in the same half-space divided by H, but $(s'w^{-1})C$ is in the other half-space. Show that $w(H) = H'$.

(f) Let H be a boundary hyperplane of C, and let $s = s_H$. Let P_s be the set of elements w of W such that C and $w^{-1}(C)$ are in the same half-space divided by H. Show that $\{P_s\}$ satisfies the three conditions of Exercise 4.

(g) Prove that (W, S) is a Coxeter system.

(h) Verify the converse: any finite Coxeter system (W, S) is isomorphic to a finite group generated by reflections in some Euclidean space.

(*Hint.* Note that we have $s_{wH} = w^{-1}s_H w$.

(a) If the points v and p belong to different half-spaces divided by a hyperplane H, then $s_H(v)$ is closer to p than v. (a) $\Rightarrow$ (b) $\Rightarrow$ (c) $\Rightarrow$ (d).

(e) $w^{-1}C$ and $(s'w^{-1})C$ have a common boundary. Consider the hyperplane containing this boundary.

(h) Let σ be the canonical representation of the Coxeter group W on a real vector space V. Since W is finite, the associated quadratic form is positive definite by Theorem 4.14. Thus, V becomes an Euclidean space, and the transformations σ_i are the usual reflections. By Corollary 1 of (4.13), σ is faithful and (h) holds.)

6. Let (W, S) be a Coxeter system of type A_n. Set

$$u_1 + u_2 + \cdots + u_n = -u_0$$

(following the notation of Definition 4.11). Show that

$$B(u_0, u_i) = 0 \quad (i \neq 1, n), \qquad B(u_0, u_j) = -1/2 \quad (j = 1, n).$$

Furthermore, show that the mapping σ_0 defined by

$$\sigma_0(x) = x - 2B(u_0, x)u_0$$

is a reflection contained in W.

(*Hint.* There is an element w of W such that $w(u_1) = u_1 + \cdots + u_n$.)

7. Suppose that the quadratic form Q associated with a Coxeter system (W, S) is positive semi-definite.

(a) Show that the set of vectors u such that $Q(u) = 0$ forms a one-dimensional subspace of V, and that it coincides with the subspace denoted by V^0 in Part (c) of the proof of Theorem 4.14.

(b) Assume that (W, S) is of type $\tilde{A}_l$ ($l = n - 1$). Show that

$$V^0 = \mathbb{R}(u_1 + u_2 + \cdots + u_n).$$

(c) Let $\bar{W}$ be the group induced by W in the factor space V/V^0. Prove that $\bar{W}$ is a Coxeter group of type A_l generated by the images of $\sigma_1, \ldots, \sigma_{n-1}$.

(d) Let $K = \operatorname{Ker}(W \to \bar{W})$. Show that K is abelian.

(*Hint.* (a) Use Lemma ((i) and (ii)) in the proof of Theorem 4.15. (c) By Exercise 6, the mapping induced by σ_n is contained in the group generated by the images of $\sigma_1, \ldots, \sigma_{n-1}$. (d) If $\tau \in K$, $\tau(u) = u + u_0$ with $u_0 \in V^0$. By Part (c) of the proof of Theorem 4.14, V^0 is centralized by $\sigma(W)$. In fact, K is a free abelian group of rank l.)

8. Show that a Coxeter group of type B_n is isomorphic to the general wreath product $Z_2 \operatorname{wr} \Sigma_n$ where the symmetric group Σ_n is considered as a permutation group on n letters. (See Exercise 4 (g), §3.)

(*Hint.* In Σ_{2n}, the elements

$$s_i = (i, i + 1)(n + i, n + i + 1) \quad (1 \le i < n) \quad \text{and} \quad s_n = (n, 2n)$$

generate a Coxeter group of type B_n.)

9. In Σ_{2n} ($n \ge 4$), define s_i ($1 \le i < n$) as above, and set

$$s_n = (n - 1, 2n)(n, 2n - 1).$$

Show that $W = \langle s_i \rangle$ and $S = \{s_i\}$ define a Coxeter group of type D_n.

Let $\bar{i}$ denote the pair $\{i, n + i\}$. Show that W induces a permutation group $\bar{W}$ on $\{\bar{1}, \bar{2}, \ldots, \bar{n}\}$ and $\bar{W} \cong \Sigma_n$.

For any integer i, $1 \le i \le n$, consider the mapping $n + i \leftrightarrow -i$, and consider W as a permutation group on $\{\pm i\}$. Show that W is isomorphic to the totality of the permutations on $\{\pm i\}$ with an even number of sign changes. Show that the order of W is $2^{n-1}(n!)$.

(*Hint.* Clearly, $\langle s_1, \ldots, s_{n-1} \rangle \cong \Sigma_n$. If W contains all the possible combinations of sign changes, then W would contain $(n, 2n)$. This is impossible since W is generated by even permutations. Thus, the order is at least $2^{n-1}(n!)$. Use the coset enumeration method to find the index of a subgroup of type D_{n-1}.)

The orders of other Coxeter groups can be computed:

$$|E_6 : D_5| = 27, \quad |E_7 : E_6| = 56, \quad |E_8 : E_7| = 240,$$
$$|F_4 : B_3| = 24, \quad |H_4 : H_3| = 120.$$

Notation : $|X_n: Y_{n-1}|$ is the index of a subgroup of type Y_{n-1} in the Coxeter group of type X_n. Even for E_8, this can be done without the aid of a computer if one can devise a clever scheme. Thus, the order of Coxeter groups of type E_6, E_7, E_8, F_4, and H_4 are

$$2^7 3^4 5, \qquad 2^{10} 3^4 5 \cdot 7, \qquad 2^{14} 3^5 5^2 7, \qquad 2^7 3^2, \qquad 2^6 3^2 5^2,$$

respectively. For the structure of these groups, the readers may consult Bourbaki [1], pp. 228–232.)

§5. Surveys of Finite Simple Groups

It should be clear by now that the structures of the composition factors of a finite group G reflect the structure of G itself. Undoubtedly then, the central problem in finite group theory is to investigate the structures of simple groups and, hopefully, to classify all of the simple groups.

The concept of a finite group appeared for the first time in the history of Mathematics in research concerning the algebraic solution of equations, and it was used very effectively, most notably in the works of Abel and Galois. In fact, Galois had already grasped the notion of a simple group, and he was aware of the nonsolvability of alternating groups, as well as of the fact that A_5 is the nonabelian simple group of the smallest order. He also knew that the linear group defined over a prime field of prime characteristic (in more modern language, the semidirect product of V and $GL(V)$) is not solvable except for a few special cases. Galois' work on group theory was later continued and developed further by Jordan, who presented many interesting results in his classic, *Traité des Substitutions* [1]. In particular, he showed that finite analogues of the classical linear groups are defined over a prime field of p elements. Dickson continued this line of investigation, and in his monograph [1] on linear groups, published around the beginning of this century, he defined the classical groups over arbitrary finite fields and proved that these groups came close to being simple groups. Thus, he showed that there are many series of finite simple groups. The so-called *classical groups* consist of the special linear groups, the symplectic groups, the unitary groups, and the orthogonal groups. In dealing with classical groups, we have already defined the special linear groups over all ground fields in Chapter 1, §9, and we have proved the main theorem on their simplicity (Theorem 9.9 of Chapter 1). For an alternate treatment of the special linear groups and for an excellent discussion of some of the other classical groups, the readers are referred to Nagao [3], Chapter 4. (There are related exercises in §3 of this chapter.)

The study of an individual simple group is not only useful for its own sake, but is also indispensable in the discovery of new results and in the development of the general theory. So, in this section, we will give a brief survey of all the simple groups known at the time of this writing.

A classical group which is not a special linear group is defined as the set of the linear transformations which leave a given form invariant. We will consider an arbitrary field F and a vector space V over F such that dim V is finite. A bilinear form f which is defined over V is said to be *nondegenerate* if the fact that $f(x, v) = 0$ for every $x \in V$ implies that $v = 0$. A linear transformation θ on V is said to leave the form f *invariant* if the following formula is satisfied for any elements $u, v \in V$:

$$f(\theta u, \theta v) = f(u, v).$$

With these definitions, we can prove a general proposition on those classical groups which are not special linear groups.

(5.1) *Let f be a nondegenerate form. If a linear transformation θ on V leaves the form f invariant, then θ is invertible. The totality of the linear transformations which leave the form f invariant forms a subgroup of $GL(V)$.*

Proof. Let θ be a linear transformation which leaves the form f invariant. Suppose that $\theta v = 0$ for some $v \in V$. Then, for any $x \in V$, we have

$$f(x, v) = f(\theta x, \theta v) = f(\theta x, 0) = 0.$$

Since f is nondegenerate, we must have $v = 0$. This means that Ker $\theta = \{0\}$. But, V is of finite dimension, so θ is an isomorphism of V and has an inverse. Thus, θ is an element of $GL(V)$. The remaining assertions are obvious. $\square$

The subgroup consisting of the linear transformations which leave a form f invariant will be denoted by $G(f)$. When f is a nondegenerate alternating form, the group $G(f)$ is called a *symplectic group*. If f is a nondegenerate symmetric form and if the characteristic of the field is not two, then the group $G(f)$ is an *orthogonal group*. In order to define a unitary group, we consider a field F which has a nonidentity automorphism τ of order 2. In this case, we can define $h(u, v)$ such that

$$h(u, v) = h(v, u)^\tau$$
$$h(\alpha u, \beta v) = \alpha \beta^\tau h(u, v)$$
$$h(u, v + w) = h(u, v) + h(u, w).$$

Such a form h is called a *Hermitian form*. Since the proposition corresponding to (5.1) holds for Hermitian forms, we can define a *unitary group* as the set of the linear transformations which leave a given nondegenerate Hermitian form invariant.

Using a nondegenerate bilinear form, we can introduce a geometric notion: that of the orthogonality of vectors. Two elements u and v of V are said to be *orthogonal* (with respect to the given form f) if $f(u, v) = 0$. For any subspace U of V, set

$$U^{\perp} = \{x \mid f(x, u) = 0 \text{ for all } u \in U\}.$$

Then, it is clear that $U^{\perp}$ is a subspace of V. Moreover, an elementary result in the theory of linear equations shows that the following dimension equation holds:

$$(5.2) \qquad\qquad \dim U + \dim U^{\perp} = \dim V.$$

For any form f and for any $\sigma \in GL(V)$, we define f^{σ} by

$$f^{\sigma}(\sigma(u), \sigma(v)) = f(u, v).$$

Then, f^{σ} is a form of the same kind as f, and we have

$$f^{\sigma\tau} = (f^{\sigma})^{\tau}.$$

Thus, the mapping $f \to f^{\sigma}$ defines an action of $GL(V)$ on the set of forms. The subgroup $G(f)$ defined earlier is the stabilizer of f with respect to the above action of $GL(V)$. So, by (7.11) of Chapter 1, we get

$$(5.3) \qquad\qquad G(f^{\sigma}) = \sigma^{-1}G(f)\sigma = G(f)^{\sigma}.$$

Let $\{u_i\}$ be a basis of V over F. A bilinear form f is uniquely determined by the values of $f(u_i, u_j)$ on the elements of the basis. If the matrix J is defined by

$$J = (f(u_i, u_j)),$$

then J is called the *matrix corresponding to* (or *representing*) the form f.

Let σ be an element of $GL(V)$. Then, σ is represented by the matrix $M_{\sigma} = (m_{ij})$ with respect to the basis $\{u_i\}$, where

$$\sigma(u_i) = \sum m_{ij} u_j.$$

Then, $\sigma \in G(f)$ if and only if

$$(5.4) \qquad\qquad M_{\sigma} J^{t}M_{\sigma} = J.$$

If the form J is nondegenerate, we have $\det J \neq 0$. So, we have

$$(5.5) \qquad\qquad \det M_{\sigma} = \pm 1.$$

In the preceding discussion, the bilinear form f can also be replaced by a Hermitian form. In this case, a few minor changes must be made. The condition (5.4) must be altered to

$$M_\sigma J\,{}^t\bar{M}_\sigma = J,$$

where the notation $\bar{M}_\sigma$ stands for $((m_{ij})^\tau)$. Also, the formula (5.5) becomes

$$(\det M_\sigma)^{1+\tau} = 1.$$

We will study each type of classical groups separately.

Symplectic Groups

For a symplectic group, the form f is alternating, so $f(u, u) = 0$ for all $u \in V$. From $f(u + v, u + v) = 0$, we get

$$f(u, v) = -f(v, u) \qquad (u, v \in V).$$

By definition, f is also nondegenerate. So, for any given element $u_0 \neq 0$, there exists element v of V such that $f(u_0, v) \neq 0$. But, since $f(u_0, \lambda v) = \lambda f(u_0, v)$, we can find an element v_0 corresponding to each given element u_0 such that

$$f(u_0, v_0) = 1.$$

Let $U = Fu_0 + Fv_0$ be the subspace spanned by u_0 and v_0. Then, we get $U \cap U^\perp = \{0\}$. So, by (5.2), V is the direct sum of U and $U^\perp$:

$$V = U + U^\perp.$$

Let f_1 be the restriction of f on U, and let f_2 be the restriction of f on $U^\perp$. Then, f_1 and f_2 are alternating forms defined on U and $U^\perp$, respectively. Suppose that $u = u_1 + u_2$ and $v = v_1 + v_2$ where $u_1, v_1 \in U$ and $u_2, v_2 \in U^\perp$. Then, we have $f(u_1, v_2) = f(u_2, v_1) = 0$ and

$$f(u, v) = f_1(u_1, v_1) + f_2(u_2, v_2).$$

Thus, we may write $f = f_1 + f_2$. Moreover, f_1 and f_2 are both nondegenerate. By using induction on dim V, we get the first half of the following proposition.

(5.6) *Let f be a nondegenerate alternating form defined on a vector space V. Then, the dimension of the space V is even. Set dim $V = 2m$. There is a basis*

$\{u_i\}$ $(i = 1, 2, \ldots, m)$ of V such that

$$f(u_i, u_{m+i}) = 1, \qquad f(u_i, u_j) = 0 \quad \text{if} \quad |i - j| \neq m.$$

Let g be another nondegenerate alternating form on V. Then, the symplectic group $G(g)$ is conjugate to the group $G(f)$, so the structure of the symplectic group does not depend on the form f. It is determined by the field F and $\dim V$.

Proof. We will prove the second part. There is a basis $\{v_i\}$ of V such that

$$g(v_i, v_{m+i}) = 1, \quad g(v_i, v_j) = 0 \quad \text{if} \quad |i - j| \neq m.$$

Let σ be the element of $GL(V)$ such that

$$\sigma(u_i) = v_i \quad (i = 1, 2, \ldots, 2m).$$

Then, by the definition of f^σ, we get $g = f^\sigma$. Hence, $G(g) = G(f)^\sigma$ by (5.3). $\square$

Unitary Groups

In this case, the field F has an automorphism τ of order 2. So, if F is a finite field, then $|F| = q^2$ where q is some prime power, and the automorphism is given by

$$\tau: \alpha \to \alpha^q$$

The fixed field F_0 of τ is the subfield of q elements, and the mappings

$$\alpha \to \alpha^{1+q} \qquad \alpha \to \alpha + \alpha^q$$

are surjective mappings from F onto F_0. In fact, the former mapping is a homomorphism from the multiplicative group F^* into F_0^*, and the order of its kernel is at most $1 + q$. Thus, the mapping $\alpha \to \alpha^{1+q}$ is surjective. Similarly, the mapping $\alpha \to \alpha + \alpha^q$ can also be proved to be surjective.

Let h be a nondegenerate Hermitian form. If we were to have $h(v, v) = 0$ for all v in V, then h would be alternating as well as Hermitian. So, we would have

$$h(u, v) = -h(v, u) = -h(u, v)^\tau$$

for all $u, v \in V$. This contradicts the assertion that there exists an element α which satisfies $\alpha + \alpha^q \neq 0$. Thus, there must be an element u of V such that $h(u, u) \neq 0$. Set $\alpha = h(u, u)$. Then,

$$\alpha^\tau = h(u, u) = \alpha.$$

So, α belongs to F_0^*, and there is an element λ of F such that $\alpha = \lambda^{1+q}$. Then, for $u_1 = \lambda^{-1}u$, we get

$$h(u_1, u_1) = \lambda^{-1-q}\alpha = 1.$$

Thus, we have the direct sum decomposition $V = U + U^\perp$ for $U = Fu_1$. The following proposition can easily be proved in the same way as (5.6).

(5.7) *Let h be a nondegenerate Hermitian form defined over a finite field F. Then, there is a basis $\{u_i\}$ of V such that*

$$h(u_i, u_i) = 1, \quad h(u_i, u_j) = 0 \quad \text{if} \quad i \neq j.$$

The structure of the unitary group $G(h)$ does not depend on the form h but is uniquely determined by the field F and dim V.

If the field F is not finite, then the structure of a unitary group may depend on the form. For example, if F is the field of complex numbers, then the form is always equivalent to those forms which satisfy

$$h(u_i, u_i) = \varepsilon_i, \quad h(u_i, u_j) = 0 \quad \text{if} \quad i \neq j,$$

where ε_i is either 1 or -1 for each value of i. For $n = 2$, it can be shown that the unitary group with $\varepsilon_1 = \varepsilon_2 = 1$ is not isomorphic to the group with $\varepsilon_1 = -\varepsilon_2 = 1$.

Orthogonal Groups

For orthogonal groups, the characteristic of the ground field F is different from 2, and the form f is symmetric. So, there is an element u of V such that $f(u, u) \neq 0$. As in the case of the unitary groups, there is a basis $\{u_i\}$ of V such that

$$f(u_i, u_i) = \alpha_i \neq 0, \quad f(u_i, u_j) = 0 \quad \text{if} \quad i \neq j.$$

Since we will assume that F is a finite field, we will be able to place some restrictions on the coefficients α_i. First, we will prove the following lemma.

Lemma. *Let F be a finite field such that char F is odd, and let K be the set of the squares of the nonzero elements of F. Then, K forms a subgroup of index 2 in F^*. For any α, β of F^*, there exists a pair (x, y) of elements of F such that*

$$x^2 + \alpha y^2 + \beta = 0.$$

Proof. By assumption, the multiplicative group F^* is a cyclic group, and its order is even. So, if γ is a generator of F^*, then the set K of the squares is equal to $\langle \gamma^2 \rangle$. Hence, the index $|F^*: K|$ of K is two. Clearly, we have

$$F^* = K \cup \gamma K.$$

If $\alpha \notin K$, then αy^2 can represent any element of γK, depending on the value of y, so the equation $x^2 + \alpha y^2 + \beta = 0$ has an obvious solution. The same is true if $-\beta = u^2 \in K$. Therefore, in proving the second half of the lemma, we will assume that $\alpha \in K$ and $-\beta \notin K$, and we will set $L = K \cup \{0\}$. If the equation $x^2 + \alpha y^2 + \beta = 0$ has a solution (x, y) for some β $(-\beta \notin K)$, then it has a solution for any β. Thus, if the equation were to have no solution, then the sum of any pair of the elements of L would also be a square. This means that L would be an additive subgroup of F, and it gives rise to a contradiction since the order q of F is not a multiple of the order $(q + 1)/2$ of the subgroup L. Hence, for any α and β of F^*, there must exist a pair (x, y) of elements of F which satisfies $x^2 + \alpha y^2 + \beta = 0$. $\square$

Corollary. *Let f be a nondegenerate symmetric bilinear form defined over a finite field F such that* char F *is odd. If* dim $V \geq 3$, *there is an element* $v \neq 0$ *such that $f(v, v) = 0$.*

Proof. Choose a basis $\{u_i\}$ of V such that

$$f(u_i, u_i) = \alpha_i \neq 0 \quad \text{and} \quad f(u_i, u_j) = 0 \quad (i \neq j).$$

Set $\alpha = \alpha_1^{-1} \alpha_2$ and $\beta = \alpha_1^{-1} \alpha_3$. By the above lemma, there are elements x and y of F such that $x^2 + \alpha y^2 + \beta = 0$. Set $v = x u_1 + y u_2 + u_3$. Then, we have $v \neq 0$, and

$$f(v, v) = \alpha_1(x^2 + \alpha y^2 + \beta) = 0. \quad \square$$

As before, let f be a nondegenerate symmetric bilinear form. If $f(v, v) = 0$ for some $v \neq 0$, then there is an element w of V such that $f(v, w) = 1$. We will show that there is in fact an element w such that

$$f(v, w) = 1 \quad \text{and} \quad f(w, w) = 0.$$

Suppose that $f(v, w) = 1$. Then, for any $\lambda \in F$, we get

$$f(w + \lambda v, w + \lambda v) = f(w, w) + 2\lambda,$$
$$f(v, w + \lambda v) = f(v, w) = 1.$$

Thus, if we replace w by $w + \lambda v$, the desired equations are satisfied for some λ.

Set $U = Fv + Fw$ for the two elements v and w which satisfy

$$f(v, v) = f(w, w) = 0, \qquad f(v, w) = 1.$$

Then, we have $U \cap U^{\perp} = \{0\}$, so the space V is the direct sum of the subspaces U and $U^{\perp}$. We have the following proposition.

(5.8) *Let f be a nondegenerate symmetric bilinear form defined on a vector space V over F. Assume that F is a finite field and that its characteristic is odd. If* dim V *is odd, let* dim $V = 2m + 1$. *Then, there is a basis $\{u_i\}$ of V such that*

$$f(u_0, u_0) = \alpha \neq 0, \qquad f(u_0, u_i) = 0,$$
$$f(u_i, u_{m+i}) = 1, \qquad f(u_i, u_j) = 0 \quad (|i - j| \neq m)$$

where the index i ranges over the integers from 1 to m. If the dimension of V is odd, the structure of the orthogonal group $G(f)$ does not depend on the form.
    If dim $V = 2m$ *for some integer m, then there is a basis $\{u_i\}$ of V such that*

$$f(u_i, u_{i+m}) = 1, \qquad f(u_i, u_j) = 0$$

for $1 \leq i < m$ and $|i - j| \neq m$. For $\{u_m, u_{2m}\}$, we have one of the following two cases :

(1) $f(u_m, u_m) = f(u_{2m}, u_{2m}) = 0$ *and* $f(u_m, u_{2m}) = 1$, *or*
(2) $f(u_m, u_m) = -\gamma f(u_{2m}, u_{2m}) \neq 0$ *and* $f(u_m, u_{2m}) = 0$,

where γ is an element of F^* which is not a square.
    If dim V *is even, there are two isomorphism classes of orthogonal groups.*

Proof. (a) Assume that dim $V = 2m + 1$. By using induction on m, we can prove the existence of a basis $\{u_i\}$ which satisfies the conditions stated in the proposition. To verify the uniqueness of the structure of the orthogonal groups, let g be another nondegenerate symmetric bilinear form. Then, there is a basis $\{v_i\}$ such that

$$g(v_0, v_0) = \beta \neq 0, \qquad g(v_0, v_i) = 0,$$
$$g(v_i, v_{m+i}) = 1, \qquad g(v_i, v_j) = 0$$

for $1 \leq i \leq m$ and $|i - j| \neq m$. Set $\gamma = \alpha^{-1}\beta$, and let σ be the linear transformation of V defined by

$$u_0 \rightarrow v_0, \qquad u_i \rightarrow v_i, \qquad v_{m+i} \rightarrow \gamma v_{m+i} \quad (1 \leq i \leq m).$$

Then, $\sigma \in GL(V)$ and we have $g = \gamma f^{\sigma}$. This implies that $G(g) = G(f)^{\sigma}$. Hence, the structure does not depend on the particular form f.

(b) Assume that dim $V = 2m$. Then, by induction on m, we can see that there is a basis $\{u_i\}$ of V such that the values of $f(u_i, u_j)$ are as given, except possibly for $\{u_m, u_{2m}\}$. To prove that this last condition is also satisfied, consider $W = Fu_m + Fu_{2m}$. If W contains an element $u \neq 0$ such that $f(u, u) = 0$, then we have the case (1). So, suppose that $f(u, u) \neq 0$ for any element $u \neq 0$ of W. In this case, we can find elements u_m and u_{2m} such that

$$f(u_m, u_m) = -\gamma f(u_{2m}, u_{2m}) \neq 0, \qquad f(u_m, u_{2m}) = 0.$$

If $u = x u_m + y u_{2m}$, then we have

$$f(u, u) = f(u_m, u_m)(x^2 - \gamma y^2).$$

Since this is not zero for $u \neq 0$, γ cannot be a square in F^*. We will also prove that in the last case, the orthogonal groups are isomorphic for any given nonsquare element γ. Suppose that we have

$$f(u_m, u_m) = \alpha, \qquad \gamma f(u_{2m}, u_{2m}) = -\alpha, \qquad f(u_m, u_{2m}) = 0.$$

If g is another form, then there is a basis $\{v_i\}$ such that

$$g(v_m, v_m) = \alpha', \qquad \gamma' g(v_{2m}, v_{2m}) = -\alpha', \qquad g(v_m, v_{2m}) = 0.$$

Since both γ and γ' are nonsquare, $\gamma \gamma'$ is a square in F. So, let $\gamma' = \lambda^2 \gamma$, and let σ be the linear transformation on V defined by

$$u_i \to v_i, \qquad u_{m+i} \to (\alpha'/\alpha) v_{m+i}, \qquad u_{2m} \to \lambda v_{2m}.$$

Then, we get $g = (\alpha'/\alpha) f^{\sigma}$. This implies that $G(g)$ is conjugate to $G(f)$.

(c) It remains to show that the orthogonal groups defined by the forms (1) and (2) are not isomorphic. For any nondegenerate form f, we have $\sigma \in G(f)$ if and only if $f(\sigma(u_i), \sigma(u_j)) = f(u_i, u_j)$ for all i and j. Using this remark, we can compute the order of $G(f)$. Set

$$Q(u) = f(u, u).$$

Then, Q is the quadratic form associated with the symmetric form f. An element u of V is said to be *isotropic* (with respect to Q) if $Q(u) = 0$. Let dim $V = n$, and write either $n = 2m + 1$ or $n = 2m$ for some integer m, depending on the parity of n. If $m \geq 2$, the element u_1 is isotropic. We can write

(*) $v = \alpha u_1 + \beta u_{m+1} + w$

where w is an element orthogonal to both u_1 and u_{m+1}. Then, we have

$$Q(v) = \alpha\beta + Q(w).$$

Let x_n be the number of isotropic elements of V. Then, the number of elements such that $Q(v) = Q(w) = 0$ is $(2q - 1)x_{n-2}$. Hence, we get

$$x_n = (2q - 1)x_{n-2} + (q - 1)(q^{n-2} - x_{n-2}).$$

For small values of n, it is easy to check that $x_1 = 1$ and $x_2 = q + \varepsilon(q - 1)$ where $\varepsilon = 1$ for the first case (1) and $\varepsilon = -1$ for the second case (2). So, by induction, we obtain

$$x_{2m+1} = q^{2m}, \qquad x_{2m} = q^{2m-1} + \varepsilon(q - 1)q^m.$$

There are x_n choices for the first element u_1. After choosing u_1, let there be y_n choices for u_{m+1}. Then, there are exactly y_n elements v such that

$$f(u_1, v) = 1 \quad \text{and} \quad f(v, v) = 0.$$

Thus, y_n is the number of isotropic elements with $\beta = 1$ in (∗). Hence, we have

$$y_n = x_{n-2} + (q^{n-2} - x_{n-2}) = q^{n-2}.$$

If we denote the form f in the n-dimensional space V by f_n, then we have $|G(f_1)| = 2$, and $|G(f_2)| = 2(q - \varepsilon)$. Starting from these values, we can compute the order of $G(f_n)$:

$$|G(f_n)| = 2(q^{2m} - 1)(q^{2m-2} - 1) \cdots (q^2 - 1)q^e$$

where $e = m^2$ and $n = 2m + 1$, and

$$|G(f_n)| = 2(q^m - \varepsilon)(q^{2m-2} - 1) \cdots (q^2 - 1)q^{m(m-1)}$$

where $n = 2m$. If dim $V = 2m$, the orthogonal groups defined by the forms (1) and (2) are not isomorphic because they have different orders. $\square$

The quadratic form Q introduced in the preceding discussion determines the symmetric form f. In fact, we have

$$Q(u + v) = Q(u) + Q(v) + 2f(u, v),$$

and the following proposition holds:

$$f^\sigma = f \Leftrightarrow Q(\sigma(u)) = Q(u) \quad \text{for all } u \in V.$$

Orthogonal Groups over the Field of Characteristic Two

If the characteristic of the ground field F is two, we can define an orthogonal group as the set of the linear transformations which leave a given quadratic form invariant. Let V be a vector space over a field F with char $F = 2$. A *quadratic form* Q defined on V is a mapping from V to F which satisfies the following two properties: for any $\lambda \in F$ and $u \in V$, we have

$$Q(\lambda u) = \lambda^2 Q(u)$$

and, for any $u, v \in V$, there is a bilinear form B such that

$$Q(u + v) = Q(u) + Q(v) + B(u, v).$$

Since the characteristic of the field F is two, we have $2u = 0$. So, the bilinear form B is *alternating*. The form B is not necessarily nondegenerate. In fact, if dim V is odd, B is always degenerate. However, we will assume in this section that B is nondegenerate. So, in particular, the dimension of V is always assumed to be even:

$$\dim V = 2m.$$

Let Q be a quadratic form defined on V. Then, for any $\sigma \in GL(V)$,

$$Q^\sigma(\sigma(u)) = Q(u)$$

defines the quadratic form Q^σ. In this way, $GL(V)$ acts on the set of the quadratic forms. The stabilizer of Q is the *orthogonal group* with respect to Q.

If $Q^\sigma = Q$, then σ leaves the associated alternating form B invariant. Hence, the orthogonal group is a subgroup of the symplectic group. There is a basis $\{u_i\}$ of V such that

$$B(u_i, u_{m+i}) = 1, \qquad B(u_i, u_j) = 0 \quad \text{if } |i - j| \neq m.$$

Then, for any element $u = \sum x_i u_i \, (x_i \in F)$ of V, we get

$$Q(u) = \sum x_i^2 Q(u_i) + \sum x_i x_j B(u_i, u_j).$$

Thus, the form Q is uniquely determined by $\{Q(u_i)\}$. We will show that we can choose a basis $\{u_i\}$ such that $Q(u_i) = 0$ except possibly when $i = m$ and $i = 2m$. We need the following lemma.

(5.9) *Let Q be a quadratic form, and let B be the associated bilinear form. If $Q(u) = 0$ for all elements u of a subspace U, then we have*

$$B(u, v) = 0 \qquad (u, v \in U).$$

Proof. If $u, v \in U$, then $u + v$ is also contained in U. Hence, we have

$$B(u, v) = Q(u + v) - Q(u) - Q(v) = 0. \quad \square$$

Suppose we choose a different basis $\{v_i\}$ of V. By (5.9), there is an element v of $Fu_m + Fu_{2m}$ such that $Q(v) \neq 0$. Then, we have

$$Q(u_i + \lambda v) = Q(u_i) + \lambda^2 Q(v) = 0$$

for some λ. Suppose that we have $Q(v_i) = 0$ for $v_i = u_i + \lambda_i v$ $(i \neq m, 2m)$. Then, we get

$$B(v_i, v_{m+i}) = 1, \qquad B(v_i, v_j) = 0$$

for $i \neq m, 2m$ and $|i - j| \neq m$. Furthermore, we can find two elements v_m and v_{2m} such that

$$B(v_m, v_i) = B(v_{2m}, v_i) = 0, \qquad B(v_m, v_{2m}) = 1$$

for all $i \neq m, 2m$. Thus, the form Q is uniquely determined by the values of $Q(v_m + \lambda v_{2m})$ when λ ranges over F. Since

(*) $$Q(v_m + \lambda v_{2m}) = Q(v_m) + \lambda^2 Q(v_{2m}) + \lambda,$$

we have one of the following two cases.

Case 1 There is an element $\lambda \in F$ such that $Q(v_m + \lambda v_{2m}) = 0$. In this case, we can choose a basis in $Fv_m + Fv_{2m}$ such that $Q(w_{2m}) = 0$ and $B(w_m, w_{2m}) = 1$. Then, (*) shows that

$$Q(w_m + \lambda w_{2m}) = Q(w_m) + \lambda.$$

Hence, we may choose a suitable element w_m so as to make $Q(w_m) = 0$.

Case 2 We have $Q \neq 0$ for all $\lambda \in F$. In this case, the quadratic polynomial

$$t^2 Q(v_{2m}) + t + Q(v_m)$$

is irreducible in F. We may change (v_m, v_{2m}) to $(\mu v_m, \mu^{-1} v_{2m})$ so as to make $Q(v_{2m}) = 1$. Set $\gamma = Q(v_m)$. Then, $t^2 + t + \gamma$ is also irreducible in F. We need to study when the above polynomial $t^2 + t + \gamma$ is irreducible. It is clear that $t^2 + t + \gamma$ is reducible in F if and only if $\gamma = \beta^2 + \beta$ for some $\beta \in F$. Let φ be the mapping defined by $\varphi(x) = x^2 + x$. Then, φ is a homomorphism of the additive group F into itself, and we have $|\operatorname{Ker} \varphi| = 2$. So, the image of φ is a subgroup F_0 such that $|F : F_0| = 2$. As we have seen before, the polynomial $t^2 + t + \gamma$ is reducible if and only if $\gamma \in F_0$. Thus, $t^2 + t + \gamma$ is irreducible if and only if γ belongs to F_1 where F_1 is the coset of F_0 other than F_0 itself. By (*), we get $Q(v_m + \lambda v_{2m}) = \gamma + \lambda^2 + \lambda$. So, if we choose $v_m + \lambda v_{2m}$ in place of v_m, we can make $Q(v_m)$ equal to any given element of F_1. Therefore, the quadratic form Q is unique when $Q \neq 0$ for all $\lambda \in F$.

As when we were considering the orthogonal groups over a field F such that char F was odd, an element u of V is said to be *isotropic* (with respect to Q) if $Q(u) = 0$. As before, the number of isotropic elements is

$$q^{2m-1} + \varepsilon(q-1)q^m$$

where $\varepsilon = 1$ in Case 1 and $\varepsilon = -1$ otherwise. We conclude that the order of an orthogonal group is

(**) $2(q^m - \varepsilon)(q^{2m-2} - 1) \cdots (q^2 - 1)q^{m(m-1)}$.

We have proved the following proposition.

(5.10) *Let V be a vector space over a finite field F of characteristic two, and let Q be a quadratic form defined on V. Suppose that the associated alternating form B is nondegenerate. Then, there is a basis $\{u_i\}$ of V such that*

$$B(u_i, u_{m+i}) = 1, \qquad B(u_i, u_j) = 0 \quad \text{if } |i - j| \neq m,$$

and, for $u = \sum x_i u_i + \sum y_i u_{m+i} \in V$, one of the following formulas holds :

$$Q(u) = \sum x_i y_i, \quad \text{or} \quad Q(u) = \sum x_i y_i + \gamma x_m^2 + y_m^2.$$

In the second formula, γ is an element of F such that the polynomial $t^2 + t + \gamma$ is irreducible in F, and the orthogonal group $G(Q)$ does not depend on γ. The order of the orthogonal group is given by the formula (**).

If the ground field is not finite, the structures of the orthogonal groups depend on the properties of the field.

Simple Groups Derived from the Classical Groups

We denote by $S_p(2m, q)$ the symplectic group defined over a $2m$-dimensional vector space over a field of q elements. Here, we will state (without proofs) some of the basic properties of the symplectic groups as well as of other classical groups.

The following properties hold:

$$S_p(2m, q) \subset SL(2m, q), \qquad S_p(2, q) = SL(2, q),$$
$$Z(S_p(2m, q)) = \{\pm 1\}.$$

(For a proof of the above propositions, see Nagao [3], pp. 170–180. Also see Exercises 3 and 4, §3 of this chapter.) The central factor group of $S_p(2m, q)$ is denoted by $PS_p(2m, q)$ or $S_{2m}(q)$, and it is a simple group except when $(2m, q)$ is $(2, 2), (2, 3)$, or $(4, 2)$. If the characteristic of the field F is two, the center of the symplectic group is $\{1\}$. We have

$$S_p(4, 2) \cong \Sigma_6$$

(See Huppert [3], p. 227.) For the order of $S_p(2m, q)$, see Exercise 1 at the end of this section.

Let $GU(n, q)$ denote the unitary group defined over an n-dimensional vector space over a field of q^2 elements, and set

$$SU(n, q) = GU(n, q) \cap SL(n, q^2).$$

By (5.5), the determinant d of a matrix in $GU(n, q)$ satisfies $d^{1+q} = 1$. Let Z be the set of elements λ of F such that $\lambda^{1+q} = 1$. Then, Z is a cyclic group of order $1 + q$, and we have

$$GU(n, q)/SU(n, q) \cong Z.$$

The center of $GU(n, q)$ is also isomorphic to the group Z, while the center of $SU(n, q)$ is a cyclic group of order $(n, q + 1)$. We have

$$SU(2, q) \cong SL(2, q).$$

The central factor group of $SU(n, q)$ is denoted by

$$PSU(n, q) \quad \text{or} \quad U_n(q),$$

and for $n > 2$, it is always simple except when $n = 3$ and $q = 2$. The group $SU(3, 2)$ of the exceptional case is a solvable group, and its order is $3^3 2^3$. Exercise 2 at the end of this section gives the order of $U_n(q)$.

If dim V is odd, the orthogonal group defined on V is unique, and it is denoted by $O(2m + 1, q)$. This group contains all of the reflections which have determinants equal to -1, so the totality of those elements with determinants equal to 1 forms a normal subgroup of index 2, and this subgroup is written as $SO(2m + 1, q)$. We remark that the commutator subgroup of $SO(2m + 1, q)$, usually denoted by $\Omega(2m + 1, q)$, coincides with the commutator subgroup of $O(2m + 1, q)$. The group $\Omega(2m + 1, q)$ is a simple group of index 4 in $O(2m + 1, q)$, and we have

$$\Omega(3, q) \cong PSL(2, q), \qquad \Omega(5, q) \cong PS_p(4, q).$$

The groups $\Omega(2m + 1, q)$ and $PS_p(2m, q)$ have the same order, but they are not isomorphic if $m > 2$ and q is odd. (See Dickson [1] for the details.)

If dim V is even there are two nonisomorphic orthogonal groups. We denote them by $O_\varepsilon(2m, q)$, where $\varepsilon = \pm 1$. In this case, the index of the commutator subgroup $\Omega_\varepsilon(2m, q)$ in $SO_\varepsilon(2m, q)$ is again 2. The order of the center of $\Omega_\varepsilon(2m, q)$ is 2 when $q^m \equiv \varepsilon \pmod 4$, and it is 1 in all other cases. The central factor group is simple if $m \geq 3$. For the orthogonal groups of low dimensions, we have

$$P\Omega_{+1}(4, q) = PSL(2, q) \times PSL(2, q)$$

$$P\Omega_{-1}(4, q) = PSL(2, q^2)$$

$$P\Omega_{+1}(6, q) = PSL(4, q)$$

$$P\Omega_{-1}(6, q) = PSU(4, q).$$

The standard reference for these results is Dieudonné [1].

Simple Groups of Lie Type

Simple groups of Lie type are groups defined analogously to the simple Lie groups. It is a well-known fact that there is a one-to-one correspondence between the simple Lie groups and the simple Lie algebras over the field of complex numbers. Lie algebras were originally studied in connection with the theory of continuous transformation groups, and the simple Lie algebras over the field of complex numbers have been completely classified by Cartan and Killing.

In 1955, Chevalley [2] published a general method of defining the subgroups of $GL(V)$ (the general linear group over an arbitrary field) which correspond to the simple Lie algebras in a manner similar to the correspondence between the complex Lie groups and the Lie algebras over the field of complex numbers. Thus, not only the classical groups but also the groups corresponding to the exceptional simple Lie algebras were defined over an arbitrary field, and Chevalley proved that with a small number of exceptions, these groups were simple groups. This method of Chevalley has an obvious advantage in that we can study all of these groups in a unified way.

We will not be able to give any of the details in this book, but we will briefly review some of the basic properties of simple Lie algebras. For the details, the readers are referred to Matsushima [1] and Bourbaki [2].

Simple Lie algebras over the field of complex numbers are uniquely determined by the corresponding *root systems* which are defined as follows. A simple Lie algebra L contains a maximal abelian subalgebra H, called a Cartan subalgebra of L, and the vector space L is decomposed into a direct sum of the eigenspaces with respect to the action of H:

$$L = H + L_\alpha + L_\beta + \cdots$$

Here, each L_α is a one-dimensional subspace. Let x_α be a nonzero element of L_α. If we denote the multiplication in L by using brackets, we have

$$[h, x_\alpha] = \alpha(h)x_\alpha,$$

where h is any element of H, α is a linear form on H (called a root), and $\alpha \neq \beta$ if $L_\alpha \neq L_\beta$. The set of all these α's is the root system associated with the simple algebra L. There is a symmetric bilinear form, called the Killing form of L, which induces a positive definite quadratic form on the real vector space $\mathbb{R}^l$ generated by all of the roots. Thus, $\mathbb{R}^l$ is in fact the usual Euclidean space. Let Σ be the root system associated with the simple Lie algebra L. For each $\alpha \in \Sigma$, let w_α be the reflection on $\mathbb{R}^l$ with respect to the hyperplane orthogonal to α. Then, it has been proved that the set of the reflections w_α $(\alpha \in \Sigma)$ generates a finite group which is the Weyl group of L. This group W is a Coxeter group (see Exercise 5, §4), and it is irreducible. Thus, W is classified according to Theorem 4.15 (i). Each type determines a unique root system, except for the graph of type B_n for $n > 2$. In this case, there are two root systems, those of type B_n and those of type C_n, and they are distinguished by the relative lengths of the roots. It has also been proved that the root systems associated with Lie algebras are never of types H_3 or H_4, but all other types can be associated with some simple Lie algebra. (The same is true for the Weyl groups associated with the buildings.)

The method of Chevalley depends on the property that for a suitably chosen basis of L, the multiplication constants among the basis elements are all integers. Such a basis is called the *canonical basis* of Chevalley. Let $\{h_1, \ldots, h_l, x_\alpha, x_\beta, \ldots\}$ be the canonical basis of Chevalley of L, and let V be a vector space on which L acts. From the properties of the Chevalley basis, it can be shown that there is a lattice M in V which is left invariant by all of the elements of the form

$$x_\alpha^n/(n!) \qquad (\alpha \in \Sigma)$$

which belong to the universal enveloping algebra of L. (See Steinberg [2] for the details, especially Theorem 2 of §2, [2].) A subset M of a vector space V is said to be a *lattice* if M contains a basis $\{u_i\}$ of V and coincides with the free abelian group generated by the set $\{u_i\}$. Furthermore, it can be shown that the set of the elements of L which leave the lattice M invariant is a lattice generated by $\{x_\alpha\}$ of the Chevalley basis together with a lattice in the Cartan subalgebra. This lattice in L is uniquely determined by the representation of L on V (and not by the lattice M), and it is denoted by $L_{\mathbb{Z}}$.

Let k be an arbitrary field, and consider the tensor products over $\mathbb{Z}$

$$V^k = M \otimes k, \qquad L^k = L_{\mathbb{Z}} \otimes k.$$

Clearly, V^k is a vector space over k, and L^k acts on V^k. For an arbitrary element $\alpha \in \Sigma$ and for any natural number n, the element $x_\alpha^n/(n!)$ acts on M. So, for any variable λ, $\lambda^n x_\alpha^n/(n!)$ acts on $M \otimes \mathbb{Z}[\lambda]$. From the basic properties of L, it follows that x_α acts on M as a nilpotent endomorphism. So, for a sufficiently large n, the element $x_\alpha^n/(n!)$ induces the zero action. Thus,

$$\exp(\lambda x_\alpha) = \sum \lambda^n x_\alpha^n/(n!)$$

acts on $M \otimes \mathbb{Z}[\lambda]$. So, $\exp(\lambda x_\alpha) \otimes 1$ acts on $M \otimes \mathbb{Z}[\lambda] \otimes k$. By the substitution $\lambda \to t$ where t is an element of k, we get an action of $\exp(t x_\alpha)$ on $V^k = M \otimes k$. We denote this automorphism of V^k by $x_\alpha(t)$, and the subgroup of $GL(V^k)$ generated by $\{x_\alpha(t)\}$ ($\alpha \in \Sigma$, $t \in k$) is called the *Chevalley group* (associated with the L-module V).

For two variables λ and μ, we get

$$\exp(\lambda x_\alpha)\exp(\mu x_\alpha) = \exp((\lambda + \mu)x_\alpha).$$

So, we have the formula

(5.11) $x_\alpha(t)x_\alpha(u) = x_\alpha(t + u) \qquad (t,\, u \in k).$

Furthermore, for any $\alpha \neq \beta$ of Σ, we have

$$[\exp(\lambda x_\alpha),\, \exp(\mu x_\beta)] = \prod \exp(c_{ij}\, \lambda^i \mu^j x_{i\alpha + j\beta}).$$

The left side is the commutator, and the product of the right side ranges over the set of integers (i, j) such that $i\alpha + j\beta \in \Sigma$. The order the products are taken in is determined by the fixed linear ordering among the roots of Σ, and the coefficients c_{ij} are integers which are determined by the roots α and β (and the ordering of the roots) and are independent of λ and μ. The above formula gives us the following:

(5.12) $[x_\alpha(t),\, x_\beta(u)] = \prod x_{i\alpha + j\beta}(c_{ij}t^i u^j)$

for any $t,\, u \in k$. The order and the range of the product in the right side, as well as the values of c_{ij}, are as before. When the field k is a finite field, the group generated by $\{x_\alpha(t)\}$ with the relations (5.11) and (5.12) is the *universal*

Chevalley group, and any Chevalley group is a homomorphic image of the universal one (Steinberg [2], §§2 and 6). The sole exception to the above statement occurs when the rank of L is one.

The original paper of Chevalley [2] deals with the adjoint representation when the vector space V is L itself. We get the groups

$$PSL(n + 1, k), \quad \Omega(2n + 1, k), \quad PS_p(2n, k), \quad \Omega_{+1}(2n, k)$$

from the simple Lie algebras of types A_n, B_n, C_n, and D_n, respectively (Ree [1]). If the characteristic of k is two, the Lie algebras of type B_n and of type C_n are isomorphic, and the associated groups are isomorphic. If the simple Lie algebra is of one of the exceptional types, then the corresponding group is simple in all but one case. The corresponding groups are said to be of type E_1, of type F_4, or of type G_2, and we use notations such as $E_6(q)$ when $|k| = q$. The only nonsimple group is $G_2(2)$ which has a normal subgroup of index 2 isomorphic to $U_3(3)$.

The orders of the Chevalley groups of exceptional types are given in the following table.

$$G_2(q): q^6(q^6 - 1)(q^2 - 1)$$

$$F_4(q): q^{24}(q^{12} - 1)(q^8 - 1)(q^6 - 1)(q^2 - 1)$$

$$E_6(q): q^{36}(q^{12} - 1)(q^9 - 1)(q^8 - 1)(q^6 - 1)(q^5 - 1)(q^2 - 1)/d$$

$$E_7(q): q^{63}(q^{18} - 1)(q^{14} - 1)(q^{12} - 1)(q^{10} - 1)(q^8 - 1)$$
$$\times (q^6 - 1)(q^2 - 1)/d$$

$$E_8(q): q^{120}(q^{30} - 1)(q^{24} - 1)(q^{20} - 1)(q^{18} - 1)(q^{14} - 1)$$
$$\times (q^{12} - 1)(q^8 - 1)(q^2 - 1)$$

The value of d in the above table is $d = (3, q - 1)$ for type E_6 and $d = (2, q - 1)$ for type E_7.

By choosing a suitable representation and the corresponding L-module V, we get the groups which are central extensions of simple groups. In this way

$$SL(n, k), \quad S_p(2n, k), \quad \text{or} \quad \text{Spin}(n, k)$$

can be considered as Chevalley groups. The last group is a central extension of the orthogonal group $\Omega_{+1}(n, k)$ (for $\text{Spin}(n, k)$, see Chevalley [1]). The above central extensions are universal Chevalley groups, and, in general, the center of a universal Chevalley group is isomorphic to the Schur multiplier of the corresponding simple Chevalley group. There are a few exceptions to this proposition, and the readers are referred to Feit [5] or Griess [1]. For example, the exceptions in the series of $PSL(n, q)$ occur when (n, q) is $(2, 4)$,

(2, 9), (3, 2), (3, 4), or (4, 2). The corresponding exceptional Schur multipliers
are

$$Z_2, \quad Z_6, \quad Z_2, \quad Z_4 \times Z_{12}, \quad \text{and} \quad Z_2.$$

The method of Chevalley produces neither the unitary groups nor the
orthogonal groups $P\Omega_{-1}(2n, k)$. However, these groups can be obtained in
the following way (Steinberg [1]). Consider the quadratic extension k_0 of k. It
can be shown that a combination of the Galois automorphism of k_0 over k
and the symmetry of the graph of type A_n (or D_n) gives rise to an automorphism σ of the Chevalley group of type A_n (or D_n) defined over the field k_0.
The set of the elements in the Chevalley group which are invariant under σ
forms the unitary group if the Chevalley group is of type A_n and the orthogonal group $\Omega_{-1}(2m, k)$ if it is of type D_n. By a similar construction using the
symmetry of the graph of type E_6, we obtain a series of simple groups. Since
the graph of type D_4 has an extra automorphism (the rotation around the
central vertex), we get another series of simple groups. These groups are
called the *Steinberg groups*, and are denoted by $^2E_6(q)$ and $^3D_4(q)$. (We fol-
lowed the notation of Feit [5] and set $q = |k|$. The group $^3D_4(q)$ is defined as
a subgroup of $D_4(q^3)$.) Similarly, we may use the following notation:

$$^2A_n(q) = U_n(q), \qquad ^2D_n(q) = P\Omega_{-1}(2n, q).$$

The orders of these groups are

$$^2E_6(q): q^{36}(q^{12} - 1)(q^9 + 1)(q^8 - 1)(q^6 - 1)(q^5 + 1)(q^2 - 1)/d$$
$$^3D_4(q): q^{12}(q^8 + q^4 + 1)(q^6 - 1)(q^2 - 1),$$

where $d = (3, q + 1)$.

Among the graphs of the irreducible Coxeter systems, there are three more
symmetric ones: B_2, G_2, and F_4. In the root systems corresponding to these
graphs, the lengths of the roots at the extremities are different. So, in general,
there is no automorphism of the corresponding Chevalley group which exchanges these roots. But, if the characteristic p of the field k is 2 for types B_2
and F_4, or 3 for the type G_2, and if the Frobenius automorphism $x \to x^p$ of k
has a square root, we can define an exceptional automorphism σ of the
Chevalley group. In each of these cases, it can be shown that the set of the
elements left invariant by σ forms a simple group except when $|k| = p$. These
groups are denoted by $^2B_2(q)$, $^2G_2(q)$, and $^2F_4(q)$. The orders are

$$^2B_2(q): q^2(q^2 + 1)(q - 1) \qquad\qquad q = 2^{2m+1}$$
$$^2G_2(q): q^3(q^3 + 1)(q - 1) \qquad\qquad q = 3^{2m+1}$$
$$^2F_4(q): q^{12}(q^6 + 1)(q^4 - 1)(q^3 + 1)(q - 1) \qquad q = 2^{2m+1}$$

where $m \geq 0$. For $m = 0$, $^2B_2(2)$ is a solvable group of order 20, $^2G_2(2)$ is isomorphic to Aut $SL(2, 8)$ and contains a normal subgroup of index 3 which is isomorphic to $SL(2, 8)$, and $^2F_4(2)$ has a normal subgroup of index 2, but this normal subgroup is a simple group which is not isomorphic to any of the simple groups we have mentioned so far (Tits [1]). The series of simple groups $^2B_2(q)$ was discovered in the study of doubly transitive permutation groups (Suzuki [3]). The book by Ito [3] is recommended for a discussion of the groups $^2B_2(q)$ and other related topics. Originally, the simple group $^2B_2(q)$ was defined as a subgroup of $S_p(4, q)$ by giving a set of generators. Soon afterwards, Ono [1] proved that it is a set of the elements left invariant under some automorphism of $S_p(4, q)$. This was discovered independently by Ree, who also constructed $^2G_2(q)$ and $^2F_4(q)$ by a similar method (Ree [2]). For this reason, these groups are often denoted by $Sz(q)$ or $Re(q)$.

The simple groups discussed above (the Chevalley groups, Steinberg groups, Suzuki groups, and Ree groups) are called *simple groups of Lie type*. It is needless to say that the method of using Lie algebras to study these groups is indispensable.

Each simple group of Lie type has a geometrical representation just like the general linear group on the flag space of a projective space. For example, consider the classical group defined as the set of the linear transformations of a vector space V which leave a nondegenerate form f invariant. The totality of the flags consisting of totally isotropic subspaces with respect to f forms a building, as defined in §3 (cf. Exercises 3 and 4 of §3). It has been proved by Chevalley, Steinberg, and Tits that any simple group of Lie type has a Tits system (for the definition, see Definition 3.24). According to a theorem of Tits mentioned in §3, a finite building whose Weyl group (W, S) satisfies $|S| \geq 3$ is isomorphic to one of the buildings associated with the simple groups of Lie type. Thus, the simple groups of Lie type are characterized as the simple groups with Tits systems. For this reason, the geometric considerations and concepts like that of buildings are significant in the study of finite simple groups.

Sporadic Simple Groups

So far, we have seen that there are two main series of finite simple groups. One is the series of alternating groups, and the other is that of the simple groups of Lie type. At present, almost all of the known finite simple groups belong to these two main series. There are only a finite number of known simple groups which are not members of these series. Those simple groups which are neither alternating groups nor groups of Lie type are collectively called *sporadic simple groups*. At the time of this writing, it is still unclear whether the sporadic simple groups are really sporadic. Thompson once compared the sporadic simple groups among the simple groups to the planets among the stars. From the beginning of group theory, the sporadic simple groups have attracted the fancy of mathematicians just as the planets interested the ancient astronomers. Around the year 1870, Mathieu published his works on quintuply transitive permutation groups of degrees 24 and 12. Witt [1] relates the amusing history of these groups and presents many of the interesting properties of the five simple groups which are called the *Mathieu groups*. Even after the nature of the Mathieu groups had been clearly determined by Witt, there was a long period during which no other sporadic simple groups were discovered, and the general feeling among the mathematicians was that the list of the simple groups of order less than a million given at the end of Dickson's book of 1901 was actually complete. The first simple group which was added to Dickson's list was $Sz(8)$ (in 1960). Now, however, this group is considered to be the simple group $^2B_2(8)$ of Lie type.

Meanwhile, during his work on Ree groups of type G_2, Janko proved that there was in fact one additional solution of the relevant Diophantine equation. He was thus able to predict the existence of a new simple group, as well as to state some of its properties. For instance, he knew that the order would be 175,560. Not only did he make this discovery, but he also actually constructed a simple group with the required properties by choosing two suitable matrices of $GL(7, 11)$ as its generators. Later, in the summer of 1967, Janko published the predicted properties of two more possibly new simple groups, and their existence was verified shortly afterwards, using the coset enumeration process performed by a computer (cf. Chapter 2, §6). This was the first of several recent constructions of simple groups using computers. The second Janko group J_2 is a simple group of order 604,800 whose existence has been reproved without the help of a computer, but as for the larger group J_3, the only proof of its existence depends on computers. At this time (January, 1977), there are twenty-six sporadic simple groups which have been discovered. The existence of two of these groups has not yet been established.

We will present a table containing the names of the people who either discovered the group or predicted its properties, the order of the group, and the name of the person who constructed it, if it is different from the discoverer. (If the simple group X contains a known subgroup Y (and if X is closely related to Y in some sense), then we will state the index $|X : Y|$ instead of the order of the group.)

Mathieu	$M_{11}: 2^4 3^2 5 \cdot 11, |M_{12} : M_{11}| = 12$
	$|M_{24} : M_{23}| = 24, |M_{23} : M_{22}| = 23,$
	$|M_{22} : PSL(3, 4)| = 22$
Janko	$J_1[1]: 2^3 3 \cdot 5 \cdot 7 \cdot 11 \cdot 19$
	$J_2[2]: |J_2 : U_3(3)| = 100$
	M. Hall–Wales [1]
	$J_3[2]: 2^7 3^5 5 \cdot 17 \cdot 19$
	G. Higman–McKay [1]

D. G. Higman– $|HS : M_{22}| = 100$
Sims [1]:
Suzuki [6]: $|S : G_2(4)| = 1782 = 2 \cdot 3^4 11$
McLaughlin [1]: $|Mc : U_4(3)| = 275 = 5^2 11$
Held [1]: $2^{10} 3^3 5^2 7^3 17$
G. Higman–McKay [2]
Conway [1] $C_1 : 2^{21} 3^9 5^4 7^2 11 \cdot 13 \cdot 23$
$C_2 : 2^{18} 3^6 5^3 7 \cdot 11 \cdot 23$
$C_3 : 2^{10} 3^7 5^3 7 \cdot 11 \cdot 23$
Fischer [1] $M'(24) : 2^{21} 3^{16} 5^2 7^3 11 \cdot 13 \cdot 17 \cdot 23 \cdot 29$
$M(23) : 2^{18} 3^{13} 5^2 7 \cdot 11 \cdot 13 \cdot 17 \cdot 23$
$M(22) : 2^{17} 3^9 5^2 7 \cdot 11 \cdot 13$
Lyons [1]: $2^8 3^7 5^6 7 \cdot 11 \cdot 31 \cdot 37 \cdot 67$
Sims [2]
Rudvalis [1]: $|R : {}^2F_4(2)| = 4060 = 2^2 5 \cdot 7 \cdot 29$
Conway–Wales [1]
O'Nan [1]: $2^9 3^4 5 \cdot 7^3 11 \cdot 19 \cdot 31$
Sims [3]
Fischer [2]: $2^{41} 3^{13} 5^6 7^2 11 \cdot 13 \cdot 17 \cdot 19 \cdot 23 \cdot 31 \cdot 47$
Leon–Sims [1]
*: $2^{46} 3^{20} 5^9 7^6 11^2 13^3 17 \cdot 19 \cdot 23 \cdot 29 \cdot 31 \cdot 41 \cdot 47 \cdot 59 \cdot 71$
(?)
Thompson [6]: $2^{15} 3^{10} 5^3 7^2 13 \cdot 19 \cdot 31$
Harada [2]: $2^{14} 3^6 5^6 7 \cdot 11 \cdot 19$
Norton [1]
Janko* [3]: $2^{21} 3^3 5 \cdot 7 \cdot 11^3 23 \cdot 29 \cdot 31 \cdot 37 \cdot 43$

The existence of the groups with the asterisks has not yet been established. For further information, the readers are advised to refer to the original papers. Feit [5] collects the properties of those groups which were known in 1970. For Conway's groups, see his paper [2]. Also, J_3, Held's group, and the groups below Lyons' have been proved to exist by using computers.

Recently, Beisiegel–Stingl [1] announced that only finite simple groups of order less than a million are cyclic groups of prime order, alternating groups, simple groups of Lie type, some Mathieu groups M_i ($i < 23$), J_1, or J_2. But, the details have not been published yet.

(* Shortly before this translation, the existence of the last Janko group was established by mathematicians at Cambridge, England, using a computer. Furthermore, Griess announced that he has constructed the so-called "monster" without the help of a computer. Thus, all twenty-six sporadic simple groups exist. Since the monster is known to contain a large number of other sporadic simple groups, Griess' work provides the computer-free proof of the existence of many other sporadic simple groups.)

Exercises

1. Let f be a nondegenerate alternating form defined in the $2m$-dimensional vector space V over a field of q elements.

(a) Show that the number of pairs $\{u, v\}$ such that $f(u, v) = 1$ is $(q^{2m} - 1)q^{2m-1}$.

(b) Show that the order of $S_p(2m, q)$ is $\prod_{i=1}^{m} (q^{2i} - 1)q^{2i-1}$.

(*Hint.* We can choose a basis of V such that $u_1 = u$ and $u_{m+1} = v$.)

2. Let h be a nondegenerate Hermitian form defined over an n-dimensional vector space V over a field of q^2 elements. Let x_n be the number of isotropic vectors in V. Show that

$$x_n = q^{2n-2}(q + 1) - qx_{n-1}.$$

Let y_n be the number of vectors of length 1. Show that

$$y_n = q^{n-1}(q^n - (-1)^n).$$

Furthermore, prove that the order of $GU(n, q)$ is

$$q^{n(n-1)/2} \prod (q^i - (-1)^i) \qquad (1 \le i \le n)$$

and that the order of $PSU(n, q)$ is

$$q^{n(n-1)/2}(q^2 - 1)(q^3 + 1) \cdots (q^n - (-1)^n)/d$$

where $d = (n, q + 1)$.

§6. Finite Subgroups of Two-dimensional Special Linear Groups

The purpose of this section is to solve the problem of finding all the possible subgroups of two-dimensional special linear groups over finite fields. This problem was solved by Dickson in his book [1], and the result has been used on various problems in finite group theory. In particular, ever since the recent works of Thompson, Gorenstein, Walter, and others indicated the importance of the concept of p-stability in the study of finite simple groups, the result of Dickson has been recognized as one of the indispensable tools in studying the basic properties of linear groups which underlie the concept of p-stability. (See Chapter 5, §4.) In this section, we will prove Dickson's theorem. The method we will use to prove the theorem is typical and is readily adaptable to further applications.

Let V be the two-dimensional vector space. Usually, we will identify the group $GL(V)$ with the totality of nonsingular 2×2 matrices with entries from the ground field E. If K is a subfield of E, the set of matrices with entries from K forms a subgroup. In this way, we get an embedding of $GL(2, K)$ into $GL(2, E)$, which we will call *standard*. Although our purpose is to find all the subgroups of $SL(2, E)$ for a finite field E, it is convenient to extend the field and to consider $SL(2, F)$ over an algebraically closed field F.

Convention 6.1. *In this section, the letter F stands for an algebraically closed field of characteristic p. Thus, the letter p stands for either a prime number or zero. If $p = 0$, an S_p-subgroup means the trivial subgroup $\{1\}$. Let V be the two-dimensional vector space over F, and consider a fixed basis $\{u, v\}$ of V. Set*

$$L = SL(V).$$

The following notation is used throughout this section :

$$d_\omega = \begin{pmatrix} \omega & 0 \\ 0 & \omega^{-1} \end{pmatrix} \quad (\omega \in F^*), \qquad t_\lambda = \begin{pmatrix} 1 & 0 \\ \lambda & 1 \end{pmatrix} \quad (\lambda \in F),$$

$$w = \begin{pmatrix} 0 & 1 \\ -1 & 0 \end{pmatrix}.$$

Furthermore, we set $D = \{d_\omega\}$ $(\omega \in F^*)$, $T = \{t_\lambda\}$ $(\lambda \in F)$, $H = DT$, and $Z = Z(L)$.

The main properties of the group L will be collected in the following lemma.

(6.2) (i) *The center Z is $\{1\}$ if $p = 2$. Otherwise, Z is a cyclic group of order 2. The group L/Z is a simple group.*

(ii) *For any $\omega, \omega' \in F^*$ and $\lambda, \mu \in F$, we have*

$$d_\omega d_{\omega'} = d_{\omega\omega'} \qquad t_\lambda t_\mu = t_{\lambda+\mu}$$
$$d_\omega^{-1} t_\mu d_\omega = t_\sigma \quad (\sigma = \omega^2 \mu), \qquad w^{-1} d_\omega w = d_\omega^{-1}.$$

(iii) *The sets D, T, and H are subgroups of L, and we have*

$$D \cong F^*, \qquad T \cong F, \qquad T \lhd H.$$

Proof. The first half of (i) follows from Cor. 2 of (9.8), Chapter 1, and the second half from Theorem 9.9 of Chapter 1 (because F is algebraically closed and infinite). The assertions in (ii) are easily proved by matrix multiplication, and (iii) follows from (ii). □

(6.3) (i) *Any element of L is conjugate to either d_ω for some $\omega \in F^*$ or to $\pm t_\lambda$ for some $\lambda \in F$.*

(ii) *If $p \neq 2$, then L contains a unique element of order 2.*

(iii) *Let x be an element of L with finite order n. Then, x is conjugate to $\pm t_\lambda$ ($\lambda \neq 0$) in L if and only if $p > 0$ and if p divides n. In this case, the order of x is either p or $2p$.*

Proof. (i) Since F is algebraically closed, the linear transformation x has at least one eigenvalue. Let ω be an eigenvalue of x, and let u_1 be the corresponding eigenvector. If we choose u_1 to be a member of a basis, then the matrix presentation of x is given by

$$\begin{pmatrix} \omega & 0 \\ \gamma & \delta \end{pmatrix}.$$

Since $x \in SL(V)$, we get $\delta = \omega^{-1}$. If $\omega = \delta$, then we have $\omega^2 = 1$ and $\omega = \pm 1$. Thus, x is conjugate to $\pm t_\lambda$. If $\omega \neq \delta$, then the eigenvector u_2 corresponding to the eigenvalue δ is linearly independent of u_1. So, $\{u_1, u_2\}$ is a basis of V, and x is conjugate to d_ω in $GL(V)$. Since we can take $\{u_1, \lambda u_2\}$ as a basis, the coordinate change can be accomplished by an element of $SL(V)$. So, x is conjugate to either d_ω or $\pm t_\lambda$ in L.

(ii) If $p \neq 2$, Z is generated by $-t_0$. There is no other element of order 2 in L. This follows easily from the following:

$$\begin{pmatrix} \alpha & \beta \\ \gamma & \delta \end{pmatrix} \in SL(V) \Rightarrow \begin{pmatrix} \alpha & \beta \\ \gamma & \delta \end{pmatrix}^{-1} = \begin{pmatrix} \delta & -\beta \\ -\gamma & \alpha \end{pmatrix}.$$

(iii) Since T is isomorphic to the additive group F^+, the element $\pm t_\lambda$ ($\lambda \neq 0$) has finite order if and only if $p > 0$. In this case, the order of $\pm t_\lambda$ is either p or $2p$. Suppose that $p > 0$ and p divides n. By (i), x is conjugate to d_ω or $\pm t_\lambda$. If x is conjugate to d_ω, then ω is a primitive n-th root of unity. This is a contradiction because the characteristic p of F cannot divide the order of a primitive n-th root of unity. Thus, x is conjugate to $\pm t_\lambda$. □

The following properties of the centralizer are basic to further investigation of the finite subgroups of $SL(V)$.

(6.4) (i) *If $x^{-1} t_\lambda x = t_\mu$ for some $x \in L$ and $\mu \neq 0$, then x is an element of H. If we have $\lambda = \mu$ in addition, then $x \in T \times Z$. Thus, if $\mu \neq 0$, we have $C_L(t_\mu) = T \times Z$.*

(ii) *If $y^{-1} d_\omega y = d_\rho$ for some $y \in L$ and $\omega \neq \pm 1$, then we have $\rho = \omega$ or ω^{-1} and $y \in \langle D, w \rangle$. If $\omega \neq \pm 1$, then we have $C_L(d_\omega) = D$. If D_1 is a subgroup of D such that $|D_1| \geq 3$, then we have $N_L(D_1) = \langle D, w \rangle$.*

Proof. Both propositions follow from easy matrix computations. For example, the equality

$$\begin{pmatrix} 1 & 0 \\ \lambda & 1 \end{pmatrix}\begin{pmatrix} \alpha & \beta \\ \gamma & \delta \end{pmatrix} = \begin{pmatrix} \alpha & \beta \\ \gamma & \delta \end{pmatrix}\begin{pmatrix} 1 & 0 \\ \mu & 1 \end{pmatrix}$$

for $\mu \neq 0$ implies that $\alpha = \alpha + \beta\mu$ and $\beta = 0$. If $\lambda = \mu$ in the above equation, we get $\alpha = \delta$. This proves (i). The proposition (ii) can be proved similarly. □

(6.5) *The centralizer of an element x in L is abelian unless x belongs to the center of L.*

Proof. By (6.3) (i), x is conjugate to either d_ω or $\pm t_\lambda$. So, it suffices to prove the proposition (6.5) for these elements. In each of these cases, (6.5) follows from (6.4) since both $T \times Z$ and D are abelian. □

In §3, we considered the projective space associated with a vector space. In the present situation, we have a projective space of dimension one, so we get a projective line $\mathscr{L}$. If we need to consider the projective lines over various fields, the projective line over the field K will be denoted by $\mathscr{L}_K$. In general, we have the following lemma.

(6.6) *Let $\mathscr{L} = \mathscr{L}_K$ be the projective line over the field K. Then, the group $PGL(\mathscr{L})$ is triply transitive on the set of the points of $\mathscr{L}$. If K is either algebraically closed or a finite field of characteristic two, then $PSL(\mathscr{L})$ is also triply transitive.*

Proof. Let P_1, P_2, and P_3 be three distinct points of $\mathscr{L}$. Let w_i be an element of V corresponding to P_i. Then, w_i and w_j are linearly independent for $i \neq j$. Thus, we have

$$w_3 = \alpha w_1 + \beta w_2 \quad \text{with} \quad \alpha\beta \neq 0.$$

Let Q_1, Q_2, and Q_3 be three distinct points. If v_i is an element of V corresponding to Q_i, there are λ and μ of K such that

$$v_3 = \lambda v_1 + \mu v_2 \quad \text{and} \quad \lambda\mu \neq 0.$$

Then, the mapping $\alpha w_1 \to \lambda v_1$, $\beta w_2 \to \mu v_2$ is extended to a linear transformation σ such that $\sigma(w_3) = u_3$. Hence, we get $P_1^\sigma = Q_1, P_2^\sigma = Q_2$, and $P_3^\sigma = Q_3$. By Definition 3.5, $PGL(\mathscr{L})$ is triply transitive.

In the above proof, we may modify σ. If ρ is the mapping which extends $\alpha w_1 \to \xi\lambda v_1$ and $\beta w_2 \to \xi\mu v_2$, then we get $\det \rho = \xi^2 \det \sigma$. So, if K is algebraically closed, or if K is a finite (perfect) field of characteristic two, then we can make $\rho \in SL(V)$. Thus, $PSL(\mathscr{L})$ is triply transitive on the set of the points of $\mathscr{L}$. □

We will return to the group L and consider its action on $\mathscr{L} = \mathscr{L}_F$.

(6.7) (i) *Any noncentral element has at most two fixed points.*

(ii) *All conjugate elements have the same number of fixed points.*

(iii) *The element d_ω ($\omega \neq \pm 1$) has exactly two fixed points.*

(iv) *Any element of the subgroup T has a common fixed point. If P is the common fixed point, then the stabilizer of P coincides with $H = TD$.*

Proof. (i) Suppose that an element x has two distinct fixed points P and Q. Let u and v be elements of V which correspond to P and Q, respectively. Since both P and Q are fixed, we get

$$x(u) = \omega u \quad \text{and} \quad x(v) = \omega^{-1} v$$

for some $\omega \in F^*$. If x has another fixed point, let it be represented by $\alpha u + \beta v$. Then, we have $x(\alpha u + \beta v) = \rho(\alpha u + \beta v)$, so we get $\rho \alpha = \alpha \omega$ and $\rho \beta = \beta \omega^{-1}$. Thus, if $\alpha \beta \neq 0$, we have $\omega = \rho = \omega^{-1}$. This implies that $\omega = \pm 1$ and $x \in Z$.

(ii) This should be obvious. (The proposition is true for any group action.)

(iii) This is also clear from the proof of (i).

(iv) Let $\{u, v\}$ be the basis of V fixed in 6.1. Every element of T fixes the point P of $\mathscr{L}$ corresponding to u. The stabilizer of P is the set of elements x such that $x(u) = \alpha u$ ($\alpha \in F^*$). Thus, we have $S_L(P) = H$. □

From now on, let G be a finite subgroup of L. It is enough to consider the case when $G \supset Z(L)$. If $Z \nsubseteq G$, then, by (6.2) and (6.3), we get $|Z| = 2$, $|G|$ is odd, and $GZ = G \times Z$. Thus, the structure of G is uniquely determined by GZ.

The key result in studying the structure of G is the following proposition about maximal abelian subgroups of G.

(6.8) *Let G be a finite subgroup of $L = SL(V)$, and let $\mathfrak{M}$ be the set of all maximal abelian subgroups of G. Assume that G contains the center Z of L.*

(i)   *If $x \in G - Z$, then we have $C_G(x) \in \mathfrak{M}$.*

(ii) *For any two distinct subgroups A and B of $\mathfrak{M}$, we have*

$$A \cap B = Z.$$

(iii) *An element A of $\mathfrak{M}$ is either a cyclic group whose order is relatively prime to p, or of the form $Q \times Z$ where Q is an S_p-subgroup of G.*

(iv) *If $A \in \mathfrak{M}$ and $|A|$ is relatively prime to p, then we have $|N_G(A): A| \leq 2$.*

If $|N_G(A): A| = 2$, there is an element y of $N_G(A)$ such that

$$y^{-1}xy = x^{-1}$$

for any element x of A.

(v) Let Q be an S_p-subgroup of G. If $Q \neq \{1\}$, then there is a cyclic subgroup K of G such that $N_G(Q) = QK$. If $|K| > |Z|$, then we have $K \in \mathfrak{M}$.

Proof. (i) By (6.5), if $x \notin Z$, then $C_L(x)$ is abelian. Thus, $C_G(x) = G \cap C_L(x)$ is also abelian. A maximal abelian subgroup of G which contains $C_G(x)$ is certainly contained in $C_L(x)$. So, $C_G(x)$ is maximal abelian, and we have $C_G(x) \in \mathfrak{M}$.

(ii) If $x \in A \cap B$, then $C_G(x)$ contains both A and B. By assumption, A and B are distinct members of $\mathfrak{M}$. So, $C_G(x)$ is not abelian. By (i), we have $x \in Z(L)$, and we get $A \cap B = Z$.

(iii) For any element x of G, $\langle Z, x \rangle$ is abelian ((2.26) of Chapter 1). Hence, if $Z \neq G$, any member A of $\mathfrak{M}$ contains an element $x \notin Z$, and $A = C_G(x)$. By (6.3) (i), the element x is conjugate to d_ω ($\omega \neq \pm 1$) or to $\pm t_\lambda$ ($\lambda \neq 0$). If x is conjugate to d_ω, $C_L(x)$ is conjugate to the subgroup D by (6.4) (ii). Since $D \cong F^*$, $C_G(x) = A$ is isomorphic to a finite subgroup of F^*. Hence, by the Cor. of (5.6), Chapter 1, A is a cyclic group such that $|A|$ is relatively prime to p. On the other hand, if x is conjugate to $\pm t_\lambda$ ($\lambda \neq 0$), by (6.4) (i), $C_G(x)$ is isomorphic to a finite subgroup of $T \times Z$. In this case, we have $T \cong F^+$, and T contains a finite subgroup $\neq \{1\}$. Thus, we get $p > 0$ and $A = Q \times Z$ where Q is an elementary abelian p-group. We will show that Q is an S_p-subgroup of G. Let S be an S_p-subgroup of G which contains Q. Then, by a fundamental theorem on p-groups, we have $Z(S) \neq \{1\}$. So, for an element $z \neq 1$ of $Z(S)$, we get

$$A = Q \times Z \subset S \times Z \subset C_G(z) \in \mathfrak{M}.$$

This proves that $C_G(z) = A$ and that $Q = S$ is an S_p-subgroup of G.

(iv) In this case, by (iii), A is a cyclic group whose order is relatively prime to p, and a generator of A is conjugate to d_ω ($\omega \neq \pm 1$). Hence, by (6.4) (ii), $N_G(A)$ is conjugate to a subgroup of $\langle D, w \rangle$. Thus, we have $|N_G(A): A| \leq 2$. The remaining assertion follows since w inverts every element of D.

(v) The proof of (iii) shows that Q is conjugate to a subgroup of T. So, without loss of generality, we may assume that $Q \subset T$. In this case, by (6.4) (i), we have $N_G(Q) \subset TD$. Since Q is an S_p-subgroup of G, we have $N_G(Q) \cap T = Q$. Thus, the group $N_G(Q)/Q$ is isomorphic to a finite subgroup of $H/T \cong F^*$. It follows that $N_G(Q)/Q$ is a cyclic group whose order is relatively prime to p. Therefore, we can choose an element x such that the order of x is prime to p and $\langle x \rangle = K$ satisfies $N_G(Q) = QK$ and $Q \cap K = \{1\}$.

We need to show that if $|K| > |Z|$, then we have $K \in \mathfrak{M}$. Let A be a member of $\mathfrak{M}$ such that $K \subset A$. By assumption, K is a cyclic group such that $|K|$ is relatively prime to p and $|K| > |Z|$. So, by (iii), A is also a cyclic group whose order is relatively prime to p. By (6.7) (iii), a generator of A has exactly two fixed points P_1 and P_2 on the projective line $\mathscr{L}$. By the same reason, a generator x of K has exactly two fixed points. Thus, P_1 and P_2 are fixed points of x, and there is no other point fixed by x. On the other hand, by (6.7) (iv), every element of T has a common fixed point. Let it be P. Then, we have $S_L(P) = H$ by (6.7) (iv). Since $K \subset H$, K must fix P. So, P is one of the fixed points $\{P_1, P_2\}$ of x. A generator of A is contained in the stabilizer of P, and we have

$$A \subset S_L(P) \cap G = N_G(Q) = QK.$$

This implies that $A = QK \cap A = (Q \cap A)K = K$. Thus, K belongs to $\mathfrak{M}$.

□

Using (6.8), we can count the number of elements of G in two different ways. This counting argument is very useful (cf. Exercise 10, §2). First, we will introduce the following notation.

As before, let G be a finite subgroup of L which contains Z. Set

$$|Z| = e.$$

Then, by (6.2), $e = 1$ or 2 according to whether or not $p = 2$. We write

$$|G| = eg,$$

so g is the order of the factor group G/Z. Let Q be an S_p-subgroup of G, and set

$$|Q| = q \quad \text{and} \quad |N_G(Q): Q| = ek.$$

Let $\mathfrak{M}$ be the totality of the maximal abelian subgroups of G. Then, $\mathfrak{M}$ contains all the conjugate subgroups of $Q \times Z$. The remaining subgroups in $\mathfrak{M}$ are cyclic groups whose orders are relatively prime to p. By (6.8) (iv), if $A \in \mathfrak{M}$ such that $|A|$ is relatively prime to p, then we have

$$|N_G(A): A| \leq 2.$$

Suppose that the members of $\mathfrak{M}$ whose orders are relatively prime to p are partitioned into conjugacy classes $C_1, C_2, \ldots, C_{s+t}$ so that, for a representative A_i from C_i, we have

$$N_G(A_i) = A_i \quad \text{for} \quad i \leq s,$$
$$|N_G(A_j): A_j| = 2 \quad \text{for} \quad s < j \leq s + t.$$

Finally, set $|A_i| = eg_i \, (i = 1, 2, \ldots, s + t)$.

By (6.8), every element of G is contained in some member of $\mathfrak{M}$, and any two different members of $\mathfrak{M}$ have only the elements of Z in common.

For each i, the number of noncentral elements which are contained in some conjugate subgroup of A_i is

$$e(g_i - 1)eg/eg_i \varepsilon = e(g_i - 1)g/g_i \varepsilon$$

where $\varepsilon = 1$ if $i \leq s$ and $\varepsilon = 2$ if $i > s$. Similarly, the conjugate subgroups of $Q \times Z$ contain

$$(eq - e)eg/eqk = e(q - 1)g/qk$$

noncentral elements. Thus, we get

$$e(g - 1) = e(q - 1)g/qk + \sum e(g_i - 1)g/g_i \varepsilon;$$

(6.9) $\qquad 1 = 1/g + (q - 1)/qk + \sum_{i=1}^{s} (g_i - 1)/g_i + \sum_{i=s+1}^{s+t} (g_i - 1)/2g_i.$

By definition, q is either 1 or a power of p, and each g_i is an integer ≥ 2. This implies that

$$(g_i - 1)/g_i \geq 1/2 \quad (i = 1, 2, \ldots, s + t).$$

Therefore, we get

$$1 > s/2 + t/4.$$

Thus, we have one of the following six cases:

Case	I	II	III	IV	V	VI
s	1	1	0	0	0	0
t	0	1	0	1	2	3

We will discuss each case separately.

(I) *In this case, the S_p-subgroup Q is different from G and is an elementary abelian normal subgroup of G. The factor group G/Q is a cyclic group whose order is relatively prime to p.*

Proof. The equation (6.9) becomes

$$1 = 1/g + (q - 1)/qk + (g_1 - 1)/g_1.$$

This implies that

(6.10) $1/qk + 1/g_1 = 1/g + 1/k.$

Thus, if $q = 1$, then we have $g = g_1$. This means that we have $G = A_1$. Therefore, G is cyclic.

Suppose that $q > 1$. Then, (6.10) shows that $k > 1$. By (6.8) (v), k must be equal to g_i for some value of i. So, in this case, we get $k = g_1$. Then, (6.10) proves that $g = qk$ and $G = N_G(Q)$. $\square$

(II) *The order $|G|$ is relatively prime to p, and G is either the group of order $4n$ defined by*

$$G = \langle x, y \,|\, x^n = y^2, \, y^{-1}xy = x^{-1} \rangle$$

where n is odd, or $G \cong SL(2, 3)$.

Proof. The equation (6.9) is reduced to

(6.11) $1/g_1 + 1/2g_2 = 1/2 + 1/g + (q - 1)/qk.$

If we had $q > 1$, then the last term would be at least $1/2k$. So,

$$1/2g_2 - 1/2k > 1/2 - 1/g_1 \geq 0.$$

This would imply that $k > g_2$. As before, by (6.8) (v), we would get $k = g_1$. Then, we would get a contradiction

$$1/2g_2 + 1/2g_1 > 1/2.$$

So, we have $q = 1$, and the order $|G|$ is relatively prime to p. Moreover, (6.11) gives us

$$1/g_1 + 1/2g_2 > 1/2.$$

This implies that $g_1 = 2$ or $g_1 = 3$.

If $g_1 = 2$, then (6.11) shows that $2g_2 = g$. Thus, we have $G = N_G(A_2)$. Let x be a generator of A_2, and let y be a generator of A_1. Then, $A_1 = N_G(A_1)$ by assumption ($s = 1$). This implies that A_1 is an S_2-subgroup of G. Thus, $g_2 = n$ is odd. By (6.8) (iv) we get the relations:

$$x^n = y^2, \qquad y^{-1}xy = x^{-1}.$$

It is easy to verify that G is indeed defined by these relations (Example 1 of §9, Chapter 2).

Suppose that $g_1 = 3$. From the formula (6.11), we get $g_2 = 2$ and $g = 12$. Then, we have $p \neq 2$ and $e = 2$. So, the order of G is 24. Let x be a generator of A_2. Then, $|N_G(A_2): A_2| = 2$ and $N_G(A_2)$ contains an element y such that

$$x^2 = y^2 \quad \text{and} \quad y^{-1}xy = x^{-1}.$$

Thus, $H = N_G(A_2)$ is the quaternion group of order 8. The element y is contained in a maximal abelian subgroup of G. In this case, $\langle y \rangle$ is conjugate to A_2. Similarly, all three cyclic subgroups of order 4 of H are conjugate, but as $|G: H| = 3$, no other conjugate subgroups of A_2 exist. Thus, we get $H \triangleleft G$, and the structure of G is uniquely determined as an extension of the quaternion group. Therefore, we have $G \cong SL(2, 3)$ (cf. Chapter 1, §9, and Chapter 2, §7). □

(III) *We have $G = Q \times Z$.*

Proof. The only elements whose orders are prime to p are elements of Z. Hence, we have $G = Q \times Z$. □

(IV) *Either $p = 2$ and G is a dihedral group of order $2n$ where n is odd, or $p = 3$ and $G \cong SL(2, 3)$.*

Proof. The equation (6.9) becomes

$$1/2 + 1/2g_1 = 1/g + (q - 1)/qk.$$

Since $g \geq 2g_1$, we get $(q - 1)/qk \geq 1/2$. Then, we have $q > 1$ and $k = 1$. The above equation is now reduced to

$$1/q + 1/2g_1 = 1/2 + 1/g.$$

We can prove (IV) by the same method as in case (II). □

(V) *We have one of the following two cases :*

(1) $g_1 = (q - 1)/d, g_2 = (q + 1)/d, g = q(q^2 - 1)/d \ (d = 1 \text{ or } 2)$.
(2) $g_1 = 2, g_2 = 5, g = 60 \ (q = 3)$.

Proof. It follows from the equation (6.9) that

(6.12) $\qquad\qquad 1/2g_1 + 1/2g_2 = 1/g + (q - 1)/qk.$

Clearly, $q > 1$ and $(q - 1)/q \geq 1/2$. Since the left side is at most $1/2$, we get $k > 1$. So, by (6.8) (v), k is equal to g_1 or g_2. Without loss of generality, we may assume that we have

$$k = g_1.$$

The equation (6.12) gives us

$$1/2g_2 = 1/g + 1/2g_1 - 1/qg_1 < 1/2g_1,$$

(6.13) $$g_1 < g_2.$$

We will prove that the following congruence holds:

(6.14) $$q \equiv 1 \pmod{g_1}.$$

The centralizer of any element x of $Q - \{1\}$ is $Q \times Z$ by (6.8) (iii). So, each element of $Q - \{1\}$ has exactly k conjugate elements in $N_G(Q)$. Thus, $q - 1$ is divisible by $k = g_1$.

We will clear the fraction in (6.12) and set

$$ag = 2g_1 g_2 q.$$

We get $g_2 q + g_1 q = a + 2(q - 1)g_2$. So, we have

(6.15) $$g_1 q = a + (q - 2)g_2.$$

By definition and by (6.15), a is a positive integer. We have

(6.16) $$g_2 \equiv a \pmod{g_1} \quad \text{and} \quad g_1 > (q - 2)g_2/q.$$

We remark that a is small. Since $g/2g_2$ is the index of a subgroup, we see that the integer a divides $g_1 q$. But, q is the order of an S_p-subgroup of G, and g_i is prime to p. Hence, the greatest common divisor of a and q is either 1 or 2. So, a divides $2g_1$.

Assume that $q \geq 4$. In this case, the inequality (6.16) shows that $2g_1 > g_2$. The congruence (6.16) tells us that $g_2 = a + lg_1$. Since a divides $2g_1$ and $2g_1 > g_2 > g_1$, we get

$$g_2 = g_1 + a.$$

Then, by (6.15), $g_1 q = a + (q - 2)(g_1 + a)$ holds. So, we get

$$2g_1 = a(q - 1).$$

This implies that $2g_2 = a(q + 1)$, $2g = a(q^2 - 1)q$, and

$$2/a = (q - 1)/g_1.$$

The last equation and (6.14) tell us that $d = 2/a$ is an integer, so we get the first case.

On the other hand, if $q \leq 3$, then we have $q = 3$ and $g_1 = 2$ by (6.14). Hence, the inequality (6.13) and (6.16) prove that $2 < g_2 < 6$. Since g_2 is relatively prime to $q = 3$, we get either $g_2 = 4$ (the first case with $d = 1$) or $g_2 = 5$ (the second case).   □

The structure of the group G will be determined later.

(VI) *The group G is isomorphic to one of the following three groups :*

$$G = \langle x, y \,|\, x^n = y^2, \, y^{-1}xy = x^{-1} \rangle \, (n \text{ is even}),$$
$$G = \hat{\Sigma}_4 \quad \text{or} \quad G = SL(2, 5),$$

where $\hat{\Sigma}_4$ is one of the representation groups of Σ_4 in which the transpositions correspond to the elements of order 4.

Proof. In this case, the equation (6.9) gives us

$$1/2g_1 + 1/2g_2 + 1/2g_3 = 1/2 + 1/g + (q - 1)/qk.$$

If $q > 1$, then the last term would be at least $1/2k$. Since k is one of the g_i by (6.8) (v), we get a contradiction. So, we have $q = 1$. Hence, we get

$$1/2g_1 + 1/2g_2 + 1/2g_3 = 1/2 + 1/g > 1/2.$$

We choose the notation so that $1 < g_1 \leq g_2 \leq g_3$. Then, we have one of the following two cases:

$$g_1 = 2, \quad g_2 = 2, \quad g = 2g_3;$$
$$g_1 = 2, \quad g_2 = 3, \quad g_3 \leq 5.$$

The first case occurs when $G = N_G(A_3)$, and we get the presentation

$$G = \langle x, y \,|\, x^n = y^2, \, y^{-1}xy = x^{-1} \rangle \qquad (n \text{ is even}).$$

In the second case, we have $3 \leq g_3 \leq 5$. But, if $g_3 = 3$, we would get $g = 12$. By Sylow's theorem, the group A_3 must be conjugate to A_2. This is a contradiction. Hence, we have two cases: $g_3 = 4, g = 24$ and $g_3 = 5, g = 60$.

Consider the case when $g = 24$. Since the index of $N_G(A_2)$ is 4, there is a homomorphism from G into Σ_4. It is easy to see that the kernel of this homomorphism is Z, so $G/Z \cong \Sigma_4$. Thus, G is a central extension of Σ_4. Since G contains a unique element of order 2, the structure of G is uniquely determined by (2.21). Thus, we have $G \cong \hat{\Sigma}_4$.

Finally, consider the case in which $g = 60$. Here, the characteristic of F does not divide $|G|$. Hence, the center Z is of order 2. Furthermore, an S_2-subgroup of G is the quaternion group (cf. the proof of (II)). Each S_2-subgroup S contains exactly 3 conjugate subgroups of A_1, and S is the normalizer of each of these three conjugate subgroups. Since G has exactly 15 conjugate subgroups of A_1, there are exactly 5 S_2-subgroups. By Sylow's theorem, there is a homomorphism from G into Σ_5. But, any element of $G - Z$ moves some S_2-subgroup. Therefore, G/Z is isomorphic to a subgroup of Σ_5. Since the order of G/Z is 60, the image of G/Z is a normal subgroup of Σ_5. We have $G/Z \cong A_5$. The structure of G is unique as the representation group A_5, so we have $G \cong SL(2, 5)$. $\square$

The preceding discussion gives us the main theorem about finite subgroups of $SL(V)$.

Theorem 6.17. *Let V be the two-dimensional vector space over an algebraically closed field F of characteristic p ($p \geq 0$). Let $L = SL(V)$. Any finite subgroup G of L is isomorphic to one of the groups in the following list.*
  Case I when $p = 0$ or $|G|$ is relatively prime to p:

  (i)   *A cyclic group.*
  (ii) *The group defined by the presentation*

$$\langle x, y \,|\, x^n = y^2, \; y^{-1}xy = x^{-1} \rangle.$$

  (iii) *The special linear group $SL(2, 3)$ over the field of 3 elements.*
  (iv) *$\hat{\Sigma}_4$, the representation group of Σ_4 in which the transpositions correspond to the elements of order 4.*
  (v)   *The special linear group $SL(2, 5)$.*

  Case II when $|G|$ is divisible by p. We will denote by Q an S_p-subgroup of G and set $|Q| = q$.

  (vi)   *$Q \triangleleft G$, Q is elementary abelian, and G/Q is a cyclic group whose order is relatively prime to p.*
  (vii) *$p = 2$ and G is a dihedral group of order $2n$ with odd n.*
  (viii) *$p = q = 3$ and $G \cong SL(2, 5)$.*
  (ix)   *The special linear group $SL(2, K)$.*
  (x)   *The group $\langle SL(2, K), d_\pi \rangle$ where K is a field of q elements and π is an element such that $K(\pi)$ is the field of q^2 elements and π^2 is a generator of the multiplicative group K^* of nonzero elements.*

Proof. First, we will consider the case when $G \not\supseteq Z(L)$. We have $p \neq 2$, and $|G|$ is odd, so $GZ = G \times Z$. Scanning through the groups in (I) to (VI), we see that G is either cyclic or $Q \triangleleft G$. Thus, we have either the case (i) or the case (vi).
    Next, assume that G contains the center Z. Then, we have one of the six

cases we have already investigated. In all the cases except for the case (V), we have determined the structure of the group G. In the second subcase of (V), we can show that $G \cong SL(2, 5)$ by the same method as the last subcase in (VI). Thus, except for the first case of (V), the structure of G is one of the listed in (i)–(ix).

In order to determine the structure of G in the first case of (V), we will prove a lemma. In the subsequent discussion, p is a fixed prime number. For any subfield K of F, there is a standard embedding of $GL(2, K)$ into $GL(2, F)$. We will identify the image with $GL(2, K)$ and call this subgroup the *standard* $GL(2, K)$. The standard $SL(2, K)$ is similarly defined. If we say that $SL(2, K)$ is contained in $SL(2, F)$, the notation $SL(2, K)$ stands for the standard subgroup in $SL(2, F)$. If q is any power of the prime p, F contains a unique subfield of q elements. So, there is a unique standard $SL(2, q)$. By our convention, we have

$$SL(2, q) \subset SL(2, q^m).$$

We will consider a fixed subfield E of F. Set $L_0 = SL(2, E)$ if E is either algebraically closed or if E is a finite field of characteristic two. Otherwise, set $L_0 = GL(2, E)$.

With these notations, the following lemma holds.

(6.18) *Suppose that G is a subgroup of L such that $|G| = iq(q^2 - 1)$ $(i = 1, 2)$ where q is a power of p. Furthermore, suppose that $G \subset SL(2, E)$. Then, the field $K = GF(q)$ is contained in E. If $i = 1$, then G is conjugate to the standard $SL(2, K)$ by some element of L_0. If $i = 2$, then we have $p > 2$ and E contains $GF(q^2)$. In this case, G is conjugate in L_0 to the subgroup*

$$\langle SL(2, K), d_\pi \rangle$$

where $\pi \in GF(q^2) - K$ and $\pi^2 \in K$.

Proof. Since $G \subset SL(2, E)$ and $|G|$ is even, the center Z is contained in G. Therefore, we have the first case of (V) (or the second case of (IV) where $q = 3$, $i = 1$ and $G = SL(2, 3)$). We will continue to use the notation of the case (V). Thus, let Q be an S_p-subgroup of G. Then, we may write

$$N_G(Q) = QM_1 \qquad (|M_1| = i(q - 1)).$$

By (6.7) (iv), every element of Q has a common fixed point on the projective line $\mathscr{L} = \mathscr{L}_E$. Let P_1 denote the common fixed point. Then, every element of M_1 fixes the point P_1. If $q > 3$ or if $i = 2$, a generator of M_1 has exactly two fixed points on $\mathscr{L}$ by (6.7) (iii), so there is exactly one more fixed point P_2.

Let u be a nonidentity element of Q. Then, $P_3 = P_2^u$ is different from P_1 and P_2. By (6.6) the group L_0 is triply transitive on $\mathscr{L}$, so there is an element v of L_0 such that

$$P_1, P_2, P_3 \rightarrow (1, 0), (0, 1), (1, 1).$$

If we consider G^v instead of G, we may assume without loss of generality that

$$Q \subset T, \qquad M_1 \subset D, \qquad u = t_1$$

where T, D, and t_1 are as defined in Convention 6.1. We will show that G contains the standard $SL(2, K)$.

We will assume that either $q > 3$ or $i = 2$. By (6.8) (iv), since $|N_G(M_1): M_1| = 2$, $N_G(M_1)$ contains an element x of order $2e$ which inverts every element of M_1. By the above assumption, we have $|M_1| > 2$, and M_1 has exactly two fixed points. Hence, the element x must interchange these two fixed points. So, we have

$$x = d_\beta w$$

for some $\beta \in E$. By assumption, M_1 is contained in D. So, an element of M_1 has the form d_ω where $\omega \in E$. Since $|M_1| = i(q - 1)$, we get $\omega^{i(q-1)} = 1$. Thus, we have

$$\omega \in K^* \ (i = 1) \quad \text{or} \quad \omega^2 \in K^* \ (i = 2).$$

If $i = 2$, the element ω belongs to $GF(q^2)$. In either case, ω or ω^2 ranges over all the elements of K^* (since $|M_1| = i(q - 1)$). Hence, E contains K if $i = 1$, while if $i = 2$, E contains $GF(q^2)$. This proves the first part. Also, it is clear that $p > 2$ if $i = 2$.

Set $\Lambda = \{\lambda \,|\, t_\lambda \in Q\}$. We will prove that $\Lambda = K$. First, notice that we have $G = N_G(Q) \cup N_G(Q)xQ$. In fact, by (3.8) of Chapter 1, $N_G(Q)xQ$ contains exactly $|Q: Q \cap N_G(Q)^x|$ cosets of $N_G(Q)$. But, $Q \cap N_G(Q)^x$ consists only of elements with at least two different fixed points on $\mathscr{L}$, so we have $Q \cap N_G(Q)^x = \{1\}$. Hence, $N_G(Q)xQ$ contains exactly q cosets. Counting the number of elements, we get

$$G = N_G(Q) \cup N_G(Q)xQ.$$

In particular, for $t_\lambda \in Q - \{1\}$, there are elements u and v such that

$$x^{-1}t_\lambda x = uxv \qquad (u \in N_G(Q), v \in Q)$$

(cf. Example 4, (e) of Chapter 2, §2). As $x = d_\beta w$, we can find elements $d_\omega \in M_1$ and $\mu, v \in \Lambda$ such that

$$w^{-1}d_\beta^{-1}t_\lambda d_\beta w = d_\omega t_\mu d_\beta wt_v.$$

This implies that $\omega\beta = -\beta^2\lambda$. Since $t_1 = u \in Q$, we have $-1 \in \Lambda$. Setting $\lambda = -1$, we get $\omega = \beta$. Thus, d_β is an element of M_1. This shows that $w = d_\beta^{-1}x \in N_G(M_1)$. We can choose w instead of x, so we may assume that $\beta = 1$. This reduces the above equation to $\lambda = -\omega$. Thus, if $i = 1, \lambda = -\omega$ ranges over K^*, and we have $\Lambda = K$. We have

$$G = \langle t_\lambda, d_\omega, w \rangle \qquad (\lambda \in K, \omega \in K^*),$$

and this is the standard $SL(2, K)$.

Assume that $i = 2$. As we have seen before, if d_ω ranges over M_1, then ω^2 ranges over K^*. By (6.2), we have

$$d_\omega^{-1}t_\lambda d_\omega = t_\mu \qquad (\mu = \omega^2\lambda).$$

So, Λ is a K-module. Since $1 \in \Lambda$, we must have $\Lambda = K$. In this case, we have

$$G = \langle t_\lambda, d_\omega, w \rangle \qquad (\lambda \in K, \omega^2 \in K^*).$$

Thus, this is the image of $\langle SL(2, K), d_\pi \rangle$ under the standard embedding, where $\pi \in GF(q^2) - K$ and π^2 is a generator of K^*. $\square$

This proves (6.18) except when $q = 3$, $i = 1$, and $G = SL(2, 3)$. Theorem 6.17 has already been proved by the portion of (6.18) we have proved so far. Before treating the exceptional case of (6.18), we will prove the following lemma.

(6.19) *Let x be an element of $SL(2, F)$ not contained in the center Z.*

(i)   *We have $x^2 \in Z$ if and only if* tr $x = 0$.
(ii) *We have $x^3 \in Z$ if and only if* tr $x = \pm 1$.

Proof. By (6.3) (i), x is conjugate to $\pm t_\lambda$ or to d_ω. If x is conjugate to $\pm t_\lambda$, then we have $x^p \in Z$ and tr $x = \pm 2$. Thus, $x^2 \in Z$ if and only if $p = 2$. So, (i) is proved in this case. The proposition (ii) can be proved similarly.

Assume that x is conjugate to d_ω. Then, we have tr $x = \omega + \omega^{-1}$. Thus, the following are equivalent:

$$x \notin Z, \ x^2 \in Z \Leftrightarrow \omega^2 = -1 \Leftrightarrow \omega + \omega^{-1} = 0.$$

Similarly, if $x \notin Z$, the following hold:

$$x^3 \in Z \Leftrightarrow \omega^3 \neq \pm 1, \ \omega^3 = \pm 1 \Leftrightarrow \omega + \omega^{-1} = \pm 1. \quad \square$$

The only thing left to prove in (6.18) is the second case of (IV): $q = 3$, $i = 1$, and $G \cong SL(2, 3)$. In this case, any element of order 3 is conjugate to t_λ for some $\lambda \in E$. So, by conjugation in L_0, we can transform G into a conjugate subgroup which contains t_1. We may, therefore, assume that $t_1 \in G$. We will show that G is conjugate to the standard embedding by some element of $T \cap L_0$.

Since $G \cong SL(2, 3)$, there is an element x of G such that $x \notin Z$, $x^2 \in Z$, and $(t_1 x)^3 \in Z$. We write

$$x = \begin{pmatrix} \alpha & \beta \\ \gamma & \delta \end{pmatrix} \qquad (\alpha, \beta, \gamma, \delta \in E).$$

By (6.19), we have $\alpha + \delta = 0$ and $\alpha + \beta + \delta = \pm 1$. So,

$$x = t_{\alpha\beta}^{-1} w^e t_{\alpha\beta} \qquad (e = \pm 1),$$
$$G = t_{\alpha\beta}^{-1} \langle t_1, w \rangle t_{\alpha\beta}.$$

Clearly, $\langle t_1, w \rangle$ is the standard $SL(2, 3)$ and $t_{\alpha\beta} \in T \cap L_0$. $\square$

The above argument also proves a conjugacy assertion that is a little stronger than the corresponding statement in (6.18). In general, we have the following proposition.

(6.20) *Let E be a subfield of F, and let G be a subgroup of $SL(2, E)$. Assume that an S_p-subgroup Q of G has order q and that $G \cong SL(2, q)$. If $t_1 \in Q$, then G is conjugate to the standard $SL(2, q)$ by an element of $T \cap SL(2, E)$.*

Proof. The last part of the proof of (6.18) shows that this proposition is valid for $q = 3$. So, assume that $q \geq 4$. Set $T_1 = SL(2, E) \cap T$. By (6.4) (i), the centralizer of t_1 in $SL(2, E)$ is $T_1 \times Z$. This implies that Q is a subgroup of T_1. By (6.7) (iv), every nonidentity element of Q has a unique fixed point on $\mathscr{L}_E$. Let P_1 be the fixed point of Q. Since $G \cong SL(2, q)$, we have $N_G(Q) = QM_1$ with $|M_1| = q - 1$. We have assumed that $q > 3$, so a generator of M_1 has exactly two fixed points. Let P be the other fixed point of M. (P_1 is also a fixed point of M.) The subgroup D has two fixed points as well. One of them is P_1. Let P_2 be the other point fixed by D. Then, by (3.6), there is an element x of $SL(2, E)$ such that $P_1^x = P_1$ and $P^x = P_2$. The element x is contained in the stabilizer of P_1. So, by (6.7) (iv), x is an element of $H = TD$. We may write $x = ty^{-1}$ with $t \in T$ and $y \in D$. Then, we have

$$P_1^t = P_1, \qquad P^t = P^{xy} = P_2^y = P_2.$$

Thus, we get $t \in T$, $Q^t \subset T$, $M_1^t \subset D$, and $t_1 \in Q^t = Q$. In the proof of (6.18) we have shown that G^t is the standard $SL(2, q)$.

It remains to show that t belongs to $SL(2, E)$. We have $t = t_\lambda$ for some λ. The point P is in $\mathscr{L}_E$, so P is represented by a vector (α, β) where both α and

β are elements of E. Since $P^t = P_2$, we get

$$(\alpha, \beta)\begin{pmatrix} 1 & 0 \\ \lambda & 1 \end{pmatrix} = (0, \gamma)$$

for some $\gamma \in F$. This implies that $\beta = \gamma \neq 0$ and $\alpha + \beta\lambda = 0$. Thus, $\lambda \in E$ and we have $t \in T_1$. $\square$

We can derive several theorems from Theorem 6.17.

Theorem 6.21 (Dickson [1]). *Let p be an odd prime number, and let λ be an element of F which is algebraic over the prime field $F_0 = GF(p)$. Set $E = F_0(\lambda)$. Let G be defined as follows:*

$$G = \left\langle \begin{pmatrix} 1 & 0 \\ 1 & 1 \end{pmatrix}, \begin{pmatrix} 1 & \lambda \\ 0 & 1 \end{pmatrix} \right\rangle$$

Then, either G is isomorphic to $SL(2, E)$, or we have $p = 3$, $\lambda^2 = -1$, $E = GF(9)$, and $G \cong SL(2, 5)$.

Proof. The group G is a subgroup of $SL(2, E)$ generated by two non-commutative elements of order p. So, by Theorem 6.17, either $G \cong SL(2, p^n)$, or $p = 3$ and $G \cong SL(2, 5)$. If $G \cong SL(2, p^n)$, then, by (6.20), there is an element α of E such that $t_\alpha^{-1} G t_\alpha$ is the standard $SL(2, p^n)$. In particular, the entries of the matrix

$$\begin{pmatrix} 1 & 0 \\ -\alpha & 1 \end{pmatrix}\begin{pmatrix} 1 & \lambda \\ 0 & 1 \end{pmatrix}\begin{pmatrix} 1 & 0 \\ \alpha & 1 \end{pmatrix} = \begin{pmatrix} 1 + \alpha\lambda & \lambda \\ -\alpha^2\lambda & 1 - \alpha\lambda \end{pmatrix}$$

belong to $GF(p^n)$. Hence, λ is an element of $GF(p^n)$. Thus, we get $E = GF(p^n)$ and $G \cong SL(2, E)$.

In the exceptional case when $p = 3$, we have $G/Z \cong A_5$. The two given generators correspond to 3-cycles of A_5. Since these 3-cycles generate A_5, they have only one letter in common. This implies that the product

$$\begin{pmatrix} 1 & 0 \\ 1 & 1 \end{pmatrix}\begin{pmatrix} 1 & \lambda \\ 0 & 1 \end{pmatrix} = \begin{pmatrix} 1 & \lambda \\ 1 & 1 + \lambda \end{pmatrix}$$

is of order 5 or 10. So, for a primitive fifth root ζ of unity over F_0, we get (by taking the trace of the matrix)

$$2 + \lambda = \lambda - 1 = \pm(\zeta + \zeta^{-1}).$$

The above equation shows that $\lambda \in GF(9)$. Furthermore, the two given generators are conjugate in G. This implies that $-\lambda$ is a square in $GF(9)$. Thus, we get $\lambda^2 = -1$.

If $p = 3$ and $\lambda^2 = -1$, then the two matrices given in Theorem 6.21 generate the group isomorphic to $SL(2, 5)$. This can be proved by noting that

$$x = \begin{pmatrix} -1 & -\lambda \\ -1 & -(1+\lambda) \end{pmatrix}, \quad y = \begin{pmatrix} 1 & 1-\lambda \\ 0 & 1 \end{pmatrix}$$

satisfy the relations of Example 4 of Chapter 2, §6 (cf. the remark after (6.24)). □

(6.22) *We have $SU(2, q) \cong SL(2, q)$.*

Proof. By definition, $SU(2, q)$ is a subgroup of $SL(2, q^2)$. So, $SU(2, q)$ is isomorphic to one of the groups in (6.17). We will compute the order of $SU(2, q)$ (cf. Exercise 2, §5).

We will choose the canonical Hermitian form corresponding to the identity matrix. Then, the condition for a matrix

$$x = \begin{pmatrix} \alpha & \beta \\ \gamma & \delta \end{pmatrix} \quad (\alpha, \beta, \gamma, \delta \in GF(q^2))$$

to be in $SU(2, q)$ is ${}^t\bar{x} = x^{-1}$ and $\det x = 1$. Thus, any element of $SU(2, q)$ has the form

$$x = \begin{pmatrix} \alpha & \beta \\ -\beta^q & \alpha^q \end{pmatrix}, \quad \alpha^{1+q} + \beta^{1+q} = 1.$$

So, the order of $SU(2, q)$ is equal to the number of solutions (α, β) of $\alpha^{1+q} + \beta^{1+q} = 1$. Clearly, if $\alpha^{1+q} = 1$, then β must be zero. If $\alpha^{1+q} \neq 1$, then for each value of α, there are exactly $1 + q$ solutions for β. Thus, we get

$$|SU(2, q)| = (1 + q) + (q^2 - q - 1)(1 + q) = q(q^2 - 1).$$

By Theorem 6.17, we get $SU(2, q) \cong SL(2, q)$. □

Finally, we will consider the structures of the subgroups of $SL(2, q)$.

(6.23) *The group $SL(2, q)$ contains a cyclic group A_1 of order $q - 1$ and a cyclic group A_2 of order $q + 1$. If the order of an element x is relatively prime to q, then x is conjugate to an element of one of the subgroups A_1 or A_2. The normalizer of A_i in $SL(2, q)$ is a group of the type (ii) of Theorem 6.17, except when $q = 3$ and $i = 1$.*

Proof. This follows immediately from (V) and Theorem 6.17. □

An element of order $q - 1$ in $SL(2, q)$ is conjugate to d_ω where ω is a generator of $GF(q)^*$. But, the elements of order $q + 1$ are 'hidden'. Since $GF(q^2)^*$ contains an element ζ of order $q + 1$, the element

$$\begin{pmatrix} \zeta & 0 \\ 0 & \zeta^q \end{pmatrix}$$

is an element of $SU(2, q)$ whose order is $q + 1$. This element and w generate a subgroup of order $2(q + 1)$ in $SU(2, q)$ which is of the type (ii) of Theorem 6.17. Thus, we can diagonalize an element of order $q + 1$ by using (6.22).

(6.24) *A necessary and sufficient condition on q for*

$$SL(2, 5) \subset SL(2, q)$$

is that q is odd and 5 divides $q(q^2 - 1)$.

Proof. If $SL(2, 5) \subset SL(2, q)$, then 5 divides the order $q(q^2 - 1)$ of $SL(2, q)$. Also, since an S_2-subgroup of $SL(2, 5)$ is a quaternion group, q must be odd.

Conversely, assume that q is odd and 5 divides $q(q^2 - 1)$. If 5 divides q, then q is a power of 5 and $SL(2, 5) \subset SL(2, 5^n)$.

Suppose that $q \equiv 1 \pmod 5$. In this case, $GF(q)^*$ has an element ζ of order 5. Set

$$x = \begin{pmatrix} \zeta & 0 \\ 0 & \zeta^{-1} \end{pmatrix}, \qquad y = \begin{pmatrix} \alpha & \beta \\ \gamma & \delta \end{pmatrix} \in SL(2, q).$$

We will determine the values $\alpha, \beta, \gamma,$ and δ so that x and y satisfy the relations of Example 4 of Chapter 2, §6. By (6.19), the relations

$$\alpha + \delta = -1, \qquad \alpha\zeta + \delta\zeta^{-1} = 0$$

give us $y^3 = (xy)^4 = 1$. The above system of linear equations has a unique solution (α, δ), and we can find β and γ such that $\alpha\delta - \beta\gamma = 1$. Thus, the group $\langle x, y \rangle$ is a homomorphic image of $SL(2, 5)$. But, by Theorem 6.17, no proper homomorphic image $\neq 1$ of $SL(2, 5)$ can be a subgroup of $SL(2, q)$. Hence, we have

$$\langle x, y \rangle \cong SL(2, 5).$$

Finally, suppose that $q \equiv -1 \pmod 5$. In this case, $GF(q^2)^*$ contains an element $\zeta \neq 1$ of order 5. We will find elements

$$x = \begin{pmatrix} \zeta & 0 \\ 0 & \zeta^q \end{pmatrix}, \qquad y = \begin{pmatrix} \alpha & \beta \\ -\beta^q & \alpha^q \end{pmatrix} \qquad (\alpha, \beta \in GF(q^2))$$

which satisfy the relations $y^3 = (xy)^4 = 1$. By (6.19) it suffices to choose

$$\alpha + \alpha^q = -1, \qquad \alpha\zeta + \alpha^q\zeta^q = 0.$$

Since $\zeta^2 \neq 1$, we get $\alpha = 1/(\zeta^2 - 1)$. With this choice of α, we can find β so as to make $y \in SU(2, q)$. We get $\langle x, y \rangle = SL(2, 5)$ as before. $\square$

As a special case of (6.24), we note that the group $SL(2, 9)$ contains $SL(2, 5)$. Thus, (6.24) gives us another way to verify that the exceptional case in Dickson's theorem really occurs. Taking the factor group by the center, we get

$$A_5 \subset PSL(2, 9).$$

So, $PSL(2, 9)$ contains a subgroup of index 6. By (7.16) of Chapter 1, there is a homomorphism from $PSL(2, 9)$ into Σ_6. Since $PSL(2, 9)$ is simple by the Corollary of (9.10), Chapter 1, we get $PSL(2, 9) \subset \Sigma_6$. By comparing the orders, we have

$$PSL(2, 9) \cong A_6.$$

Subgroups of $PSL(2, q)$

In the remainder of this section, we will study the subgroups of $PSL(2, q)$ where q is a power of p. Since $PSL(2, q)$ is isomorphic to the factor group of $SL(2, q)$ by its center Z, the Correspondence Theorem asserts that every subgroup of $PSL(2, q)$ has the form G/Z, where G is a subgroup of $SL(2, q)$ which contains Z. Thus, by Theorem 6.17, we can find all the subgroups of $PSL(2, q)$.

Theorem 6.25. *Let q be a power of the prime p. Then, a subgroup of $PSL(2, q)$ is isomorphic to one of the following groups.*

(a) *The dihedral groups of order $2(q \pm 1)/d$ and their subgroups $(d = (2, q - 1))$.*

(b) *A group H of order $q(q - 1)/d$ and its subgroups. An S_p-subgroup Q of H is elementary abelian, $Q \lhd H$, and the factor group H/Q is a cyclic group of order $(q - 1)/d$.*

(c) *A_4, Σ_4, or A_5*

(d) *$PSL(2, r)$ or $PGL(2, r)$, where r is a power of p such that $r^m = q$.*

Proof. By (6.23), if the order of an element of $SL(2, q)$ is relatively prime to p, then the order is a divisor of $q + 1$ or of $q - 1$. Thus, the subgroups of types (i), (ii), and (vii) in Theorem 6.17 give us the subgroups of $PSL(2, q)$ referred to in (a).

Clearly, an S_p-subgroup Q of $PSL(2, q)$ is isomorphic to the corresponding S_p-subgroup of $SL(2, q)$. Hence, Q is elementary abelian. The normalizer H of Q is the factor group of the corresponding normalizer in $SL(2, q)$ by the center Z. Thus, the order of H is $q(q - 1)/d$. Let Q_1 be any nonidentity subgroup of Q. Then, by (6.4), the normalizer of Q_1 is contained in H. Thus, the subgroups of type (vi) in Theorem 6.17 correspond to the subgroups of type (b) of Theorem 6.25.

It is clear that the subgroups of types (iii), (iv), (v), and (viii) give rise to the groups of type (c).

For a subgroup of type (ix) or (x) in Theorem 6.17, we set $|K| = r$. (Note the change of symbols.) Then, by (6.18), K is contained in $GF(q)$. So, we have $q = r^m$. It is clear that $SL(2, r)$ corresponds to the subgroup $PSL(2, r)$, so we need only to show that

$$G/Z \cong PGL(2, r)$$

where $G = \langle SL(2, r), d_\pi \rangle$. Set $E = GF(r^2)$. Then, the element π generates E over K: $E = K(\pi)$. By definition, we get

$$PGL(2, r) = GL(2, r)/Z_K$$

where Z_K is the center of $GL(2, r)$. Consider $GL(2, E)$ and its center Z_E. Clearly, we have $Z_K = GL(2, K) \cap Z_E$. So,

$$PGL(2, r) \cong GL(2, K)Z_E/Z_E.$$

The definition of G gives us $G \cap Z_E = Z$. Hence, we have

$$G/Z \cong GZ_E/Z_E.$$

The decomposition

$$d_\pi = \begin{pmatrix} \pi & 0 \\ 0 & \pi^{-1} \end{pmatrix} = \begin{pmatrix} \pi^2 & 0 \\ 0 & 1 \end{pmatrix} \begin{pmatrix} \pi^{-1} & 0 \\ 0 & \pi^{-1} \end{pmatrix}$$

and the fact that $\pi^2 \in K$ show that $GZ_E \subset GL(2, K)Z_E$. So,

$$GZ_E/Z_E \subset GL(2, K)Z_E/Z_E.$$

Thus, the group G/Z is isomorphic to a subgroup of $PGL(2, r)$. By comparing the orders, we get

$$G/Z \cong PGL(2, r). \quad \square$$

In the last part of the preceding proof, we used the formula for the order of $PGL(2, r)$. In general, an element x of $GL(2, K)$ is contained in $SL(2, K)Z_K$ if and only if det x is a square in K^*. Thus, we get

$$PGL(2, K)/PSL(2, K) \cong K^*/(K^*)^2$$

where $(K^*)^2$ denotes the subgroup of K^* consisting of all the squares of the elements of K^*. If K is a finite field, then

$$|K^*/(K^*)^2| = 1 \quad \text{or} \quad 2$$

according to whether the characteristic of K is two or odd. In particular, note that there is no difference between $PGL(2, 2^n)$ and $PSL(2, 2^n)$.

We remark that there are some subgroups which are always contained in $PSL(2, q)$, while others are not. For example, $PSL(2, q)$ does not contain A_4 for some values of q. We will prove the following theorem which gives a complete survey of the subgroups of $PSL(2, q)$.

Theorem 6.26. Set $L = PSL(2, q)$ and $q = p^n$.

(i)   The group L always contains subgroups of types (a) and (b) of Theorem 6.25.

(ii) The group L does not contain any subgroup isomorphic to A_4 if and
only if $p = 2$ and n is odd.

(iii) The group L contains a subgroup isomorphic to Σ_4 if and only if
$q^2 \equiv 1 \pmod{16}$.

(iv) The necessary and sufficient condition for L to contain a subgroup
isomorphic to A_5 is

$$q(q^2 - 1) \equiv 0 \pmod 5.$$

(v)   If $q = r^m$, then L contains a subgroup isomorphic to $PSL(2, r)$. Assume
that q is odd. Then, L contains a subgroup isomorphic to $PGL(2, r)$ if
and only if m is even and $q = r^m$.

Proof. (i) This follows immediately from (6.23) and the proof of Theorem 6.25.

(ii) Suppose that $p = 2$. In this case, a subgroup isomorphic to A_4 in $PSL(2, q)$ comes from a subgroup of type (vi) in Theorem 6.17. Thus, the proof of Theorem 6.25 shows that the normalizer of an S_2-subgroup of L must contain a subgroup of this type. This implies that $q = 2^n \equiv 1 \pmod 3$. Hence, n must be even. Conversely, if n is even, then $PSL(2, 2^n)$ contains $PSL(2, 4)$. But, $PSL(2, 4) \cong A_5$, and it contains a subgroup isomorphic to A_4.

If $p = 3$, then $SL(2, 3) \subset SL(2, q)$. Hence, the central factor group contains $SL(2, 3)/Z \cong A_4$.

If $p > 3$, we get $q^2 \equiv 1 \pmod 3$. As in the proof of (6.24), $SL(2, q)$ contains elements x and y such that

$$x^3 = y^3 = 1, \qquad xy \notin Z, \qquad (xy)^2 \in Z.$$

Then, we get $\langle x, y \rangle \cong SL(2, 3)$ by Example 3 of Chapter 2, §6. Thus, $PSL(2, q)$ does not contain any subgroup isomorphic to A_4 only if $p = 2$ and n is odd.

(iii) If L contains a subgroup isomorphic to Σ_4, then the S_2-subgroups are not abelian. Hence, we have $p \neq 2$ and $q^2 \equiv 1 \pmod{16}$. Conversely, assume that $q^2 \equiv 1 \pmod{16}$. Then, the order of an S_2-subgroup of L is at least 8. By (ii), L contains a subgroup A isomorphic to A_4. Let B be an S_2-subgroup of A. Then, B is an elementary abelian normal subgroup of A. On the other hand, B is contained in an S_2-subgroup of L as a proper subgroup. Hence, the normalizer of B in L is strictly larger than A. Scanning the table of subgroups in Theorem 6.25, we conclude that $N_L(B)$ is isomorphic to Σ_4.

(iv) If q is odd, then the assertion follows from (6.24). If q is even, then $q^2 = 2^{2n} \equiv 1 \pmod 5$ if and only if n is even. In this case, we get $SL(2, 4) \subset SL(2, 4^m)$. Since $SL(2, 4) \cong A_5$, the assertion is proved.

(v) If $q = r^m$, then $SL(2, q)$ contains the standard $SL(2, r)$ because $GF(r)$ is a subfield of $GF(q)$. Assume that $q = r^m$ is odd. In this case, we have

$$|PGL(2, r) : PSL(2, r)| = 2.$$

Hence, if L contains a subgroup isomorphic to $PGL(2, r)$, then the index

$$|L : PSL(2, r)|$$

is even. On the other hand, $q = r^m$ and

$$|L : PSL(2, r)| = r^{m-1}(r^{2(m-1)} + r^{2(m-2)} + \cdots + r^2 + 1).$$

The right side is even if and only if m is even.

Conversely, suppose that m is even. Then, L contains $PSL(2, r^2)$. So, L contains

$$\langle SL(2, r), d_\pi \rangle / Z \cong PGL(2, r). \quad \square$$

(6.27) *The smallest index of a proper subgroup G of $PSL(2, q)$ is $q + 1$ except for $q = 2, 3, 5, 7, 9, 11$. In these exceptional cases, the smallest index is q except when $q = 9$. If $q = 9$, then the smallest index is 6.*

Proof. The subgroup G is of one of the four types listed in Theorem 6.25. The largest order among the subgroups of type (a) is $2(q + 1)$. So, the smallest index is at least $q(q - 1)/2 > q + 1$ if $q > 3$.

The index of the normalizer of an S_p-subgroup is equal to $q + 1$.

For the subgroups of type (d), the smallest index is at least

$$r^{m-1}(r^{2(m-1)} + \cdots + r^2 + 1)/2 > q + 1.$$

If the index of one of the subgroups of type (c) is less than $q + 1$, then we get

$$q(q^2 - 1)/d < 60(q + 1)$$

where $d = (2, q - 1)$. This gives us an upper bound on $q: q \leq 11$. For each prime power $q \leq 11$, we apply Theorem 6.26 to get all the possible subgroups, and we check that $q = 4$ and 8 do not give exceptional subgroups and that the smallest index is as stated in (6.27). $\square$

It is amazing when we realize that in his famous last letter, Galois could state with certainty all of the special cases of Proposition (6.27) for a prime q.

Exercises

1. Let i be the number of conjugacy classes of the subgroups which are isomorphic to $SL(2, q)$ in the special linear group $SL(2, q^n)$. Show that if either q is even or n is odd, then we have $i = 1$. On the other hand, if q is odd and n is even, show that $i = 2$. Also prove that all subgroups isomorphic to $SL(2, q)$ are conjugate in $GL(2, q^n)$ for any q^n.

(*Hint.* Use (6.20) and Exercise 1 of Chapter 1, §9.)

2. Suppose that the group $PSL(2, q)$ contains a subgroup isomorphic to A_5. Show that the number of conjugacy classes of the subgroups isomorphic to A_5 is 1 if q is even or if $q = 5^{2m+1}$, and that otherwise, the number of conjugacy classes is 2.

(*Hint.* Use $A_5 \cong SL(2, 4) \cong PSL(2, 5)$ and Exercise 1 for the cases when q is a power of 2 or 5. In all other cases, show that an element of order 5 is conjugate to

$$x = \begin{pmatrix} \zeta & 0 \\ 0 & \zeta^{-1} \end{pmatrix} \quad \text{or} \quad \begin{pmatrix} \zeta & 0 \\ 0 & \zeta^q \end{pmatrix}$$

within $SL(2, q)$ or $SU(2, q)$, and then show that there are two conjugacy classes of elements y such that $y^3 = 1$ and $(xy)^2 \in Z$. All subgroups isomorphic to A_5 are conjugate in $PGL(2, q)$.)

3. Let j be the number of conjugacy classes of the subgroups isomorphic to $PSL(2, q)$ in the group $PSL(2, q^n)$. Show that $j = 2$ if q is odd and n is even, while $j = 1$ otherwise.

4. Suppose that $PSL(2, 2^n)$ contains a subgroup H isomorphic to A_4. Show that H is conjugate to a subgroup of the standard $PSL(2, 4)$. Furthermore, prove that any subgroup of $PSL(2, 2^n)$ which is isomorphic to A_4 is conjugate to H.

(*Hint.* Some conjugate subgroup H_1 of H contains t_1. Let $H_1 = \langle t_1, x \rangle$ with the relations $x^3 = (t_1 x)^3 = 1$. Show that some conjugate of x lies in the standard $PSL(2, 4)$.)

5. Let q be a power of an odd prime number, and set $L = SL(2, q)$ and $G = L/Z(L)$. Prove the following propositions.

(a) The subgroups of order 4 in L are all conjugate to each other, and the order of the normalizer of a subgroup of order 4 is equal to $2(q - \varepsilon)$ where ε is $+1$ or -1 and $q \equiv \varepsilon \pmod 4$.

(b) All elements of order 2 of G are conjugate.

(c) G contains exactly $|G|/12$ subgroups which are elementary abelian and of order 4.

(d) Let e be the number of conjugacy classes of the elementary abelian subgroups of order 4 in G. Then, $e = 2$ if $q \equiv \pm 1 \pmod 8$, and $e = 1$ if $q \equiv 3$ or $5 \pmod 8$.

(*Hint.* Use (6.23). An elementary abelian subgroup of order 4 is generated by two distinct commuting elements u and v of order 2. Count the number of such pairs (u, v) using the fact that the centralizer of u is a dihedral group of order $q - \varepsilon$. If V is an elementary abelian subgroup of order 4, then $N_G(V)$ is isomorphic to A_4 or Σ_4. Thus, (d) holds.)

6. Show that if $q \equiv 3$ or $5 \pmod 8$, all subgroups isomorphic to A_4 in $PSL(2, q)$ are conjugate. Also prove that if $q \equiv \pm 1 \pmod 8$, then $PSL(2, q)$ has two conjugacy classes of subgroups which are isomorphic to A_4, as well as two classes of conjugate subgroups which are isomorphic to Σ_4.

7. Show that a maximal subgroup of $PSL(2, q)$ is one of the following groups.

(i) A dihedral group of order $2(q - \varepsilon)/d$ where $d = (2, q - 1)$. (Exceptions occur when $\varepsilon = 1, q = 3, 5, 7, 9, 11$ and $\varepsilon = -1, q = 2, 7, 9$.)

(ii) A solvable group of order $q(q - 1)/d$.

(iii) A_4 when q is a prime number > 3 and

$$q \equiv 3, 13, 27, 37 \pmod{40}.$$

(iv) Σ_4 when q is an odd prime number and $q \equiv \pm 1 \pmod 8$.

(v) A_5 when q is of one of the following forms: $q = 5^m$ or 4^m where m is a prime, q is a prime number congruent to $\pm 1 \pmod 5$, or q is the square of an odd prime number which satisfies $q \equiv -1 \pmod 5$.

(vi) $PSL(2, r)$ when $q = r^m$ and m is an odd prime number.

(vii) $PGL(2, r)$ when $q = r^2$.

(*Hint.* Use Theorem 6.25, 6.26, and the above exercises. When $r = 3$ in (vi) and (vii), the linear groups are reduced to A_4 and Σ_4, respectively. So, Σ_4 is also maximal in $PSL(2, 9) = A_6$. Similarly, A_4 is maximal in $PSL(2, 3^m)$ if m is an odd prime number.)

8. A nonabelian simple group is said to be a *minimal simple group* if all proper subgroups are solvable. Show that $PSL(2, q)$ is a minimal simple group if and only if q is of one of the following forms: $q = 2^m$ where m is a prime; $q = 3^l$ where l is an odd prime number; q is a prime number greater than 5 which is congruent to $\pm 2 \pmod 5$; or $q = 5$.

(*Hint.* Use Exercise 7 and (9.10) of Chapter 1, §9. According to Thompson [4], all minimal simple groups other than $PSL(2, q)$ are either $PSL(3, 3)$ or $S_z(2^m)$ where m is an odd prime number.)

9. (a) Suppose that E is a subfield of F. Show that the standard embedding $GL(2, E) \to GL(2, F)$ induces an injection from $PGL(2, E)$ into $PGL(2, F)$.

(b) Prove that a finite subgroup of $PGL(2, E)$ is isomorphic to one of the groups listed in Theorem 6.25.

(c) Determine the type of the subgroups of $PGL(2, q)$ which are not contained in $PSL(2, q)$.

(*Hint.* Note that $PSL(2, F) = PGL(2, F)$.)

Bibliography

The main references on algebra are the following two books, referred to in the text as [IK] and [AS], respectively:

Iyanaga, S. and Kodaira, K.: *Gendai Sûgaku Gaisetsu* (Introduction to Modern Mathematics), **I**. Gendai Sûgaku vol. 1. Tokyo: Iwanami Shoten 1961.

Akizuki, Y. and Suzuki, M.: *Kôtô Daisûgaku* (Higher Algebra), **I** and **II**. Iwanami Zensho. Tokyo: Iwanami Shoten 1952 and 1957, Revised editions 1980.

For the representation theory of finite groups, the readers are referred to:

Curtis, C. and Reiner, I.: *Representation Theory of Finite Groups and Associative Algebras*. New York: Wiley–Interscience 1962.

For the theory of permutation groups, the book by Wielandt is the standard reference:

Wielandt, H.: *Finite Permutation Groups*. New York and London: Academic Press 1964.

There are many textbooks on group theory. We will mention just one:

Asano, K. and Nagao, H.: *Gunron* (Group Theory). Iwanami Zensho. Tokyo: Iwanami Shoten 1965. (Referred as [AN] in text.)

Recently, Kondo [1] appeared. The books, Ito [3], Nagao [3], and Tsuzuku [1], contain topics which were not discussed in this book, and they are highly recommended as supplements to this book.

The following is a partial list of papers on group theory and is intended as a guide for those who are interested in further investigating some of the topics beyond the level on which they are discussed in this book.

Alperin, J. L.: [1] Centralizers of abelian normal subgroups of *p*-groups. *J. Algebra* **1**, 110–113 (1964).

[2] Sylow intersections and fusion. *J. Algebra* **6**, 222–241 (1967).

Alperin, J. L., Brauer, R., and Gorenstein, D.: [1] Finite groups with quasi-dihedral and wreathed Sylow 2-subgroups. *Trans. Amer. Math. Soc.* **151**, 1–261 (1970).

[2] Finite simple groups of 2-rank two. *Scripta Math.* **29**, 191–214 (1973).

Artin, E.: [1] The orders of the linear groups. *Comm. Pure Appl. Math.* **8**, 355–365 (1955).

[2] The orders of the classical simple groups. *Comm. Pure Appl. Math.* **8**, 455–472 (1955).

[3] *Geometric Algebra*. New York: Interscience Publishers 1957.

Aschbacher, M.: [1] A condition for the existence of a strongly embedded subgroup. *Proc. Amer. Math. Soc.* **38**, 509–511 (1973).

[2] Finite groups with a proper 2-generated core. *Trans. Amer. Math. Soc.* **197**, 87–112 (1974).

[3] On finite groups of component type. *Illinois J. Math.* **19**, 87–115 (1975).

[4] A homomorphism theorem for finite graphs. *Proc. Amer. Math. Soc.* **54**, 468–476 (1976).

[5] A characterization of Chevalley groups over fields of odd characteristic, I and II. *Ann. of Math.* (2) **106**, 353–398, 399–468 (1977). Correction. *Ann. of Math.* (2) **111**, 411–414 (1980).

[6] On finite groups in which the generalized Fitting group of the centralizer of some involution is extraspecial. *Illinois J. Math.* **21**, 347–364 (1977).

[7] Thin finite simple groups. *J. Algebra* **54**, 50–152 (1978).

[8] The uniqueness case for finite groups. To appear.

Aschbacher, M., Gorenstein, D., and Lyons, R.: [1] The embedding of 2-locals in finite groups of characteristic 2-type. To appear.

Aschbacher, M. and Seitz, G. M.: [1] Involutions in Chevalley groups over fields of even order. *Nagoya Math. J.* **63**, 1–91 (1976), *Nagoya Math. J.* **72**, 135–136 (1978).

[2] On groups with a standard component of known type. *Osaka J. Math.* **13**, 439–482 (1976).

Auslander, L.: [1] On a problem of Philip Hall. *Ann. of Math.* (2) **86**, 112–116 (1967).

Baer, R.: [1] Representations of groups as quotient groups, I, II, and III. *Trans. Amer. Math. Soc.* **58**, 295–347, 348–389, 390–419 (1945).

Baer, R. and Levi, F.: [1] Freie Produkte und ihre Untergruppen. *Compositio Math.* **3**, 391–398 (1936).

Baumann, B.: [1] Endliche nichtauflösbare Gruppen mit einer nilpotenten maximalen Untergruppen. *J. Algebra* **38**, 119–135 (1976).

Baumslag, G.: [1] Wreath products and *p*-groups. *Proc. Cambridge Philos. Soc.* **55**, 224–231 (1959).

Beisiegel, B.: [1] Über einfache endliche Gruppen mit Sylow-2-gruppen der Ordnung höchstens 2^{10}. *Commun. Algebra* **5**, 113–170 (1977).

Beisiegel, B. and Stingl, V.: [1] *Announcement* (Mainz, Dec. 1974).

Bender, H.: [1] On the uniqueness theorem. *Illinois J. Math.* **14**, 376–384 (1970).

[2] On groups with abelian Sylow 2-subgroups. *Math. Z.* **117**, 164–176 (1970).

[3] Transitive Gruppen gerader Ordnung, in denen jede Involution genau einen Punkt festlässt. *J. Algebra* **17**, 527–554 (1971).

[4] Finite groups with large subgroups. *Illinois J. Math.* **18**, 223–228 (1974).

[5] Goldschmidt's 2-signalizer functor theorem. *Israel J. Math.* **22**, 208–213 (1975).

[6] On the normal *p*-structure of a finite group and related topics, I. *Hokkaido Math. J.* **7**, 271–288 (1978).

Blackburn, N.: [1] On a special class of *p*-groups. *Acta Math.* **100**, 45–92 (1958).

[2] Generalizations of certain elementary theorems on *p*-groups. *Proc. London Math. Soc.* **11**, 1–22 (1961).

Bourbaki, N.: [1] *Groupes et algèbres de Lie*, Chap. 4, 5 et 6. Paris: Hermann 1968.

[2] *Groupes et algèbres de Lie*, Chap. 7 et 8. Paris: Hermann 1975.

Brauer, R.: [1] On groups whose order contains a prime number to the first power, I and II. *Amer. J. Math.* **64**, 401–420, 421–440 (1942).

[2] Zur Darstellungstheorie der Gruppen endlicher Ordnung. *Math. Z.* **63**, 406–444 (1956).

[3] Zur Darstellungstheorie der Gruppen endlicher Ordnung, II. *Math. Z.* **72**, 25–46 (1959/60).

[4] Some applications of the theory of blocks and characters of finite groups, I and II. *J. Algebra* **1**, 152–167, 307–334 (1964).

Brauer, R. and Fowler, K. A.: [1] On groups of even order. *Ann. of Math.* (2) **62**, 565–583 (1955).

Brauer, R. and Suzuki, M.: [1] On finite groups of even order whose 2-Sylow group is a quaternion group. *Proc. Nat. Acad. Sci. U.S.A.* **45**, 1757–1759 (1959).

Brauer, R., Suzuki, M. and Wall, G. E.: [1] A characterization of the one-dimensional unimodular projective groups over finite fields. *Illinois J. Math.* **2**, 718–745 (1958).

Bruhat, F.: [1] Représentations induites des groupes de Lie semi-simples complexes. *C. R. Acad. Sci.*, Paris **238**, 437–439 (1954).

Buekenhout, F. and Shult, E. E.: [1] On the foundations of polar geometry. *Geometriae Dedicata* **3**, 155–170 (1974).

Burnside, W.: [1] On groups of order $p^{\alpha}q^{\beta}$, II. *Proc. London Math. Soc.* (2) **2**, 432–437 (1904).

[2] *Theory of Groups of Finite Order*. Second edition. Cambridge: University Press 1911.

Camm, R.: [1] Simple free products. *J. London Math. Soc.* **28**, 66–76 (1953).

Carter, R.: [1] Nilpotent self-normalizing subgroups of soluble groups. *Math. Z.* **75**, 136–139 (1960/61).

[2] *Simple Groups of Lie type*. London-New York: Wiley 1972.

Chevalley, C.: [1] *The Algebraic Theory of Spinors*. New York: Columbia University Press 1951.

[2] Sur certains groupes simples. *Tôhoku Math. J.* (2) **7**, 14–66 (1955).

Chunikhin, S. A.: [1] *Subgroups of Finite Groups*. Minsk 1964. (Translation. Groningen: Wolters-Noordhoff 1969).

Conway, J. H.: [1] A group of order 8,315,553,613,086,720,000. *Bull. London Math. Soc.* **1**, 79–88 (1969).

[2] Three lectures on exceptional groups. In: *Finite Simple Groups. Proceedings of an Instructional Conference*, pp. 215–247. London and New York: Academic Press 1971.

Conway, J. H. and Wales, D. B.: [1] Construction of the Rudvalis group of order 145,926,144,000. *J. Algebra* **27**, 538–548 (1973).

Coxeter, H. S. M. and Moser, W. O. J.: [1] *Generators and Relations for Discrete Groups*. Ergeb. Math. 14 Second edition. Berlin: Springer-Verlag 1964.

Dade, E. C.: [1] Lifting group characters. *Ann. of Math.* (2) **79**, 590–596 (1964).

[2] On normal complements to sections of finite groups. *J. Austral. Math. Soc.* **19**, 257–262 (1975).

Dickson, L. E.: [1] *Linear Groups with an Exposition of the Galois Field Theory*. Leibzig: Teubner 1901 (New York: Dover Publ. 1958).

Dieudonné, J.: [1] *La géométrie des groupes classiques*. Ergeb. Math. 5, Second edition. Berlin: Springer-Verlag 1963.

[2] Sur les générateurs des groupes classiques. *Summa Brasil. Math.* **3**, 149–179 (1955).

Dornhoff, L.: [1] *Group Representation Theory*. New York: Marcel Dekker, Inc. 1971–72.

Feit, W.: [1] On a class of doubly transitive permutation groups. *Illinois J. Math.* **4**, 170–186 (1960).

[2] On groups with a cyclic Sylow subgroup. In: *Proceedings of the International Conference on the Theory of Groups*, pp. 85–88. New York: Gordon and Breach 1967.

[3] *Characters of Finite Groups*. New York: W. A. Benjamin, Inc. 1967.

[4] Lectures on representations of finite groups. (Mimeographed Notes, Yale University 1968–74).

[5] The current situation in the theory of finite simple groups. In: *Actes Congr. Internat. Math.* Nice 1970, I. 55–93 (1971).

Feit, W. and Thompson, J. G.: [1] Groups which have a faithful representation of degree less than $(p - 1)/2$. *Pacific J. Math.* **11**, 1257–1262 (1961).

[2] Finite groups which contain a self-centralizing subgroup of order 3. *Nagoya Math. J.* **21**, 185–197 (1962).

[3] Solvability of groups of odd order. *Pacific J. Math.* **13**, 775–1029 (1963).

Fischer, B.: [1] Finite groups generated by 3-transpositions. *Invent. Math.* **13**, 232–246 (1971). (Remaining portion in preprint.)

[2] Lectures on a new (3,4)-transposition group at University of Illinois at Urbana–Champaign, 1974.

Fischer, B., Gaschütz, W. and Hartley, B.: [1] Injektoren endlicher auflösbarer Gruppen. *Math. Z.* **102**, 337–339 (1967).

Fuchs, L.: [1] *Abelian Groups*. New York: Pergamon Press 1960.

[2] *Infinite Abelian Groups, I*. Pure and Applied Mathematics vol. 36. New York-London: Academic Press 1970.

Gaschütz, W.: [1] Zur Erweiterungstheorie der endlichen Gruppen. *J. Reine Angew. Math.* **190**, 93–107 (1952).

[2] Zur Theorie der endlichen auflösbaren Gruppen. *Math. Z.* **80**, 300–305 (1962/63).

[3] Nichtabelsche p-Gruppen besitzen äussere p-Automorphismen. *J. Algebra* **4**, 1–2 (1966).

Gaschütz, W., Neubüser, J. and Yen, T.: [1] Über den Multiplikator von p-Gruppen. *Math. Z.* **100**, 93–96 (1967).

Glauberman, G.: [1] Fixed points in groups with operator groups. *Math. Z.* **84**, 120–125 (1964).

[2] Central elements in core-free groups. *J. Algebra* **4**, 403–420 (1966).

[3] On the automorphism group of a finite group having no non-identity normal subgroups of odd order. *Math. Z.* **93**, 154–160 (1966).

[4] A characteristic subgroup of a p-stable group. *Canad. J. Math.* **20**, 1101–1135 (1968).

[5] Prime power factor groups of finite groups, II. *Math. Z.* **117**, 46–56 (1970).

[6] Global and local properties of finite groups. In: *Finite Simple Groups. Proceedings of an Instructional Conference*. pp. 1–64. London: Academic Press 1971.

[7] A sufficient condition for p-stability. *Proc. London Math. Soc.* (3) **25**, 253–287 (1972).

[8] On groups with a quaternion Sylow 2-subgroup. *Illinois J. Math.* **18**, 60–65 (1974).

[9] On solvable signalizer functors in finite groups. *Proc. London Math. Soc.* **33**, 1–27 (1976).

[10] Factorizations in local subgroups of finite groups. CBMS Regional conference series in Math. 33. Amer. Math. Soc. 1977.

Goldschmidt, D. M.: [1] A group theoretic proof of the $p^a q^b$ theorem for odd primes. *Math. Z.* **113**, 373–375 (1970).

[2] 2-signalizer functors on finite groups. *J. Algebra* **21**, 321–340 (1972).

[3] 2-fusion in finite groups. *Ann. of Math.* (2) **99**, 70–117 (1974).

[4] Strongly closed 2-subgroups of finite groups. *Ann. of Math.* (2) **102**, 475–489 (1975).

Golod, E. S. and Šafarevič, I. R.: [1] On the class field towers. *Izv. Akad. Nauk SSSR* **28**, 261–272 (1964). (Amer. Math. Soc. Translations **48**, 91–102 (1965).)

Gomi, K.: [1] Finite groups with a standard subgroup isomorphic to $S_p(4, 2^n)$. *Japanese J. Math.* **4**, 1–76 (1978).

[2] Finite groups with a standard subgroup isomorphic to *PSU*(4, 2). *Pacific J. Math.* **79**, 399–462 (1978).

Gorenstein, D.: [1] *Finite Groups*. New York: Harper and Row 1968.

[2] On the centralizers of involutions in finite groups. *J. Algebra* **11**, 243–277 (1969).

[3] The classification of finite simple groups, I. Simple groups and local analysis. *Bull. Amer. Math. Soc.* (New series) **1**, 43–199 (1979).

Gorenstein, D. and Harada, K.: [1] Finite groups whose 2-subgroups are generated by at most 4 elements. *Memoirs Amer. Math. Soc.* **147** pp. 1–464 (1974).

Gorenstein, D. and Lyons, R.: [1] On the structure of finite groups of characteristic 2 type, I and II. (To appear)

Gorenstein, D. and Walter, J. H.: [1] The characterization of finite groups with dihedral Sylow 2-subgroups. *J. Algebra* **2**, 85–151, 218–270, 354–393 (1965).

[2] The π-layer of a finite group. *Illinois J. Math.* **15**, 555–565 (1971).

[3] Centralizers of involutions in balanced groups. *J. Algebra* **20**, 284–319 (1972).

[4] Balance and generation in finite groups. *J. Algebra* **33**, 224–287 (1975).

Green, J. A.: [1] On the number of automorphisms of a finite group. *Proc. Roy. Soc. London Ser. A.* **237**, 574–581 (1956).

Griess, R. L.: [1] Schur multipliers of the known finite simple groups. *Bull. Amer. Math. Soc.* **78**, 68–71 (1972).

[2] Schur multipliers of some sporadic simple groups. *J. Algebra* **20**, 320–349 (1972).

[3] Schur multipliers of finite simple groups of Lie type. *Trans. Amer. Math. Soc.* **183**, 355–421 (1973).

[4] A construction of F_1 as automorphisms of a 196883 dimensional algebra. (To appear).

Gruenberg, K. W.: [1] Resolutions by relations. *J. London Math. Soc.* **35**, 481–494 (1960).

[2] A new treatment of group extensions. *Math. Z.* **102**, 340–350 (1967).

Hall, M.: [1] On the number of Sylow subgroups in a finite group. *J. Algebra* **7**, 363–371 (1967).

Hall, M. and Senior, J. K.: [1] *The Groups of Order 2^n ($n \leq 6$)*. New York: Macmillan 1964.

Hall, M. and Wales, D. B.: [1] The simple group of order 604,800. *J. Algebra* **9**, 417–450 (1968).

Hall, P.: [1] A note on soluble groups. *J. London Math. Soc.* **3**, 98–105 (1928).

[2] A contribution to the theory of groups of prime power order. *Proc. London Math. Soc.* (2) **36**, 29–95 (1933).

[3] A characteristic property of soluble groups. *J. London Math. Soc.* **12**, 198–200 (1937).

[4] On the Sylow systems of a soluble group. *Proc. London Math. Soc.* (2) **43**, 316–323 (1937).

[5] On the system normalizers of a soluble group. *Proc. London Math. Soc.* (2) **43**, 507–528 (1937).

[6] The classification of prime-power groups. *J. Reine Angew. Math.* **182**, 130–141 (1940).

[7] The construction of soluble groups. *J. Reine Angew. Math.* **182**, 206–214 (1940).

[8] Finiteness conditions for soluble groups. *Proc. London Math. Soc.* (3) **4**, 419–436 (1954).

[9] Theorems like Sylow's. *Proc. London Math. Soc.* (3) **6**, 286–304 (1956).

[10] Some sufficient conditions for a group to be nilpotent. *Illinois J. Math.* **2**, 787–801 (1958).

[11] The Frattini subgroups of finitely generated groups. *Proc. London Math. Soc.* (3) **11**, 327–352 (1961).

Hall, P. and Hartley, B.: [1] The stability group of a series of subgroups. *Proc. London Math. Soc.* (3) **16**, 1–39 (1966).

Hall, P. and Higman, G.: [1] On the p-length of p-soluble groups and reduction theorems for Burnside's problem. *Proc. London Math. Soc.* (3) **6**, 1–42 (1956).

Harada, K.: [1] Finite simple groups with short chains of subgroups. *J. Math. Soc. Japan*, **20**, 655–672 (1968).

[2] On the simple group F of order $2^{14}3^6 5^6 7 \cdot 11 \cdot 19$. In: *Proceedings of a conference on finite groups*, pp. 119–276. New York: Academic Press 1976.

[3] On finite groups having self-centralizing 2-subgroups of small order. *J. Algebra* **33**, 144–160 (1975).

[4] On Yoshida's transfer. *Osaka J. Math.* **15**, 637–646 (1978).

Harris, M.: [1] On balanced groups, the $B(G)$-conjecture, and Chevalley groups over fields of odd order. (To appear).

Hattori, A.: [1] On exact sequences of Hochschild and Serre. *J. Math. Soc. Japan* **7**, 312–321 (1955).

Held, D.: [1] The simple groups related to M_{24}. *J. Algebra* **13**, 253–296 (1969).

Higgins, P. J.: [1] Presentations of groupoids, with applications to groups. *Proc. Cambridge Philos. Soc.* **60**, 7–20 (1964).

Higman, D. G.: [1] Focal series in finite groups. *Canad. J. Math.* **5**, 477–497 (1953).

[2] Remarks on splitting extensions. *Pacific J. Math.* **4**, 545–555 (1954).

[3] Finite permutation groups of rank 3. *Math. Z.* **86**, 145–156 (1964).

Higman, D. G. and Sims, C. C.: [1] A simple group of order 44,352,000. *Math. Z.* **105**, 110–113 (1968).

Higman, G.: [1] A finitely generated infinite simple group. *J. London Math. Soc.* **26**, 61–64 (1951).

[2] Complementation of abelian normal subgroups. *Publ. Math. Debrecen* **4**, 455–458 (1956).

[3] Enumerating p-groups, I. Inequalities. *Proc. London Math. Soc.* (3) **10**, 24–30 (1960).

[4] Odd characterizations of finite simple groups. (Mimeographed lecture notes, University of Michigan 1968.)

[5] Construction of simple groups from character tables. In: *Finite Simple Groups. Proceedings of an Instructional Conference.* pp. 205–214. London: Academic Press, 1971.

Higman, G. and McKay, J.: [1] On Janko's simple group of order 50,232,960. *Bull. London Math. Soc.* **1**, 89–94 (Correction: 219) (1969).

[2] On Held's simple group. (Unpublished)

Higman, G., Neumann, B. H. and Neumann, H.: [1] Embedding theorems for groups. *J. London Math. Soc.* **24**, 247–254 (1949).

Hirsch, K. A.: [1] On infinite soluble groups, I and II. *Proc. London Math. Soc.* (2) **44**, 53–60, 336–344 (1938).

[2] On infinite soluble groups, III. *Proc. London Math. Soc.* (2) **49**, 184–194 (1946).

Hochschild, G. and Serre, J.-P.: [1] Cohomology of group extensions. *Trans. Amer. Math. Soc.* **74**, 110–134 (1953).

Huppert, B.: [1] Normalteiler und maximale Untergruppen endlicher Gruppen. *Math. Z.* **60**, 409–434 (1954).

[2] Subnormale Untergruppen und p-Sylowgruppen. *Acta Sci. Math.* (Szeged) **22**, 46–61 (1961).

[3] *Endliche Gruppen I.* Die Grundlehren der Mathematischen Wissenschaften, Band 134. Berlin: Springer-Verlag 1967.

Isaacs, I. M.: [1] An alternate proof of the Thompson replacement theorem. *J. Algebra* **15**, 149–150 (1970).

[2] Some of Yoshida's transfer results. (Unpublished)

Ito, N.: [1] On the degrees of irreducible representations of a finite group. *Nagoya Math. J.* **3**, 5–6 (1951).

[2] On π-structures of finite groups. *Tôhoku Math. J.* (2) **4**, 172–177 (1952).

[3] *Theory of Finite Groups.* Modern Mathematics vol. 7. Tokyo: Kyoritsu Shuppan 1970. (English version: Mimeographed Notes, University of Illinois at Chicago Circle 1970.)

Iwahori, N.: [1] Centralizers of involutions in finite Chevalley groups. In: *Seminar on Algebraic Groups and Related Finite Groups*, pp. F1–F29. Lecture Notes in Mathematics 131. Berlin: Springer-Verlag 1970.

Iwasawa, K.: [1] Über die Einfachheit der speziellen projektiven Gruppen. *Proc. Imp. Acad. Tokyo* **17**, 57–59 (1941).

[2] Einige Sätze über freie Gruppen. *Proc. Imp. Acad. Tokyo* **19**, 272–274 (1943).

Iyanaga, S.: [1] *The Theory of Numbers.* Gendai Sûgaku 10. Tokyo: Iwanami Shoten 1969. (Translation: North-Holland Math. Library vol. 8. Amsterdam-Oxford: North-Holland Publ. Co. 1975.)

Janko, Z.: [1] A new finite simple group with abelian Sylow 2-subgroups and its characterization. *J. Algebra* **3**, 147–186 (1966).

[2] Some new simple groups of finite order, I. In: *Symposia Mathematica* (INDAM, Rome 1967/68) Vol. **1**, pp. 25–64. London: Academic Press 1969.

[3] A new finite simple group of order 86,775,571,046,077,562,880 which possesses M_{24} and the full cover of M_{22} as subgroups. *J. Algebra* **42**, 564–596 (1976).

Jónsson, B.: [1] On direct decompositions of torsion-free abelian groups. *Math. Scand.* **5**, 230–235 (1957).

Jordan, C.: [1] *Traité des substitutions et des équations algébriques.* Paris: Gauthiers-Villars 1870. (Librairie Scientifique et Technique Albert Blanchard 1957.)

Kaloujnine, L. and Krasner, M.: [1] Le produit complet des groupes de permutations et le problème d'extension des groupes. *C. R. Acad. Sci. Paris* **227**, 806–808 (1948).

Kantor, W. M.: [1] Rank 3 characterizations of classical geometries. *J. Algebra* **36**, 309–313 (1975).

Kegel, O. H.: [1] Die Nilpotenz der H_p-Gruppen. *Math. Z.* **75**, 373–376 (1960/61).

Kondo, T.: [1] *Group Theory*. Iwanami Kôza, Kiso Sûgaku. Tokyo: Iwanami Shoten 1976/77.

Krasner, M. and Kaloujnine, L.: [1] Produit complet des groupes de permutations et problème d'extension de groupes, I, II, and III. *Acta Sci. Math.* (Szeged) **13**, 208–230 (1950); **14**, 39–66, 69–82 (1951).

Kuhn, H. W.: [1] Subgroup theorems for groups presented by generators and relations. *Ann. of Math.* (2) **56**, 22–46 (1952).

Kulikov, L. J.: [1] On the theory of Abelian groups of arbitrary power. *Mat. Sb. 16* (**58**), 129–162 (1945).

Leech, J.: [1] Coset enumeration on digital computers. *Proc. Cambridge Philos. Soc.* **59**, 257–267 (1963).

Leon, J. S. and Sims, C. C.: [1] The existence and uniqueness of a simple group generated by {3, 4}-transpositions. *Bull. Amer. Math. Soc.* **83**, 1039–1040 (1977).

Lyndon, R. C.: [1] The cohomology theory of group extensions. *Duke Math. J.* **15**, 271–292 (1948).

[2] Cohomology theory of groups with a single defining relation. *Ann. of Math.* (2) **52**, 650–665 (1950).

Lyons, R.: [1] Evidence for a new finite simple group. *J. Algebra* **20**, 540–569 (1972). Errata: *J. Algebra* **34**, 188–189 (1975).

MacLane, S.: [1] A proof of the subgroup theorem for free products. *Mathematica* **5**, 13–19 (1958).

MacWilliams, A. P.: [1] On 2-groups with no normal abelian subgroups of rank 3 and their occurrence as Sylow 2-subgroups of finite simple groups. *Trans. Amer. Math. Soc.* **150**, 345–408 (1970).

Magnus, W.: [1] Beziehungen zwischen Gruppen und Idealen in einem speziellen Ring. *Math. Ann.* **111**, 259–280 (1935).

Matsumoto, H.: [1] Générateurs et relations des groupes de Weyl généralisés. *C. R. Acad. Sci. Paris* **258**, 3419–3422 (1964).

Matsushima, Y.: [1] *The Theory of Lie Rings*. Gendai Sûgaku Kôza 3-A. Tokyo: Kyoritsu Shuppan 1956.

Matsuyama, H.: [1] Solvability of groups of order $2^a p^b$. *Osaka J. Math.* **10**, 375–378 (1973).

McLaughlin, J.: [1] A simple group of order 898,128,000. In: *Theory of Finite Groups. A Symposium.* pp. 109–111. New York: W. A. Benjamin, 1969.

Mendelsohn, N. S.: [1] An algorithmic solution for a word problem in group theory. *Canad. J. Math.* **16**, 509–516 (1964). Correction: *Canad. J. Math.* **17**, 505 (1965).

Miyamoto, I.: [1] Finite groups with a standard subgroup isomorphic to $U_4(2^n)$. *Japanese J. Math.* **5**, 209–244 (1979).

Nagao, H.: [1] A note on extensions of groups. *Proc. Japan Acad.* **25**, 11–14 (1949).

[2] A proof of Brauer's theorem on generalized decomposition numbers. *Nagoya Math. J.* **22**, 73–77 (1963).

[3] Groups and Designs. Tokyo: Iwanami Shoten 1974. (Translation to appear)

Nagata, M.: [1] Note on groups with involutions. *Proc. Japan Acad.* **26**, 564–566 (1952).

Nakayama, T.: [1] A theorem on modules of trivial cohomology over a finite group. *Proc. Japan Acad.* **32**, 373–376 (1956).

[2] On modules of trivial cohomology over a finite group. *Illinois J. Math.* **1**, 36–43 (1957).

[3] On modules of trivial cohomology over a finite group, II. Finitely generated modules. *Nagoya Math. J.* **12**, 171–176 (1957).

Nakayama, T. and Hattori, A.: [1] *Homological Algebra*. Gendai Sûgaku Kôza 10-B. Tokyo: Kyoritsu Shuppan 1957.

Neumann, B. H.: [1] Adjunction of elements to groups. *J. London Math. Soc.* **18**, 4–11 (1943).

[2] Twisted wreath products of groups. *Arch. Math.* (Basel) **14**, 1–6 (1963).

Neumann, B. H. and Neumann, H.: [1] Embedding theorems for groups. *J. London Math. Soc.* **34**, 465–479 (1959).

Neumann, P. M.: [1] On the structure of standard wreath products of groups. *Math. Z.* **84**, 343–373 (1964).

[2] An enumeration theorem for finite groups. *Quart. J. Math. Oxford* (2) **20**, 395–401 (1969).

Nielsen, J.: [1] A basis for subgroups of free groups. *Math. Scand.* **3**, 31–43 (1955).

Norton, S.: [1] Construction and properties of a new simple group. Thesis, Cambridge University 1975.

Novikov, P. S. and Adjan, S. I.: [1] Infinite periodic groups, I, II, and III. *Izv. Akad. Nauk SSSR Ser. Mat.* **32**, 212–244, 251–524, 709–731 (1968). (*Math. USSR Izv.* **2**, 209–236, 241–479, 665–685 (1968).)

O'Nan, M. E.: [1] Some evidence for the existence of a new simple group. *Proc. London Math. Soc.* (*2*) **32**, 421–479 (1976).

Ono, T.: [1] An identification of Suzuki groups with groups of generalized Lie type. *Ann. of Math.* (*2*) **75**, 251–259 (1962).

Patterson, A. P.: [1] On Sylow 2-subgroups with no normal abelian subgroups of rank 3 in finite fusion-simple groups. *Trans. Amer. Math. Soc.* **187**, 1–67 (1974).

Pontrjagin, L. C.: [1] The theory of topological commutative groups. *Ann. of Math.* **35**, 361–388 (1934).

Rado, R.: [1] A proof of the basis theorem for finitely generated Abelian groups. *J. London Math. Soc.* **26**, 74–75; erratum, 160 (1951).

Ree, R.: [1] On some simple groups defined by C. Chevalley. *Trans. Amer. Math. Soc.* **84**, 392–400 (1957).

[2] A family of simple groups associated with the simple Lie algebra of type (F_4). *Amer. J. Math.* **83**, 401–420 (1961).

[3] A family of simple groups associated with the simple Lie algebra of type (G_2). *Amer. J. Math.* **83**, 432–462 (1961).

Robinson, D. S.: [1] Joins of subnormal subgroups. *Illinois J. Math.* **9**, 144–168 (1965).

[2] A property of the lower central series of a group. *Math. Z.* **107**, 225–231 (1968).

Romanovskii, A. V.: [1] On Čunikin's theory of indexials. *Dokl. Akad. Nauk BSSR* **18**, 297–299, 379 (1974).

Roquette, P.: [1] Über die Existenz von Hall-Komplementen in endlichen Gruppen. *J. Algebra* **1**, 342–346 (1964).

Rudvalis, A.: [1] A new simple group of order $2^{14}3^35^37 \cdot 13 \cdot 29$. *Notices Amer. Math. Soc.* **20**, A–95 (1973).

Schenkman, E.: [1] The splitting of certain solvable groups. *Proc. Amer. Math. Soc.* **6**, 286–290 (1955).

[2] *Group Theory*. Princeton: D. Van Nostrand 1965.

Schmidt, O.: [1] Über unendliche spezielle Gruppen. *Mat. Sb.* 8(50), 363–375 (1940).

Schreier, O.: [1] Über die Erweiterung von Gruppen, I. *Monatsh. Math. Phys.* **34**, 165–180 (1926).

[2] Über die Erweiterung von Gruppen, II. *Abh. Math. Sem. Univ. Hamburg* **4**, 321–346 (1926).

[3] Die Untergruppen der freie Gruppen. *Abh. Math. Sem. Univ. Hamburg* **5**, 161–183 (1927).

[4] Über den Jordan-Hölderschen Satz. *Abh. Math. Sem. Univ. Hamburg* **6**, 300–302 (1928).

Schur, I.: [1] Über die Darstellung der endlichen Gruppen durch gebrochene lineare Substitutionen. *J. Reine Angew. Math.* **127**, 20–50 (1904).

[2] Untersuchungen über die Darstellungen der endlichen Gruppen durch gebrochene lineare Substitutionen. *J. Reine Angew. Math.* **132**, 85–137 (1907).

[3] Über die Darstellungen der symmetrischen und alternierenden Gruppen durch gebrochene lineare Substitutionen. *J. Reine Angew. Math.* **139**, 155–250 (1911).

Scott, W. R.: [1] *Group Theory*. Englewood Cliffs, N.J.: Prentice-Hall 1964.

Serre, J.-P.: [1] Cohomologie des groupes discrets. *C. R. Acad. Sci. Paris Ser. A-B* **268**, A268–A271 (1969).

Shult, E. E.: [1] On the fusion of an involution in its centralizer. *Notices Amer. Math. Soc.* **17**, 548 (1970) (Full paper unpublished).

[2] On a class of doubly transitive groups. *Illinois J. Math.* **16**, 434–455 (1972).

[3] Groups, polar spaces and related structures. Combinatorics Part 3. *Proc. Advanced Study Institute*, Breukelen, pp. 130–161, 1974.

[4] Disjoint triangular sets. *Ann. of Math.* (*2*) **111**, 67–94 (1980).

Sibley, D. A.: [1] Finite linear groups with a strongly self-centralizing Sylow subgroup. *J. Algebra* **36**, 158–166 (1975).

[2] Coherence in finite groups containing a Frobenius section. *Illinois J. Math.* **20**, 434–442 (1976).

Sims, C. C.: [1] Enumerating p-groups. *Proc. London Math. Soc.* (*3*) **15**, 151–166 (1965).

[2] The existence and uniqueness of Lyons' group. In: *Finite Groups '72. Proceedings of the Gainsville Conference*. Pp. 138–141. Amsterdam-London: North-Holland 1973.

[3] The existence and uniqueness of O'Nan's group. Unpublished.

Smith, F.: [1] On the centralizers of involutions in finite fusion-simple groups. *J. Algebra* **38**, 268–273 (1976).

[2] On finite groups with large extra-special 2-subgroups. *J. Algebra* **44**, 477–487 (1977).

Smith, S. D.: [1] A characterization of orthogonal groups over $GF(2)$. *J. Algebra* **62**, 39–60 (1980).

[2] Large Extraspecial subgroups of widths 4 and 6. *J. Algebra* **58**, 251–281 (1979).

Stallings, J. R.: [1] On torsion-free groups with infinitely many ends. *Ann. of Math.* (2) **88**, 312–334 (1968).

Steinberg, R.: [1] Variations on a theme of Chevalley. *Pacific J. Math.* **9**, 875–891 (1959).

[2] Lectures on Chevalley groups. Lecture notes, Yale University 1967–68.

Suzuki, M.: [1] A characterization of simple groups $LF(2, p)$. *J. Fac. Sci. Univ. Tokyo Sect. I.* **6**, 259–293 (1951).

[2] The nonexistence of a certain type of simple groups of odd order. *Proc. Amer. Math. Soc.* **8**, 686–695 (1957).

[3] A new type of simple groups of finite order. *Proc. Nat. Acad. Sci. U.S.A.* **46**, 868–870 (1960).

[4] On a class of doubly transitive groups. *Ann. of Math.* (2) **75**, 105–145 (1962).

[5] On a class of doubly transitive groups, II. *Ann. of Math.* (2) **79**, 514–589 (1964).

[6] A simple group of order 448,345,497,600. In: *Theory of Finite Groups. A Symposium.* pp. 113–119. New York: W. A. Benjamin 1969.

[7] Characterizations of linear groups. *Bull. Amer. Math. Soc.* **75**, 1043–1091 (1969).

[8] Recent development in the theory of finite simple groups. *Sûgaku* **27**, 99–110 (1975). (In Japanese)

Swan, R.: [1] Groups of cohomological dimension one. *J. Algebra* **12**, 585–610 (1969).

Takahashi, M.: [1] Bemerkungen über den Untergruppensatz in freie Produkte. *Proc. Imp. Acad. Tokyo* **20**, 589–594 (1944).

Takeuti, G.: [1] Mathematical Logic. Tokyo: Baihukan 1973.

Tate, J.: [1] Nilpotent quotient groups. *Topology* **3**, suppl. 1 109–111 (1964).

Thompson, J. G.: [1] Finite groups with fixed-point-free automorphisms of prime order. *Proc. Nat. Acad. Sci. U.S.A.* **45**, 578–581 (1959).

[2] Normal p-complements for finite groups. *J. Algebra* **1**, 43–46 (1964).

[3] Factorizations of p-solvable groups. *Pacific J. Math.* **16**, 371–372 (1966).

[4] Vertices and sources. *J. Algebra* **6**, 1–6 (1967).

[5] Nonsolvable finite groups all of whose local subgroups are solvable. *Bull. Amer. Math. Soc.* **74**, 383–437 (1968); *Pacific J. Math.* **33**, 451–536 (1970); *Pacific J. Math.* **39**, 483–534 (1971); *Pacific J. Math.* **48**, 511–592 (1973); *Pacific J. Math.* **50**, 215–297 (1974); *Pacific J. Math.* **51**, 573–630 (1974).

[6] A replacement theorem for p-groups and a conjecture. *J. Algebra* **13**, 149–151 (1969).

[7] A simple subgroup of $E_8(3)$. In: *Finite Groups. Symposium.* pp. 113–116. Tokyo: Japan Soc. for Promotion of Sci. 1976.

[8] Simple groups of order prime to 3. (Unpublished: cf. in *Symposia Math.* **XIII** pp. 517–530. London: Academic Press 1974.)

[9] Notes on the *B*-conjecture (unpublished).

Timmesfeld, F.: [1] Groups generated by root involutions, I. *J. Algebra* **33**, 75–135 (1975).

[2] Groups with weakly closed T.I. subgroups. *Math. Z.* **143**, 243–278 (1975).

[3] On elementary abelian TI-subgroups. *J. Algebra* **44**, 457–476 (1977).

[4] Finite simple groups in which the generalized Fitting group of the centralizer of some involution is extraspecial. *Ann. of Math.* (2) **107**, 297–369 (1978). Correction: *Ann. of Math.* **109**, 413–414 (1979).

Tits, J.: [1] Algebraic and abstract simple groups. *Ann. of Math.* (2) **80**, 313–329 (1964).

[2] *Buildings of Spherical Type and Finite BN-Pairs.* Lecture Notes in Math. 386. Berlin: Springer-Verlag 1974.

Tsuzuku, T.: [1] *Finite Groups and Finite Geometries.* Tokyo: Iwanami Shoten 1976. (Translation to appear)

Van der Waerden, B. L.: [1] Free products of groups. *Amer. J. Math.* **70**, 527–528 (1949).

Walter, J. H.: [1] The characterization of finite groups with abelian Sylow 2-subgroups. *Ann. of Math.* (2) **89**, 405–514 (1969).

[2] The *B*-conjecture; 2-components in finite simple groups. *Proc. Symp. Pure Math.* **37**, 57–66 (1980). *Amer. Math. Soc.*

Weir, A. J.: [1] The Reidemeister–Schreier and Kuroš subgroup theorems. *Mathematika* **3**, 47–55 (1956).

Wiegold, J.: [1] Multiplicators and groups with finite central factor-groups. *Math. Z.* **89**, 345–347 (1965).

Wielandt, H.: [1] Ein Verallgemeinerung der invarianten Untergruppen. *Math. Z.* **45**, 209–244 (1939).
[2] *p*-Sylowgruppen und *p*-Faktorgruppen. *J. Reine Angew. Math.* **182**, 180–193 (1940).
[3] Zum Satz von Sylow. *Math. Z.* **60**, 407–408 (1954).
[4] Zum Satz von Sylow, II. *Math. Z.* **71**, 461–462 (1959).
[5] On the structure of composite groups. In: *Proceedings of the International Conference on the Theory of Groups*, pp. 379–388. New York: Gordon and Breach 1967.
Wielandt, H. and Huppert, B.: [1] Arithmetical and normal structure of finite groups. In: *Proc. Sympos. Pure Math.* vol. VI, pp. 17–38. Providence, RI: Amer. Math. Soc. 1962.
Witt, E.: [1] Die 5-fach transitiven Gruppen von Mathieu. *Abh. Math. Sem. Univ. Hamburg* **12**, 256–264 (1938).
[2] Spiegelungsgruppen und Aufzählung halbeinfacher Liescher Ringe. *Abh. Math. Sem. Hansischen Univ.* **14**, 289–322 (1941).
Yamada, H.: [1] Finite groups with a standard subgroup isomorphic to $G_2(2^n)$. *J. Fac. Sci. Univ. Tokyo* Sect. I, **26**, 1–52 (1979).
[2] Finite groups with a standard subgroup isomorphic to $^3D_4(2^{3n})$. *J. Fac. Sci. Univ. Tokyo* Sect. IA **26**, 255–278 (1979).
Yamazaki, K.: [1] On projective representations and ring extensions of finite groups. *J. Fac. Sci. Univ. Tokyo* Sect. I **10**, 147–195 (1964).
Yoshida, T.: [1] Character-theoretic transfer. *J. Algebra* **52**, 1–38 (1978).
Zassenhaus, H.: [1] *The Theory of Groups*. Second edition. New York: Chelsea 1958.

Index